THE MANUFACTURE OF SOAPS, OTHER DETERGENTS AND GLYCERINE

ELLIS HORWOOD SERIES IN APPLIED SCIENCE AND INDUSTRIAL TECHNOLOGY

Series Editor: Dr D. H. SHARP, OBE, lately General Secretary, Society of Chemical Industry; formerly General Secretary, Institution of Chemical Engineers; and former Technical Director, Confederation of British Industry

Published and in active publication

PREPARATION, PROPERTIES AND INDUSTRIAL APPLICATIONS OF ORGANOFLUORINE COMPOUNDS
Editor: R. E. BANKS, Department of Chemistry, University of Manchester Institute of Science and Technology
CERAMIC WHITEWARES
A. DINSDALE, lately Director of Research, British Ceramic Research Association
MATCHMAKING: Science, Technology and Manufacture
C. A. FINCH, Managing Director, Pentafin Associates, Chemical, Technical and Media Consultants, Stoke Mandeville, and S. RAMACHANDRAN, Senior Consultant, United Nations Industrial Development Organisation for the Match Industry
WATER – ABSTRACTION, STORAGE, TREATMENT, DISTRIBUTION
J. JEFFERY, General Manager, North Surrey Water Company
PAINT AND SURFACE COATING: Theory and Practice
Vol. 1: Colloid, Surface and Polymer Chemistry of Modern Surface Coatings
Vol. 2: Physical Properties of Liquid Coating Compositions and of Derived Films
Editor: R. LAMBOURNE, lately Section Manager, Research Department, ICI plc Paints Division
CROP PROTECTION CHEMICALS
B. G. LEVER, Development Manager, ICI plc Plant Protection Division
HANDBOOK OF MATERIALS HANDLING
Translated by R. G. T. LINDKVIST, MTG, Translation Editor: R. ROBINSON, Editor, Materials Handling News. Technical Editor: G. LUNDESTO, Rolatruc Limited
FERTILIZER TECHNOLOGY
G. C. LOWRISON, MSc, FIChemE, MBIM, lately of Fison's Fertilisers Ltd.
NON-WOVEN BONDED FABRICS
Editor: J. LUNENSCHLOSS, Institute of Textile Technology of the Rhenish-Westphalian Technical University, Aachen and W. ALBRECHT, Wuppertal
MICROCOMPUTERS IN THE PROCESS INDUSTRY
E. R. ROBINSON, Head of Chemical Engineering, North East London Polytechnic
POLYMERS IN PACKING TECHNOLOGY
J. STEPEK, Czechoslovakia. Translation Editor: G. E. J. REYNOLDS, Surrey
REFRACTORIES TECHNOLOGY
C. STOREY, Consultant, Durham, One time General Manager, Refractories, British Steel Corporation
PETROLEUM TECHNOLOGY
A. H. SWEATMAN, Visiting Professor, Imperial College and G. A. HOGG, Natural Gas Consultant, lately of British Petroleum Company plc
PERFUMERY TECHNOLOGY 2nd Edition
F. V. WELLS, Consultant Perfumer and former Editor of Soap, Perfumery and Cosmetics, and M. BILLOT, former Cheif Perfumer to Houbigant-Cheramy, Paris, President d'Honneur de la Societe Technique des Parfumeurs de la France
THE MANUFACTURE OF SOAPS, OTHER DETERGENTS AND GLYCERINE
E. WOOLLATT, Consultant, British Executive Service Overseas

THE MANUFACTURE OF SOAPS, OTHER DETERGENTS AND GLYCERINE

EDGAR WOOLLATT, B.Sc. C.Eng., F.I.Chem.E.
Consultant, British Executive Service Overseas
formerly Development Manager, Lever Brothers Limited, Cheshire

ELLIS HORWOOD LIMITED
Publishers · Chichester

Halsted Press: a division of
JOHN WILEY & SONS
New York · Chichester · Brisbane · Toronto

First published in 1985 by
ELLIS HORWOOD LIMITED
Market Cross House, Cooper Street, Chichester, West Sussex, PO19 1EB, England

The publisher's colophon is reproduced from James Gillison's drawing of the ancient Market Cross, Chichester.

Distributors:

Australia, New Zealand, South-east Asia:
Jacaranda-Wiley Ltd., Jacaranda Press,
JOHN WILEY & SONS INC.,
G.P.O. Box 859, Brisbane, Queensland 4001, Australia

Canada:
JOHN WILEY & SONS CANADA LIMITED
22 Worcester Road, Rexdale, Ontario, Canada.

Europe, Africa:
JOHN WILEY & SONS LIMITED
Baffins Lane, Chichester, West Sussex, England.

North and South America and the rest of the world:
Halsted Press: a division of
JOHN WILEY & SONS
605 Third Avenue, New York, N.Y. 10158 U.S.A.

British Library Cataloguing in Publication Data
Woollatt, Edgar
The manufacture of soaps, other detergents and glycerine. –
(Ellis Horwood series in applied science and industrial technology)
1. Surface, active agents
I. Title
668'.1 TP994

Library of Congress Card No. 85–14048

ISBN 0–85312–567–8 (Ellis Horwood Limited)
ISBN 0–470–20234–3 (Halsted Press)

Typeset by Ellis Horwood Limited
Printed in Great Britain by Unwin Brothers of Woking

Table of Contents

Preface

This book is intended for students, both inside and outside the soaps, other detergents, and glycerine industries, and for those who work in these industries. It is concerned primarily with manufacturing processes and equipment, but it also considers ancillary topics needed for a full understanding of the industries. These non-specialist discussions, which are likely to be of particular interest to those who do not have access to specialist departments normally available in large organisations, include:

- a simplified, non-mathematical, discussion of the theory of detergency and related matters.
- the properties and roles of the various ingredients used in detergent products; and some principles of formulation.
- a review of world markets, including consideration of the rise and fall of various types of product; and some recent trends, particularly in the UK.
- environmental problems.
- the interpretation and use of analytical information, although details of analytical procedures are generally outside the scope of the book.
- factors which affect vacuum maintenance and the performance of fans.

Every effort is made to establish the principles which control the various elements into which a process of manufacture can be divided; for example, saponification, washing, and fitting in the case of soap production by direct saponification. Alternative types of equipment which can be used for these elements are then reviewed before complete plants are considered. In many cases the historical evolution of modern plants is described, as this widens understanding; and because some types of equipment no longer used by major manufacturers may still be useful in smaller factories, particularly when large capital expenditure is not worthwhile, or even feasible.

As far as possible, approximate calculations, involving only arithmetic, are used to illustrate various process elements. Like other writers, the present author is handicapped by lack of data available only in company files, but, despite this limitation, the rough quantitative approach is considered worthwhile because:

- it provides a deeper insight into the various operations.
- it may encourage readers to analyse the performance of their own plants, which is often a necessary first step towards improved efficiency, or modification to meet changed circumstances.

The author would like to thank the many manufacturers of plant and ingredients who have helped with literature and other information. Many of the illustrations are taken from the brochures, or from drawings, or photographs, provided for the purpose. Special thanks are due to Albright and Wilson Ltd and the Boots Company plc for most useful discussions.

The author has drawn heavily upon his long experience with Unilever companies, and upon information gained during visits to various companies overseas since his retirement; but it is emphasized that opinions expressed are those of the author and do not necessarily represent the views of any organization. No attempt is made to indicate possible patent restrictions upon the use of information provided in this book.

Finally the author would like to thank Mrs. M. Honor and Mrs. A. Simpson for the careful way in which they typed the script.

To: *G. R. ('Bob') Elliott; mentor and colleague for many years*

Units and factors

In this book units have been chosen which, in the opinion of the author, are convenient for each topic; and likely to be familiar to many readers. SI equivalents are given in this section, together with other factors and some notes which may be useful.

A high degree of precision is not required for the simple calculations in this book; and some fine points are ignored in the compilation of definitions and factors.

TIME

The SI unit is the second (s).

1 minute (min) = 60s; and 1 hour (h) = 3600s

LENGTH

The SI unit is the metre (m).

1 foot (ft) = 12 inches (in) = 0.305 m
1 inch = 2.54 cm = 25.4 mm

AREA

The SI unit is the square metre (m^2)

1 square foot (ft^2) = 144 square inches (in^2)
= 0.093 m^2
1 in^2 = 645 mm^2 = $6.45 \times 10^{-4} m^2$

VOLUME AND CAPACITY

The SI unit is the cubic metre (m^3). The litre (l), originally the volume occupied by 1 kg of cold water (strictly at a defined temperature having a defined density), has been redefined as equal exactly to 1 cubic decimetre ($10^{-3} m^3$). For almost all purposes the difference is negligible.

1 cubic foot (ft^3) = 0.0283 m^3 = 28.3 l
1 Imperial gallon (Imp. gal.) = 0.161 ft^3 = 4.55 l
= 0.00455 m^3

1 Imp. gal. of cold water weighs 10 lb
1 US gallon (US gal.) = 0.833 Imp. gal.
= 3.79 l
= 0.00379 m^3
1 US gal. of cold water weighs 8.33 lb

In this book, the Imp. gal. and the US gal. are used only in the conversion of units of water hardness.

MASS, OR QUANTITY OF MATTER

The SI unit is the kilogram (kg), equal to 1000 grams (g).

1 pound (lb) = 0.454 kg
1 ton (Imperial) ≑ 2240 lb = 1016 kg
= 1.016 tonne (t)

In the USA, masses are commonly quoted in lb, but the 'short ton' of 2000 lb is sometimes used. The word 'weight', which is the force with which a mass is attracted by gravity and hence the quantity measured by a scale, is often used loosely when 'mass' would be more correct.

Amount of substance is also measured in terms of the mole (mol), defined as the number of elementary particles the amount contains. The mole has now been broadened to apply to atoms, molecules, ions, or equivalents, with the relevant particle specified unless it is obvious from the context.

In scientific work the gram mol is commonly used; in technical calculations this is often replaced by the lb mol, or kg mol

Some approximate atomic masses (formerly called atomic weights) are:

Hydrogen, H	1.0	Chlorine, Cl	35.5
Boron, B	10.8	Potassium, K	39.1
Carbon, C	12.0	Calcium, Ca	40.1
Nitrogen, N	14.0	Chromium, Cr	52.0
Oxygen, O	16.0	Manganese, Mn	54.9
Sodium, Na	23.0	Iron, Fe	55.9
Magnesium, Mg	24.3	Nickel, Ni	58.7
Aluminium, Al	27.0	Copper, Cu	63.5
Silicon, Si	28.0	Zinc, Zn	65.4
Phosphorus, P	31.0	Bromine, Br	79.9
Sulphur, S	32.0	Silver, Ag	107.9
		Iodine, I	126.9
		Barium, Ba	137.4

DENSITY, RELATIVE DENSITY, AND SPECIFIC GRAVITY

The SI unit of density is the kg/m^3.

1 lb/ft^3 = 16.0 kg/m^3 = 0.016 g/cm^3 = 0.016 g/ml
62.4 lb/ft^3 = lg/cm^3

Specific gravity, now usually called relative density, is density relative to that of water (or that of air in the case of gases). Temperatures, and in the case of gases also pressures, usually need to be defined.

The old Twaddell and Baumé scales are widely used in the detergents industry.

$$\text{Degrees Twaddell } (^\circ\text{Tw}) = 200\ (\text{specific gravity} - 1)$$

$$\text{or SG} = \frac{^\circ\text{Tw}}{200} + 1.$$

For liquids heavier than water, degrees Baumé (°Bé) $= 145 - \dfrac{145}{\text{SG}}$, so that a liquid of SG = 1.5 has a density of 48.3°, say 48°Bé.

For liquids lighter than water $^\circ\text{Bé} = \dfrac{140}{\text{SG}} - 130$. It will be noted that water of SG 1.0 is equivalent to 0°Tw and to 0°Bé on the heavier than water scale. On the lighter than water scale water of SG 1.0 is equivalent to 10°Bé. Minor variations in the heavier than water scale have been used in different countries.

TEMPERATURE

The SI unit of temperature is the degree Kelvin (K). This is the same as the degree Celsius (°C), often called the degree Centigrade, except that the Kelvin scale starts from absolute zero (− 273.15°C, taken here as − 273°C) and the Celsius scale at the melting point of ice, 0°C. That is K = 273 + °C.

The Fahrenheit scale is defined by the melting point of ice 32°F and the boiling point of water under normal atmospheric pressure 212°F. Hence, temperature in °F = (°C x 1.8) + 32, or 1.8 °F = 1 °C.

Degrees Rankine (°R), or Fahrenheit absolute, run from the absolute zero on the Fahrenheit scale which is − 460°F.

FORCE

The SI unit of force is the Newton (N) which is the force which when acting on a mass of 1 kg gives it an acceleration of 1 m/s^2.

The pound force (lbf) is the force with which a mass of 1 lb is attracted by gravity, that is its weight.

1 lbf = g poundals	= 32.2 poundals approximately
	= 4.45 N
1 kgf = g dynes	= 981 dynes approximately
	= 9.81 N
1 dyne	= 10^{-5}N

g, the accleration due to gravity varies slightly from place to place on the earth's surface.

PRESSURE AND STRESS

Pressure and stress are force per unit area, so that the SI unit is N/m^2.

The bar (approximately one atmosphere at sea level) = $10^5 N/m^2$.

Gauge pressures are pressures above atmospheric (as measured by many types of gauge which zero at atmospheric pressure); and absolute pressures are measured from zero pressure (a full vacuum).

Various units are used in industry, and in this book, to suit different situations.

1 lbf/in^2 (pounds per square inch, psi)
$= 6890\ N/m^2 = 0.0689$ bar
1 bar $= 14.5\ lbf/in^2$
1 in water $= 249\ N/m^2$
1 mmHg (mercury) $= 133\ N/m^2$

Several slightly different values are taken for one atmosphere.

760 mmHg $= 1.01 \times 10^5\ N/m^2$
$= 1.01$ bar
1 bar $= 753$ mmHg.

HEAT, WORK, AND OTHER FORMS OF ENERGY

The SI unit is the joule (J) = N.m

1 footpound (ft . lbf.) = 1.356J

For thermal calculations it is often convenient to use the British thermal unit (Btu), or the corresponding centrigrade heat unit (CHu) or kilogram calorie (kg . cal). These units vary with the specific heat of water; and hence with the temperature which is assumed, or defined. For this book a specific heat of 1 is adequate. Then 1 Btu equals the heat required to raise 1 lb water 1°F = 1.06 kJ (1.06×10^3J). Similarly, 1 CHu, the heat to raise 1 lb water 1°C, = 1.91 kJ, and 1 kg . cal, the heat to raise 1 kg water 1°C = 4.19 kJ.

Energy is also expressed in terms of power × time.

1 horsepower hour (hph) = 2.68 MJ (2.68×10^6 J)
1 kilowatt hour (kWh) = 3.6 MJ

POWER

Power is rate at which work is done, so that the SI unit, the watt (W) = J/s.

1 horsepower (hp) = 746 W, or 0.746 kW
= 33 000 ft . lbf/min
= 550 ft . lbf/s

LATENT HEAT

1 Btu/lb = 2326 J/kg
= 0.556 CHu/lb
1 CHu/lb = 4187 J/kg

SPECIFIC HEAT

1 Btu/lb . °F = 1 CHu/lb . °C = 4187J/kg . °C

THERMAL CONDUCTIVITY

$$1\ \text{Btu . ft/ft}^2\text{ . h . °F} = 1\text{CHu . ft/ft}^2\text{ . h . °C} = 1.73\ \text{W/m . °C} = 1.73\ \text{J . m/m}^2\text{ . s . °C}$$

COEFFICIENT OF HEAT TRANSFER

$$1\ \text{Btu/ft}^2\text{ . h . °F} = 1\ \text{CHu/ft}^2\text{ . h . °C} = 5.68\ \text{W/m}^2\text{ . °C} = 5.68\ \text{J/m}^2\text{ . s . °C}$$

SURFACE TENSION

The SI unit is N/m.

$$1\ \text{dyne/cm} = 10^{-5}\text{N/m}\ 10^{-2} = 10^{-3}\text{N/m}$$

DYNAMIC, OR ABSOLUTE, VISCOSITY

The SI unit is N . s/m^2

$$1\ \text{centipoise (cP)} = 10^{-3}\text{N . s/m}^2$$
$$1\ \text{lbf . s/ft}^2 = 47.9\ \text{N . s/m}^2$$

KINEMATIC VISCOSITY

The SI unit is m^2/s.

$$1\ \text{centistoke (cSt)} = 10^{-6}\text{m}^2\text{/s}$$
$$1\ \text{ft}^2\text{/s} = 0.093\ \text{m}^2\text{/s}$$

NOTE ON THE UNITS OF VISCOSITY

The SI system avoids a complication in older systems which can result from confusion between the lb mass and the lb force; but the relationship between dynamic viscosity, kinematic viscosity and fluid density can conveniently be explained in terms of the older English system. It can also be discussed in terms of SI units when the acceleration is unity.

Considering only dimensions and substituting lbf = lb mass × ft/s^2 in the expression for absolute viscosity lbf . s/ft^2, it is found that absolute viscosity also has the dimensions

$$\frac{\text{lb mass . ft . s}}{\text{s}^2\qquad \text{ft}^2} = \frac{\text{lb mass}}{\text{ft . s}}$$

Kinematic viscosity, which is absolute viscosity divided by density, thus has the dimensions

$$\frac{\text{lb mass}}{\text{ft . s}} \times \frac{\text{ft}^3}{\text{lb mass}} = \text{ft}^2\text{/s},$$

as before.

1

Preliminary information including some definitions and some elementary chemistry

1.1 HISTORICAL OUTLINE

Although some cleaning operations can be carried out with organic solvents and other materials, most are done with water; and the detergents (cleaning agents) discussed in this book are intended to improve the cleaning performance of water.

Not surprisingly, the early history of detergent materials is somewhat obscure. Crude alkaline carbonates became available at an early date from several sources:

- Natural deposits of sodium carbonate from some arid regions of North East Africa.
- Barilla, the ash of seaweeds, which contains sodium carbonate.
- The ashes of wood, and other land plants, which contain potassium carbonate (potash).

It is not known when the process of converting these alkali carbonates to the corresponding caustic alkalis (the hydroxides) by the action of slaked lime was discovered, although this is sometimes said not to have been until towards the end of the Middle Ages.

Crude soaps seem to have first been made some 2000–3000 years ago in the Nile valley and in early centres of civilization around the Mediterranean sea, but it is also said that the art was learnt from Celtish peoples further north. The Romans are known to have been considerable users of soap; and during the period of the Roman Empire stale urine was collected as a source of ammonium carbonate for cleaning purposes.

Fats from domestic animals which had been killed for food, and olive oil (which could be obtained more readily than other vegetable oils), were used for the manufacture of soap. Until caustic alkalis became available, saponification must have been difficult and probably incomplete. The fats may have been allowed to deteriorate with the development of free fatty acids which react readily with alkali carbonates; and the soap so formed would have helped the reaction. The process has been said to have been one in which the fats and alkali carbonate solution were boiled to dryness in a pot over an open fire. More water was added, and the mixture boiled down again; this process being repeated as many times as necessary. The first reaction may have been slow hydrolysis of the fat to give free fatty acid which was then neutralized by the alkali carbonate to give soap. If sodium carbonate was used, a 'hard soap' was produced, but with potassium carbonate the product would have been some sort of 'soft soap', unless a further stage of salting out with common salt (sodium chloride) was employed to convert the potassium soap to sodium soap.

During the Middle Ages soap was made at various places in Italy, France, Spain, England, and possibly other countries. Marseilles, in France, became particularly famous and many small factories were established there. In some countries soap manufacture was taxed and this tax was not repealed in England until 1853. Following a period of importation from Europe, soap manufacture was started in North America early in the history of the European colonies and of the Republic. Some American companies with currently well-known names were started 150–200 years ago.

In England, and probably many other countries, soap manufacture was carried out in a large number of very small factories; and in farm houses and other homes. Gradually the trade became concentrated in a smaller number of larger factories; and this process has continued and accelerated with developments in the technology and marketing of detergent products.

As the chemical industry developed, alkalis (sodium carbonate and sodium hydroxide) made from salt have largely replaced the sources mentioned earlier. Local tallow was the major fat used in England, but it became necessary to augment it with imported tallow and other oils and fats, such as coconut oil. Export of soap also became an important trade; and these two factors caused areas near to a port to be particularly suitable for soap factories.

It so happened that some major centres of soap manufacture in England, including London and Bristol, were also ports. Merseyside was already an important soapmaking area supplying soft and hard soap to the textile mills, as well as for other uses, when the port of Liverpool and the rise of the alkali industry adjacent to the salt deposits in Cheshire made Merseyside predominant in the soapmaking field. Similar factors doubtless influenced the location of soap manufacture in other countries.

Soap cannot be used in acid solution (because it is decomposed with the liberation of fatty acids); and it forms a precipitate with the calcium and magnesium in hard water. Hence, although soap is a good all-round detergent, it has limitations, particularly for some industrial purposes in the textile field. About 1835, Fremy and Runge reacted olive oil with sulphuric acid; and in

1874 Turkey Red oil was made by the action of sulphuric acid on castor oil. These products, and other similar ones, were found useful in certain dyeing and other textile processes. As explained later, castor oil contains a large proportion of an hydroxy acid which can be sulphated, but with both castor and olive oils the reactions with sulphuric acid are complex.

During the first and second world wars, and the period between, various alternatives to soap were developed, particularly in Germany and later in the U.S.A. In part, this was to obtain detergent products from petroleum, and other non-fatty materials, because of shortages of oils and fats, although some soap alternatives were, and still are, made from fats. By the mid-1950s, products based on non-soapy detergents had obtained a major share of the market in many developed countries. Soap powders continued to be very important in the United Kingdom and are still significant there, although they are declining steadily. In most developed countries soap products are now relatively insignificant in the domestic market, except for toilet soaps used for personal washing. In less developed countries, household soaps continue to be the main detergent. The rise and fall of different types of product, and some recent trends, are discussed later.

1.2 SURFACTANTS, BUILDERS AND FILLERS

1.2.1

Soaps, and the alternative non-soapy detergents, have a number of important properties which are discussed in Chapter 2. One such property is the reduction of the surface energy, or surface tension (more strictly interfacial tension), of solutions which contain them, so that they are called surface active agents. This is commonly contracted to *surfactants,* the term used in this book. On the continent of Europe, the equivalent term *tensides* is often used. Surfactants contain a hydrophobic (water hating, or repelling) and a hydrophilic (water loving, or attracting) end. Alternatively, this can be expressed as a lipophilic (fat loving, equals hydrophobic) and a lipophobic (fat hating, equals hydrophilic) end. Hartley described such chemical compounds as having amphipathic properties. Non-soapy surfactants are also sometimes called *syndets.*

1.2.2

Solutions of acids, bases (alkalis) and salts contain charged particles called ions. Ions which carry positive electrical charges are called cations and others which carry negative charges are called anions. Ions can carry one unit of charge (called monovalent or univalent), two units (divalent), and so on. Examples are:

hydrochloric acid which ionizes $HCl = H^+ + Cl^-$
sulphuric acid which ionizes $H_2SO_4 = H^+ + HSO_4^-$ and $HSO_4^- = H^+ + SO_4^{2-}$
sodium hydroxide which ionizes $NaOH = Na^+ + OH^-$
ammonium hydroxide which ionizes $NH_4OH = NH_4^+ + OH^-$
sodium carbonate which ionizes $Na_2CO_3 = 2Na^+ + CO_3^{2-}$
magnesium sulphate which ionizes $MgSO_4 = Mg^{2+} + SO_4^{2-}$
trisodium orthophosphate which ionizes $Na_3PO_4 = 3Na^+ + PO_4^{3-}$

All normal chemical compounds are electrically neutral, so that the charges on the positive and negative ions which they contain, or produce in aqueous solution, balance. That is sodium chloride and magnesium sulphate are Na^+Cl^- and $Mg^{2+}SO_4^{2-}$ respectively, but sodium sulphate is $Na_2^+SO_4^{2-}$.

It should be noted that chloride ions (Cl^-) have completely different properties from those of chlorine gas (Cl_2). Similarly, the sodium ion (Na^+) differs from metallic sodium (Na).

Many surfactants ionize; and those with the hydrophobic-hydrophilic portion in the anion are called *anionic surfactants.* Examples are:

- soap, which is a mixture of salts (usually sodium) of long chain fatty acids, $R\cdot COO^-Na^+$, where R represents a long hydrocarbon chain.
- alkyl benzene sulphonates, $R-C_6H_4-SO_3^-Na^+$.

(the difference between sulphonates and sulphates is discussed later).

Similarly, surfactants with the hydrophobic-hydrophilic portion in the cation are described as *cationic.* Examples are:

- Quaternary ammonium salts, such as alkyl trimethyl ammonium chloride, $R-N(CH_3)_3^+Cl^-$
- Alkyl pyridinium salts, such as cetyl pyridinium bromide, $C_{16}H_{33}-C_5H_4N^+Br^-$.

The balancing ions, such as Na^+ or Cl^-, in anionic and cationic surfactants respectively, are sometimes called *gengenions.* Unlike the anionics, which are the most important group of surfactants used in detergent products, the cationics are used mainly in fabric softeners and as germicides.

A third group of surfactants, which is becoming increasingly important in the detergents field is the *nonionic* type. By definition, these compounds do not ionize, but they contain a hydrophobic and a hydrophilic end to make them surfactants. There are naturally occurring nonionic surfactants, such as saponin, but those of importance in the detergents industry are synthetic. The hydrophobic portion is normally an alkyl chain, or an alkyl substituted aromatic hydrocarbon. The hydrophilic portion is usually a polyglycol chain produced by the condensation of ethylene oxide with itself and with an active hydrogen atom in the hydrophobe molecule. In the most important sub-group, the active hydrogen atom is in an hydroxyl group of a long chain alcohol, or an alkyl phenol. The reactions can be illustrated by reference to a long chain alcohol, R–OH.

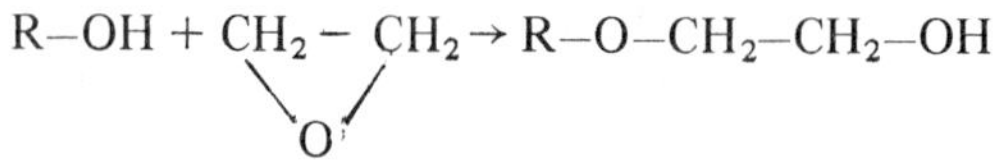

This compound, in which one molecule of ethylene oxide has been attached, is itself an alcohol; and more ethylene oxides can be added similarly to form a polyglycol chain containing n $-O-CH_2-CH_2-$ groups.

$$R-O-CH_2-CH_2-OH + n(CH_2)_2O \rightarrow R-O-CH_2-CH_2-(O-CH_2-CH_2)_{n-1} -O-CH_2-CH_2-OH$$

Other compounds to which ethylene oxide can be added to form nonionic surfactants are fatty acids (in which the carboxyl group, –COOH, contains an –OH), long chain amines and long chain amides, in which there is an active hydrogen atom in the $-NH_2$ group.

The ethylene oxide condensates just considered end with an alcohol group; and this can be sulphated to produce another type of anionic surfactant.

Certain nonionic surfactants are made with propylene oxide in conjunction with ethylene oxide; these will be considered later.

There is also a fourth group of surfactants known as the *amphoterics*. These can ionize as either anionic or cationic substances, depending upon the alkalinity, or acidity, of the solution in which they are present. Amphoteric surfactants are generally expensive and used only in rather specialized products.

A very large number of surfactants have been developed and the more important types, with their properties and uses, are discussed later. Extensive lists of surfactants, including their trade names, are given in various books, particularly Sisley and Wood (1972).

1.2.3

It will be appreciated that soap is a detergent, but the word 'detergent' is frequently confined to the non-soapy surfactants and products based upon them. In this book the term 'detergent' will be taken to include soap; and non-soapy types will be called non-soapy detergents (abbreviated to NSDs). NSDs are also called soapless detergents, or synthetic detergents. The latter term is not very appropriate, as soaps can also be made from fatty acids produced by the oxidation of petroleum-based hydrocarbons.

1.2.4

Soaps and non-soapy surfactants may require the addition of other substances to make them perform well. This is particularly so with the NDSs, which need additives to make them effective heavy duty fabric washing products capable of dealing with really dirty clothes and other articles. Such additives are called *builders*. The overall cleaning effect of a detergent solution is the net result of soil removal and soil redeposition; and some authors say that only substances which perform both functions should be called builders. *Fillers* are substances added to a detergent to cheapen the product. In some cases the distinction between fillers and builders is blurred. For example, sodium silicate and sodium carbonate which are, or were, added to household soaps to reduce cost also have some beneficial effects.

1.3 TYPES OF SOAPS AND SOAPMAKING TERMS

1.3.1

The oils and fats used in the manufacture of soap are essentially triglycerides, that is esters of the trihydric alcohol, glycerol. In this book the common, but not universal, convention which reserves the word *'glycerol'* for the pure compound, $CH_2OH–CHOH–CH_2OH$ (1,2,3 trihydroxypropane or 1,2,3 propanetriol), and calls the actual materials *'glycerines'*, is followed. For example, a soap lye crude glycerine may be found to contain 83.5% glycerol and a distilled glycerine 99.1% glycerol. The relevant British Standard specifications BS2621-5:1979 use the word glycerol for both the pure compound and the actual materials. The fatty acids which are combined with the glycerol in a fat are almost exclusively straight chain monocarboxylic acids with an even number of carbon atoms, including the one in the carboxylic group. For example:

stearic acid	$C_{17}H_{35}\cdot COOH$
oleic acid	$C_{17}H_{33}\cdot COOH$
lauric acid	$C_{11}H_{23}\cdot COOH$

The triglycerides, which comprise the bulk of an oil or fat which is fresh, have the formula $CH_2{-}O{-}OCR_1{-}CH{-}O{-}OCR_2{-}CH_2{-}O{-}OC{-}R_3$ where R_1, R_2 and R_3 are long chain alkyl groups which may be the same, or different. The oils and fats, and their fatty acids and minor constituents, are discussed in Chapter 3. Here it is only necessary to add that an oil is a fat which is liquid under the prevailing conditions, but the distinction is somewhat arbitary and a matter of convention. Coconut oil is normally solid in temperate areas, but it is called an oil because it is liquid in its tropical countries of origin. Tallow, which is a typical fat, is now commonly transported as an oil in heated road and ships' tanks. In this book it will frequently be convenient to use the single word fat to include both oils and fats.

1.3.2

The soaps used as detergents are the soluble salts of mixtures of fatty acids. The cation is usually sodium, but potassium, or a mixture of sodium and potassium, is used in soft and shaving soaps. Ammonia (giving ammonium soaps) and various organic nitrogenous bases are sometimes used to neutralize the fatty acids in specialist products. The fatty acid salts of other metals are substantially insoluble. Those of calcium and magnesium are formed as undesirable precipitates when soap is used in hard water, but they, and some other insoluble soaps, are used for various industrial purposes outside the scope of this book.

Household sodium soaps are called *'hard soaps'* to distinguish them from jelly-like *'soft soaps'*, which were at one time used extensively in the textile industry and the home. As already mentioned, soft soaps are commonly made with potassium as the cation; and from liquid oils. The term hard soap can be confusing in that these soaps are normally softer when made than the other main group of solid soaps, the toilet soaps.

1.3.3

Sodium soaps can be made by:

- *direct saponification*, almost always with caustic soda.

$$\underset{\text{fat}}{CH_2\cdot O\cdot OCR\cdot CH\cdot O\cdot OCR\cdot CH_2\cdot O\cdot OCR} + \underset{\text{caustic soda}}{3NaOH} \rightarrow$$

$$\underset{\text{glycerol}}{CH_2OH{-}CHOH{-}CH_2OH} + \underset{\text{soap}}{3R\cdot COO\,Na}$$

- *splitting,* or *hydrolysis,* of the fat with water to fatty acids plus glycerol; followed by separation of the glycerol, and neutralization of the fatty acids with caustic soda, or sodium carbonate, to form soap.
- *neutralization of fatty acids produced in other ways,* particularly by the oxidation of a petroleum hydrocarbon.

A major distinction is between soaps made by an *'all-in'* process and soaps made by the *'fully-boiled'* process. All-in soaps are so-called because the fat (or fatty acid) and the alkali are just mixed, so that everything put into the pot goes into the product. Glycerol is not separated; and it is important to avoid any undue excess of alkali. All-in processes are divided into *'semi-boiled',* in which external heat is applied, and *'cold-process'* in which it is not, the only heat coming from the heat of the saponification reaction. All-in processes can be carried out very simply; and they are still used by some small local manufacturers in some of the less-developed countries, where these manufacturers are sometimes known as traditional soapmakers. Soaps which include potassium are made by an all-in process, because graining with salt, which is an inherent part of the fully-boiled process, would displace potassium by sodium.

Most soap made by direct saponification of fats is produced by some version of the fully-boiled process. As discussed in detail in Chapter 4, this process includes three main stages, although other operations are sometimes added:

- *saponification,* the chemical reaction which forms soap and liberates glycerol. This is sometimes called *killing* the fat.
- *washing,* in which the soap mass is *grained,* or *salted-out,* with salt (or other electrolyte) and, on standing, separates into a lower aqueous layer called *lye,* and an upper layer of *soap curd,* also called *kettle wax.* The soap curd contains a proportion of enmeshed lye. A washing operation transfers part of the glycerol to the lye; and also removes dirt from the soap.
- *fitting* or *finishing.* In this operation the electrolyte content of the soap mass is adjusted so that on prolonged standing (or passage through a centrifuge) an upper layer of *neat soap,* or *soap base,* and a lower layer of *nigre,* or *nigre lye,* is obtained. The nigre is reprocessed. Fitting achieves a further purification of the soap and produces the neat soap in a desirable physicochemical state. In America, fitting is commonly called *pitching.*

Removal of the neat soap from a vessel in which it is settled is called *cleansing.* Soapmakers use the term *closing* to describe a reduction in electrolyte content (with consequent reduction in tendency to separate) and *opening* for the converse operation. Those terms, however, have rather different connotations in relation to the washing and the fitting stages.

The large vessels in which soap is made in the traditional form of the process are often called *pans* or *kettles.*

Bar soaps, known as *curd soaps,* were made at one time by an elaboration of the washing stage just described; but neat soap is the normal starting material for the production of most soap products (hence the name 'soap base'). If the fitting stage is carried out properly, and the soap is well-settled, the neat soap has a fairly constant composition, although this can be varied somewhat as discussed in Chapter 4. Usually neat soap contains about 61–64% *TFM* (*total fatty matter*), equivalent to about 67–69% anhydrous soap and 30–32% water.

When neat soap is cooled with no more than minor additions (if any), a *genuine soap* is produced. It is also possible to reduce the soap content of the product by addition of suitable alkaline solutions (or dilute brine) to give what are called *filled,* or *run,* soaps. A special type of filled soap is described as *glued-up.*

Genuine soaps made by the older processes using frames, or water coolers, have substantially the same composition as the base from which they are produced. With the modern spray-chilling and extrusion processes, the distinction between genuine and filled soaps is obscured by the fact that evaporation occurs during the process and dilute electrolyte has to be added if it is required to bring the TFM back to 62–64%. Genuine and filled soaps can be sold as *bars,* or cut into *billets* and stamped into *tablets.* These soaps are known by various names: *laundry, common,* or *household,* as well as *hard* or *bar.* No one term is entirely satisfactory to cover all variations from relatively crude bars to perfumed and packaged tablets which approach toilet soaps, but the word *household* is adopted in this book.

1.3.4

To a soapmaker a *lye* is an aqueous liquor which normally contains only a negligible concentration of soap, the exception being a *thick*, or *nigre lye* which may carry about 1–3% soap. The term lye is used in several ways. It came originally from the alkali industry, where it means a solution of caustic soda (or possibly other alkali); and it is sometimes used in this sense in the soap industry, particularly by older soapmakers. The lye separated from soap curd in the washing stage of soapmaking is called *spent lye* provided it contains little caustic soda. If it contains a substantial concentration of caustic soda (and hence requires to be processed further in the soapmaking department), it is called *half-spent lye.* The glycerol concentration in the spent lye produced is very important; but when a soapmaker speaks of the strength of a lye (or a soap mass), the reference is normally to caustic soda rather than glycerol content.

1.3.5

When fats are hydrolysed with water, with or without a catalyst, the process is often called *splitting,* and the aqueous glycerine separated *sweet water.* In some countries, soap lyes are called sweet water, and some glycerine distillers use the term sweet water for the dilute liquor produced in some types of still, which is otherwise called *weaks.*

In this book the term *lye* is used for any liquor separated after soap has been grained, and *sweet water* for the glycerol bearing liquor from a splitting operation.

1.4 ACIDS, ALKALIS, AND BUFFERS. pH AND INDICATORS

1.4.1

Acids and bases are now defined as proton donors and proton receivers respectively, but for present purposes, the older and narrower definition of an acid as a substance which yields hydrogen ions in solution and a base (alkali) as one which yields hydroxyl ions is adequate. Actually, all aqueous solutions contain both hydrogen and hydroxyl ions, and the definitions should be in terms of an excess of either hydrogen or hydroxyl ions.

Pure water itself ionizes reversibly to a very small extent, $H_2O \rightleftharpoons H^+ + OH^-$, and the equilibrium constant is equal to $\frac{[H^+] \times [OH^-]}{[H_2O]}$,

where $[H^+]$ = concentration of hydrogen ions in gram equivalents (moles) per litre†

$[OH^-]$ = concentration of hydroxyl ions in gram equivalents (moles) per litre†

and $[H_2O]$ = concentration of molecular water in gram equivalents (moles) per litre†

As the extent of the ionization is very slight, the concentration of molecular water is constant, and $[H^+] \times [OH^-] = K$, the ionic product for water. At 25°C, $K = 10^{-14}$, so that $[H^+] \times [OH^-] = 10^{-14}$. To avoid the need to consider concentrations in terms of 10 raised to various negative powers, the pH and pOH scales have been devised. pH is defined as $-\log_{10}[H^+]$ and pOH as $-\log_{10}[OH^-]$. The ionic product equation can then be rewritten pH + pOH = 14. From the definitions of pH and pOH it follows:

- the lower the pH the higher the concentration of hydrogen ions, that is the more acidic the solution; and similarly for pOH.
- because the scales are logarithmic, a solution with a pH of 3 contains ten times the hydrogen ion concentration of one with a pH of 4.

From the relationship pH + pOH = 14, pOH = 14 − pH, and for most purposes the pOH scale is unnecessary, as the whole range of acidity and alkalinity can be expressed in terms of pH.

In solutions which are neutral (neither acidic nor alkaline) pH = pOH = 7; and those with a pH below 7 are acidic and those with a pH above 7 are alkaline, or basic.

A decinormal (N/10) solution of any strong acid, assumed to be fully ionized, contains 0.1 gram equivalents, or moles, of hydrogen ions per litre, that is $[H^+] = 0.1$; and its pH $= -\log_{10}[0.1] = 1$. More concentrated solutions of strong acids can have a negative pH, as well as one between 1 and 0. Similarly, an N/10 solution of a strong base, say caustic soda, has $[OH^-] = 0.1$ and a pOH of 1. Its pH is, therefore, 14 − 1 = 13. Again, more concentrated solutions of caustic soda have a pH above 13.

†Strictly activities should be used instead of concentrations, but this elaboration is beyond the scope of this treatment.

To set N/10 caustic soda in context with other alkalis to be discussed later, it contains 4 g/l of NaOH, or 0.4% m/m as its specific gravity is approximately 1.

Strong lavatory cleaners are commonly acids, but most other cleaning with aqueous solutions requires alkaline conditions. Both pH (alkaline potential) and quantity of alkali present are important, but should be distinguished. From the previous discussion it will be noted that an N/10 solution of caustic soda has a high alkaline potential, but only a small alkaline capacity (that is, it will only neutralize a small quantity of acid).

1.4.2

Four classes of salts, those from strong acid-strong base, weak acid-strong base, strong acid-weak base, and weak acid-weak base are now considered.

Salts of a strong acid and a strong base

Solutions of salts of strong acids and strong bases, such as sodium chloride and sodium sulphate, are substantially neutral. The alkylbenzene sulphonic acids are strong, so that the sodium alkylbenzene sulphonates are neutral.

Salts of a weak acid and a strong base

The alkalis used in detergent products are nearly always salts of this type, so that their properties are important to the detergents technologist.

Consider the addition to pure water of sodium acetate, $CH_3{\cdot}COONa$, a substance similar in many ways to a soap, although acetic acid is soluble in water and stronger than the fatty acids in soap (but still weak compared with hydrochloric acid). The anion of the weak acid, the acetate ion in this example, reacts with water to form molecular acetic acid and hydroxyl ions

$$CH_3{\cdot}COO^- + H_2O \rightarrow CH_3{\cdot}COOH + OH^-$$

so that the solution becomes alkaline. A more complete way to consider the situation is to say that the solution contains sodium ions (Na^+) and that all the following reactions must reach and maintain equilibrium:

$$CH_3{\cdot}COO^- + H_2O \rightleftharpoons CH_3{\cdot}COOH + OH^- \quad (a)$$
$$H_2O \rightleftharpoons H^+ + OH^- \quad (b)$$
$$CH_3{\cdot}COOH \rightleftharpoons H^+ + CH_3COO^- \quad (c)$$

The sequence of events can be indicated as follows:

(1) reaction (a) forward increases $[OH^-]$.
(2) the increase in $[OH^-]$ from (1) causes a decrease in $[H^+]$ from (b) in reverse.
(3) the decrease in $[H^+]$ from (2) leads to an increase in $[CH_3{\cdot}COO^-]$ from (c) forward.
(4) the increase in $[CH_3{\cdot}COO^-]$ from (3) causes further increase in $[OH^-]$ from (1) above, and so on until all the equilibria are established.

The sodium ion in the sodium acetate solution does not react significantly with water, $Na^+ + H_2O \rightarrow NaOH + H^+$ because NaOH is a strong base which ionizes

$NaOH \rightarrow Na^+ + OH^-$, followed by $H^+ + OH^-$ -
to $Na^+ + H_2O$.

The weaker the acid, the higher the co
the lower that of its ions; hence the higher t
produced and the more alkaline the solution
a pH around 9, depending upon the type of s
acid and the silicic acids are very weak, s
alkaline solutions than soap. Sodium carbo
bicarbonate (which can be regarded as carbonic acid half
caustic soda) is mildly alkaline.

The hydrolysis reaction for the bicarbonate ion is $HCO_3^- + H_2O \rightleftharpoons H_2CO_3 + OH^-$. That for the carbonate ion takes place in two stages

$$CO_3^{2-} + H_2O \rightleftharpoons HCO_3^- + OH^-$$
$$HCO_3^- + H_2O \rightleftharpoons H_2CO_3 + OH^-$$

Salts of a strong acid and a weak base

In this case the cation of the weak base reacts with water leaving the solution acidic. For example with ammonium chloride $NH_4^+ + H_2O \rightleftharpoons NH_4OH + H^+$.

Salts of a weak acid and a weak base

In a solution of a salt of a weak acid and a weak base both cation and anion react with water. For ammonium acetate

$$NH_4^+ + CH_3{\cdot}COO^- + H_2O \rightleftharpoons NH_4OH + CH_3{\cdot}COOH$$

The pH of the solution then depends upon the relative strengths of the acid and base.

This discussion of the acidity or alkalinity of various salts shows that the presence of a replaceable hydrogen in a formula does not necessarily make a compound acidic in reaction. The different strengths of sulphuric and carbonic acids cause sodium bisulphate ($NaHSO_4$) to be a strong acid whereas sodium bicarbonate ($NaHCO_3$) is a weak alkali, although both are their respective acids half neutralized with caustic soda.

1.4.3

Fig. 1.1, courtesy of PQ International Inc., shows how the pH of some important alkalis varies with concentration. It will be noted that the pH of solutions of the less strong alkalis tends to level off as concentration increases and never reaches the levels produced by caustic soda and the very alkaline silicates. In some cases it even falls with an increase in concentration. Table 1.1 gives information concerning the pH of 1% m/m solutions of various alkalis. The different types of sodium silicate are described in section 1.5.7; and the types of sodium phosphate in section 1.5.10. Reference is also made later to the other alkalis listed.

1.4.4

A buffer, or buffer solution, is one in which the pH changes relatively little when an acid or alkali is added to it. Many 'soils' (i.e. the matter to be removed by

...ashing process) (particularly those which contain sweat) include acidic ...stituents, and it is important that detergent solutions should remain at near ...he selected pH throughout the washing process; this requires a buffering effect. Buffers are also used to produce solutions of known pH, for example, in the preparation of a set of comparator tubes which show the colours developed by an indicator at a series of known pH values.

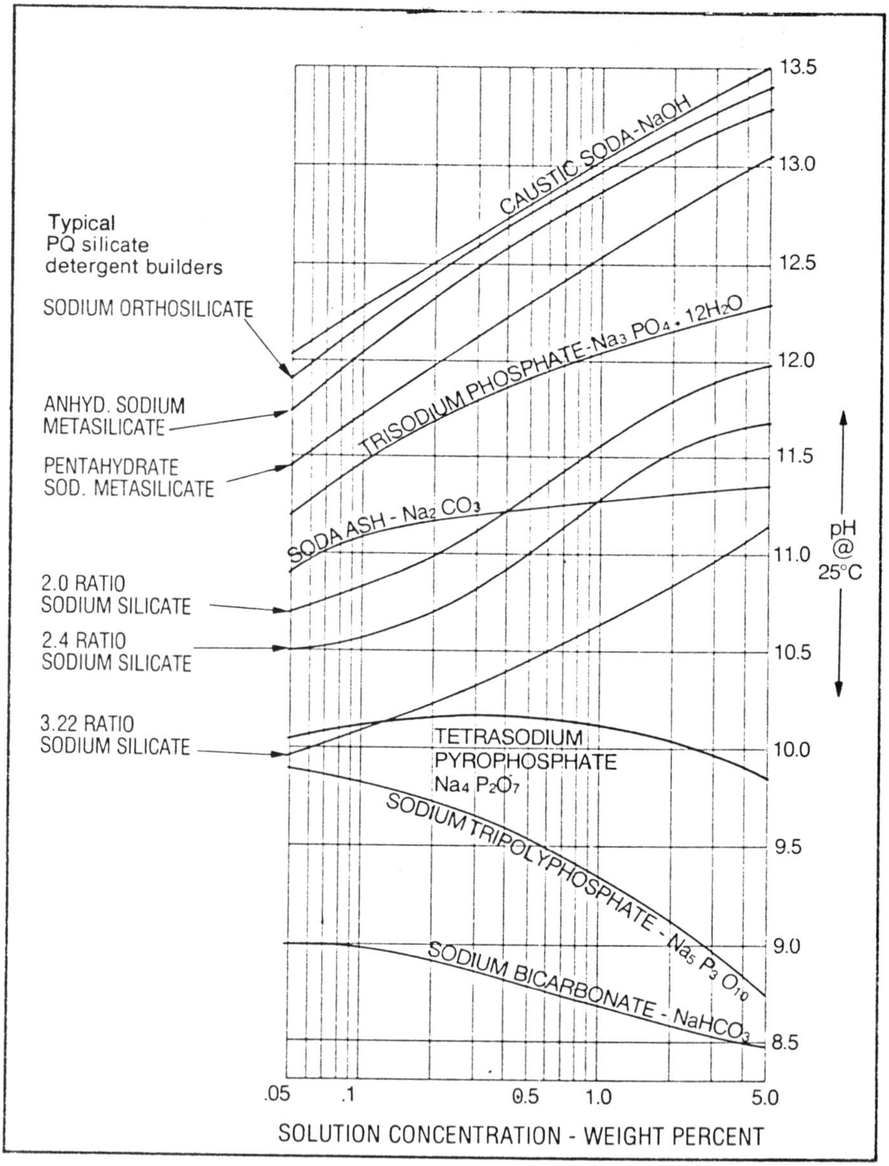

Fig. 1.1 – pH/concentration curves for some important alkalis.

Table 1.1 – Approximate pH values for 1% m/m solutions of various alkalis. (Figures quoted by different authors vary slightly, and the figures in this table do not all agree exactly with Fig. 1.1).

Alkali	pH 1% m/m solution
Caustic soda	13.2
Sodium orthosilicate, anhydrous	12.9
Sodium sesquisilicate, anhydrous	12.6
Sodium metasilicate, anhydrous	12.3
Trisodium orthophosphate, anhydrous	12.0
Ammonium hydroxide	11.4
Sodium carbonate, anhydrous	11.4
Alkaline sodium silicate, anhydrous	11.2
Sodium percarbonate	10.6
Neutral sodium silicate, anhydrous	10.5
Sodium pyrophosphate, anhydrous	10.5
Sodium perborate, $NaBO_2 \cdot H_2O_2 \cdot 3H_2O$	10.2
Sodium sesquicarbonate, $Na_2CO_3 \cdot NaHCO_3 \cdot 2H_2O$	10.0
Sodium tripolyphosphate, anhydrous	9.5
Borax, $Na_2B_4O_7 \cdot 10H_2O$	9.2
Disodium hydrogen phosphate, anhydrous	8.8
Sodium bicarbonate	8.4
Graham's salt (Calgon), sometimes called sodium hexametaphosphate	6.9

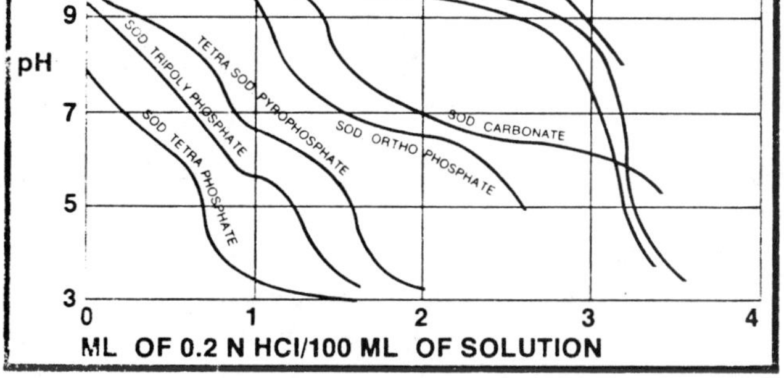

Fig. 1.2– Buffering capacity of various alkali solutions.

Unless there are other differences in composition, as may occur in commercial materials, a hydrate should show the same pH as its anhydrous salt after adjustment of concentration to the same level of anhydrous salt in the two solutions.

Effective buffers are produced by a combination of a weak acid and its salt, or a weak base and its salt. Examples are:

Boric acid (HBO_2) and borax ($Na_4B_2O_7{\cdot}10H_2O$),
Sodium bicarbonate ($NaHCO_3$) and sodium carbonate (Na_2CO_3), and
sodium dihydrogen orthophosphate (NaH_2PO_4), and
disodium hydrogen phosphate (Na_2HPO_4).

In the second and third examples, the sodium bicarbonate and the sodium dihydrogen phosphate act as weak acids. The sodium silicates work in a similar way and are very effective buffers.

The buffering action of a weak acid (HA) and its sodium salt (NaA) can be explained as follows.

The solution contains Na^+ and A^- ions from the salt, together with the unionized acid, its ions and water. The equilibria

$$HA \rightleftharpoons H^+ + A^- \qquad (a)$$

and $$H_2O \rightleftharpoons H^+ + OH^- \qquad (b)$$

have to remain satisfied.

If acid is added to the system $[H^+]$ is increased, but it is then reduced nearly to its original value by reaction (a) in reverse. If alkali is added, OH^- is temporarily increased, but is reduced by reaction (b) in reverse, leaving less $[H^+]$. HA then ionizes according to (a) to restore equilibrium at near its original value. Fig. 1.2, courtesy PQ International Inc., shows the buffering capacity of various salts in terms of pH after various additions of strong acid.

1.4.5

The changes in pH as an alkali is added to an acid, as in an analytical titration, depend upon the strengths of the acid and the alkali. Sections of the curves for the four cases are shown approximately in Fig. 1.3 (a), (b), (c) and (d). The end point in a titration can be determined using an electrical pH meter, or from conductivity measurements, but indicators which change colour at certain values of pH are commonly used, particularly in factory practice. Many indicators are available and their colour changes occur at different pH ranges; this makes the proper selection of indicator important in many cases. Some common indicators with the pH range over which the colour change occurs are shown in Table 1.2. In order to obtain the correct end point in a volumetric analysis it is necessary to select an indicator which changes colour at near the equivalence point. Suitable indicators can be selected as shown in Table 1.3 by combining the information in Table 1.2 with that from Fig. 1.3.

Some important applications in the detergents industry are:

- The estimation of *free fatty acid (FFA)* in a fat is carried out in alcoholic solution (FFA is soluble in alcohol), by titration with standard caustic soda, or caustic potash, to the phenolphthalein end point (estimation of a weak acid with a strong base).

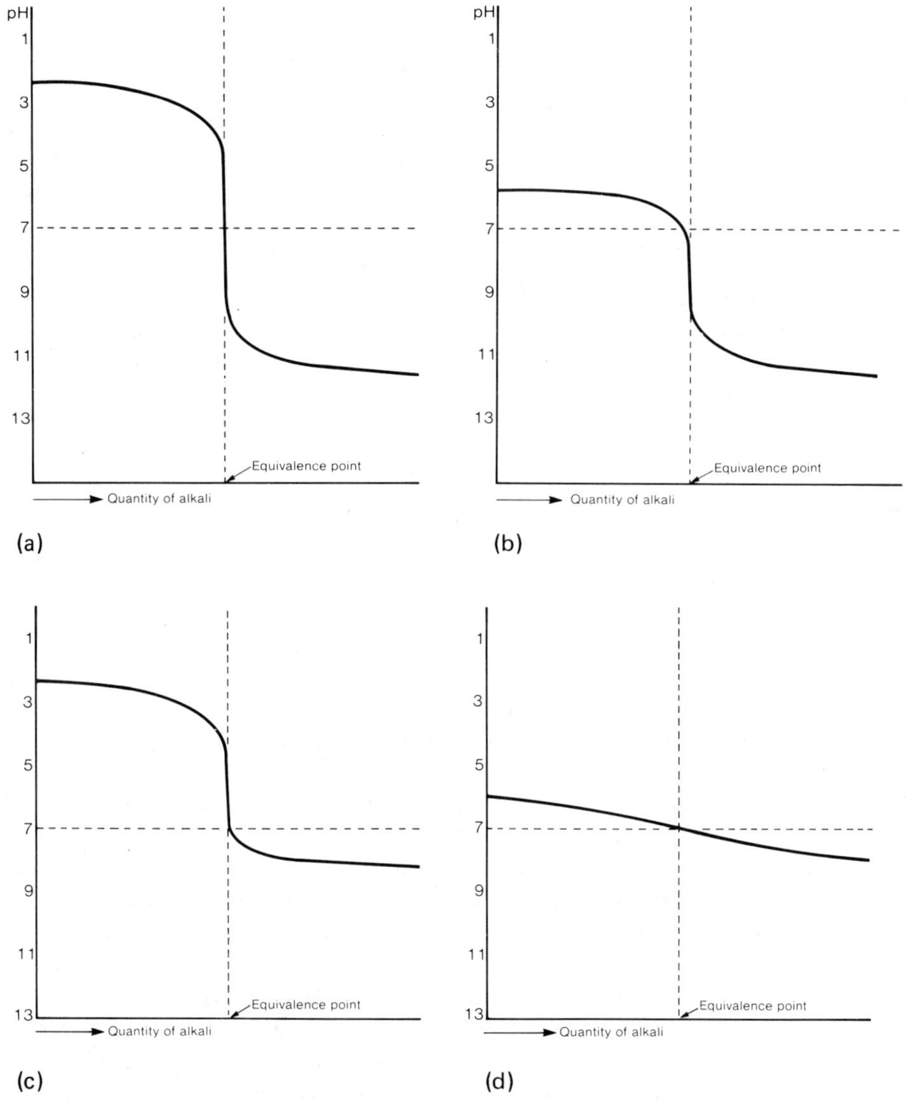

Fig. 1.3 – Titration curves.

Table 1.2 – Some common indicators

Indicator	Colour change	pH range
Methyl orange	red – yellow orange	3.1– 4.5
Bromcresol green	yellow – blue	3.8– 5.4
Methyl red	red – yellow	4.7– 6.4
Litmus	red – blue	5.0– 7.0
Cresol red	yellow – red	7.2– 8.8
Thymol blue	yellow – blue	8.0– 9.6
Phenolphthalein	colourless – purple	8.5–10.5
Thymolphthalein	colourless – blue	9.4–10.6

Table 1.3 – Indicator selection

Titration	Approx. pH range at equivalence point	Suitable indicator
Strong acid – strong base	4–10	Almost any
Weak acid – strong base	7–10	Phenolphthalein
Strong acid – weak base	4– 7	Methyl orange, methyl red
Weak acid – weak base	very narrow	No indicator satisfactory

- To estimate *free caustic alkali in a soap,* the sample is dissolved in hot alcohol (neutralized if necessary), barium chloride solution is added to precipitate any sodium carbonate present, and the caustic soda is titrated with standard sulphuric acid to the phenolphthalein end point. Phenolphthalein is chosen because it is required to estimate only the strong alkali, caustic soda, and not to include mild alkalis such as soap. It is necessary to remove sodium carbonate from solution because it is alkaline to phenolphthalein (in the titration of a sodium carbonate solution the phenolphthalein end point is at the bicarbonate stage).
  Soaps are often checked for caustic soda content by 'spotting' a freshly cut surface with phenolphthalein solution, noting whether a purple colour develops, and, if so, how quickly. With experience it is possible to judge roughly the free caustic soda content of the sample. Soapmakers now use this test, instead of the traditional testing on the tongue, at relatively high levels of caustic soda. For the routine examination of production samples of toilet soap, which should contain little, or no, free caustic soda, spotting is useful to decide whether, or not, a quantitative estimation is necessary.
- If it is required to estimate the *total free alkali* (hydroxide plus carbonate) in a soap, an excess of standard sulphuric acid is added to the sample in alcoholic solution, the mixture is boiled to remove carbon dioxide liberated from the carbonate, and the excess acid (actually fatty acid liberated from

the soap by the stronger sulphuric acid) is back titrated with standard caustic soda to the phenolphthalein end point.

- An important sequence of tests is used to determine *total fatty matter (TFM), total alkali* and *'soda as soap'*. The sample of soap is decomposed by the addition of a known quantity of standard acid, the fatty matter liberated is separated (by ether extraction) and weighed. The excess acid is back titrated with standard alkali to the methyl orange end point. Methyl orange is selected for this estimation, because it is required to include alkali combined with weak acids, such as silicic and boric. Hence weak acids left in the aqueous liquid from which the fatty matter has been separated should not be allowed to act as acids in the analysis. The alkali which is combined with the fatty acids to form soap, known as 'soda as soap', is determined by dissolving the extracted fatty matter in neutral alcohol and titrating to the phenolphthalein end point with standard caustic soda. In many routine tests, it is only required to estimate TFM. It is then not necessary to use standard acid, and it is possible in many cases to add a weighed piece of beeswax and to separate the fatty matter which has been liberated from the soap in the form of a wax cake. The weight of TFM is found by the deduction of the weight of the added beeswax from that of the cake. Some sample calculations using the results of tests such as those just outlined are given section 1.7, after various materials have been discussed.

1.4.6

The term *'active alkali'* is used by some authors to denote that portion of the alkalinity of a material which is titratable to phenolphthalein, that is the portion which is available at a pH above about 8.5. In caustic soda the whole of the alkalinity is active, in sodium carbonate half, and in sodium bicarbonate none.

1.5 INORGANIC MATERIALS IMPORTANT IN THE DETERGENTS INDUSTRY

1.5.1 Common salt, or sodium chloride (NaCl)

Salt is used mainly to grain soap. It can be purchased in crystalline form, but it is now often bought as a saturated brine containing about 25% NaCl. When crude glycerine is being recovered from soap lyes, a major part of the salt in the lyes is recovered in the evaporation process and is returned to soapmaking as either solid salt, or brine. As explained in Chapter 8, salt is now often recovered directly as brine. In any case some new salt has to be purchased to make up for losses.

Solid salt is used in some soapmaking procedures, but modern processes usually require saturated brine, made by dissolving salt in water, or in a liquor containing a small concentration of glycerol. A convenient system for the preparation of brine on the large scale is shown diagrammatically in Fig. 1.4. The salt is fed in a stream (usually by a worm conveyor) to the top of a small tower to the base of which water is introduced at the required rate. The slurry overflows to one of a group of tanks in which dissolution is completed by compressed air introduced through a perforated pipe. With experience, the flows of water and salt can be adjusted so that saturated brine is produced, leaving only a small excess of salt in the tanks. This has to be dissolved occasionally by addition of water to the tank and blowing with air. If the salt is slightly dirty, or if it contains calcium or magnesium, it is often useful to make the brine slightly alkaline and to filter it before it is sent to the soapmaking department. This also reduces the content of heavy metals, such as copper and iron.

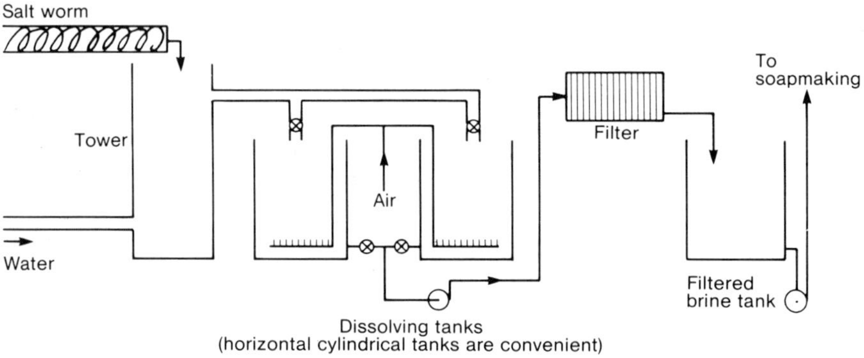

Fig. 1.4 – Brine preparation.

1.5.2 Caustic soda, or sodium hydroxide (NaOH)

Caustic soda is sold as powder, flake, solid blocks produced by running molten, anhydrous, material into lightweight steel drums, and also as a concentrated solution containing 45–50% NaOH. Detergent manufacturers normally need to dissolve solid to produce a concentrated solution before use and, when possible, there are considerable advantages in buying solution, provided transport costs are not excessive. Caustic soda is a dangerous chemical and adequate precautions, including the wearing of goggles and other protective clothing, are important when it is being handled. For reasons that are discussed in Chapter 4, efficient soapmaking normally requires the use of a concentrated solution containing 45–50% NaOH which is subsequently diluted with lyes. The heat evolved when solid caustic soda is dissolved to produce these concentrated solutions is sufficient to raise the temperature to near boiling point. Solid blocks in steel drums are normally the cheapest form of solid, and they can be dissolved by cutting off the ends of the drums with hammer and chisel (and sometimes perforating the cylinders with a pick) before placing the material in the dissolving tank with suitable lifting tackle. The tank should be sized so that the rather small quantity of water needed to produce a solution of the required concentration covers the drum, or drums. Some operators use a tank with a perforated plate, or grid, near the base. The rise in temperature and density gradients provide circulation and reasonably rapid solution. The empty drum shells must be rinsed with a little water from a hose to prevent loss of alkali and to make the metal safe to handle. More complicated dissolving plants have also been described.

It is often convenient to express quantities and concentrations of alkaline materials in terms of Na_2O. The conversion factors can be calculated by the usual methods of elementary chemistry. Using a table of atomic masses, as given in the summary of units and factors at the beginning of this book, and considering caustic soda as an example,

$$2\ NaOH = Na_2O + H_2O$$

$$2 \times (23 + 16 + 1) = (46 + 16) + (2 + 16)$$

$$\text{or } 2 \times 40 = 80 \qquad 62 \qquad 18$$

The equivalent mass of Na_2O = 62/2 = 31; and x% of NaOH is equivalent to $x \times 31/40 = 0.775x\%$ Na_2O

The concentration of a caustic soda solution is often expressed, and determined approximately, from its density measured at a known temperature around ambient. Density is mass per unit volume (that is g/ml, or lb/ft^3), but it is often expressed in terms of specific gravity, or relative density, which is density relative to that of water. Some factories and published literature still use the old Twaddell (Tw), or Baumé (Bé), scales defined in the table of symbols and units. Table 1.4 gives approximate relationships between concentration and density for some caustic soda solutions.

Table 1.4 – Approximate concentrations and densities of caustic soda solutions at 15.6°C (60°F)

% NaOH	% Na_2O	S.G.	°Tw	°Bé
1	0.78	1.01	2	1.4
2	1.55	1.02	4	2.8
4	3.10	1.04	8	5.6
8	6.20	1.09	18	12.0
10	7.75	1.11	22	14.4
20	15.5	1.22	44	26.1
25	19.4	1.28	56	31.7
30	23.3	1.33	66	36.0
35	27.1	1.38	76	39.9
40	31.0	1.43	86	43.6
45	34.9	1.48	96	47.0
50	38.8	1.53	106	50.2

The concentrated solution containing 45–50% NaOH is commonly called 100°Tw liquor.

1.5.3 Caustic potash, or potassium hydroxide (KOH)

Considerations similar to those for caustic soda apply to caustic potash. In most factories much less caustic potash than caustic soda is used, and somewhat less concentrated solutions are usually adequate. The heat of solution is even greater than that for caustic soda and solutions are not normally produced at more than about 40% KOH. The equivalent mass of caustic potash is $39.1 + 16 + 1 = 56.1$ and that of K_2O is $[(2 \times 39.1) + 16]/2 = 47.1$, so that y% KOH is equivalent to $y \times (47.1/56.1) = 0.84y$% K_2O, $y \times (40/56.1) = 0.713y$% NaOH, or $y \times (31/56.1) = 0.553y$% Na_2O.

1.5.4 Sodium carbonate (Na_2CO_3)

Anlydrous sodium carbonate, commonly known as soda ash, is available in light and heavy forms, which essentially differ only in bulk density. Most sodium carbonate is produced from salt, but natural deposits are found in certain places, including California and Lake Magadi in Kenya. Care is necessary in selecting natural sodium carbonate for use in detergent products because it is liable to contain iron and organic dirt, and that from Lake Magadi also contains some 1% of sodium fluoride (NaF) which is toxic.

Soda ash is commonly handled in bags, but large users normally find bulk handling worthwhile.

Sodium carbonate forms hydrates, the decahydrate, $Na_2CO_3 \cdot 10H_2O$, being traditional washing soda once widely found in households. Other hydrates are $Na_2CO_3 \cdot H_2O$ and $Na_2CO_3 \cdot 7H_2O$. Detergent manufacturers seldom need to buy hydrated sodium carbonate, but the formation of these hydrates is very important in the manufacture of certain types of soap and NSD powders as described later.

1.5.5 Sodium bicarbonate ($NaHCO_3$)

Sodium bicarbonate is a fine white powder which is only mildly akaline, so that it is little used in detergent formulations. Mixtures of pure soap powder and sodium bicarbonate have been made for special purposes, and the buffering action of sodium bicarbonate may be useful in reducing the pH of certain high carbonate formulations.

1.5.6 Sodium sesquicarbonate ($Na_2CO_3 \cdot NaHCO_3 \cdot 2H_2O$)

Sodium sesquicarbonate, also known as 'modified alkali', exists in the form of needle-shaped crystals. It is intermediate in alkalinity between sodium carbonate and sodium bicarbonate, and has a number of uses. In particular:

- as a mild alkali in commercial laundries, although this use is declining.
- as a major constituent of certain floor and other hard surface cleaners.

1.5.7 The sodium silicates

A number of different types of sodium silicate are available and are important in many industries including detergents. They can be regarded as combinations of SiO_2 (silicon dioxide), Na_2O and, in some cases, H_2O. There are two major types:

- *crystalline silicates* which are definite chemical compounds.
  sodium orthosilicate, $SiO_2 \cdot 2Na_2O$ (Na_4SiO_4),
  sodium sesquisilicate, $2SiO_2 \cdot 3Na_2O$ ($Na_6Si_2O_7$),
  and, sodium metasilicate, $SiO_2 \cdot Na_2O$ (Na_2SiO_3).
- *colloidal silicates* which are not definite compounds and which can have any ratio of SiO_2/Na_2O between about 1.6 and 3.85. The ratio is normally expressed by mass, but as the molecular mass of SiO_2 (60) is nearly the same as that of Na_2O (62), the composition by mass is nearly the same as that on a molar basis (mass ratio $SiO_2/Na_2O \times (62/60)$ = molar ratio). In the detergents industry, the usual ratios, by mass, are about 3.2, known as 'neutral silicate', and 2.0, known as 'alkaline silicate'. Despite these names, all the sodium silicates are alkaline, as shown in Table 1.1 and Fig. 1.1, but the higher the ratio the less alkaline the material. The alkalinity of a sodium silicate solution is directly related to its Na_2O content, the SiO_2 having little, if any, effect.

Orthosilicate and sesquisilicate are very alkaline and are used for specially difficult jobs in some commercial laundries, Metasilicate, sometimes known by the proprietary *brand* name 'Metso', is widely used in laundries and as an ingredient in some detergent formulations. It is available anhydrous and as the pentahydrate $Na_2O \cdot SiO_2 \cdot 5H_2O$. Both are white solids, and various granularities are commonly sold. Both materials are delivered in paper sacks.

Sodium silicates are normally made by fusing together pure quartz sand (more than 99% SiO_2) and soda ash which provides the Na_2O; carbon dioxide is evolved. The colloidal types leave the cooler which follows the furnace as a 'soluble glass'. This glass is not readily soluble and has to be dissolved in a digester under pressure. The solution is settled to remove impurities and concentrated by evaporation. This is normally done by the silicate manufacturer, but some large users may find it economical to instal the equipment needed to handle soluble glass, particularly if the silicate has to be transported long distances. Suppliers of soluble glass, such as Joseph Crosfield & Sons Ltd., or PQ International Inc., can provide details. The silicate solutions produced from soluble glass are sometimes called 'water glass', an old name for sodium silicate solutions.

Concentrated sodium silicate solutions are viscous, and the viscosity is greater for those with a high SiO_2/Na_2O ratio than for those with a lower ratio (above about 1.7 where a minimum is found). For this reason, alkaline silicate (ratio 2) is commonly purchased at 100°Tw or 120°Tw (120°Tw alkaline silicate contains 32.15% SiO_2 plus 16.07% Na_2O giving a total solids content of 48.2%), whereas neutral silicate (ratio 3.2) is usually bought at around 80°Tw (28.9% SiO_2 plus 9.0% Na_2O giving 37.9% total solids).

Sodium silicate solutions can be handled in mild steel, or cast iron, equipment. Tanks should be covered to avoid formation of hard skins by evaporation of water from the surface. Suitable heating facilities are necessary in tanks subject to cold, or temperate, climates to avoid excessive viscosity at low temperatures.

Sodium silicate solutions can be drum dried to produce soluble powders containing 15–20% of water. Crosfields sell 'C' soluble powder (an alkaline silicate of ratio 2.0, containing 56.0% SiO_2, 28.0% Na_2O and 16.0% H_2O) and 'M' soluble powder (a neutral silicate of ratio 3.3, containing 64.5% SiO_2, 19.5% Na_2O and 16.0% H_2O). Other silicate manufacturers produce similar products. 'C' soluble powder will dissolve readily in cold water, but 'M' soluble requires water at 75–80°C for easy dissolution. Detergent manufacturers who have to import sodium silicates may find soluble powders particularly suitable; they are also bought for use in some dry mixed powders.

1.5.8 Potassium silicates

Potassium silicates are sometimes used in soft soaps, liquid soaps and heavy duty liquid NSD products.

1.5.9 Borax

Borax is available as the decahydrate, $Na_2B_4O_7 \cdot 10H_2O$, and as the pentahydrate, $Na_2B_4O_7 \cdot 5H_2O$. It is a white powder which is a weak alkali used in some mild soap and NSD products.

1.5.10 Sodium phosphates

Sodium phosphates are very important constituents of detergent products, although now sometimes restricted for environmental reasons (as discussed in section 1.11). A number of different types of sodium phosphates are, or have been, used in detergent products, and it is important to distinguish between them. As described in Van Wazer (1958), the chemistry of the phosphates is complicated and the following account is somewhat simplified.

The anhydride, phosphorus pentoxide (P_2O_5) hydrates fully to orthophosphoric acid, $P_2O_5 + 3H_2O = 2H_3PO_4$. Orthophosphoric acid, commonly called phosphoric acid, is a tribasic acid which forms three series of salts; for sodium these are:

sodium dihydrogen orthophosphate, NaH_2PO_4
disodium hydrogen orthophosphate, Na_2HPO_4
trisodium orthophosphate, Na_3PO_4
(the syllable 'ortho' is often omitted).

These series of salts result from the ionization of the acid in three stages:

$$H_3PO_4 \rightleftharpoons H^+ + H_2PO_4^-$$
$$H_2PO_4^- \rightleftharpoons H^+ + HPO_4^{2-}$$
and $$HPO_4^{2-} \rightleftharpoons H^+ + PO_4^{3-}$$

In the first stage phosphoric acid is strong, in the second moderately weak, and in the third stage very weak. In consequence, NaH_2PO_4 is acidic, Na_2HPO_4 is mildly alkaline, and Na_3PO_4 is strongly alkaline. These salts form hydrates, and both hydrates and anhydrous salts are used for various purposes in detergent products. The most important orthophosphates are anhydrous trisodium phosphate, Na_3PO_4, trisodium phosphate dodecahydrate, $Na_3PO_4 \cdot 12H_2O$, and disodium hydrogen phosphate dodecahydrate, $Na_2HPO_4 \cdot 12H_2O$.

Most orthophosphates, other than those of the alkali metals except lithium, are substantially insoluble, and the sodium salts soften water by the precipitation of calcium and magnesium phosphates.

The condensed phosphates can be regarded as derived from acids formed by the elimination of water from two, or more, molecules of orthophosphoric acid; $m\ H_3PO_4 - (m-1)\ H_2O$, leading to the following types:

m	empirical formula	name
1	H_3PO_4	orthophosphoric acid
2	$H_4P_2O_7$	pyrophosphoric acid
3	$H_5P_3O_{10}$	tripolyphosphoric acid
4	$H_6P_4O_{13}$	tetrapolyphosphoric acid

As m approaches infinity, $mH_3PO_4 - (m-1)\ H_2O$, tends to $(H_3PO_4 - H_2O)_m = (HPO_3)_m$. Graham's salt, which is a hygroscopic glass, commonly known by its proprietary brand name, 'Calgon', is the sodium salt derived from an acid of this type. According to Van Wazer it should not be called sodium hexametaphosphate, which name correctly applies to a ring compound with quite different properties. All the sodium salts of the chain polyphosphoric acids are sequestering, complexing, or chelating, agents, that is they take alkaline earth and heavy metal ions into complexes which leave only a very low concentration of the ions of the complexed metal free in the solution. For example, sodium pyrophosphate complexes calcium from a solution of calcium chloride by the reaction

$$4Na^+ + P_2O_7^{4-} + Ca^{2+} + 2\ Cl^- \rightarrow 4\ Na^+ + [CaP_2O_7]^{2-} + 2\ Cl^-$$

Various other compounds chelate Ca^{2+} and Mg^{2+} as discussed later.

There are also a number of ring sodium phosphates with the general formula $(NaPO_3)_n$, of which the only one significant in the detergent industry is sodium trimetaphosphate, $Na_3P_3O_9$. Unlike the chain polyphosphates, this compound is not a chelating agent, but, as discussed in Chapter 10, it can be used as precursor of sodium tripolyphosphate.

The various condensed phosphates can be produced by elimination of H_2O from a phosphate containing a higher ratio of H_2O (or Na_2O) to P_2O_5, or by addition of H_2O (or a base such as NaOH) to one with a lower ratio. The most important condensed phosphates are tetrasodium pyrophosphate and pentasodium tripolyphosphate. The pyrophosphate is formed by heating disodium hydrogen orthophosphate

$$2\ Na_2HPO_4 = Na_4P_2O_7 + H_2O,$$

and the tripolyphosphate by heating a mixture of disodium hydrogen orthophosphate and sodium dihydrogen orthophosphate

$$2\ Na_2HPO_4 + NaH_2PO_4 = Na_5P_3O_{10} + 2H_2O.$$

In both cases small proportions of other sodium phosphates are found in the product.

The very important pentasodium tripolyphosphate, often called polyphosphate or TPP, exists in two forms or phases. Form 1 is the form stable at high temperatures and it hydrates rapidly to the hexahydrate, $Na_5P_3O_{10}\cdot 6H_2O$. Form 2, the low temperature form, hydrates more slowly. In practice, the sodium tripolyphosphate of commerce is a mixture of the two forms, but manufacturers can produce grades with various Form 1 contents. The approximate Form 1 content is commonly measured by the Temperature Rise Test; this is carried out by measuring the rise in temperature when water is added to the sample of tripolyphosphate in glycerine under standard conditions. Under the test conditions about 90% of the potentially available heat of hydration is released by Form 1 and 10% by Form 2. From this, Form 1 content = (TR − 6) x 4%. Three grades produced by one manufacturer have:

T.R.°C	Form 1 from formula %	Description
16.4	42	Fast hydrating
12.8	27	Fast hydrating
6.5	2	Slow hydrating

Particle size also influences rate of hydration; fine powders hydrating more rapidly than coarser powders of the same Form 1 content. The material with TR = 16.4°C just mentioned is a granular grade, whereas the other two are relatively fine powders. Some practical implications of rate of hydration are considered in Chapter 10.

The condensed phosphates hydrolyse slowly in solution, and tripolyphosphate changes to ortho and pyrophosphate. The rate of hydrolysis increases with temperature and under acidic conditions. If the hydrate $Na_5P_3O_{10}\cdot 6H_2O$ is heated sufficiently to remove water, decomposition to ortho and pyrophosphate occurs and little, or no, tripolyphosphate remains.

The various sodium phosphates are white powders which can be handled in bags, or in bulk.

1.5.11 Potassium phosphates

Potassium phosphates corresponding to the sodium salts can be produced; as in the case of the silicates, potassium pyrophosphate is sometimes selected for use in liquid detergents because of its superior solubility.

1.6 BLEACHES, INORGANIC AND ORGANIC

1.6.1 General

Chemical bleaching (as distinct from adsorption bleaching with activated earths or activated carbon) is used to improve the colour of ingredients in the manufacture of detergent products, and chemical bleaches are also used in formulations to improve performance by the removal of organic colour from fabrics or other articles. Chemical bleaching is usually carried out with oxidizing agents, but reducing agents have sometimes been employed.

1.6.2 Sodium and other hypochlorites

These so-called chlorine bleaches are made by reacting chlorine gas with an alkali, or alkaline earth. At one time slaked lime was used to give bleaching powder (also called 'chloride of lime', a misnomer); but the alkali is now usually caustic soda which gives sodium hypochlorite, or 'bleach'.

$$2\ NaOH + Cl_2 = NaOCl + NaCl + H_2O$$

To achieve adequate stability it is necessary to leave an excess of 1 or 2% of caustic soda, so that bleach liquor is strongly alkaline (hypochlorous acid is also weak, so that its sodium salt is itself alkaline).

It will be noted that sodium chloride equivalent to the hypochlorite is formed and is left in the bleach liquor. This is significant in that the chloride ion, Cl^-, as well as the hypochlorite ion, OCl^-, is necessary for the production of chlorine gas when bleach liquor is made acidic. Hydrochloric acid provides its own chloride ion, but with other acids, such as sulphuric, the sodium chloride plays an essential part in the reaction

$$NaOCl + 2\ HCl = NaCl + Cl_2 + H_2O$$
$$NaOCl + NaCl + H_2SO_4 = Na_2SO_4 + Cl_2 + H_2O$$
$$\text{or, ionically, } OCl^- + Cl^- + 2H^+ = Cl_2 + H_2O$$

Because of the toxic nature of chlorine gas it is important to avoid its evolution into the atmosphere, particularly in buildings. Accidents have occurred where acid has been used to adjust the pH when sodium hypochlorite is employed to sterilize water in swimming baths, etc., and too much has been added. Similarly, when sodium hypochlorite and acidic lavatory cleaners have been mixed in error. The same type of situation could occur in the detergents factory unless proper precautions are taken. A subtle variation of the problem can arise with lavatory cleaners which include trichlorcyanuric acid (see section 1.6.3), and the acid, sodium bisulphate. Hypochlorous acid is generated from the trichlorcyanuric acid, but no significant evolution of chlorine occurs when normal supplies of water are used to flush the lavatory. If, however, the products are used with sea, or other salt, water an unpleasant amount of chlorine can be produced.

In the treatment of fabrics (and in the factory), hypochlorite bleach is normally used in alkaline solution. The active agent is then really active, or nascent, oxygen (O). If oxidizable materials, such as most organic coloured compounds, are present, the active oxygen will react; if not, the atoms may combine slowly to produce gaseous oxygen (O_2), which is gradually released. Although the effective bleaching agent is active oxygen, it is usual to determine the concentration of hypochlorite present in terms of available chlorine, that is the quantity of chlorine liberated when the solution is made acidic. One method is to add potassium iodide, acidify, and titrate the iodine liberated with standard sodium thiosulphate.

$$\begin{array}{l} NaOCl + NaCl + H_2SO_4 \rightarrow Na_2SO_4 + Cl_2 + H_2O \\ \quad\downarrow \\ NaCl + O \end{array}$$

Hence one atom of active oxygen, atomic mass 16, is equivalent to two atoms of chlorine, atomic mass $2 \times 35.5 = 71$ (or to one molecule of chlorine, molecular mass 71). The molecular mass of sodium hypochlorite, NaOCl, is 74.5. Therefore, 74.5 parts of NaOCl are equivalent to 71 parts of available chlorine; so that a solution containing 10.0% of NaOCl contains $10.0 \times (71/74.5) = 9.53\%$ available chlorine. This close numerical similarity may seem surprising.

When used to bleach fabrics in a washing process, sodium hypochlorite is normally added to a rinse in a separate operation after the main wash, otherwise part of the available chlorine would be mopped up by oxidizable matter in the soil. Another very important use of sodium hypochlorite is as a disinfectant. It is effective against all types of micro-organisms, but some types, such as bacterial spores, require longer times and higher concentrations than more delicate types of organism. The active chemical species is understood to be hypochlorous acid and disinfectant performance increases with fall in pH. The main limitation of sodium, and other, hypochlorites is again that they can be used up by other oxidizable matter often present with the organisms; this means that sufficient must be used to maintain the necessary concentration of active bacteriocide for the required time.

Sodium hypochlorite solution is produced with about 13–14% available chlorine content, and it is commonly received in factories at this strength. Even when strongly alkaline it tends to decompose slowly on storage. Decomposition can be with evolution of oxygen gas as explained earlier, but it also occurs by the formation of sodium chlorate,

$$3NaOCl = NaClO_3 + 2NaCl$$

Loss of available chlorine is greater when the concentration of sodium hypochlorite is high than when it is low. Bleach liquids for sale for domestic purposes in temperate climates do not normally start with more than 10% available chlorine and often have less. For export to hotter countries 5% is usually the maximum. Decomposition, particularly with evolution of oxygen gas, is increased by the presence of heavy metal catalysts. Iron, or steel, particles in a solution containing traces of copper are particularly potent 'fizzers', that is centres from which the gas is evolved. Probably metallic copper 'plates out' to produce active spots on the iron. For this reason, sodium hypochlorite solution is often handled in non-metallic tanks, such as stoneware, concrete, certain plastics, or steel lined with a suitable rubber or enamel. Some special metals are also satisfactory. To minimize problems with fizzers, particles should be removed by filtration at appropriate stages. High concentration (10%) bleach for the retail trade may be sold in bottles (glass or suitable plastic) fitted with 'venting caps'. These caps have a spongy, or other, wad which, when tightened to the correct torque, allow gas to escape if the pressure in the bottle rises, but retain liquid.

In situations, such as some tropical countries, where sodium hypochlorite solution cannot conveniently be purchased and stored, it can be made by:

- blowing chlorine gas into a liquid containing caustic soda to produce hypochlorite *in situ* (care must be taken to use a suitable diffuser and to control the process so as to avoid dangerous escape of the gas).

- reacting solid bleaching powder, approximately $Ca(OCl)Cl\ H_2O$, or the more concentrated calcium hypochlorite, $Ca(OCl)_2$, with sodium carbonate,

$$Ca(OCl)Cl + Na_2CO_3 = CaCO_3 + NaOCl + NaCl$$

  The insoluble calcium carbonate can then be separated, leaving a solution of sodium hypochlorite.

Lithium hypochlorite is also available in powder form.

1.6.3 Substances which produce available chlorine when dissolved

When it is desired to add a substance to a dry detergent product which will produce available chlorine when water is added for use, there are two major types of compound available:

- chlorinated trisodium orthophosphate, or sodium phosphate hypochlorite (SPH), described as $(Na_3PO_4 \cdot 12H_2O)_5\ NaOCl$ or $(Na_3PO_4 \cdot 11H_2O)_4\ NaOCl$ approximately. This contains about 3.5% available chlorine.
- an extensive range of organic compounds which contain N–Cl groups in which the chlorine is 'available'. The most important are probably trichlorcyanuric acid (TCCA), with about 90% available chlorine, and potassium dichlorcyanurate, with 59% available chlorine.

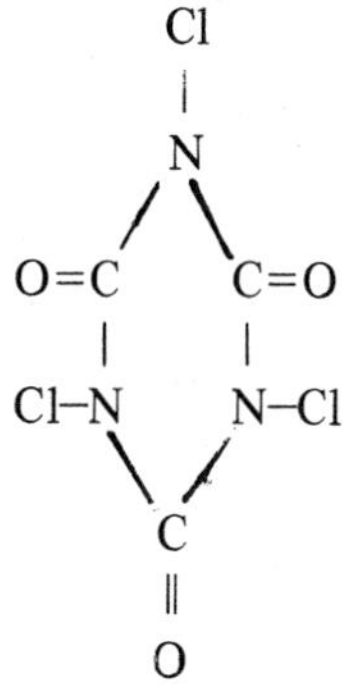

trichlorcyanuric acid

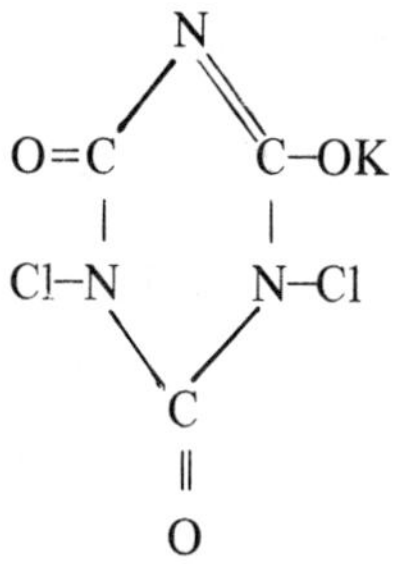

potassium dichlorcyanurate

Others include Chloramine T(25% available chlorine), Chloroxal (66%), and 1.3 dichloro 5.5 dimethyl hydantoin (66%).

An important practical difference between the two groups of compounds is their behaviour in relation to the water content of the product. With SPH, the water shown in the formula is necessary for stability, and if the product dries out on storage available chlorine is likely to be lost. On the other hand, the chlorcyanurates and similar compounds are readily decomposed with loss of available chlorine if the product picks up water. As discussed in Chapter 11, scouring powders containing TCCA, etc. usually need an ingredient, such as anhydrous trisodium orthophosphate, which will mop up water vapour which leaks into the pack, and so protects the TCCA. These products also commonly include a stabilizing olefine, usually as a perfume ingredient.

1.6.4 Sodium chlorite, $NaClO_2$

Commercial sodium chlorite containing about 80% of $NaClO_2{\cdot}2H_2O$, with sodium chloride, sodium hydroxide and sodium carbonate (of which the alkalis are stabilizers) is available. A material is also sold which contains about 48% $NaClO_2{\cdot}2H_2O$, plus 48% sodium nitrate which is added as a corrosion inhibitor. Sodium chlorite is used in some processes for the chemical bleaching of fats. Solutions containing up to about 35% chlorite can be kept at ambient temperatures in buildings in the UK; more concentrated solutions (up to 50%) require to be heated and may be less convenient.

1.6.5 Hydrogen peroxide and the per-salts

These compounds liberate active, or nascent, oxygen which is the bleaching agent. They are often considered milder on dyes and fabrics than sodium hypochlorite, but at concentrations which give equal stain removal this is not necessarily true.

Hydrogen peroxide solution is sometimes used in industry, but it cannot be included in detergent products, as can certain 'per-salts' which are widely included in detergent powders sold in many countries. The most important per-salt is sodium perborate, $NaBO_3{\cdot}4H_2O$, or $NaBO_2{\cdot}H_2O_2{\cdot}3H_2O$, containing about 10% available oxygen. The mono-hydrate, containing about 16% available oxygen is also sometimes used. As shown in Table 1.1, sodium perborate is also an alkali. Sodium perborate is generally considered ineffective at temperatures below 60°C and it works best at higher temperatures of 80–100°C, although some benefits from the inclusion of sodium perborate at temperatures as low as 35°C have been claimed. A number of compounds described as *bleach activators* have been found which react with perborate in the wash liquor to form peracetic acid (or its salt) which bleaches at lower temperatures than perborate. The most important of these activators is TAED, tetra-acetyl ethylene diamine. Others are TAGU, tetra-acetyl glycol uril, tetra-acetyl methylene diamine, and p-acetoxybenzene sulphonate.

The principal other per-salt used in detergent formulations is sodium percarbonate, $2Na_2CO_3{\cdot}3H_2O_2$. This compound is less stable than sodium perborate in detergent powders, but it may have some advantage in products to be used at relatively low temperatures, and it avoids the inclusion of boron to which there are some environmental objections when the waste water containing it may be used for irrigation. Persulphates are mentioned in Chapter 11 in relation to scouring powders.

1.6.6 Reducing bleaches

Reducing bleaches are of little importance in the detergent industry, but sodium hyposulphite, $Na_2S_2O_4$ (also known as sodium hydrosulphite or blankite), is sometimes used to improve the colour of soap during the boiling process.

1.7 A SCHEME FOR THE ANALYSIS OF SOAP PRODUCTS

1.7.1 General

In section 1.4.5 some analytical tests used for soap products are outlined. Calculations using the results of such tests, and other information presented in subsequent sections, will now be given for two typical examples, a household soap and a soap powder. First some additonal tests will be outlined.

Chlorides (salt) can be determined as follows. A 10 g sample of soap is dissolved in hot distilled water and transferred to a 250 ml narrow necked graduated flask. Sufficient calcium nitrate is added to precipitate the soap (about 15 ml of a 20% solution), and the flask is filled to the 250 ml mark with distilled water. The mixture is filtered and 100 ml taken in a porcelain bowl where it is titrated with standard silver nitrate solution using a little potassium chromate as indicator. The reactions are:

$$NaCl + AgNO_3 \rightarrow \underset{\substack{\text{silver chloride}\\ \text{(insoluble white)}}}{AgCl} + NaNO_3$$

$$K_2CrO_4 + 2AgNO_3 \rightarrow \underset{\substack{\text{silver chromate}\\ \text{(insoluble brick red)}}}{Ag_2CrO_4} + 2KNO_3$$

Because silver chloride is less soluble than silver chromate, there is no permanent change of colour from the yellow due to potassium chromate to brick red until all the chloride has been precipitated. A blank estimation, without soap, should be carried out and the figure deducted from the test titration. If 0.1 N silver nitrate solution is used, 1 ml is equivalent to $(58.5/10) \div 1000 = 0.00585$ g NaCl, or 0.00355 g chloride expressed as Cl. With the quantities just specified, the salt found came from $10 \times (100/250) = 4$ g soap, and, if the titration after deduction of the blank is y ml, the salt content of the soap is $y \times 0.00585 \times (100/4) = 0.146\, y\%$ as NaCl.

Silica, SiO_2. To separate silica, a sample of the soap, or soap powder, is ashed in a platinum dish with fusion mixture (potassium carbonate plus sodium carbonate) and the mix is heated to melting point. When cool, the melt is dissolved in distilled water, transferred to a porcelain basin and carefully acidified with hydrochloric acid. It is then evaporated to dryness, moistened with hydrochloric acid and again evaporated to dryness after any large lumps have been broken with a glass rod. The basin and contents are baked in an oven at 135°C, or gently on a sand bath, cooled, and then a little hydrochloric acid and water added. This process renders the silica insoluble, so that it can be filtered and weighed after the filter paper has been ashed. If insolubles other than silica may be present, the silica can be vaporized as silicon tetrafluride, SiF_4, by evaporating two or three times after additions of hydrofluoric acid and a few drops of sulphuric acid. This second residue is not silica and should be deducted from the crude residue first obtained.

Available oxygen can be estimated by acidification of a solution of a weighed sample with sulphuric acid and titration with standard potassium permanganate.

The estimation of glycerol is discussed in Chapter 8. Some analytical methods are described in the British Standard Specifications mentioned in section 1.8, and many others are available from books on the analysis of detergent products, such as those listed in the bibliography.

1.7.2 Household soap

The main constituents of a genuine household soap are:

- soap, comprising fatty matter (mainly fatty acids) and the alkali combined with it
- free caustic soda
- sodium carbonate
- sodium chloride
- glycerol
- water, possibly plus colour and perfume.

As an example, assume that the following figures are obtained:

total fatty matter (TFM) = 63.30% (this is often loosely called fatty acids (FA)%)
total alkali = 7.70% as Na_2O
soda as soap = 7.58% as Na_2O
free caustic soda = 0.02% as Na_2O, equivalent to 0.03% as NaOH
chloride = 0.63% as NaCl
glycerol = 0.35%.

The neutralization of FA reaction, assuming the hypothetical alkali, Na_2O, can be written with the appropriate molecular masses,

$$\begin{array}{cccc} 2R\cdot COOH & + \; Na_2O & \rightarrow 2R\cdot COONa & + \; H_2O \\ x & 62 & (x + 62 - 18) & 18 \\ & & = x + 44 & \end{array}$$

62 parts of Na_2O used in neutralization add 44 parts (Na_2O – H_2O) to the fatty acids to form soap, so that the soda as soap, estimated to be 7.58% as Na_2O, adds 7.58 × (44/62) = 5.38 parts to the 63.30 parts of TFM, and the anhydrous soap in the sample is 63.30 + 5.38 = 68.68%.

$$\frac{\text{TFM\%}}{\text{soda as soap\%}} = \frac{63.30}{7.58} = 8.35$$

is defined as the fatty acid ratio and it varies with the composition of the fatty acids in the fat blend. Fatty acid ratio and related figures, such as saponification value, are discussed in Chapter 3. Residual alkali, assumed to be sodium carbonate, = 7.70 − 7.58 − 0.02 = 0.10% as Na_2O, equivalent to 0.17% as Na_2CO_3.

A small residual alkali which is obtained in this way as a difference between two relatively large quantities is subject to substantial error. If the figure for total free alkali (hydroxide plus carbonate) is important, it should be determined directly as explained in section 1.4.5. The composition of the household soap can now be tabulated as shown in Table 1.5.

Table 1.5 – Composition of a household soap calculated from analytical data

	% TFM	% Na_2O	% actual constituent
TFM	63.30	–	–
Total alkali	–	7.70	–
Soda as soap	–	7.58	–
Anhydrous soap	–	–	68.68
Free caustic soda	–	0.02	0.03
Residual alkali, assumed Na_2CO_3	–	0.10	0.17
Salt, NaCl	–	–	0.63
Glycerol	–	–	0.35
Water etc. (by difference)	–	–	30.14
Total	–	7.70	100.00

1.7.3 Soap Powder

For the soap powder assume the following analytical results:

TFM = 44.30%
total alkali = 16.38% as Na_2O
soda as soap = 5.37% (giving a fatty acid ratio of 8.25)
silica = 1.70% as SiO_2
available oxygen = 0.97%
boron as B_2O_3 = 2.34%
phosphorus = nil.

As in the previous example, anhydrous soap = 44.30 + [5.37 x (44/62)] = 48.11%. If it is assumed that neutral silicate was used in the manufacture of the powder, the Na_2O as silicate = 1.70/3.2 = 0.53%, and the anhydrous silicate = 1.70 + 0.53 = 2.23%. From the presence of available oxygen and B_2O_3, it is reasonable to assume that the powder contains sodium perborate. The usual form, $NaBO_2{\cdot}H_2O_2{\cdot}3H_2O$, contains about 22.7% B_2O_3, 20.2% Na_2O, and 10.4% available oxygen. Available oxygen is lost slowly during storage leaving a borate which will be assumed to be $NaBO_2$ ($Na_2O{\cdot}B_2O_3$). 2.34% B_2O_3 is equivalent to [(2.34 x 100)/22.7] = 10.3% sodium perborate added originally, 0.97% available oxygen is equivalent to [(0.97 x 100)/10.4] = 9.3% perborate. It is deduced that the sample tested contains 9.3% sodium perborate, containing 1.88% Na_2O, and that about 1% perborate has changed to 0.43% $NaBO_2$, containing 0.20% Na_2O. The residual alkali = 16.38 – 5.37 – 0.53 – 1.88 – 0.20 = 8.40% Na_2O equivalent to 14.36% Na_2CO_3. The estimated composition of the powder can be tabulated as shown in Table 1.6.

It is possible to estimate the carbon dioxide content of a sample by the use of a train in which a sample of the powder is acidified, the carbon dioxide swept out in a stream of carbon dioxide free air, water vapour separated and the carbon dioxide collected in potassium hydroxide and weighed. This is necessary if the sample may contain sodium bicarbonate as well as carbonate.

Table 1.6 – Composition of a soap powder calculated from analytical data

	% TFM	% Na_2O	% actual ingredient
TFM	44.30	–	–
Total alkali	–	16.38	–
Soda as soap	–	5.37	–
Anhydrous soap	–	–	48.11
Anhydrous sodium silicate (neutral)	–	0.53	2.23
Sodium perborate, $NaBO_2{\cdot}H_2O_2{\cdot}3H_2O$	–	1.88	9.30
Sodium borate, $NaBO_2$	–	0.20	0.43
Sodium carbonate	–	8.40	14.36
Water, salt, glycerol etc. (by difference)	–	–	25.57
Total	–	16.38	100.00

If a powder contains phosphates, the P_2O_5 content can be determined by a gravimetric method, and various more complicated procedures are available for the estimation of different types of phosphate. If only a figure for P_2O_5% is available, it is necessary to make assumptions as to the type of phosphate, and/or to use trial and error methods to make the figures balance in a way which seems reasonable. There are also complications with regard to the determination of the total alkali.

1.8 SPECIFICATIONS

Detergent products sold directly to housewives are usually made to the manufacturer's own formulations, rather than to any official specification, although such specifications might perhaps be invoked if a question of trade description arises. Transactions for products and intermediates, such as dried soap noodles, between companies (and/or other organizations) require some agreed specification which may, or may not, be an official published one. Some large purchasing organisations, such as USA government departments, issue detailed specifications for a wide range of cleaning products. It is outside the scope of this book to survey such documents, but the main features of the relevant British Standard Specifications are given in Tables 1.7 and 1.8 by permission of the British Standards Institution, 2 Park Street, London W1A 2BS from which complete copies of the documents can be obtained. British Standards are revised at regular intervals, and some of the documents mentioned in this book are currently being reviewed. Care should be taken to ensure that the latest versions are consulted.

Table 1.7 – British Standard Specifications for solid soap products. The following table summarizes the main requirements of BS 1910, 1911, 1912 and 1914. Apart from the latest (1984) edition of BS 1914, these specifications also include general preambles along the lines that the product should be of firm texture, free from objectionable odour and have good lathering and cleaning properties.

	BS 1910:1966	BS 1911:1966	BS 1912:1973	BS 1914:1984		
	Carbolic soap	Genuine hard soap	Soap flakes	Toilet soap*		
				Type A	Type B	Type S
TFM, including rosin, % m/m	>59	>62	>80	>78.0	>76.5	>79.0
Fatty acids derived from lauric oils as % on TFM m/m	–	–	–	>13.0†	–	>13.0
Rosin, % m/m on TFM	<7.5	<7.5	<3.0	<2.0	<2.0	<2.0
Ethanol insoluble % m/m	<2.0	<2.0	<2.0	<1.25	<2.5	<1.25
Free caustic soda % m/m as Na_2O	<0.10	<0.05	<0.05	<0.05	<0.05	not applicable
Total free alkali % m/m as Na_2O	<0.30	<0.25	<0.30	–	–	–
Chloride as NaCl % m/m	<1.0	see below	<1.0	<0.7	<0.8	<1.0
Chloride as NaCl % m/m hard soap type 1 (batch)	–	<0.75	–	–	–	–
Chloride as NaCl % m/m hard soap type 2 (continuous)	–	<1.25	–	–	–	–
Phenols	>1.5	–	–	–	–	–
Thickness, mm	–	–	<0.20	–	–	–
Free fatty acid content, expressed as fatty acid of mean molecular mass 248, % (m/m)	–	–	–	–	–	2.0–10.0

†See BS 684 for method.

*In BS 1914, type A is a more strictly specified material, and type B is suitable for all general purposes. Type S (first introduced in the 1984 edition) contains a specified proportion of its total fatty matter as free fatty acids and therefore contains no free caustic alkali. Types A and B usually contain very little free fatty acid.

Table 1.8 – British Standard Specifications for liquid and soft soaps. This table summarizes BS 1913, 1545, and 4405. In addition to general statements concerning odour, etc., BS 1545:1969 (liquid toilet soaps) says that these products shall not contain non-aqueous solvents or synthetic detergents.

	BS 1913:1966 soft soap type A	BS 1913:1966 soft soap type B	BS 1545:1969 liquid toilet soap	BS 4405:1969 liquid soap for domestic and industrial purposes
TFM, including rosin, %	>38	>28	>15	>10[†]
Rosin, % on TFM	Nil	<15	<2	<3
Free caustic alkali as K_2O	<0.1	<0.5	<0.03	<0.04
Total free alkali as K_2O	<1.0	<1.0	–	–

†The TFM usually contains more than 20% of fatty acids of molecular weight of 200, or above.

BS 1545:1969 and BS 4405:1969 also include the requirement that the products shall remain clear and bright, shall not deposit sediment, and shall show no gelling after 24 hours at 3° ± 1°C. It is also noted that liquid soaps may absorb carbon dioxide from the air and lose water by evaporation, which may affect the behaviour of the products when chilled.

BS 1715:1963 gives the methods of analysis of soaps and soap powders to be used in conjunction with the various specifications, and BS 1910:1966 includes some additional procedures peculiar to carbolic soaps. It is recognized that soap products dry out on storage, or may be made with higher TFMs than the minima specified, and it is permissible to recalculate the various analytical figures to correspond to the minimum specified TFM before making comparison with maxima quoted in the specification. For example, if when analysed a genuine hard soap has a TFM of 74.3% as against a specified 62% and a total free alkali content of 0.28% as Na_2O, the recalculated total free alkali is $0.28 \times (62/74.3) = 0.23\%$ which is below the limit of 0.25%.

BS 3762:1964, 'Methods of sampling and testing detergents', contains useful information on these topics. BS 5610:1978, 'Surface-active agents – Detergents for washing fabrics – Guide for comparative testing of performance' is mentioned in section 2.2.1. Presumably because of the very wide range of formulations used for NSD products, there are no general specifications for these materials. BS 2756:1970, 'Recommendations for the use of detergents in the dairy industry', provides a review of cleaning and sanitizing procedures and products for hard surfaces which is of general interest outside the dairy field.

Reference will be made later to specifications related to the oils and fats, and to glycerines.

1.9 SOME TECHNICAL IMPLICATIONS OF WEIGHT CONTROL LEGISLATION

In many countries it is required by law to declare the weights of certain products, but the way in which this is defined can vary. In the United Kingdom prior to 1st January 1980, the declared weight of a detergent product had to be exceeded when the item was finally sold at retail. Obviously some packs are stored much longer than others, and the storage conditions can vary widely, so that allowance had to be made for a considerable range of weight losses in the case of products which tend to lose weight, which includes soaps and soap powders. Manufacturers found it necessary to carry out extensive statistical studies of the weights of samples purchased from retail shops of various types in order to decide the minimum weight which could safely be packed. The problem could be minimized by the use of packaging which reduced the rate of moisture loss. Since January 1980, the law in the UK has been brought into line with that in other EEC countries. The regulations are rather complex and should be consulted by those concerned. In outline, the main control is now average weight in the factory. This is based upon samples of household soap seven days old and of toilet soap 24 hours old, but considerable attention is paid to the production records, provided the inspector is satisfied by the procedures adopted. The inspector can also carry out a reference test using laid down methods. Other inspectors can monitor samples in retail shops, but, if not satisfied, they refer the matter to the inspector responsible for the factory of origin, who proceeds as just discussed. This new procedure is much less onerous than the former one, and drying out is much less important relative to weight control legislation. In some countries the weights of soaps have to be declared in terms of TFM, or anhydrous soap content, which does not change with loss of weight as the tablets dry out. With the modern vacuum chilling and extrusion processes this may make it technically convenient, and commercial satisfactory, to extrude and pack at, say 67–68% TFM, rather than to add liquor to compensate for the evaporation which occurs in the vacuum chilling unit. On the other hand, with the weight-when-packed system, there may be some marketing inducement to pack as near as possible to the permitted minimum TFM (that is 62% in BS 1911:1966).

Ways in which tablet weights can be controlled are discussed later, but it may be of interest to explain how the weight at which a competitor's soap was packed can be estimated from an old tablet, such as one purchased at retail:

- weigh the tablet, say y grams.
- take an average sample by cutting a representative eighth. Weigh. Dissolve in distilled water and estimate TFM in the whole, or a known fraction, of the solution to give the average TFM of the sample, say $a\%$.
- estimate the TFM of a sample taken from the inside of the tablet, say $b\%$. This will closely approximate the TFM% at which the tablet was packed, unless the sample is very old and dried out to an abnormal extent.
- original weight $= y \times (a/b)$ grams.

1.10 WATER HARDNESS AND WATER SOFTENING

Water used for washing, and other purposes, in home and factory contains impurities to a degree which varies with its source. In the detergent field, the most important impurities are the calcium and magnesium salts which constitute water hardness. Traces of iron can lead to iron stains (sometimes known as iron mould), but this is more common in laundries than in the home.

The effect of water hardness is obvious with soap in that precipitates are formed; and no lather is produced until the calcium and magnesium in the water have been removed as insoluble soaps. Although it does not prevent lather, hardness, particularly calcium hardness, does have an adverse effect upon the performance of NSD products, at least in the washing of textiles, unless builders are also present in sufficient quantity. Unbuilt non-soapy surfactants are used successfully for dishwashing in hard as well as soft water. These matters are discussed further in Chapter 2.

Water hardness is divided into:

- *temporary,* or *alkaline, hardness* due to bicarbonates. This hardness was called temporary because it is largely removed by boiling (or heating to near boiling point), when the bicarbonates are converted to relatively insoluble carbonates which precipitate on kettles and other heating surfaces. Actually, the alkaline earth carbonates are slightly soluble, so that the hardness originally due to bicarbonates is not completely removed by boiling. The modern term is alkaline hardness, so called because this hardness can be determined by titration with standard acid to the methyl orange end point.
- *Permanent hardness,* due to other salts, such as sulphate, chloride and nitrate.

Total hardness can be estimated by:

- titration with Wanklyn's (standard soap) solution to a permanent lather (that is one which persists after the bottle in which the estimation is carried out has rested on its side for five minutes). Particularly with water high in magnesium, a false end point can occur when the calcium, but not the magnesium, has been precipitated. This lather does not last for five minutes.
- titration with standard EDTA (ethylene diamine tetra-acetic acid) solution, using as indicator a black dye, known variously as Eriochrome, Solochrome, etc. EDTA is a sequestering agent which, like the condensed phosphates, complexes the calcium and magnesium in the water (see also Chapter 2).

Permanent hardness is not estimated directly, but is found as the difference between total and alkaline hardness.

Hardness, calcium plus magnesium, is usually expressed in terms of $CaCO_3$, although CaO has been used in some countries. Degrees Clark are grains of $CaCO_3$ per Imperial gallon (or parts per 70,000). Degrees in terms of parts per 100,000, and of grains per US gallon (parts per 58,400), are also used, but the modern method of expressing hardness in parts per million of $CaCO_3$ avoids any ambiguity. The conversion factors are:

$1°$ US = 17.1 ppm
$1°$ UK (or Clark) = 14.3 ppm
$1°$ parts per 100,000 = 10.0 ppm

The demarcations between waters classed as soft, moderately hard, etc., are necessarily arbitary, but one American author (Levitt, 1967) suggests about 0–60 ppm soft, 60–120 moderately hard, 120–200 hard, and above 200 very hard.

Laundries usually soften the water to be used in washing operations, but unsoftened water is normally used in the home. Dishwashing machines often have inbuilt water softeners, but, so far as is known, this is not the case with clothes washing machines. Detergent manufacturers therefore need to formulate their products to perform well in the wäters available to the housewives in their markets. Attempts have been made to offer hard and soft water versions of certain brands, but one difficulty is that, at least in the UK, there are pockets of hard water in predominantly soft water areas, and conversely.

There are three main methods to soften water:

- the *lime-soda process* in which calcium hydroxide (slaked lime) and sodium carbonate are added to the raw water. The main reactions are:

  $$Ca(HCO_3)_2 + Ca(OH)_2 \rightarrow 2CaCO_3\downarrow + 2H_2O$$
  $$Mg(HCO_3)_2 + 2Ca(OH)_2 \rightarrow Mg(OH)_2\downarrow + 2CaCO_3\downarrow + 2H_2O$$

  (magnesium hydroxide is considerably less soluble than magnesium carbonate).

  $$CaSO_4 + Na_2CO_3 \rightarrow CaCO_3\downarrow + Na_2SO_4$$

  $CaCO_3$ and $Mg(OH)_2$ are relatively insoluble, and are separated by settling and decantation, followed, if necessary, by filtration. Because of the slight solubility of $CaCO_3$ and $Mg(OH)_2$, the water retains a little hardness.

- *base exchange.* Various materials, including natural zeolites (types of silicate minerals), synthetic zeolites, and certain synthetic resins, are capable of softening water by an exchange of sodium for calcium and magnesium. The bed of base exchange granules is then regenerated by the action of a sodium chloride (salt) solution which reverses the softening reaction. Calling any base exchange material NaZ, the reactions are roughly:

  Softening $2NaZ + Ca\ salt \rightarrow CaZ_2 + 2Na\ salt$

  Regeneration $CaZ_2 + 2NaCl \rightarrow 2NaZ + CaCl_2$

  The excess brine (or other sodium salt used for regeneration), carrying with it the soluble calcium and magnesium salts (usually chlorides), is run to waste, and the regenerated bed is rinsed before being used for the next cycle. Alkaline hardness is converted to sodium bicarbonate in the softened water. This is a disadvantage in some situations, for example, it can reduce the pH of a wash system comprising soap and sodium sesquicarbonate. A suitable addition of caustic soda converts the bicarbonate to sodium carbonate. Water softened by base exchange may not be suitable for boiler feed, particularly if the raw water is high in alkaline hardness. One procedure is to mix a little raw water with the softened water. A base exchange bed filters out any particulate dirt from the water. It also removes any soluble iron, but the treatment of raw water high in iron content can lead to problems in regeneration.

- *de-mineralization,* which is a development of the base exchange process. A wide range of anionic and cationic exchange resins is now available, and it is possible to remove substantially all the ions from a water to give the equivalent of distilled water. The deminerlization process can be outlined by saying:

  (a) the various cations, Ca^{2+}, Mg^{2+}, Na^{+}, etc., are exchanged for H^{+}, leaving a liquor containing the anions as the equivalent acids. This resin is regenerated with mineral acid.
  (b) the various acids are removed with an anion exchange resin, regenerated with caustic soda, to leave a water which is nearly pure after carbon dioxide has been removed.
  (c) more than one stage is usually used for each operation, and/or the process is completed with a final mixed bed containing both cationic and anionic resins.

Good quality condensed steam is often available in factories and is a good source of nearly pure water.

1.11 ENVIRONMENTAL PROBLEMS

1.11.1 General

As in other industries, some problems of an environmental nature can arise in, and in the environs of a factory, due to noise, gaseous and liquid effluents, and solid wastes. This section is concerned, however, with two problems, lack of adequate biodegradability and eutrophication, which can occur when waste waters containing certain detergents are discharged from the homes and factories in which they have been used. The two problems are quite distinct. Taylor (1980) provides a useful review.

1.11.2 Biodegradability

When waste soapy water is discharged into a sewage system, or directly into a river or lake, no real problems normally arise. Excess soap is precipitated by the hardness in the receiving water, and it is also readily broken down by micro-organisms such as those which play a major role in sewage treatment. Soap is, therefore, described as biodegradable. This is not the case with certain non-soapy surfactants. Problems arose with foam on small rivers, for example in Yorkshire, England, due to the discharge of waste liquors from textile factories. The problem developed into one of major importance in the 1950s and early 1960s when in many countries the discharge of domestic waste waters containing NSD products had serious adverse effects on sewage treatment plants and on the rivers, etc. into which the effluents, treated or untreated, were discharged. Even water from wells near places where these effluents were discharged was sometimes found to foam. At that time the trouble was traced to alkyl benzene sulphonates with highly branched alkyl chains, in particular those derived from propylene tetramer (the so-called PT alkyl benzenes, or alkylates). These surfactants do not break down sufficiently readily in sewage treatment plants and they are described as biologically 'hard', or as not biodegradable. The micro-organisms attack side chains two carbon atoms at a time and the process is stopped if a tertiary carbon (one to which only alkyl groups and no hydrogen atoms are attached) is reached. This led to the linear alkyl benzene sulphonates, often called LABS or LAS, in which the side chains are derived from suitable, substantially straight chain, hydrocarbons. In fact, this was a reversion to improved forms of the 'keryl' benzenes which were used before the PT type was developed and came to be the dominant anionic surfactant. LABS surfactants are accepted as adequately biodegradable, although they are not broken down as readily and completely as soaps and other surfactants derived from fats, or synthesized to contain a completely unbranched chain with an even number of carbon atoms and no benzene ring. Around 1965, the use of PT type sulphonates was stopped in the more developed countries by voluntary agreement by manufacturers, or by legislation. They then became relatively cheap and continued to be used in some parts of the world.

Nonionic surfactants can also lack adequate biodegradability, but problems of analysis of low concentrations in effluent waters make regulation more difficult than for anionics. The alkyl phenol ethylene oxide types, once widely used, are being, or have been, phased out by voluntary agreement. There is still some argument, but they are probably not sufficiently degradable to be acceptable. The octyl and nonyl side chains in these compounds are highly branched. Other nonionics based on straight chain, or relatively straight chain, alcohols are currently acceptable, but there is some uncertainty regarding the fate of the polygylcol chains. Surfactants with long polyethylene oxide, and/or polypropylene oxide, chains are particularly suspect.

The biodegradability of cationic surfactants has been little discussed, but it is probably not important, because of the mutual precipitation when most anionic and cationic surfactants are mixed.

1.11.3 Eutrophication

It is a natural process for lakes and ponds to silt up gradually. When little nutrient is present, and hence there is little plant life, the condition is called oligotrophic. When rivers and rains bring in nutrients, algae and other plants grow and the condition becomes eutrophic. 'Cultural' eutrophication can occur if human activities, usually the discharge of large quantities of sewage into relatively small lakes, or into lakes from which little, or no, water is discharged to the sea, result in the presence in the water of excessive quantities of nutrients. This abundance of nutrients leads to a massive explosion of the population of algae, with production of oxygen. At the end of the growing season most of the algae die and bacteria feed on the dead plant material. This process requires oxygen, that is the destruction of the dead algae imposes a biological oxygen demand (BOD), similar to that caused by the discharge of milk, or any other destructible organic matter, including normal sewage. In extreme cases, this can eventually turn the lake waters from an aerobic to an anaerobic condition (one lacking dissolved oxygen), which kills fish and other animals. Under anaerobic conditions, anaerobic bacteria flourish, and may produce toxins and evil-smelling sulphurous compounds which can make the water foul and objectionable. Interestingly, it is understood that where algae grow all the year round, as in the tropics, it is possible to use them to help in sewage treatment. There have been arguments as to whether phosphorus, nitrogen or, even, carbon, is the element which controls the excessive growth, but phosphorus has received the main blame. Phosphorus is introduced into rivers and lakes from human excrement, and from drainings from agricultural land, as well as from the phosphates in detergent products, but detergents are known to be a major source and this source is the only one readily subject to regulation. The inclusion of phosphates in detergent products is now banned in certain areas of the USA. In some other countries there are restrictions on the percentage of phosphorus which is permitted. The maximum permitted level varies from country to country; typically, it is between 5% and 6% as P, which is equivalent to 20% to 25% sodium tripolyphosphate, but there are higher and lower limits. In the UK there are no limits and no serious eutrophication problems. It is understood that, in 1985, the restricted level of phosphorus in Switzerland is to be replaced by a complete ban. Much work has been carried out on replacements for phosphates in NSD heavy duty products, but, so far at least, no fully satisfactory alternative for the complete replacement of sodium tripolyphosphate has been found. Possible solutions to the formulation problem are discussed in Chapter 2; here it will only be said that some alternative builders may cause other serious ecological difficulties. Technically, the best solution to the eutrophication problem caused by phosphorus would seem to be tertiary treatment of sewage, as is being carried out in Sweden and other areas where eutrophication is a serious problem. This would obviate the need for restriction of the use of phosphates in detergents, but full sewage treatment, including removal of phosphorus, is unlikely to become universal in the near future.

1.11.4 Boron

In some parts of the world there are restrictions on the use of sodium perborate and other boron compounds, because boron compounds in water used for irrigation can have an adverse effect upon crops.

2

What detergents have to do and how they do it. The roles of various ingredients and some principles of formulation. Market survey and recent trends

2.1 DETERGENT DUTIES

Detergents are used for many purposes, and there are a wide range of substrates (surfaces to be cleaned) and of soils. The main substrates are:

- textile fabrics
- human skin
- cooking utensils, crockery, glasses and cutlery
- sinks, draining surfaces and baths
- floors and walls
- lavatory bowls.

Textiles are used for many purposes, but the most important classes which are washed are clothing, bedclothes, towels and cleaning cloths. Many kinds of fibres are used in these textiles, in particular wool, cotton, linen, silk, rayon (regenerated cellulose), cellulose acetate, polyamide (nylon), polyester (terylene, Dacron, etc.), and polyacrylate. Wool is now often treated to make it less liable to shrink and felt, and so to be machine washable. Blends of different fibres, particularly polyester/cotton, are becoming increasingly important, and many fabrics are given special crease-resistant and other treatments by the manufacturer. The type of fibre and other differences between fabrics, such as type of weave, or knit, and type of dyestuff, can affect the adhesion of the soil and permissible conditions of washing (that is conditions which will minimize damage to the article).

The soils to be removed are also of many types, but most include both particulate and greasy elements. Particulate soils can be general household dusts, or dusts of more specialized kinds (clay, coal, iron oxide, etc.), depending upon the circumstances under which the article became dirty. Sebum, an exudate from the skin which contains glycerides, free fatty acids, waxes (esters) and sterols, is normally part of the soil on clothing and other articles which come into contact with the human body, such as towels and bedclothes. Soils on fabrics can also include:

- hydrocarbon oils and greases, particularly on mechanics overalls
- spilt food and drink
- blood
- urine and faeces
- grass stains
- cosmetic residues
- paint and varnish.

Hence a great variety of situations can arise from different combinations of soils and fabrics in the textile washing field alone. Some soils produce stains which are not normally removable by ordinary washing, and which require special treatments with solvents, bleaches or acids. It is not necessary to enumerate the soils and stains which can be found on other substrates, but it will be mentioned that, in addition to food residues, lipstick stains often need to be removed from cups and drinking glasses.

Products are sold to housewives for all the main purposes just indicated, as well as for more specialized duties, such as the cleaning of ovens and windows. Products are also required by commercial and other organizations for:

- duties similar to those in the home, although the equipment and detergent products used may be somewhat different.
- operations of a non-domestic type, such as the cleaning of bottles, process plant and metal parts.

2.2 HOW DETERGENTS WORK

2.2.1

The theory of detergency is complex and still not fully understood, partly because of the very wide range of situations which can arise from different combinations of substrate, soil, detergent solution and physical conditions, such as temperature and agitation. The literature is enormous and some requires a rather specialized knowledge of physical chemistry. Some generalizations reported conflict with others, again in part because the experiments were carried out with different systems under different conditions, and it is now thought that there is no single mechanism of cleaning. The final assessment of detergent performance is that achieved by the housewife in the home (or by other users under their conditions). Methods of washing, and the dirtiness of the articles to be washed, can vary widely, and manufacturers make much use of consumer and other in-home tests. Systematic tests under controlled conditions in the laboratory are also needed, particularly during the earlier stages of development work and for control purposes. Much effort has been devoted to the preparation of soiled test pieces which give results which correlate with statistical analyses of in-home tests. A 'half-way house' which is often useful is to collect the laundry from a suitable group of households and to select reasonably matched articles (or even half articles) for washing machine, or other, tests in the laboratory. BS 5610:1978 (which is the same as the international document ISO 4319 – 1977) provides valuable guidance on the setting up and interpretation of tests in which detergents are used to wash fabrics. Points include:

- the human eye is often better than instruments in comparing results, although instrumental readings are necessary for correlation, with due precautions, of one set of results with others.
- the hardness of a water supply can vary from day to day, or even within a day, and synthetic hard water may be needed for comparative tests. When synthetic hard water is used, the calcium/magnesium ratio, as well as total hardness, should be specified.

A comprehensive discussion of the theory of detergency is outside the scope of this book, but a simplified, non-mathematical, treatment of the major aspects is included to help readers to understand important points in relation to formulation. Those requiring a fuller, although concise, modern discussion are referred to the chapter on detergency in Rosen (1978). The following discussion refers directly to fabrics, but many points are also relevant to other substrates.

2.2.2

The washing process involves two stages, the removal of soil from the substrate and its disposal in the wash liquor without redeposition. In the case of fabrics, it is also essential for the detergent solution to penetrate between the fibres into the fabric; this is related to surface tension as discussed below. The shower proofing of rainproof garments, so that they permit a flow of air but not penetration of liquid water, is the opposite of penetration by a washing solution to achieve detergency.

2.2.3

The molecules in a liquid attract one another. At the surface there are no forces attracting from above, and this leaves a net inward force acting on the molecules in the surface. This is called surface tension and it acts something like a skin which tends to contract to make the area of the surface as small as possible. Surface tension is demonstrated by the way in which a falling droplet of liquid tends to a spherical shape (a sphere having the smallest surface area per unit volume). Similarly, together with gravity, it causes a liquid at rest to have a level surface. There are interfacial tensions at all interfaces, liquid–gas, liquid–

liquid, liquid–solid, and gas–solid. Surface tension is the interfacial tension at a liquid–gas interface.

Water has a high surface tension, but when a surfactant is dissolved in it, the surface tension of the solution is progressively lowered until the concentration reaches a value dependent upon the particular surfactant used and then remains constant with further increases in concentration. The concentration at which the surface tension ceases to fall is called the cmc (critical micelle concentration). It has been found that when a small quantity of surfactant is added to water, the surfactant ions (or molecules in the case of a nonionic surfactant) concentrate more in the surface layer than in the bulk of the liquid. At the liquid surface the surfactant ions are orientated so that the hydrophilic ends are in the water and the hydrophobic ends point out to the air. Similarly, if a suitable surfactant is dissolved in oil, the hydrophobic ends point inwards to the oil and the hydrophilic outwards to the air. Reverting to the aqueous system, at the cmc the solution becomes saturated in the normal sense, but more surfactant can be introduced into the liquid by the formation of what are known as micelles. These are bunches of surfactant ions (or molecules), probably roughly spherical in shape, with the hydrophilic ends pointing outward to the aqueous solution and the hydrophobic ends inward to form a more, or less, oily globule. In the case of the ionized surfactants, some of the gegenions (such as Na^+ with a sodium salt, or Cl^- in the case of a cationic surfactant) may be included in the micelles. Fig. 2.1 illustrates diagrammatically the main points in this section.

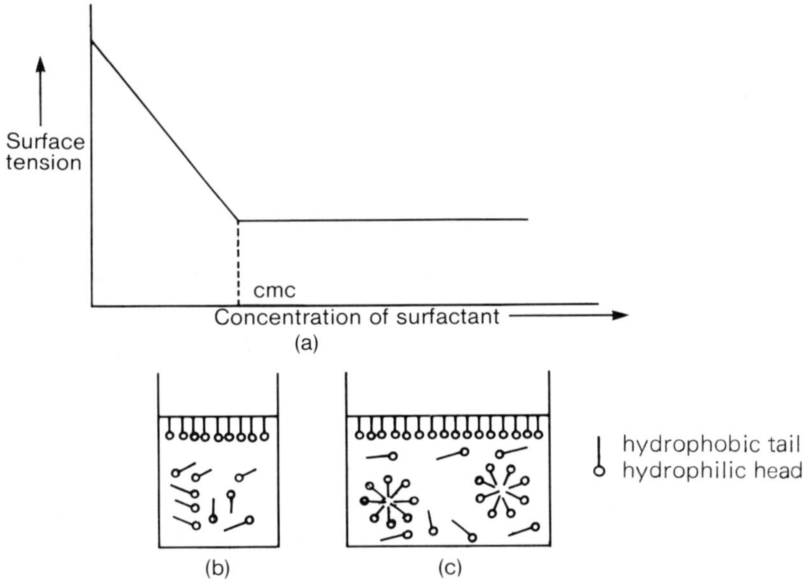

(a) surface tension — concentration diagram
(b) portion of liquid just below cmc
(c) portion of liquid just above cmc

Fig. 2.1 – Surface tension and micelle formation.

The solubility of a surfactant varies with temperature and an important temperature, known as the Krafft point, is that at which solubility rises rapidly with the formation of micelles. This is also the cmc for the temperature concerned (see also section 2.2.12 for some special features of the solubility of nonionic surfactants).

Micelles, which give surfactant solutions their colloidal nature, are very important in relation to detergent properties because they:

- provide a reservoir from which the liquid surfaces are kept saturated and the surface tension of the solution at its minimum.
- can solubilize oily matter. The interior of a micelle is almost like a hydrocarbon solvent which can 'dissolve' oily soil and carry it away in the wash liquor. This property is also employed when a soap solution is used to solubilize cresols to produce the lysol type of disinfectant, and in similar applications.

2.2.4

Water will not wet a surface which is even slightly greasy, as can be seen by the way in which a small quantity of water forms droplets, rather than an even film, when placed on such a surface. Addition of surfactant to the water lowers the surface tension and provides a solution which has improved wetting properties. Changes in wetting power can be measured by the contact angle, θ, between the surface and the aqueous solution, as shown in Fig. 2.2. θ equal to 180° represents complete non-wetting and $\theta = 0^\circ$ complete wetting. For most practical purposes, an angle of 90° is regarded as the boundary between wetting and non-wetting, A liquid making a contact angle of less than 90° will flow into a capillary tube having the surface properties under consideration without the application of external pressure. This explains the role of the surface tension of a washing solution in its penetration of the interstices of a soiled fabric, which is an essential first step towards the removal of the soil.

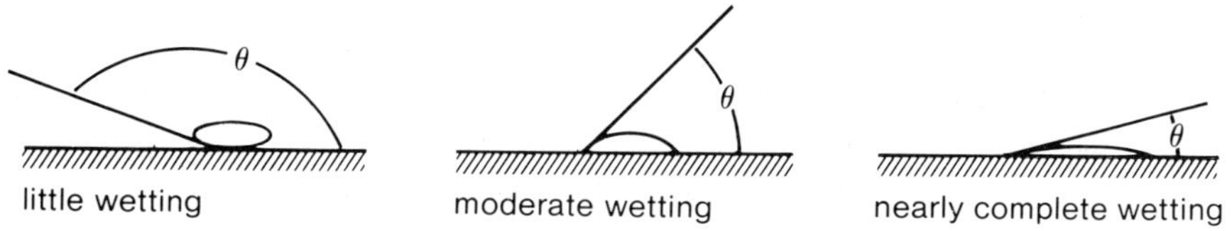

Fig. 2.2 – Behaviour of aqueous droplets.

2.2.5

Somewhat similar considerations apply to the removal of fatty soil from a substrate by what is known as the 'rolling-up' mechanism. The sequence of events is shown diagrammatically in Figs. 2.3 and 2.4. A solution which has sufficiently strong wetting properties in relation to the substrate can penetrate under the soil and gradually ease it away as a droplet. If the oily droplet is then solubilized in a micelle, it can be carried away without being re-deposited.

Fig. 2.3 shows complete removal; whereas Fig. 2.4 indicates what can happen when a droplet is only partially removed, leaving a small portion on the substrate. This mechanism of fatty soil removal requires a temperature high enough to render the soil mobile, that is substantially liquid.

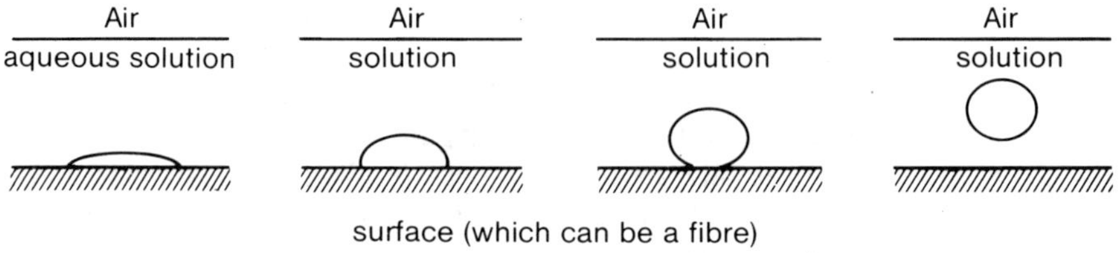

Fig. 2.3 – Removal of an oily droplet from a surface.

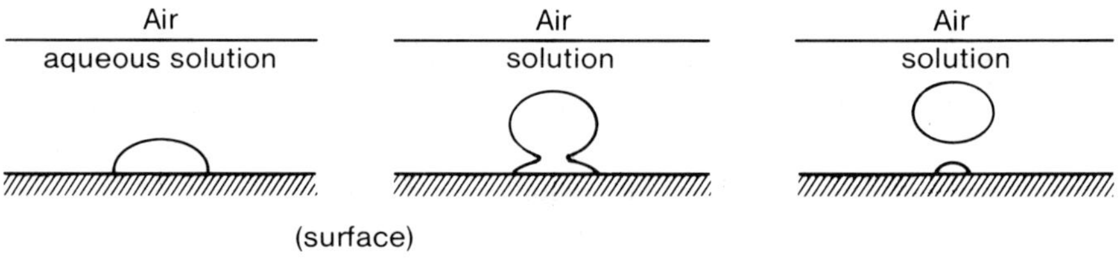

Fig. 2.4 – Removal of an oily droplet. (cont.)

There is another mechanism for the removal of fatty, or oily, soil from a substrate which involves the formation of a mesomorphic (or liquid crystal) phase which is considered to be more important than the roll-up mechanism under some circumstances.

2.2.6

The development of electrical charges on soil and substrate, which can result from the behaviour of the various ions present, is important in the removal of particulate soils by anionic detergents. Nonionic detergents, which are generally less effective than anionics against particulate soils but good against oily soils, do not produce electrical effects in the same way as anionics, and some other mechanism is probably involved. When, as is commonly the case, soils are both oily and particulate, removal of the oily soil helps removal of the particulate. These rather complex matters are beyond the scope of this elementary discussion.

2.2.7

Although articles can sometimes be cleaned by just soaking in a detergent solution, mechanical action is usually an important element in the cleaning process. This may be applied in various ways, with the hands, by jets of liquid, by the agitators, or the revolving and tumbling action, of a washing machine, or even by ultrasonics.

2.2.8

Thermal energy is also important and, broadly speaking, an increase in washing temperature improves performance, or makes the operation easier, although it may increase the damage done to some articles being washed. Boiling is particularly effective. The subject is, however, not straightforward, and it has become controversial with suggestions that much energy used to heat water might be saved by the use of cold water, or at least lower temperatures than have been normal in some countries. In tropical countries, much washing is done with unheated water, but this is considerably warmer than that from a cold water tap in a temperate area, and standards of performance may be different. There is little doubt that an increase in temperature up to about 40°C (that is to above the melting point of most fats) is beneficial, as will be appreciated by anyone who has tried to wash greasy plates in cold water. A number of factors need to be considered in the selection of the best temperature to use, and there is now a substantial increase in the amount of washing carried out at 40°C, or below, using detergents formulated to suit these conditions.

2.2.9

The main source of Ca^{2+} and Mg^{2+} ions in the wash liquor is the hardness of the water used, but these ions can also come from the soil, or from deposits which have been formed in a fabric during previous washes or rinses.

It has long been known that when the detergent used is soap, adverse effects result from the hardness present. Calcium and magnesium soaps are precipitated, and are liable to form scums, 'bath-tub rings', or deposits in a fabric. Such deposits in fabrics are liable to carry particulate dirt and to result in a greying of the articles. Soap is wasted, because no effective detergent (soluble soap) is present until virtually all the calcium and magnesium present have been precipitated and a sufficient concentration of soluble soap is achieved to give a good lather. NSD products lather as well, or sometimes better, in hard than in soft water, and at one time magnesium alkylbenzene sulphonates were sometimes used in detergents instead of the sodium salts. Various authorities, including the authors of BS 2756:1970 (concerned with cleaning in the dairy industry), say that hardness of water has little effect on the performance of NSD products. On the other hand, Rosen (1978) says that the presence in the wash liquor of polyvalent cations, notably Ca^{2+} and Mg^{2+}, is invariably detrimental to the cleaning process for a number of reasons. It has been known from the earliest days of NSD products for domestic use that non-soapy surfactants do not perform well as heavy-duty fabric washing products unless they are properly built; and the conflict is probably due to the adverse effect being mainly, if not entirely, in the fabric washing field, to which most of the literature on detergency refers.

The prime function of a builder is to remove calcium and magnesium ions from the solution, although builders also have other functions discussed later. Removal of Ca^{2+} and Mg^{2+} ions from the wash liquor can be achieved in one of three ways:

- by precipitation, using salts such as sodium carbonate or sodium orthophosphate. Soap is its own precipitative builder, which is one major reason why soap alone is a more effective detergent than other surfactants alone, although it also provides mild alkalinity. Additional builders can, however, be useful in soap products.

- by sequestration, in which the calcium and magnesium are taken into complex ions, as explained in section 1.5.10 in relation to sodium tripolyphosphate and other condensed phosphates. There are other types of sequestering agents, such as nitrilotriacetic acid (NTA) and ethylene diaminetetra-acetic acid (EDTA). Small concentrations of Ca^{2+} and Mg^{2+} ions remain in equilibrium with the complex ions, in concentrations which depend upon the stability of the various complexes. In this respect EDTA is better than NTA, which, in turn, is better than sodium tripolyphosphate. The way in which sequestering agents affect the build-up of calcium on cotton test pieces is shown in Fig. 2.5, courtesy of Unilever Educational Publications. The test pieces of cotton fabric were successively soiled with sebum and washed at 25°C in 12.5° hard water containing 0.1% anionic detergent, with and without 0.1% builder. The curves show the pick-up of calcium (in mmol/g × 10^3) after various numbers of soil/wash cycles.

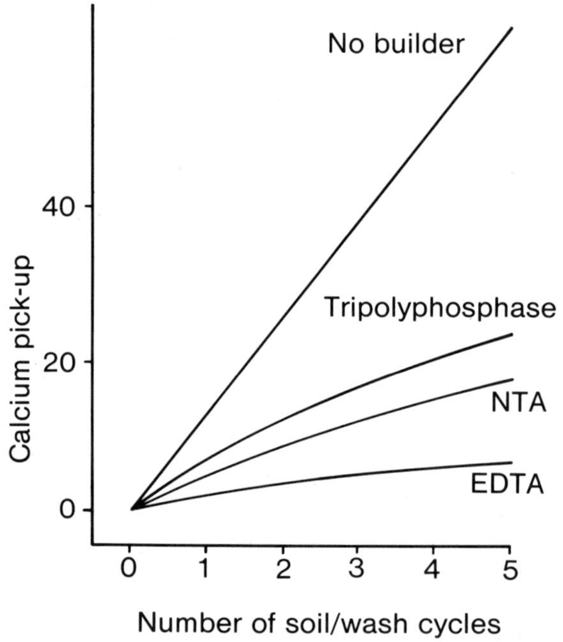

Fig. 2.5 – Build-up of calcium on cotton test pieces.

- by ion exchange, using synthetic zeolites, similar to those used in ordinary water softening. These zeolites do not exchange significantly with magnesium, and they need to be used in conjunction with other builders which may include sodium citrate.

It is often thought that anionic non-soapy surfactants do not precipitate insoluble calcium and magnesium salts, but this is not necessarily correct. Recent papers, such as Hollingsworth (1978) and the same author (1976) explain how the calcium (and similarly, but to a lesser extent, magnesium) salts of anionic surfactants of the alkylbenzene sulphonate type can precipitate, and so be inactivated, if the concentration of Ca^{2+} ions is above a certain limit. Rosen (1978) gives examples where the lack of solubility of some anionic surfactants has an adverse effect on performance.

Ways in which calcium and magnesium ions interfere with the washing process, in addition to the precipitation of anionic surfactants, include their tendency to:

- cause particulate dirt to flocculate and redeposit on a fabric
- combine with the free fatty acid in sebum and other fatty soils, and so to interfere with the removal of these soils. Fig. 2.5 shows how a cotton test piece soiled with sebum can pick-up calcium.

Although de-activation of calcium and magnesium is the major function of builders, they are also needed to:

- be a source of alkalinity
- provide a buffering action to maintain the pH of the wash liquor near to the required value throughout the wash
- to deflocculate and disperse particulate soils.

Sodium tripolyphosphate provides all four actions and is generally considered the best builder so far discovered; it is also a useful constituent for the production of detergent powders with good physical properties. It is possible to use a combination of builders, for example, a colloidal sodium silicate is often used to augment the relatively poor buffering capacity of sodium tripolyphosphate, as well as for its own deflocculating and other properties. Fig. 1.2 in the previous chapter shows the relative buffering capacity of the sodium silicates and other alkaline salts including the phosphates.

Figs. 2.6, 2.7 and 2.8 from Hollingsworth (1976), courtesy of Ciba-Geigy Ltd., illustrate several important points in relation to the building of a commercial C_{12} alkylbenzene sulphonate (0.5g/l) with sodium tripolyphosphate at pH 10 and 50°C. At the equivalence points shown in Figs. 2.6 and 2.7, there are equimolecular concentrations of sodium tripolyphosphate and Ca^{2+}; at concentrations of sodium tripolyphosphate above this level there is sufficient to sequester most of the Ca^{2+} ions, leaving about 10 ppm $CaCO_3$. Figs. 2.6 and 2.7, which show results averaged from tests with various test cloths, indicate how detergent efficiency is much improved by the presence of the phosphate, but falls off rapidly if the concentration of phosphate falls below that at the equivalence point. They also show that additional improvement occurs if the phosphate concentration is increased further. This may be because sodium tripolyphosphate alone has considerable cotton washing power. Fig. 2.8 shows how detergent efficiency varies with the concentration of residual free Ca^{2+} ions (that is those

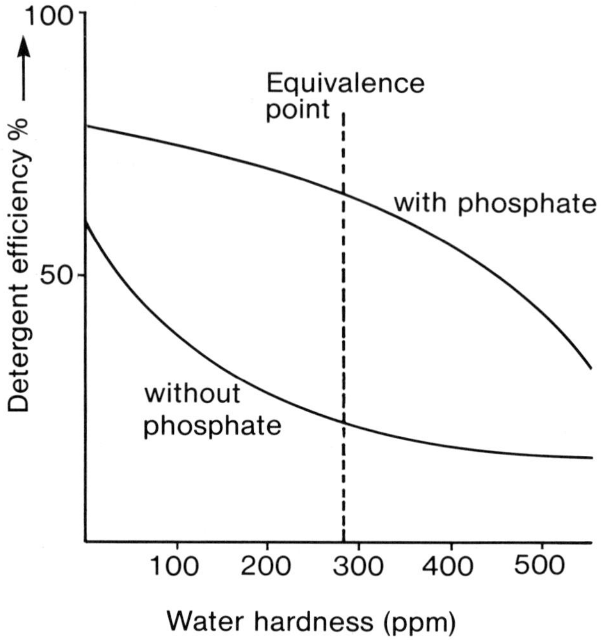

Fig. 2.6 – The effect of water hardness on detergent efficiency in the presence and absence of 1 g/l of sodium tripolyphosphate.

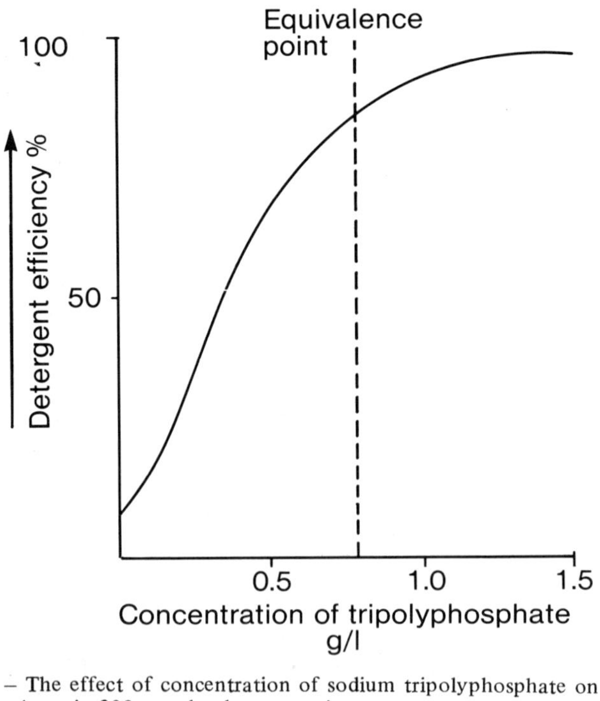

Fig. 2.7 – The effect of concentration of sodium tripolyphosphate on detergent efficiency (tests in 200 ppm hardness water).

which are not sequestered). The free Ca^{2+} ion concentration is shown in terms of pCa, which corresponds exactly with pH, so that the higher the value of pCa the lower the concentration. Curves 1, 2 and 3 in Fig. 2.8 are for three types of test cloth (made with different soils and/or fabrics). It will be seen that results can vary widely from one test cloth to another, but that the point at which detergent efficiency starts to fall off sharply is at nearly the same pCa for all three.

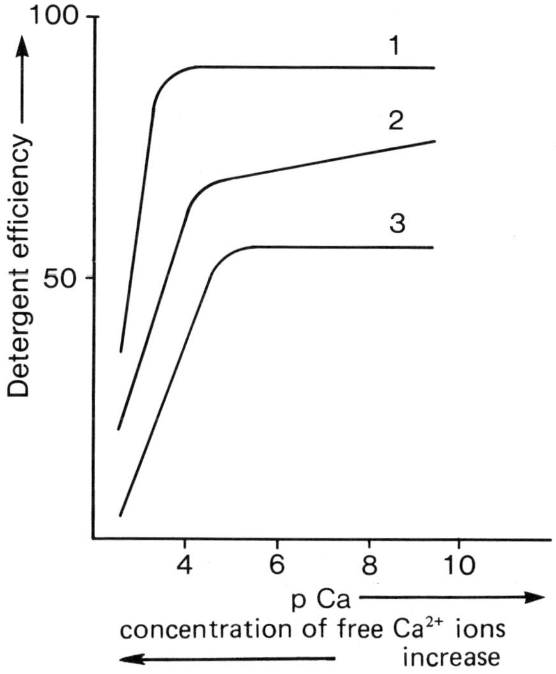

Fig. 2.8 – The effect of concentration of residual free calcium ions on detergent efficiency.

A system is said to be fully built when sufficient builder is present to cope with all the calcium and magnesium. The achievement of this state in the domestic situation depends upon the hardness of the water and the concentration of product used by the housewife, as well as upon the formulation of the product.

All salts, including non-builders such as sodium chloride and sodium sulphate, can be beneficial under some circumstances by decreasing the solubility of the surfactant and promoting its adsorption onto the soil and substrate.

2.2.10

Detergency normally improves with increase in pH. Caustic soda and the very alkaline silicates are used for oven cleaning and for some industrial purposes, such as bottle and dairy plant washing. For most domestic purposes, and for fabric washing generally, it is necessary to avoid too high a pH, and a value around 9.5–10.5 is usually desirable. Cotton can stand higher pHs than wool, and the human skin can be sensitive to solutions which are too alkaline.

Pure soaps are slightly hydrolysed when dissolved in water to produce an acid soap (or a fatty acid) and a slightly alkaline solution.

$$2\ R{\cdot}COO\ Na + H_2O = (R{\cdot}COO)_2H\ Na + Na^+ + OH^-.$$

In practice, acid soap formation is minimized by finishing the soap with a slight excess of alkali and by the use of alkaline builders. In addition to the prevention of the formation of acid soap, some alkaline builder helps the washing process by raising the pH of the wash liquor and by reacting with saponifiable matter in the soil (including sebum). The soap so generated in the soil itself is particularly useful in soil removal. 'An introduction to laundary chemistry', published by the British Guild of Cleaners and Launderers, points out that the use of modified alkali (sodium sesquicarbonate), which has in the past been thought adequate to protect the soap when woollens and other delicate articles are being washed, is not sufficiently alkaline if base-exchange softened water containing a substantial concentration of sodium bicarbonate is used.

2.2.11

The number of carbon atoms, including any in a benzene nucleus, in the hydrophobic part of an anionic surfactant is significant and is known as its carbon number, a better term than chain length, because the chain is sometimes branched. Fig. 2.9 shows diagrammatically surface tension curves for some sodium soaps

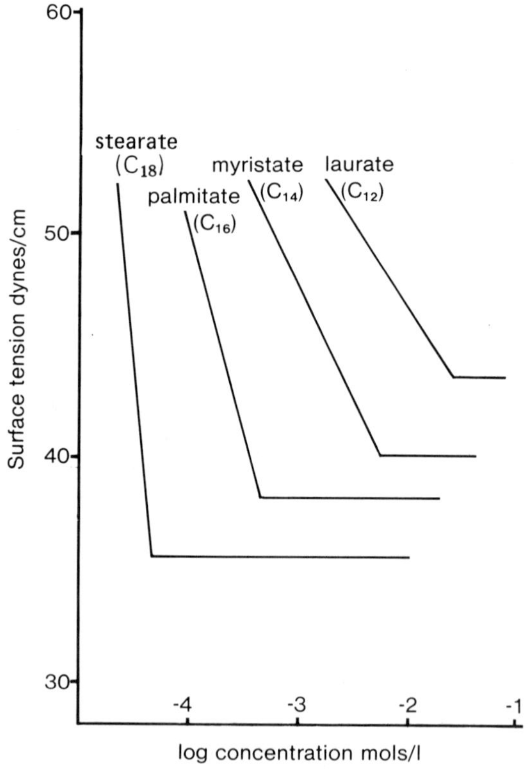

Fig. 2.9 – Surface tensions of sodium soap solutions.

of pure fatty acids, and Fig. 2.10 similar information for some sodium alkylbenzene sulphonates, both courtesy Unilever Educational Publications. As indicated on Fig. 2.10, the numbers identifying the curves are the numbers of carbon atoms in the alkyl chains, so that 6 must be added to obtain the carbon numbers. Soaps with less than about 8 carbon atoms do not show detergent properties. It will be seen from Fig. 2.9 that the cmc becomes lower, as does the minimum surface tension, as the carbon number increases. The solubility of the soaps falls with increase in carbon number, and there is an optimum carbon number depending upon the conditions of use. Soaps with a high proportion of C_{12} and C_{14} acids made from coconut or palm kernel oil, have good solubility and are particularly useful at low temperatures, or when the water is salty. Soaps made from tallow and palm oil, which contain high proportions of C_{16} and C_{18} acids, are less soluble, but give superior detergency at relatively high temperatures. As discussed later, most commercial soaps are made from fat blends.

Similar considerations apply to non-soapy anionic surfactants as shown in Fig. 2.10 for the alkylbenzene sulphonates, and similar information is available

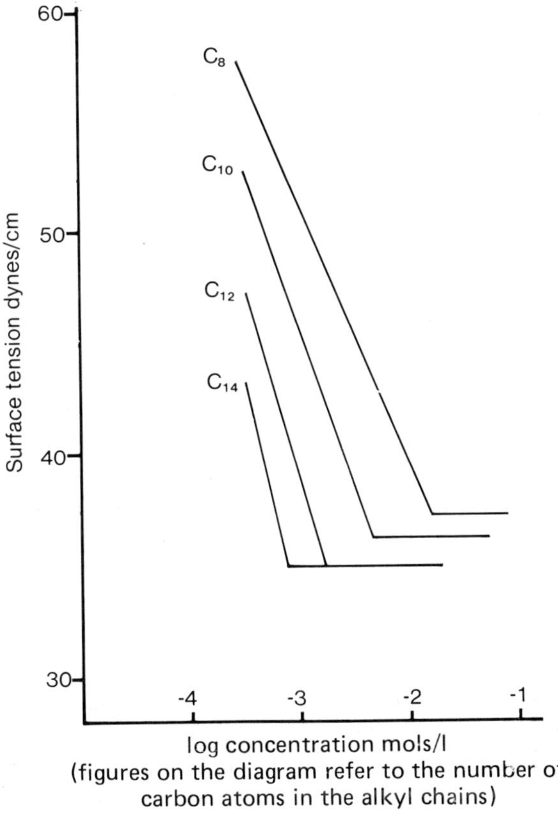

Fig. 2.10– Surface tensions of sodium alkylbenzene sulphonate solutions.

for the alkyl sulphates and other groups. In general, it is found that the best detergency is obtained by selecting a surfactant with the highest carbon number which does not lead to precipitation on the substrate under the conditions of use; important factors are concentration of Ca^{2+} ions and temperature. As for soaps, the optimum is usually with a carbon number of about 18. Commercial surfactants are seldom, if ever, pure compounds, and blends with a slightly higher average carbon number may be selected when detergency is particularly important, and conversely when the emphasis is on solubility, as for concentrated liquid products.

In the case of the important linear alkylbenzene sulphonates, the position of the phenyl group is also important. It can be attached to any carbon in the main chain, numbered as shown below, which is drawn for the 2-phenyl isomer.

$$\begin{array}{ccccccc} 7 & 6 & 5 & 4 & 3 & 2 & 1 \\ -CH_2- & CH_2- & CH_2- & CH_2- & CH_2- & CH- & CH_3 \\ & & & & & | & \\ & & & & & C_6H_4-SO_3Na & \end{array}$$

The 1–phenyl isomer shows excellent detergency in distilled water, but is too sensitive to calcium ions to be usable under ordinary conditions. Commercial LABS contains 2–, 3–, 4– and other isomers; 5– and 6– being said by some authors to be particularly suitable for heavy duty products. Other types of isomerism are also possible, including the position of the sulphonic group in the benzene nucleus.

2.2.12

As outlined in section 1.2.2, the most important nonionic surfactants used in detergent products contain a hydrocarbon hydrophobic and a polyglycol (or polyethylene oxide) hydrophilic end. Alkylphenols, such as nonylphenol (carbon number 15) and octylphenol (carbon number 14), both with highly branched alkyl chains, with a suitable ethylene oxide chain, were widely used in liquid detergent formulations, but as mentioned in 1.11.2, they are no longer employed in some countries because of insufficient biodegradability. The alternative hydrophobic end is usually some type of long chain alcohol. The properties and uses of these nonionic surfactants vary widely with the length of the ethylene oxide chain as well as with the carbon number of the hydrophobic end, and there tends to be a balance between the two. Hydrophilic chains with an average of 7–15 ethylene oxide units are usual for detergents 9–10 being particularly common. These chain lengths are averages, because a range of lengths is obtained in the production of commercial materials. They are described as, for example nonylphenol $\overline{10}$EO, the bar over the figure indicating that it is an average.

An important property of the ethylene oxide condensates is that above a certain temperature, known as the cloud point, they cease to be miscible with water and come out of solution. This is because the compounds hydrate at lower temperatures, which facilitates solution in water, but cease to have a shell of water molecules as the cloud point is reached. Another criterion sometimes used in relation to nonionic surfactants is the HLB (hydrophilic-lipophilic balance), defined as HLB = (content of ethylene oxide%)/5.

Griffin (1949 and 1955) used HLB to indicate the uses of various types of nonionic surfactant as follows:

Range HLB	Application
3– 6	Water-in-oil (W/O) emulsifier
7– 9	wetting agent
8–15	oil-in-water (O/W) emulsifier
13–15	detergent
15–18	solubilizer

The variation in properties with length of EO chain is illustrated in Table 2.1, courtesy of Dow Chemical Company, for their range of Dowfax 9N surfactants (nonylphenol ethylene oxide condensates). It will be noted that as the ethylene oxide content increases and the material becomes more hydrophilic, the cloud point rises. Fig. 2.11, courtesy of Unilever Educational Publications, shows

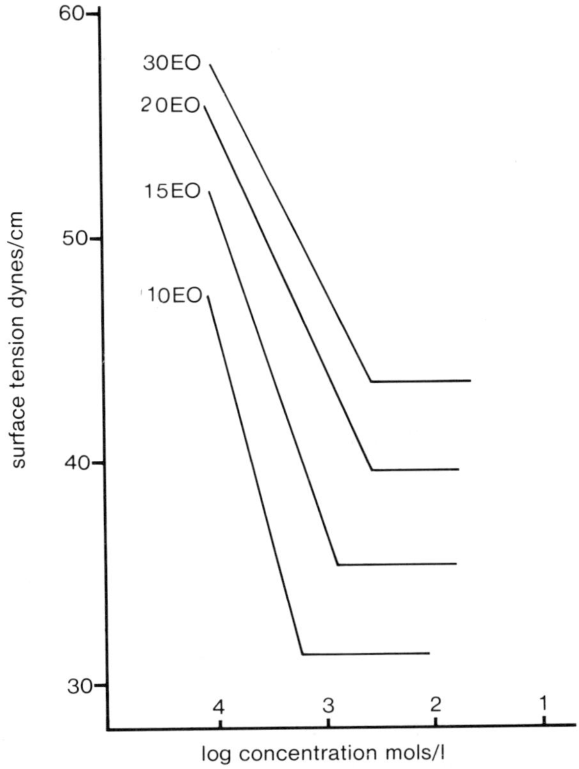

Fig. 2.11 – Surface tensions of solutions of nonylphenol ethylene oxide condensates.

Table 2.1 – Properties of Dowfax 9N condensates

Type	9N3	9N5	9N7	9N7B	9N7.5	9N8	9N9	9N10	9N12	9N14/15	9N20	9N25	9N30	9N50
Appearance	L	L	L	L	L	L	L	L	SS	SS	WS	WS	WS	WS
Cloud point °C (10% in 25% aqueous butyl diglycol solution)	46	62	71	–	–	–	–	–	–	–	–	–	–	–
Cloud point °C (1% aqueous solution)	–	–	–	15	32	45	50	56	84	91	–	–	–	–
Cloud point °C (1% in 10% NaCl solution)	–	–	–	–	–	–	–	–	–	–	72	74	75	78
HLB	7.5	10.0	11.7	11.9	12.1	12.3	12.9	13.3	14.1	14.3	16.0	16.7	17.1	18.0

L = clear liquid.
SS = semi solid.
WS = waxy solid.
The number following the N is the average number of E0 units in the hydrophilic chain. Dow make some other types not included in this table.

diagrammatically the surface tension-concentration curves for various nonyl phenol ethylene oxide condensates.

Generally, the detergency of a nonionic surfactant is at its optimum at just below its cloud point in the wash liquor, because solubilization of oily soils by the micelles of the surfactant increases markedly at this point. Surfactants with cloud points below the wash liquor temperature will separate out of solution and consequently show poor detergency, whereas those with cloud points much above that temperature may show relatively poor detergency because their greater solubility in the wash liquor decreases both solubilization of oily soil and adsorption of the surfactant onto soil and substrate.

One important property of nonionic surfactants is that they tend to be low, or very low, foaming.

2.2.13

Rosen (1978) summarizes the results of a number of investigations, including some which compare the performance of different detergents against various soils on various fabrics. These show that nonionics are superior to anionics under some circumstances and the converse in other situations. It is difficult to summarize further without risk of misleading, but it would seem that nonionics are particularly good for the removal of some greasy soils from some synthetic fabrics, particularly at relatively low temperatures, and anionics for handling particulate soils and when the fabric is cotton. There are, however, exceptions.

2.2.14

The circumstances in which improved foam, or less foam, are required are discussed later. Foaming involves a large increase in the liquid-air interface and this is helped by the presence of surfactants in the water; pure liquids do not foam. Soaps foam well, provided sufficient is present to ensure soluble soap in the solution. The stability of a foam, or lather, is related to the properties of the films of solution in the foam, particularly their plastic, and/or slow draining characteristics. Although too much fatty material can 'kill' a lather (as in hand dishwashing), certain limited additions of fatty materials can improve the quality of a lather. Examples are:

- fatty acid molecules formed by the hydrolysis of soaps interpose themselves between the soap ions in the surface zones of a soap solution and improve the lather. An addition of free fatty acid makes the lather of a soap more creamy, but, by itself, also reduces lather volume; this volume can be increased by the use of a fat blend which contains a high proportion of lauric (coconut or palm kernel) oil to give an important type of toilet soap discussed later.
- lather boosters, such as lauric monoethanolamide, which act similarly, are added to NSD products when it is required to improve lather.

Alkaline builders can also influence the lather characteristics of a detergent product.

Addition of soap to another anionic surfactant normally reduces the lather which would be produced by either separately.

As mentioned in the previous section, nonionic surfactants are low foaming, and they are valuable constituents of formulations intended to perform without the development of much lather.

2.3 ADDITIONAL INFORMATION CONCERNING INGREDIENTS AND THEIR ROLES

2.3.1 Anionic surfactants

2.3.1.1

Soap in its various forms is the subject of several later chapters. The other anionic surfactants of practical importance in the detergents industry are mainly sulphonates and sulphates. Sulphonation and sulphation are often carried out by the manufacturers of detergent products and these processes are the subject of Chapter 9. Some large detergent manufacturers may produce intermediates such as fatty alcohols from fats, and may make ethylene oxide condensates of various kinds, but surfactants, or the intermediates from which they can be made by sulphonation or sulphation, are usually purchased from petrochemical, or other chemical, companies. The present section will only list, with brief notes, the more important types of non-soapy anionic surfactants.

2.3.1.2

Alkylbenzene sulphonates, $R - C_6H_4 - SO_3Na$, sometimes known as dodecylbenzene sulphonates, or by various proprietary names, are still the most important type of surfactant used in the detergents industry. As mentioned earlier, the original types had a side-chain made from a paraffin fraction and were known as kerylbenzene sulphonates (from kerosene). It was then found that highly branched material produced by the polymerization of propylene, C_3H_6, provided a very satisfactory side chain when attached to benzene. Polymers from pure propylene contain only C_6, C_9, C_{12} and C_{15} hydrocarbons, but some ethylene C_2H_4, and butylene C_4H_8, may also be present in the propylene to be polymerized, giving hydrocarbons with other carbon numbers. The propylene polymer, and/or the alkylbenzene made from it, is fractionated to the required extent. The so-called propylene tetramer (PT) type of side chain material is mainly C_{12}, with a molecular mass of the alkylbenzene near to the theoretical 246 for $C_{12}H_{25}-C_6H_5$. A slightly higher molecular mass type, sometimes called tridecylbenzene, with an average molecular mass around that for $C_{13}H_{27}-C_6H_5$ (260), contains substantial proportions of the C_{12} and C_{15} materials. Lower molecular mass types with a considerable content of C_9 side chain have also been made. As will be appreciated from section 2.2.11, the sodium sulphonates from the lower molecular mass hydrocarbons are more soluble than those of higher molecular mass, which makes them particularly suitable for liquid products, but less effective in heavy duty products to be used at high washing temperatures for which the 260 MM materials are especially good.

For some years alkylbenzene sulphonates made with polypropylene side chains dominated the field, but, as explained in section 1.11.2, they are no longer used in the developed countries because of lack of biodegradability. They have been replaced by the linear alkylbenzene types (coded LABS or LAS) which are refined versions of the original kerylbenzene sulphonates; these are also available with a range of average molecular mass and of narrowness of 'cut' (that is range of carbon numbers in a particular product).

The alkylbenzenes (often called alkylates) are mobile liquids which can readily be transported to the detergent manufacturer, in drums or in bulk. The sulphonic acids (in two slightly different forms discussed in Chapters 9 and 10), or the sodium sulphonates, can also be purchased. The latter are usually sold as powders or flakes, containing around 80–85% of active surfactant, the rest being mainly sodium sulphate plus a little water.

2.3.1.3

Alkyl sulphates, $R - O - SO_2 - ONa$, are another important type of anionic surfactant. Long chain alcohols to be sulphated to produce sodium salts for use in detergents can be obtained from several sources:

- from fats, or fatty acids, by reduction catalytically with gaseous hydrogen, or by the action of sodium metal on the fatty acid esters. The basic reaction is $R{\cdot}COOH + 2H_2 \rightarrow R{\cdot}CH_2 - OH + H_2O$. With the sodium method, unsaturated fatty acids are converted into the corresponding unsaturated alcohols, but with the gaseous hydrogen method, the double bonds are hydrogenated to give only saturated alcohols. Various blends of alcohol are possible, but those of commercial importance are derived mainly from coconut (or palm kernel) oil and from tallow. In the case of the 'lauric oils' (coconut and palm kernel oils), alcohols produced without fractionation can be used, or lauryl alcohols of varying purity (broad, medium or narrow cut) can be derived prior to sulphation to give 'sodium lauryl sulphates' of different qualities. Tallow alcohol sulphates are usually made without fractionation of the alcohols.
- from the waxes in the oil of the sperm whale, but this oil is no longer generally available. Alcohols from this source were used for the production of some early surfactants.
- from the Ziegler process for the polymerization of ethylene, which produces linear primary alcohols with an even number of carbon atoms like those made from fats. The Alfols (registered trade mark of the Continental Oil Company, USA) are of this type and products with various chain lengths are available.
- from olefins and carbon monoxide using variations of the OXO process. These processes give alcohols with both odd and even numbers of carbon atoms, and they are often slightly branched.
- by a process developed by the Union Carbide Corporation, thought to involve oxidation of a paraffinic feedstock, giving products sold under the trade name Tergitol.

The sulphates made from these, or any other, alcohols are sulphuric esters, in which the alkyl group from the alcohol is attached to the sulphur atom through an oxygen. This makes the sulphates readily hydrolysable at high temperatures, or particularly, under acidic conditions. so that the acid produced in the sulphation reaction must be neutralized immediately and not transported as such. This contrasts with the sulphonic acids and the sulphonates, in which the sulphur atom is attached directly to a carbon one (in the benzene ring in the case of the alkyl benzene types), and the bond is not broken under normal conditions.

Another type of alkyl sulphate, the *secondary alcohol sulphate, Teepol 10,* was made from olefins derived from cracked waxes, and it was first sold by Shell in 1942. Teepol is a registered trade mark used generically by Shell for their range of detergent products. Teepol 10 was the first liquid soapless detergent, at least on general sale. Its formula is

$$R - \underset{}{\overset{CH_3}{\overset{|}{CH}}} - O - SO_2 - ONa$$

Teepol 10, and similar products made by other companies, were very widely used for a number of years. As a liquid Teepol 10 was used for dishwashing and general cleaning. It was also spray dried mixed with sodium sulphate to produce a powder which was an ingredient in some early NSD powder products. Again, it was blended with soap to produce tablets to be used for toilet purposes in sea water, for example by the USA armed forces during the 1939–45 war. These secondary alcohol sulphates have a strong, not very pleasant, odour, and hydrolyse readily, particularly if allowed to become acidic. The hydrolysis reaction is

$$R - CH(CH_3) - O - SO_2 - ONa + H_2O \rightarrow R - CH(CH_3) - OH + NaHSO_4$$

The sodium bisulphate produced is strongly acidic, so that if the decomposition starts under near neutral conditions it accelerates as the pH falls; that is it is autocatalytic.

2.3.1.4

The *alkyl ether sulphates* are important anionic surfactants used in liquid products. They are produced by the reaction of a long chain alcohol (usually some grade of lauryl alcohol) with 2–4 moles of ethylene oxide, and sulphation of the alcohol so formed. The formula for lauryl ether sulphate is thus

$$C_{12}H_{25} - [O - CH_2 - CH_2 -]_n - O - CH_2 - CH_2 - O - SO_2 - ONa$$

The lauryl ether sulphates are commonly available as aqueous solutions at about 28% AD (active detergent). By using the physical chemistry of the systems, Albright and Wilson have recently developed certain alkyl ether sulphates, and alkyl sulphates, in the form of very high active mobile liquids. Fig. 2.12, courtesy of Albright and Wilson Ltd, shows phase boundaries for their Empicol ESB70 (sodium lauryl 2E0 sulphate). Phase G is the high active mobile liquid. The product ESB70 contains 70% anionic active, which minimizes transport costs, and it also includes lower percentages of sulphate and chloride calculated on active matter than the 28% material.

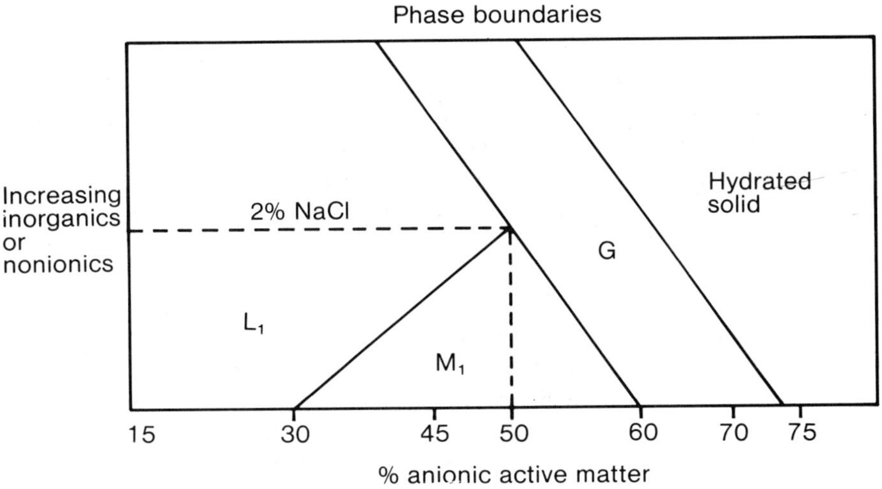

Fig. 2.12 – Effect of inorganics and nonionics on Empicol ESB70.

2.3.1.5

The *alkylphenol ethylene oxide sulphates* are another group of products which have been used in liquid detergents, but, as they are based on nonylphenol or octylphenol, they are now probably objectionable in some countries because of insufficient biodegradability. The alkylphenol ethylene oxide sulphates are made by light ethoxylation of the alkylphenol followed by sulphation. In order to avoid sulphonation of the benzene ring, sulphation is usually done with sulphamic acid

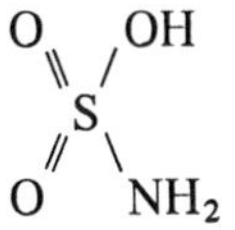

which yields the ammonium salt, for example,

$$C_9H_{19} - C_6H_4 - [O - CH_2 - CH_2]_n - O - CH_2 - CH_2 - O - SO_2 - ONH_4.$$

2.3.1.6

Alpha-olefin sulphonates have aroused considerable interest in recent years, but, because of cost and technical problems, it is not thought that they are very widely used. Anionic surfactants made from long chain alpha-olefins include two main constituents:

- alpha-olefin sulphonate $R - CH = CH - SO_3Na$ and
- alkyl hydroxysulphonate $R - \underset{\displaystyle OH}{\underset{|}{CH}} - CH_2 - SO_3Na$.

Special points in relation to their manufacture as discussed in Chapter 9.

2.3.1.7

A fairly recent development is the use of *fatty methyl ester sulphonates, MES.* Methyl esters are produced by replacement of glycerol in fairly saturated fats, such as tallow, palm oil, and the lauric oils, by reaction with methyl alcohol. The methyl esters are then sulphonated under suitable conditions, when the $-SO_3H$ group is attached to the carbon next to the ester group. The sodium MES materials are good surfactants, 100% biodegradable, and of interest as scum dispersants.

2.3.1.8

Many other types of anionic surfactant have been made and used for various purposes at different times. The benzene nucleus in the alkylaryl sulphonates serves only to facilitate sulphonation, and the *Mersolates,* which are primary alkyl (or alkene) sulphonates, $R - SO_3 - Na$, were an early type used in detergent products. Interest in this type of anionic has revived recently, but, so far as is known, they are not used extensively.

Other early anionic surfactants were the *Igepons.* Igepon A, an alkyl isethionate,

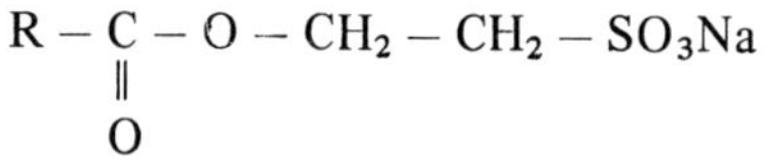

and Igepon T, an alkyl methyl tauride,

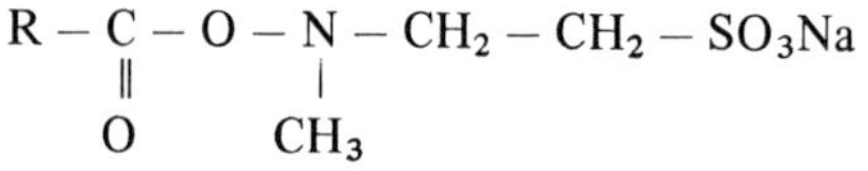

were first made by I. G. Farben Industries in the 1930s, but have been used more recently as surfactants in non-soapy toilet tablets.

2.3.1.9

The following figures, courtesy of Shell International Petroleum Company Ltd, give consumptions of alkylates (alkylbenzenes) and of long chain alcohols, used as detergent intermediates in the non-Communist world in 1981.

	Alkylate ('000 tonnes)		Long chain alcohols ('000 tonnes)
	Hard	Soft	
West Europe	–	358	158
North America	–	249	237
Asia	107	161	55
South America	152	43	8
Middle East	23	39	–
Africa	30	27	1
Oceania (Australia, New Zealand and Pacific Islands)	–	19	3

2.3.2 Nonionic surfactants, including lather boosters

The ethylene oxide condensates with alkyl phenols have been discussed in section 2.2.12. Similar nonionic surfactants can be made from the various types of high carbon number alcohols mentioned in section 2.3.1.3. Of these, materials such as *tallow alcohol* $\overline{10}$*EO* have become important as surfactants in controlled and low sudsing products.

Other types of nonionic surfactant include:

- *lauryl,* and similar, *monoethanolamides,* $R - CO - NH - C_2H_4 - OH$. Narrow cut lauryl monoethanolamide (LME or LEA), broader cut types, whole coconut monoethanolamide (CME or CEA), and whole palm kernel monoethanolamide (PKEA) are widely used as lather boosters and lather stabilizers in powders and liquids, and as viscosity modifiers in liquids. Corresponding isopropanolamides are also sometimes used.
- *lauryl,* and similar, *diethanolamides,*

$$R - CO - N \begin{array}{l} \diagup C_2H_4OH \\ \diagdown C_2H_4OH \end{array}$$

  are used for broadly similar purposes to those for which the monoethanolamides are employed, particularly in certain liquid products.
- *alkyl,* usually lauryl/myristyl (C_{12}/C_{14}), *di-methylamine oxides,*

$$\begin{array}{c} \quad CH_3 \\ \quad | \\ R - N \rightarrow O \\ \quad | \\ \quad CH_3 \end{array}$$

  are used in liquid products as foam boosters, foam stabilizers and viscosity modifiers. Because of their mildness to the skin, they are used particularly in shampoos, foam baths and the like.
- *polypropylene condensates* of at least two major types have been developed. In the first, a molecule of propylene oxide is condensed onto a conventional hydrophobe and a suitable number of ethylene oxide molecules then added. Propylene oxide chains are less hydrophilic than ethylene oxide chains with the same number of units. These mixed condensates are said to provide good detergency. The second type is the group of very low foaming nonionics sold under the trade name of *Pluronic* by Wyandotte Chemical Corporation†. These are especially suitable for machine dishwashing detergents for which foam suppression rather than foam generation is required. When propylene oxide is condensed by itself polypropylene glycols are formed. With a relatively small number of units they are water soluble, but with increasing complexity they become water insoluble and can be used as hydrophobes. These glycols carry hydroxyl groups at each end to which hydrophilic ethylene oxide chains can be added to produce the nonionic surfactants.

†now part of BASF.

The foaming characteristics of nonionic surfactants vary considerably, but, as already indicated, some are of great, and increasing, importance for their low foaming, as well as other, properties.

2.3.3 Cationic surfactants

Cationic surfactants are relatively expensive and not very good detergents. In general, they are not compatible with anionic surfactants and the two kinds usually precipitate each other, but there can be exceptional combinations.

Because of their activity against various micro-organisms, some types, such as alkylbenzyldimethyl ammonium chloride, are used as detergent-sanitizers in dairy and other food industries.

Very large quantities of types such as di-hardened tallow dimethyl ammonium chloride,

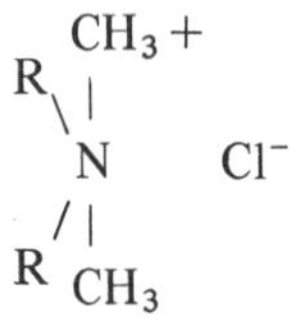

are used as the major constituent of fabric softeners.

2.3.4 Hydrotropes

Hydrotropes are substances which can act as solubilizers and viscosity modifiers. They are used to lower the temperature at which liquid detergent products separate on cooling (the lower cloud point) and to modify the viscosity of these products. They are also used to reduce the viscosity of slurries to be fed to spray drying towers, and to improve the physical properties of the blown powders produced from these slurries.

The main types are:

- *aromatic sulphonates with a short side chain,* particularly sodium (or other) toluene sulphonate (STS, $CH_3 - C_6H_4 - SO_3Na$), xylene sulphonate (SXS, $(CH_3)_2 - C_6H_3 - SO_3Na$), and, less usually, cumene sulphonate ($(CH_3)_3 - C_6H_2 - SO_3Na$).
  This is the type of hydrotrope used in slurries to be blown in spray drying plants, as well as in liquid detergent products. The materials are commonly available as dried powders. In some cases it is convenient to co-sulphonate the toluene, or xylene, with the alkylate intended to produce the anionic surfactant.
- *urea,* $NH_2 - CO - NH_2$, a white powder.
- *industrial ethyl alcohol.* Ethyl alcohol is used in many liquid detergents, but it is not permitted in some countries. Methyl alcohol is a satisfactory hydrotrope, but is little used because of its toxicity. Iso-propyl alcohol may be an acceptable alternative to ethyl alcohol.

2.3.5 Builders and sequestrants

2.3.5.1 Precipitative builders

The main precipitative builders, sodium carbonate, sodium orthophosphates and sodium silicates, provide alkalinity and buffering power, as well as remove calcium and magnesium from the wash liquor. The precipitated calcium and magnesium salts can, however, cause fabrics to become harsh. Calcium and magnesium soaps can also result in undesirable precipitates in fabrics, but, in contrast to the carbonates and silicates, the soaps have some lubricating and softening effect.

The colloidal sodium silicates also help to:

- deflocculate particulate soils, and to disperse oily soils
- suspend soils and reduce redeposition
- protect metals, particularly aluminium
- improve the crispness and storage properties of blown NSD powders.

2.3.5.2 Sequestratnt builders and sequestering agents for other purposes

As explained in sections 1.5.10 and 2.2.9, the condensed phosphates sequester calcium and magnesium in complex ions, and have other properties required of a builder. Sodium tripolyphosphate is generally considered the best all-round builder, and it has been, and is, used in enormous quantities all round the world. Sodium pyrophosphate has also been used, but it has tended to be replaced by sodium tripolyphophate in NSD powders. Potassium pyrophosphate is used in some heavy duty liquids, because the pyrophosphates are less prone to hydrolysis than the tripolyphosphates, and the potassium salts are more soluble than the sodium. Graham's salt (so-called sodium hexametaphosphate) is an excellent sequestering agent (better than pyro- or tripoly-phosphate), but it is hygroscopic and generally not suitable for use in powder products.

The rise in concern with regard to eutrophication and the consequent restrictions on the use of phosphorus compounds in detergent products (see section 1.11.3) has led to much effort to find satisfactory alternatives. The most important sequestrant alternatives which have been seriously considered, or used, are:

- *aminocarboxylic acids* and their sodium salts, in particular ethylene diamine tetra-acetic acid (EDTA),

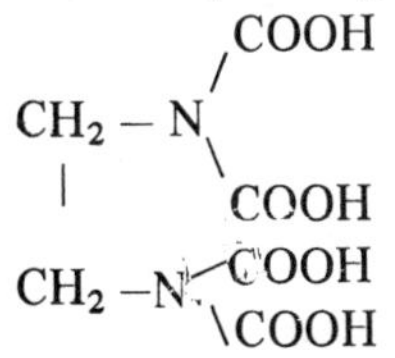

and nitrilotriacetic acid (NTA),

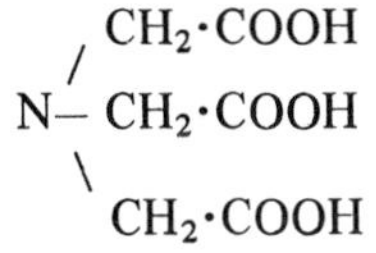

These compounds can be bought as the acids, or as one of their salts, but they are converted into the fully neutralized sodium salts in alkaline detergents. The differences in the molecular masses of the acids and their salts are sufficient to make the way in which concentrations are expressed important, for example, 30% of the tetrasodium salt of EDTA is equivalent to about 22% of the free acid.
EDTA is an excellent sequestrant, as shown in Fig. 2.5, but it has not been used as a builder because it is expensive and not biodegradable; it would also be difficult to include in large concentrations in powders. EDTA is widely used in small concentration in soaps and other products to sequester heavy metals, such as copper, which catalyse deterioration reactions. At one time, NTA was expected to become a major replacement for phosphates, and plants were projected to produce it in large quantities at an acceptable cost. Apart from technical problems in the manufacture of good powders, the environmental acceptability of NTA is uncertain. One problem is that it can sequester metals such as cadmium and mercury, and introduce these toxic elements into water supplies. In 1983, NTA was used as a detergent builder only in Canada. NTA cannot replace sodium tripolyphosphate completely, and it is best used in conjunction with sodium tripolyphosphate; a mixture of NTA and borax has been suggested, but boron is not universally acceptable.

- *hydroxycarboxylic acids* and their salts, which have some sequestrant powers. Compounds of this type include citric acid,

$$\begin{array}{ccccc} CH_2 & - & CH & - & CH_2 \\ | & & | & & | \\ COOH & & OH & & COOH \end{array}$$

tartaric acid,

$$\begin{array}{l} HO - CH - COOH \\ \qquad\;\; | \\ HO - CH - COOH \end{array}$$

gluconic acid, $CH_2OH - (CH - OH)_4 - COOH$, and di-glycollic acid,

$$\begin{array}{ccccc} CH_2 & - & O & - & CH_2 \\ | & & & & | \\ COOH & & & & COOH \end{array}$$

but none so far seems to provide a satisfactory solution to the general heavy duty detergent problem. Gluconic acid is useful in very alkaline solutions containing free caustic soda. Citrates are used in conjunction with zeolites.

Other materials used to inactivate heavy metals for preservative reasons will be mentioned later.

2.3.5.3

Certain inorganic base-exchange materials, that is Type A synthetic zeolites (sodium aluminium silicates), are not satisfactory as the sole agent for dealing with calcium and magnesium, but they are becoming of increasing importance as partial replacements for sodium tripolyphosphate. Stavitsky (1977), explains how the zeolites need to be very fine (less than 10μ) in order to exchange cations rapidly, and not to show as visible deposits on cloths. Zeolites need the support of a soluble complexing agent to remove calcium and magnesium from solid substrates, and to deal satisfactorily with the magnesium. Sodium tripolyphosphate may be used if this is permissible, and sodium citrate is said to be useful to handle magnesium.

Other interesting recent papers concerning non-phosphate builders are Schweiker (1978, and 1981) Berth (1978), Crutchfield (1978) and Borchert and Marino (1980).

2.3.6 Anti-redeposition agents

Sodium silicates have useful anti-redeposition properties, but special ingredients are commonly added, especially to NSD formulations, to reduce the redeposition of soil onto certain fabrics, particularly untreated cotton. A sodium carboxymethyl cellulose (SCMC or CMC) is nearly always chosen for this purpose. Cellulose is a natural polymer made up from a large number of glucose units, and by the action of chloracetic acid and alkali, a number of $-\ CH_2{\cdot}COONa$ groups are attached by reaction with hydroxyl groups in the cellulose,

$$\begin{array}{l}
-\underset{\substack{|\\ OH}}{CH}-\underset{\substack{|\\ OH}}{CH}-\underset{\substack{|\\ OH}}{CH}-+\ Cl-CH_2{\cdot}COOH+2NaOH\rightarrow-\underset{\substack{|\\ O\\ |\\ CH_2-COONa}}{CH}-\underset{\substack{|\\ OH}}{CH}-\underset{\substack{|\\ OH}}{CH}+NaCl+2H_2O
\end{array}$$

Grades of SCMC differ one from another in two respects:

- the length of the polymer chain, called the degree of polymerization (DP). This affects the viscosity of the solutions and about 200–500 glucose units is generally desirable.
- degree of substitution (DS), defined as the number of sodium carboxymethyl units per glucose unit.
  The following table indicates how the properties of SCMC vary with DS:

Value of DS	Properties of SCMC
Below 0.4	not sufficiently soluble
0.4 – 0.8, particularly 0.6 – 0.8	usually optimum
Above 0.8	too water soluble and not sufficiently adsorbed onto cotton

SCMC is a fibrous material, something like a crude cotton wool in texture.

It can be dispersed in the slurries used in the production of spray dried powders, but if it is to be included in a dry mixed powder it should usually be pre-ground in a mill with an inorganic salt to ensure proper distribution and an acceptable appearance of the powder.

Other cellulosic anti-redeposition agents, such as methyl cellulose, methyl hydroxypropyl cellulose and methyl hydroxybutyl cellulose have been tested on polyester fabrics and appear to show considerable benefit in nonionic detergent systems built with sodium carbonate. Some of these materials are also used to thicken liquids.

Other anti-redeposition agents, such as polyvinyl alcohol (PVA) and polyvinylpyrrolidone (PVP) are sometimes used, particularly in liquid detergents intended for the washing of fabrics.

2.3.7 Optical brighteners (OBs), also known as fluorescent whitening agents (FWAs), fluorescers, optical dyes, or (incorrectly) optical bleaches

For many years, housewives and commercial laundries used laundry blue to improve the colour of washed white articles. This blue, which is still available in many countries, helped to neutralize the yellow/red discolouration which develops on fabrics and thus gave the articles a better white/blue shade. It, however, did this by reducing the total reflectance from the fabric, and hence its brightness. A better solution to the problem has been achieved during the last 40 years by the use of optical brighteners, which are also used in the paper and fabric manufacturing industries, which absorb ultra-violet light and emit the energy so gained in the form of longer wavelength, visible, light. They improve the brightness, as well as the shade, of white articles, and justify the advertising claim to add brightness to whiteness. The way in which optical brighteners affect fabrics is shown dramatically by the way in which clothing on a darkened theatre stage can glow when illuminated with invisible, ultra-violet, light. Nearly all white fabrics now on sale have been treated with optical brighteners during manufacture, and it is difficult to buy white cloths which do not fluoresce. The function of optical brighteners in detergents is to maintain, or build-up, this fluorescent whitening effect.

Optical brighteners have been developed to improve the appearance of white textiles made with various types of fibre. Brighteners for wool were used at one time, but the most effective, and far the most important, optical brighteners are those which affect cotton and other cellulosic fibres (viscose and linen).

For many years the predominant types were the substituted di-triazinyl aminostilbene di-sulphonates,

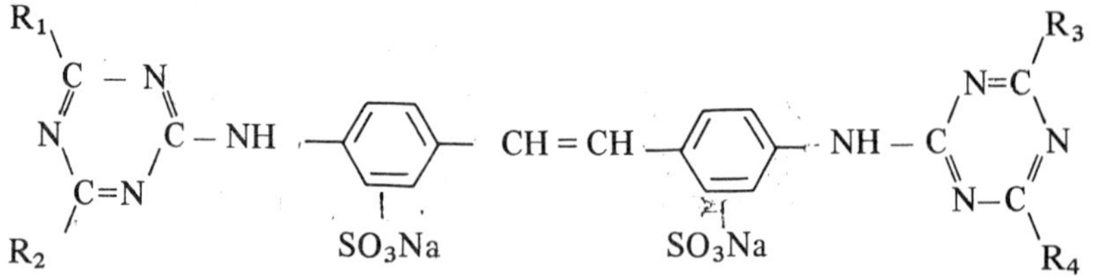

and these are still of major importance.

Symmetrical compounds in which $R_1 = R_3$ and $R_2 = R_4$ are normally, but not necessarily, used. The Rs can be aniline, morpholine, diethanolamine, or other groups.

More recently, other types of optical brighteners for cotton have been developed, in particular a series derived from a di-styryl biphenyl skeleton. These are understood to have superior properties to justify their high cost. The various manufacturers of optical brighteners usually have a generic name for their products, such as Photine from Hickson and Welch, Tinopal from Ciba-Geigy, and Leucophor from Sandoz. They then identify particular types by a suffix, such as Tinopal 5MBS-X, Photine CBUS-B, and Leucophor DU. Thus, for example, one Tinopal may be quite different from another Tinopal, whereas one Photine may be very similar to one Leucophor, or to one Tinopal. The various manufacturers issue useful literature about their products and, in some cases (Ciba-Geigy and Hickson and Welch, for instance) about various aspects of optical brightening in general.

The optical brighteners added to detergents to improve washing performance often also improve the colour of nearly-white powdered products. Special types, such as Tinopal GS (described as a naphtho-triazole-stilbene derivative), are available specifically to improve the colour of white products such as toilet soaps.

Points to be considered in the selection and use of an optical brightener are:

- the way in which the optical brightener is taken up by the fibres from the wash liquor. Figs. 2.13, 2.14 and 2.15, courtesy of Hickson and Welch, illustrate a number of points. Fig. 2.13 shows how the fluorescence of a cotton cloth increases with the number of washes at various wash liquor

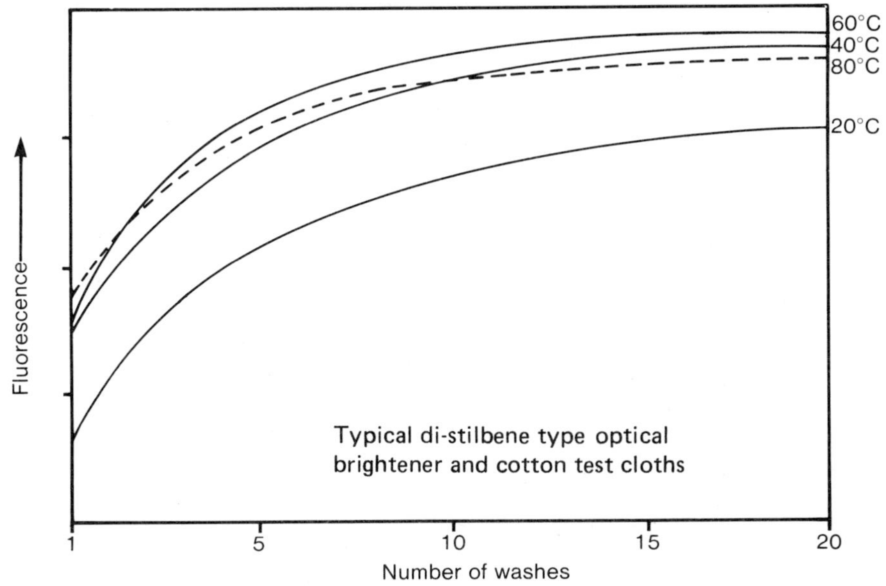

Fig. 2.13 – Effect of wash temperature on fluoresence.

temperatures; concentrations, ratio of wash liquor to cloth, and washing time kept constant. Fig. 2.14 shows how fluorescence builds up with

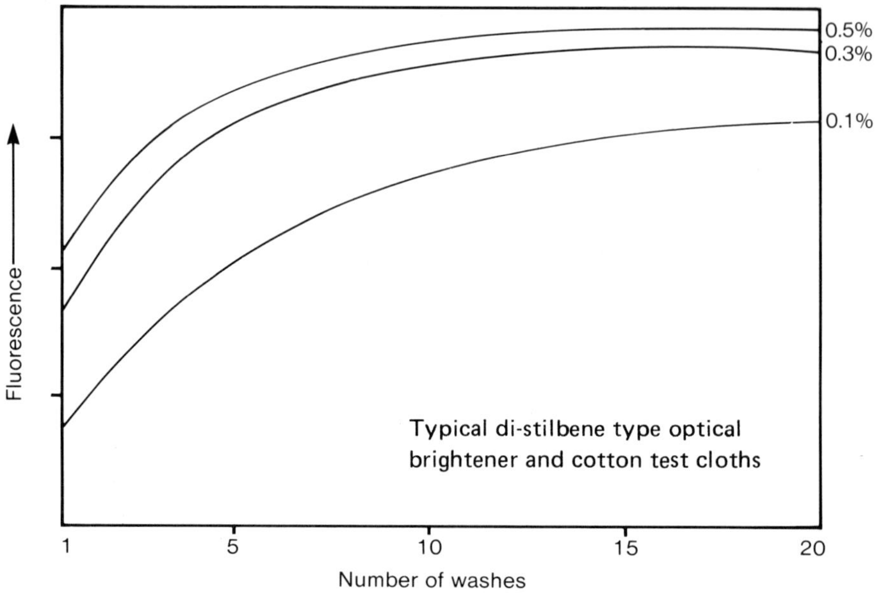

Fig. 2.14 – Effect of concentration in product on fluorescence.

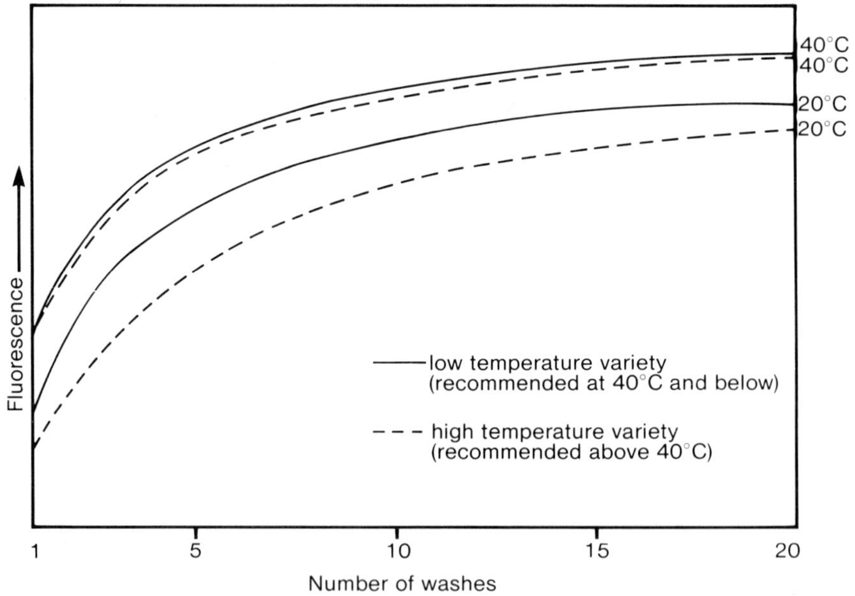

Fig. 2.15 – Effect of number of washes on fluorescence.

number of washes using various concentrations of optical brightener in the product, other factors being kept constant. Fig. 2.15 shows the merits of a low-temperature version of an optical brightener when used at temperatures below 40°C. When, as in an automatic washing machine, the wash liquor is used only once, it is important that as much as possible of the optical brightener is taken up from the wash liquor, as the rest goes to drain and is wasted. If, however, the wash liquor is used for several loads of articles (as is sometimes the case with non-automatic machines and with hand washing), a somewhat slower take-up may be useful, so that some optical brightener remains to be effective with the later loads.

- different optical brighteners can produce slightly different shades of near-white; bluish tints are often preferred. Some types have a tendency to produce yellowish, or greenish, tints when used at high concentrations.
- in many countries, the use of white articles is declining and that of coloured ones is increasing. Optical brighteners, particularly when included at high concentration in the product, can affect the colour of pastel shaded articles, but, in general, the effect is not serious. Detergent products without optical brighteners are sometimes sold for use with coloured articles.
- adequate stability to light is important, particularly in countries where articles are likely to be dried, or worn, in strong sunlight.
- optical brighteners used must be compatible with other ingredients in the formulation. Available chlorine should normally be avoided in blowing slurries to which optical brighteners are to be added. Little difficulty is experienced with sodium perborate, even if activators are also included. Most optical brighteners are anionic, so that they are compatible with anionic and nonionic surfactants, but many varieties are not suitable for use in cationic fabric softeners. See Stensby (1965) for types which can be used in fabric softeners.
- optical brighteners can be added as the final solid, or as a pre-slurry in an electrolyte solution such as sodium silicate, in the preparation of a slurry to be spray dried. In the production of dry mixed powders, it is often desirable to grind the optical brightener with, say, SCMC and a powder such as sodium sulphate in a premixing operation in order to ensure good dispersion.
- concentrations of 0.05 – 0.6% of optical brightener are usual in heavy duty powders, with around 0.3% now common, but the concentration required varies with type of OB and other circumstances. Nonionic surfactants in the formulation tend to reduce the take-up of optical brightener by fabrics.

As mentioned earlier, the main optical brighteners just discussed are substantive only to cotton and other cellulosic fibres. The use of pure cotton is declining, but the cotton in the increasingly common blends, such as polyester/cotton, is whitened in a way which improves the appearance of the blended fabric. Expensive optical brighteners intended for synthetic fibres have been used, but they are not very effective and are now of little importance. One company has withdrawn the nylon fluorescer from its range because less and less white nylon fabric is being washed.

Some non-detergent products intended to be used to augment fluorescent whitening in a separate operation are available; these may contain optical brighteners of special types, such as amino-coumarins.

2.3.8 Enzymes

Enzymes are complex nitrogenous substances of proteinaceous type which catalyse various reactions of a biochemical nature. Some of these reactions take place in living organisms, but others occur in non-living systems, as when starches are being broken down to sugars in the production of alcoholic beverages. There has been considerable interest from around the 1960s in the inclusion of certain enzymes in detergent products, although they had been tried much earlier in countries of continental Europe.

There are three types of enzyme which might be of use in detergents:

- proteases which break down proteins to amino acids, or intermediate compounds.
- amylases which break down starches.
- lipases which attack fats.

Practical interest has centred on the proteases, although most commercial concentrates also include some amylase. Enzymes require time to act and only operate at below about 50°C: at higher temperatures they are destroyed. They are particularly suitable when the wash process includes a cool soak, and the modern interest started in specialist products intended for a preliminary soak prior to the main wash. Early types of protease enzymes were sensitive to alkalinity above about pH9, and, during storage of the powders, to moisture and oxidizing agents such as sodium perborate. More robust types have now been developed which are more suitable for incorporation in heavy duty detergent products. Useful information is given in Sterace (1981). Three generations of enzymes are mentioned, powders, prills (agglomerates with a nonionic surfactant of melting point about 38°C), and encapsulates (in which the enzymes are granulated with an organic binder and coated with inorganic salts). The high sodium tripolyphosphate powders used in the USA resulted in a wash liquor pH of about 9.2–9.5, so that the earlier proteases could be included. Types of protease are now available which can be used in phosphate free, high sodium carbonate, powders which give a wash liquor of pH 10 to 11. Adverse effects of moisture pick-up during storage can be mitigated by protective packaging.

Protease enzymes in detergent products help the removal of stains which include proteins, such as blood, gravy, milk, and faeces, and it has been found that they also tend to improve slightly the overall colour of washed articles.

Some problems have been encountered in detergent factories due to reactions by employees to traces of enzymatic dust. If enzymes are to be used, particular care is needed with regard to dust collection and product handling, but some of the modern encapsulated types are probably less likely to cause trouble than the earlier powders. A relatively small number of people are allergic to enzymes and have encountered problems from the use of enzymatic powders in the home.

2.3.9 Anti-caking agents

The flow and storage properties of spray-dried powders are controlled primarily by the basic formulation and manufacturing procedure, but powders, such as some silica gels and calcium silicates, which have a large surface area and can adsorb substantial quantities of water, are sometimes added to spray dried, and other, powders to reduce caking on storage under humid conditions. Their benefits are sometimes limited.

2.4 TYPES OF DETERGENT PRODUCTS AND THEIR FORMULATION

2.4.1 General

Detergent products must be formulated to give a good cleaning performance under the conditions in which they are to be used, but, to be commercially successful, they must also meet other criteria; they must:

- be capable of satisfactory manufacture with the equipment available
- be safe during manufacture and in use. This includes freedom from damage to the articles to be cleaned and from harmful, or unpleasant, effects on the user (irritancy, allergy, unpleasant dustiness, etc.)
- have secondary characteristics, such as colour, perfume, and flow properties, which appeal to the user
- store and transport without undue deterioration
- be capable of being sold at a profit in what is often a highly competitive market. If a product is to be sold at a higher price than other brands in the same field, its superiority needs to be established clearly and demonstrated to the public.

Some of these points will be discussed more specifically in relation to various types of product in subsequent sections.

Most detergent products sold at retail are made to the manufacturer's own formulae, rather than to public specifications. These formulae are not generally disclosed, but, in most cases, their main features can be determined by a competent analyst. The typical formulations quoted in subsequent sections are only indications of some important types of product and many variations are usually possible.

2.4.2 Foam, lather, or suds

Depending upon circumstances, lather may be desirable, neutral, or positively objectionable. Some examples are:

- in the case of soap products, a good lather is an indication that soluble soap has dissolved and remains present in the wash liquor.
- possibly because of their experience with soap products, housewives originally liked NSD heavy duty powders (as well as soap powders) to show a copious, creamy, lather in the hand washing bowl, or in the washing machine. Special efforts were made to provide this lather.
- Most horizontal drum type automatic washing machines do not work satisfactorily with high lathering products. With them, overflow may occur

and too much lather reduces the effectiveness of the mechanical action. This tempts the user to dose too little powder, so that detergency suffers, particularly in hard water, because the wash liquor contains too little builder. Special low, or controlled, sudsing products have been developed to meet this need, particularly in Europe (now including the UK), where low sudsers have become dominant in the market. Many manufacturers have aimed to produce products which give nearly as much lather as the machines will take, partly because housewives tend to prefer controlled to very low lathers.

- with toilet soaps a lather is commonly worked up, particularly on the hands when difficult soil has to be removed. This lather provides a concentrated soap solution (6 – 8%) which is a very effective cleaning agent, and the lather helps to carry away the soil when it is rinsed off. The character of the lather, particularly its creaminess, can affect consumer preference.
- in hand dishwashing, for which a liquid product is normally used, good lather is also important. Fatty soil from the articles being washed is dispersed in the lather as well as in the solution, rather than left as an oily layer, and the lather reduces any tendency for redeposition as the plates and other articles are withdrawn from the wash liquor. A good layer of lather also minimizes the unpleasant appearance of the dirty liquor.
- on the other hand, dishwashing machines require products which produce virtually no lather and preferably which suppress any tendency for proteinaceous soils to foam. Foaming can upset the effectiveness of the sprays, cause spotting on the articles being washed, and, possibly, lead to an overflow from the machine.
- floor cleaning products should usually not produce much lather. They are often sold on a 'no-rinse' advertising story, although this should, perhaps, more correctly be called 'easy rinse', as it is necessary to transfer the soil which has been separated from the floor to the mop, or bucket.

2.4.3 Heavy duty fabric washing powders

2.4.3.1

Various market factors which influence product requirements need to be considered before a formulation is selected. In the less developed tropical countries, the majority of the population continue to rely heavily on pieces of household soap for fabric washing. When they start to buy NSD powders they tend to use them at low concentration in unheated water to soak the articles, and then rub dirty parts of the articles with a piece of soap to complete the job (apart from rinsing). More wealthy people in these countries, and most of the inhabitants of many more developed countries, use washing powders (soap or NSD) at higher concentrations alone, either for washing by hand or in machines. Important differences between different machines and washing processes are:

- as mentioned in the previous section, horizontal drum automatic machines are increasingly common in Europe. These machines need a low, or controlled, sudsing product and, as the wash liquor is used only once, rapid exhaustion of the optical brightener onto the fabrics is desirable.

- in the USA vertical spindle machines, commonly automatic, are normal. These, like the vertical spindle machines (single tub and twin-tub, manual as well as automatic) used in many other countries, can usually take medium or high lathering products. American machines are commonly equipped with dispensers for sodium hypochlorite bleach, which is unusual in Europe.
- hand washing cannot be carried out in water which is hotter than about 40 – 45°C, and in tropical countries it is often done in unheated, but slightly warm, water. American washing machines do not usually include heaters and normally use water from the domestic system at 45 – 60°C. In Europe, machines have in-built heaters and offer a range of washing temperature between about 40° and 90 – 95°C. At one time, hot programmes were commonly used for robust and dirty articles, which often also required to be sanitized. In the UK, and other countries, there is now a move towards lower temperature washing at around 40 – 60°C, and in Japan the low temperatures of 25 – 30°C are common. A recent survey stated that in 1983 over 40% of washes in the UK were carried out at 40°C or below.
- ranges of product concentrations and of the ratio of wash liquor to mass of fabric are also encountered.

As indicated earlier, the types of fabrics to be washed are important. Detergent manufacturers need to assess this and other factors for their own markets, but the following trends are of interest.

Per capita consumptions of different textiles worldwide

	1930	1950	1970	1980
Cotton %	83	71	53	30
Wool %	14	11	7	5
Synthetics, including blends %	3	18	40	65
Total kg/capita	1.6	3.7	6.0	–

Proportion of white textiles in households, 1982

	Japan	Europe	USA/Canada	Africa	Asia	Central/Latin America
White %	50	28	25	40	35	30
Coloured %	50	72	75	60	65	70

2.4.3.2

A number of important points related to formulation stem from the previous section and from the earlier technical discussions.

Sodium perborate is not included in USA products because housewives normally do any bleaching that is required with sodium hypochlorite; also the washing temperatures are not high enough for perborate without an activator to be really effective (although, contrary to general opinion, some Japanese authors claim that sodium perborate can show beneficial effects at temperatures as low as 35 – 50°C). Similarly, sodium perborate is not usual in those tropical countries where much of the washing is done in unheated water. About 10% of sodium perborate ($NaBO_3 \cdot 4H_2O$) has been common in the UK for many years, and higher levels of up to 20 – 25% have been used in continental Europe, where high temperatures up to near boiling have been common for the washing of heavily soiled articles. Bleach activators were not commercially successful when first tried, but they are becoming of increasing importance in products intended for use at medium temperatures. Heavy metals, particularly copper, catalyse the decomposition of perborate, and other per-salts, and reduce their efficiency. A little magnesium silicate, which adsorbs these metals, or a sequestrant such as EDTA or NTA, is usually included in powders which contain perborate.

The surfactant in high lathering powders is usually sodium alkylbenzene sulphonate, although others, such as tallow alcohol sulphate, can be used. About 20% surfactant is frequently used in powders with perborate, but higher levels up to 30% may be included in non-perborate powders, particularly for tropical markets. A high proportion of anionic surfactant helps to give an adequate lather when the product is used at low concentration in the wash bowl, but it accentuates the chance that the wash system will not be fully built. In order to achieve the foam characteristics desired by housewives, 2 – 3% of an ethanolamide type lather booster is usual. As in other cases, the narrow cut LEA is usually more effective on a weight-for-weight basis than a broader cut material, or CEA. It is also more expensive, and an interesting 'best value for money' decision has to be taken.

Various surfactant systems can be selected for controlled and low sudsing powders. Soaps and other anionic surfactants mutually reduce lather, and soaps of high titre (high melting point) fatty acids, giving relatively insoluble sodium soaps, have been used to moderate the lather of alkylbenzene sulphonates. Suitable nonionic surfactants, such as tallow alcohol $\overline{10}$EO, are very important in this field, in ternary mixtures with LABS and soap, alone, or in combination with LABS. Foam suppressors, such as phosphate esters, have been considered. There is now a move away from high titre soaps towards higher concentrations of nonionics. This is associated with the desire to use higher concentrations of nonionics because of the tendency towards lower washing temperatures.

When the use of phosphates is not restricted, or prohibited, it is usual to include 30 – 50% of sodium tripolyphosphate and some sodium silicate. Other possible builders have already been discussed. Sodium carboxymethyl cellulose (SCMC) is included at 1 – 2% as an antiredeposition agent.

The makers of optical brighteners refer to the use of these substances at concentrations of about 0.05 – 0.6% in the product. The higher levels were used at one time, particularly to justify competitive advertising claims, but levels now seem to be decreasing. Makers will provide recommendations for their various products, but about 0.2 – 0.3% of a typical di-stilbene type optical brightener is probably a reasonable level at which to start trials. Newer types of optical brightener understood to be better in some respects are mentioned in section 2.3.7.

It is necessary to leave about 10% of water in many spray-dried powders. Sodium sulphate is commonly included as a filler to complete the formula; it also helps powder characteristics. Some ingredients, such as sodium perborate, perfume and enzymes, cannot be spray dried and are added to already blown base powders.

Important physical properties of powders, such as particle size distribution, bulk density, and solubility (strictly rate of solution) are discussed in Chapter 10, although they can be affected by composition as well as by method and conditions of processing.

A number of typical formulations from various sources are shown in Table 2.2.

Table 2.2 – Compositions of some heavy duty fabric washing powders

Formula	1	2	3	4	5	6	7	8	9
LABS (linear alkylbenzene sulphonate)	30	20	8	6	–	14	14	20	–
Tallow alcohol sulphate	–	–	12	10	–	–	–	–	–
Alpha-olefin sulphonate	–	–	–	–	–	–	4	–	–
Alkylpolyethoxysulphate	–	–	–	–	–	–	5	–	–
Soap	–	–	–	–	–	5	–	–	–
Nonionic (such as tallow alcohol 10EO)	–	–	–	–	10	1	1	–	10
LEA or CEA (lather booster)	3	3	3	–	–	–	–	3	–
Sodium tripolyphosphate	40	35	50	45	45	48	–	–	35
Sodium silicate (as solids)	6	6	2	6	6	2	20	10	6
Sodium carbonate	–	–	–	–	–	–	20	30	–
Sodium bicarbonate	–	–	–	–	–	–	–	10	–
Sodium sulphate	8[a]	13[a]	13[a]	10[a]	16[a]	17[a]	23[a]	22	34[a]
STS or SXS (toluene or xylene sulphonate)	?[a]	?[a]	?[a]	?[a]	?[a]	?[a]	?[a]	–	?[a]
SCMC (antiredeposition agent)	2	2	1	2	2	2	2	2	2
Sodium perborate, $NaBO_3 \cdot 4H_2O$	–	10	–	10	10	–	–	–	–
Optical brightener, perfume, etc.	1	1	1	1	1	1	1	1	1
Water	10	10	10	10	10	10	10	2	12
	100	100	100	100	100	100	100	100	100

[a]STS or SXS is added as required to adjust slurry consistency and to make the powder crisp. Sodium sulphate is included as necessary to bring total to 100.

Formula 1: a high active no perborate powder giving high lather, particularly for tropical markets

Formula 2: a typical, high lather, phosphate powder containing perborate, for the UK and other European markets. As mentioned earlier, some powders for continental Europe contain, or contained, up to, say, 25% perborate

Formula 3: a high lather, high phosphate powder without perborate of the USA type

Formula 4: a medium high lather formulation with phosphate and perborate

Formula 5: a low lather formulation with phosphate and perborate

Formula 6: a low lather formulation with soap, phosphate and no perobrate

Formula 7: a medium/high lather formulation without phosphate or perborate

Formula 8: a high lather non-phosphate and non-perborate formulation intended to be dry mixed; with sodium bicarbonate to reduce the pH to 10 – 10.5

Formula 9: a low lather formulation with phosphate and no perborate.

2.4.4 Light duty powders

This type of powder is relatively unimportant. Light duty powders often contain 20 – 25% LABS, with the rest mainly sodium sulphate, but about 10% of sodium tripolyphosphate may be included. Sodium lauryl sulphate has been favoured as the surfactant in specialist wool washing products.

2.4.5 Dishwashing and other light duty liquids

2.4.5.1

Dishwashing liquids are generally solutions of surfactants with other ingredients added only to give the desired colour and perfume, and, when needed, to adjust viscosity and storage properties. A good cleaning and emulsifying performance is necessary to ensure that the articles are washed clean and drain free from streaks or spots, but dishwashing liquids are normally judged using tests in which the disappearance of the lather as soil is added (dirty articles washed) is taken as the end point. The shape of the lather – soil curve, Fig. 2.16, is important. Curve A, in which a good lather is maintained nearly to the end is usually preferred by housewives to curve B, even though solution B may emulsify grease as well as, or even better than, solution A.

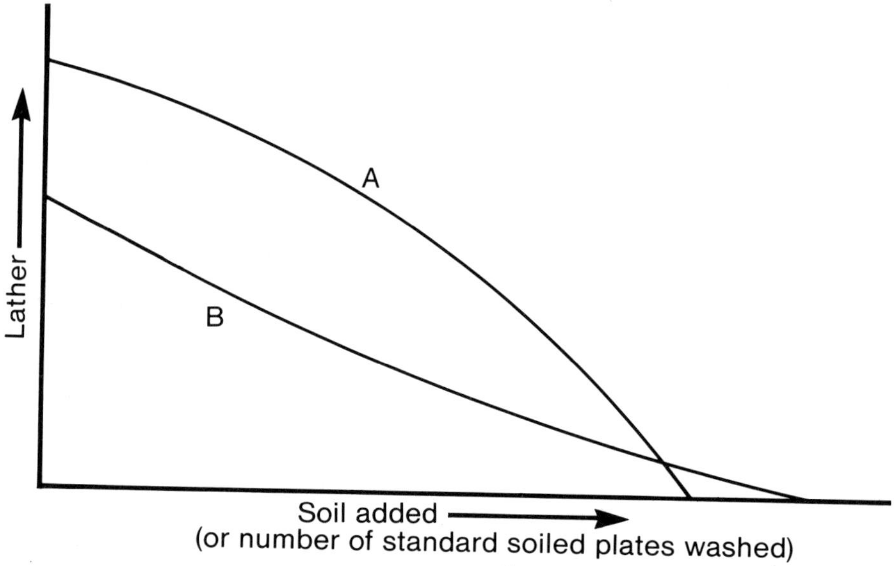

Fig. 2.16 – Performance of dishwashing liquids.

Various secondary characteristics are also important:

- some surfactant systems, particularly those containing a substantial proportion of lauryl ether sulphate, give a dishwashing solution with a soft feel which is generally preferred to the harsh feel given by some other surfactant systems.
- dishwashing liquids are naturally a pale straw, or amber, colour, but they can be given various attractive shades. Tests in the UK some years ago

showed that green and yellowish colours were generally acceptable and had good connotations of efficiency. Green was also associated in the housewives' minds with mildness. A pink opaque colour supported an image of mildness rather than efficiency, and tended to polarize opinions. Blue had clinical associations and was little liked. Housewives in other countries, and at other times, may have different opinions and feelings regarding colour.

- the perfume added to a dishwashing liquid needs to strike a balance between being attractive to the user and being liable, or thought liable, to scent the washed articles. A lemon type of odour is often considered fresh and appropriate, but certain brands have been very successful with a strong, characteristic, perfume. Many people have said that they would like an unperfumed dishwashing liquid, but consumer tests have shown that products with almost any reasonably selected perfume are preferred to one which is unperfumed, but has the smell of its detergent ingredients. The alcohol commonly included in the formulation has a significant, and generally liked, effect on the odour.

Important marketing decisions have to be taken in selecting the best concentration and viscosity at which to make a product. Some cheap brands have quite low concentrations, but others, selling at a higher unit price, contain 30–40% surfactants, and often also use a superior blend. Factors to be considered include:

- concentrated liquids save packaging and transport costs, and help to engender an image of quality. They also encourage housewives to use sufficient product to ensure a good result. On the other hand, too much surfactant in the dishwashing bowl can product a troublesome amount of foam and seem extravagant, and housewives have been known to dilute dishwashing liquids in the bottle after purchase. Some potential customers are attracted by a large bottle at a low unit cost, even though they are probably aware that the quality is relatively low.

- although, other factors equal, there is some broad relationship between viscosity and concentration, it is possible to vary the viscosity quite widely at a given concentration. Again, conflicting considerations affect the choice. A viscous liquid tends to give an impression of concentration, but may reduce the quantity dosed and hence the performance. The size of the hole in the nozzle through which the product is transferred to the bowl also tends to affect dosage.

2.4.5.2

Following Teepol 10, which was at one time widely used for dishwashing, blends of surfactants have become usual. Mixtures of alkylbenzene sulphonate (ABS) and alkylphenol ethylene oxide condensate (APEO) were common, particularly in cheaper brands. Blends of ABS and lauryl ether sulphate (LES), with or without a small proportion of lauryl monoethanolamide (LEA), have been found superior, and are now normal in the UK, although the proportions can vary. The latter type of formulation containing a substantial proportion of LES gives a dishwashing liquor with a good feel and a lather/soil curve of type A in Fig. 2.16.

As mentioned in section 1.11.2, the alkylphenol ethylene oxide condensates are no longer used in many developed countries because of lack of biodegradability, but they can, if required, be replaced by similar condensates based on high carbon number alcohols. Other ingredients, such as alkylphenol ethylene oxide sulphates (alkylphenol ether sulphates (APES)), amine oxide surfactants and lauryl diethanolamide have also been used. As mentioned earlier, compounds loosely described as 'lauryl' can be of different grades depending upon the narrowness of cut, and, generally speaking, the narrower cut versions are superior on an equal weight basis but more expensive. It is not easy to decide the qualities of ingredients, the ratios in which they are used and the total concentration of surfactants, which will give the best value for money and the greatest appeal in the market place. One top quality formulation has contained about 16% LABS, 16% LES and 3% LEA.

Large manufacturers normally blend their liquid products from individual ingredients, and the ABS at least is commonly made in the factory. ABS can be transferred from the sulphonation plant to the liquid making section as sulphonic acid, or as neutralized paste; transfer as paste has advantages if bleaching with hypochlorite is used. Other manufacturers who do not make surfactants, may find it better to buy concentrated blends from chemical manufacturers which only require addition of minor ingredients, such as colour and perfume, and of water if the product is to be diluted before sale.

Whichever method is adopted, one important point to be watched is that detergent liquids (unless extremely concentrated) need to be preserved against bacterial decay which can produce most objectionable odours, malodorous sulphur compounds being formed from the sodium sulphate. About 0.2% of formalin (a 37% aqueous solution of formaldehyde) is normally used as the preservative, but this is prohibited in some countries, so that manufacturers need to consider the legal position in markets they propose to enter. The suppliers of liquid ingredients often, but not always, include a little formalin in their liquid materials. In addition to the selection of a surfactant system which will give the required performance, it is necessary to ensure adequate temperature stability, particularly with regard to the low temperatures which may be encountered during storage, and to achieve the desired, or at least an acceptable, viscosity. Hydrotropes are commonly required, and blends selected from alcohol, SXS and urea are usual. All normally help to improve low temperatures stability; alcohol, and to a lesser extent SXS, tend to reduce viscosity, whereas urea tends to increase it. Other factors which affect viscosity, and cloud and clear points (the respective temperatures at which separation starts when the product is cooled and at which it becomes clear again when it is re-warmed) are:

- the nature of the surfactants. Linear ABSs are more soluble than branched, and relatively low molecular mass types are more soluble than those of higher molecular mass.
- the electrolyte content of the product, in ways which can be complex, as explained below. ABS made by sulphonation with oleum normally contains more sodium sulphate than that sulphonated with gaseous sulphur trioxide.

- the cation used to neutralize the sulphonic acid. Special bases, such as potassium hydroxide or an ethanolamine, give more soluble sulphonates than the usual sodium hydroxide.

The effects of dilution, and of electrolyte concentration, on the viscosity of a detergent liquid can vary widely with the surfactant system. Some points are illustrated by Figs. 2.17, 2.18, 2.19 and 2.20, courtesy of Shell International

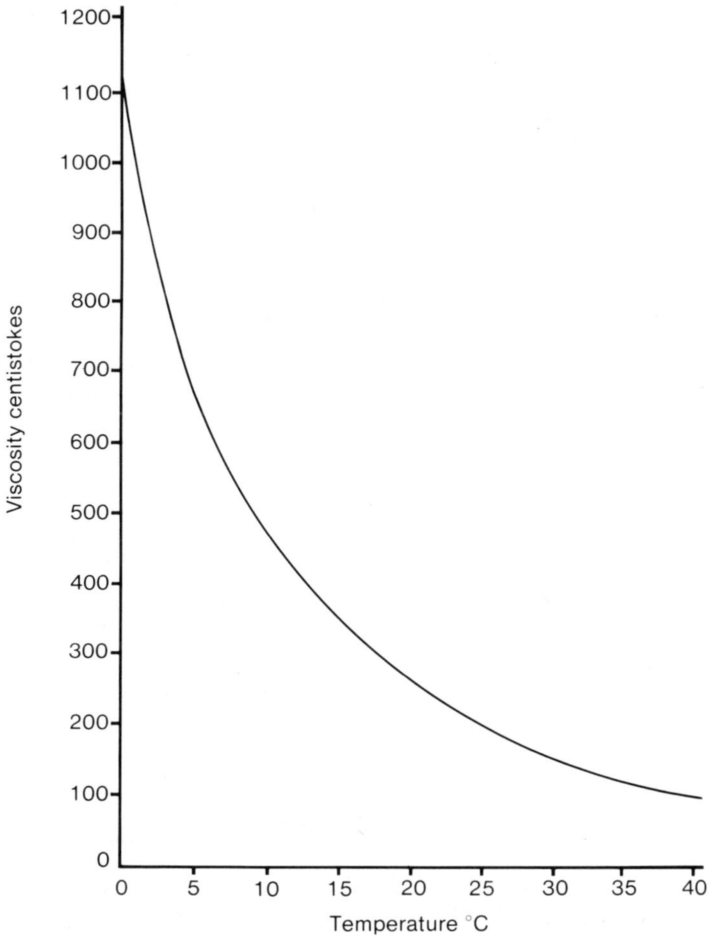

Fig. 2.17 – Viscosity of Teepol GD53 at various temperatures.

Chemical Co. Ltd., related to Teepol GD53 (based on 18% LABS, 6% LES and 6% alkyl ethoxylate), Teepol CH53 (based on LABS plus APEO), and a binary blend of 80 parts LABS plus 20 parts LES (these ingredients having the same molecular masses as those in Teepol GD53. The clear points of these blends are 0 – 1°C. Fig. 2.17 shows the kinematic viscosity (absolute viscosity/density) of

Teepol GD53 at various temperatures. Fig. 2.18 gives viscosity–dilution data for the three blends. It will be seen that whereas the viscosity of the 80/20 binary blend falls very rapidly on dilution from 30% to 18% active matter (surfactant),

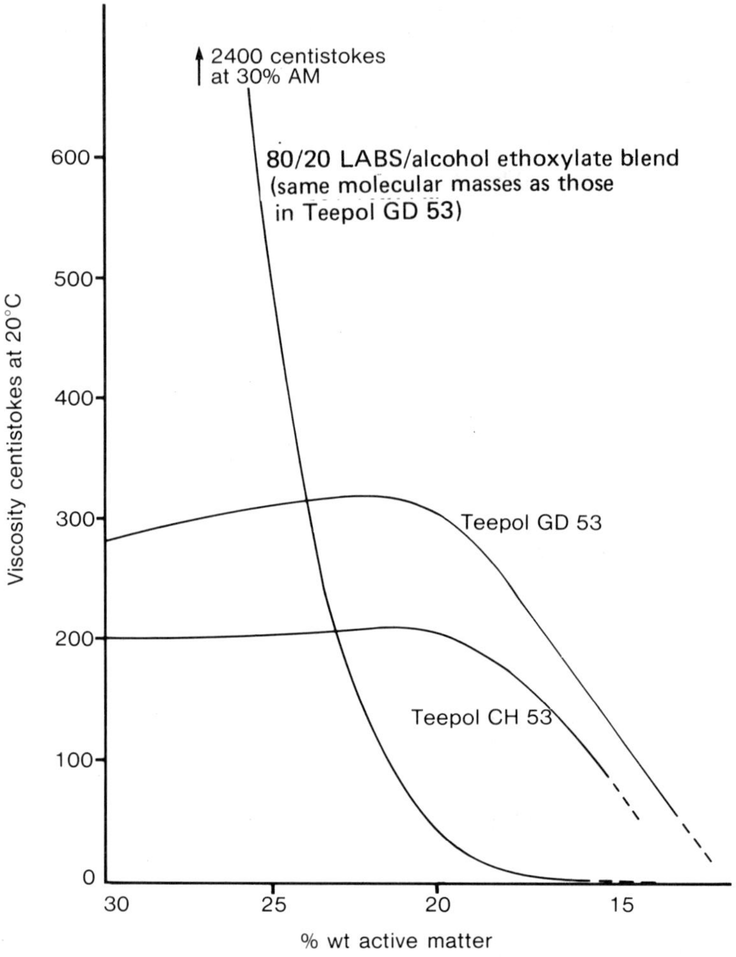

Fig. 2.18 – Viscosity/dilution relationships.

that of Teepol GD53 and Teepol CH53 changes little between 30% and 20% active matter. Figs. 2.19 and 2.20 show respectively the effects on the viscosity of Teepol GD53 of small additions of sodium chloride and of magnesium sulphate (expressed in terms of the anhydrous salt). It will be seen that whereas small additions of electrolyte reduce the viscosity at active matter concentrations of 20% and above, they increase viscosity at lower active matter concentrations. Although sodium chloride is more effective than magnesium sulphate as a viscosity modifier, Shell recommend the use of magnesium sulphate because:

- it gives a better feel to the skin in use
- it is not corrosive to metals as is sodium chloride.

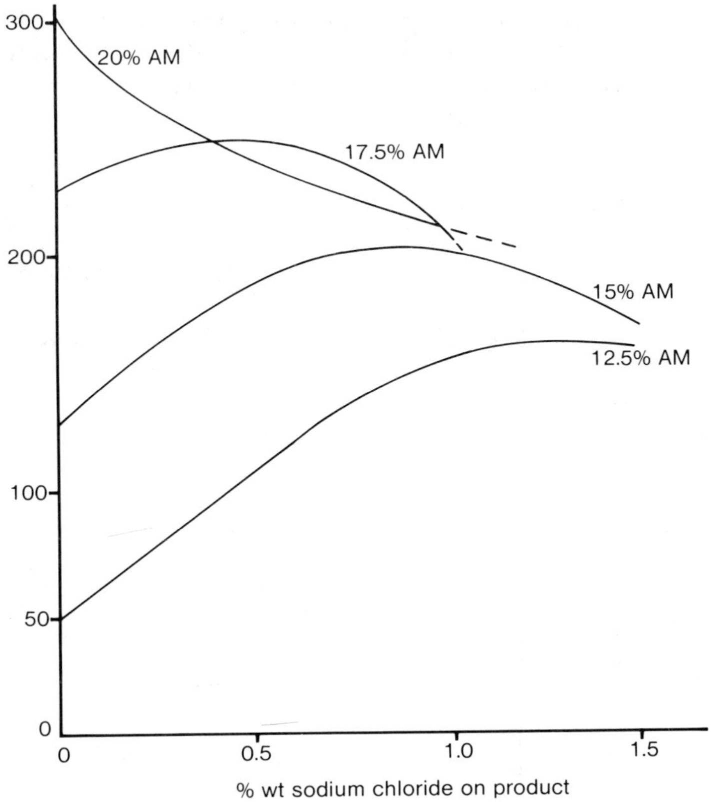

Fig. 2.19 – Effect of sodium chloride additions on the viscosity of Teepol GD53.

It will be evident from Figs. 2.18, 2.19 and 2.20 that experimental data are needed to decide the effects of concentration and electrolyte upon the viscosity of a surfactant system as used in liquids.

Dishwashing liquids and other mobile, unbuilt, liquids are sometimes used for light duty fabric washing and general cleaning. Brands sold for the washing of wool and other delicate fabrics may be formulated specially to minimize risk of damage, and selection of surfactant can be important in this respect.

2.4.6 Floor and other hard surface cleaning products

Types of product in this category include:

- powders based on sodium sesquicarbonate needles, usually with small additions of a nonionic surfactant and an available chlorine release agent, such as trichlorcyanuric acid
- mobile liquids containing alkalis and surfactants, usually low foaming.
- thick liquids containing surfactants and sodium tripolyphosphate. British Patent 882 569 (November 1961) to T. G. Jones and D. W. Stephens (and Unilever) describes systems which will suspend insoluble powdered abrasives

to give abrasive creams. Similar systems, containing ABS, groundnut oil soap, lauric diethanolamide, and sodium tripolyphosphate, will suspend additional sodium tripolyphosphate to give non-abrasive floor and general hard surface cleaners.

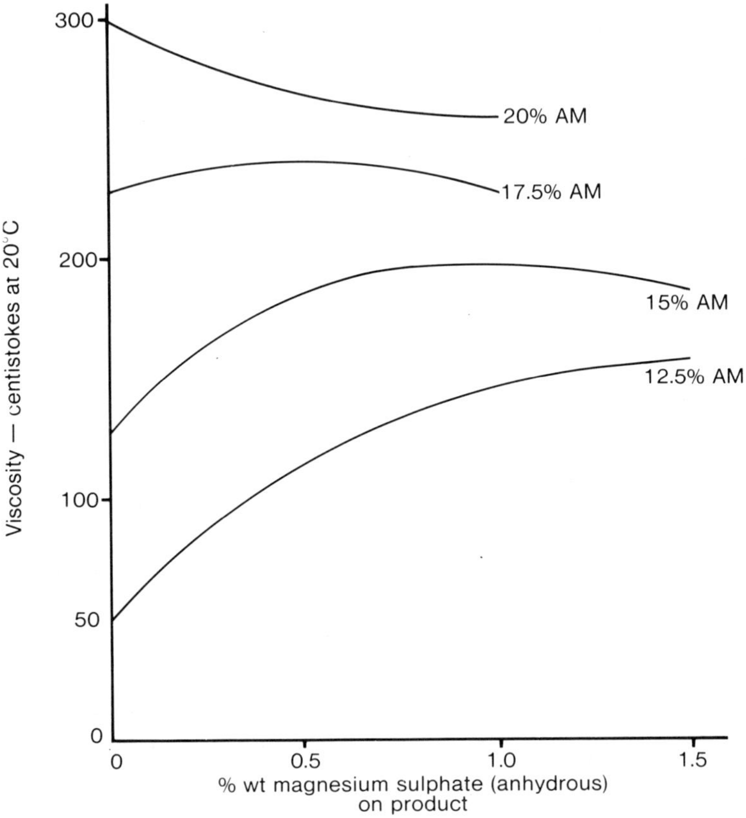

Fig. 2.20 – Effect of magnesium sulphate additions on the viscosity of Teepol GD53.

2.4.7 Heavy duty liquid fabric washing detergents

True heavy duty liquids need to include builders for the same reasons as for heavy duty powders, but some unbuilt liquids have attempted to enter the field without much success. Their performance is generally poor relative to that of built products.

One approach to the formulation of a true heavy duty liquid is to use potassium and triethanolamine salts because of their greater solubility compared with their sodium counterparts. The phosphate is usually potassium pyrophosphate because it is less readily hydrolysed than the tripolyphosphate, and hydrotropes may be needed. Selection and proper dispersion of the cellulosic anti-redeposition agent may be difficult. Three published formulae are:

	1	2	3
Triethanolamine linear alkyl benzene sulphonate	–	10	15
Fatty alcohol ethylene oxide condensate	13	–	28
Soap	–	–	2
Potassium pyrophosphate	25	20	–
Triethanolamine	–	–	14
Potassium silicate	3	2	–
Potassium xylene sulphonate	–	10	–
Alcohol	–	–	7
Optical brightener, cellulosic, perfume, etc.	2	5	1
Water	57	53	33
	100	100	100

The triethanolamine in non-phosphate formula 3 is intended to provide the necessary buffered alkalinity.

An alternative approach is to use about 25% of sodium tripolyphosphate, some of which is suspended by a technique similar to that just described for thick liquids in section 2.4.6.

2.4.8 Miscellaneous products

The formulations of various other types of product are discussed later in the sections dealing with their manufacture.

2.5 MARKET SURVEY. THE RISE AND FALL OF PRODUCT TYPES, AND SOME RECENT TRENDS

2.5.1

For many years, household soaps dominated the detergents market and this is still the case in many of the less developed countries. In the more developed countries other products, such as soap powders, soap flakes, toilet soaps, scouring powders and various more specialized cleaning agents, increased rapidly from about 1900. More recently, from the 1940s and 1950s, products based on non-soapy surfactants have emerged and have largely displaced soap products in the domestic market, except in the toilet soap field. Some completely new types of product, such as dishwashing liquids and fabric conditioners, have emerged, as well as replacements for the more traditional soap products.

Market evolution in developing countries seems to follow a similar pattern to that which occurred earlier in the developed countries, except that modern types of product are now available and intermediate types, such as soap powders, are often by-passed.

Broadly speaking what occurred in the developed countries and is now occurring as other countries develop, is:

- as standards of living rise, the per capita consumption of household soap rises. This soap is used for most purposes.
- the better-off members of a community tend to buy toilet soaps for personal washing.
- they also move to washing powders and other specialized products, particularly as they acquire washing machines. As mentioned previously, many people in developing countries start by buying relatively small quantities of NSD powders which they use in conjunction with household soap.
- as a large proportion of the population moves to toilet soaps, powders and other products, the sales of household soap fall. Very little household soap has been used in the USA for some time. Consumption is now very low in the UK and Japan, and is falling in other developed countries.
- new types of product come into use, such as dishwashing liquids (now standard items in many countries), fabric softeners, floor and window cleaners. In some countries, particularly the USA, bath additives have to some extent replaced toilet soaps.

Types of product have a life cycle in which they increase, stabilize and then decline. Often they continue in a small way long after a major decline has occurred, although the increasing dominance of supermarkets, which are reluctant to stock slow-selling lines, now tends to curb this.

Very substantial quantities of household soap, and other detergents, have been exported from Eurpoe and North America to less developed countries. A common pattern is for local manufacture to be established at a certain stage, first for household soaps and then for more sophisticated products as the market develops. Government actions, in the form of tariffs, quotas and currency restrictions, sometimes accelerate, and possibly distort, the natural development of local manufacture.

2.5.2

Most governments issue statistics for imports, productions and/or consumptions of soaps, other detergents and raw materials. These are very valuable in assessing market size and trends, but they must sometimes be interpreted with care, if only because clerks can be confused by brand and chemical names and do not always classify products correctly. Other bodies, such as the AIS (Association Internationale de la Savonnerie et de la Detergence), publish very useful statistics, but the most comprehensive data on a world-wide basis are probably those provided at regular intervals by the detergent manufacturing company, Henkel KGaA, of Düsseldorf, West Germany. The present author is indebted to Henkel for permission to quote some of their figures, converted in Tables 2.5 and 2.6 to per capita consumptions. Henkel point out that, like figures from other sources, their data may contain some inconsistences and are incomplete for some countries. Additional information concerning the UK market, including consumption figures from another source, is considered in section 2.5.3. Table 2.3 shows world productions and consumptions of cleaning products and Table 2.4 corresponding per capita consumptions for various years between 1960 and 1978.

Table 2.3 – World production and consumption of soaps, detergents and cleaning compounds (1000 tonnes)

Continent	1970		1974		1978		1980	
	Production	Consumption	Production	Consumption	Production	Consumption	Production	Consumption
West Europe, including Yugoslavia	4 905	4 767	6 145	5 993	6 732	6 562	7 402	7 050
East Europe	2 190	2 253	2 277	2 283	2 999	3 019	3 187	3 268
North America	4 598	4 574	6 253	6 118	7 354	7 268	7 706	7 564
Central America	607	617	740	752 }	2 466	2 519	2 728	2 757
South America	868	887	1 268	1 240 }				
Oceania (Australia, New Zealand and Pacific Islands)	228	237	309	276	277	286	296	300
Africa	724	780	857	877	1 182	1 308	1 463	1 494
Asia	2 257	2 324	2 637	2 670	4 936	4 877	5 212	5 241
World	16 377	16 439	20 486	20 209	25 946	25 839	27 994	27 674

Table 2.4 – World per capita consumption, soaps, detergents and other cleaning materials kg/capita/annum

Continent	1960	1966	1968	1970	1974	1976	1978	1980
West Europe	9.7	11.5	12.7	13.8	16.9	16.9	17.8	18.9
East Europe	6.5	7.5	6.2	6.5	6.2	8.1	8.1	8.6
North America	12.8	18.2	19.2	20.2	26.3	26.8	30.1	30.1
Central America	4.1	4.5	6.6	6.9	7.7 }	6.2	7.5	7.8
South America	4.6	4.7	4.3	4.7	6.0 }			
Oceania	13.1	14.1	14.3	15.4	16.2	15.3	15.6	16.1
Africa	2.3	2.2	2.4	2.3	2.3	2.5	2.9	3.2
Asia	1.0	1.1	1.3	1.2	2.1	1.9[a]	2.0[c]	2.1[e]
World	3.8	4.4	4.8	4.6	7.0	5.8[b]	6.1[d]	6.3[f]

[a]excluding China 2.5
[b]excluding China 7.3
[c]excluding China 2.6
[d]excluding China 7.6
[e]excluding China 2.5
[f]excluding China 7.7

It will be noted that over the large areas of the continents, net imports and exports are not very large; this is not always true for individual countries and types of product. Increases in consumptions are due to a combination of increases in population and in per capita consumptions. Table 2.5 shows approximate per capita productions and consumptions of various types of product in some countries of Western Europe in 1972. Table 2.6 shows per capita production for the years 1972, 1979 and 1982 for countries throughout the world. It will be noted that the information is incomplete, and some figures are surprising, but the table does provide a useful broad survey of the detergents industry. It is of interest to consider some trends and comparisons in the light of the generalizations in section 2.5.1. Apart from errors due to difficulties in classification, per capita consumptions may differ from figures given in Table 2.6 for productions because of significant exports and/or imports of certain types of product. For example, there were substantial exports of toilet soaps from the UK; and the more sophisticated types of product were imported by some less well-developed countries, although often only in small quantities.

2.5.3

To conclude this chapter the main product areas are reviewed briefly, and recent trends in the UK market are illustrated by figures courtesy of AGB/TCA – Lever Bros.

2.5.3.1 Household soaps

The rise and fall of the UK household soap market is shown by the following approximate figures:

Year	Tonnes/annum	Consumption kg/annum/capita
1801	24 000	1.6
1861	over 100 000	3.6
1900	320 000	7.9
1979	13 000	0.2
1983	5 700	0.1

Table 2.5 – Approximate per capita productions and consumptions of some types of detergent products in various countries of Western Europe in 1972

Country	Household soap		Toilet Soap		Soap Flakes		Soap Powders		NSD Solid (mainly powder)		NSD Liquid	
	Prod.	Cons.	Prod.	Cons.	Prod.	Cons.	Prod.	Cons.	Prod.	Cons.	Prod.	Cons.
Austria	0.3	0.3	0.4	0.6	–	–	0.1	0.1	6.9	6.3	2.2	2.2
Belgium/Luxemburg	2.3	2.5	0.6	0.8	0.1	0.1	0.5	0.4	11.0	7.0	2.8	2.4
Denmark	0.9	0.4	1.2	0.7	0.7	0.2	–	0.4	19.2	8.6	2.1	1.5
Finland	0.9	0.9	0.2	0.5	–	–	6.0	6.0	0.1	0.1	1.4	1.4
France	1.2	1.1	0.8	0.7	0.1	0.1	–	–	11.4	9.0	1.6	1.7
West Germany	0.5	0.5	1.1	1.0	0.1	0.1	0.4	–	15.2	10.9	2.2	–
Greece	1.5	1.5	0.5	0.5	–	–	–	–	2.9	2.9	–	–
Italy	1.6	1.5	0.8	0.7	–	–	–	–	9.2	8.9	0.3	0.3
Netherlands	0.7	0.6	0.8	0.8	–	–	0.7	0.7	12.8	8.6	3.6	3.0
Norway	1.2	1.2	1.0	0.8	–	–	–	–	7.9	7.9	2.3	3.2
Portugal	6.4	6.4	0.3	0.3	0.1	0.1	–	–	3.4	3.3	1.0	1.0
Spain	2.4	2.4	0.6	0.5	0.4	0.4	0.2	0.2	5.6	5.5	0.3	0.3
Sweden	0.5	0.5	0.8	1.0	0.1	0.1	0.6	0.8	6.0	6.2	2.1	3.2
Switzerland	0.3	0.3	1.0	1.1	–	–	–	–	13.9	10.2	2.3	2.3
United Kingdom	0.5	0.4	1.6	1.3	0.2	0.2	2.0	2.0	5.9	4.7	2.2	2.2

Table 2.6 – Approximate per capita productions of various product types in different countries kg/capita/annum

Country	Household soaps			Toilet soaps			Soap powders and flake			Syndets (NSDs) Solid			Syndets (NSDs) Liquid		
	1972	1979	1982	1972	1979	1982	1972	1979	1982	1972	1979	1982	1972	1979	1982
North America															
Canada	0.2	–	0.1	1.0	–	1.1	0.5	–	–	5.2	–	9.3	2.5	–	0.8
United States	0.1	0.2	0.2	1.7	1.8	1.8	0.7	0.5	0.5	12.9	11.8	12.2	8.6	9.3	10.3
Western Europe															
Austria	0.3	0.2	0.1	0.4	0.2	0.2	0.1	–	–	6.9	8.1	8.5	2.2	2.1	2.3
Belgium/Luxemberg	2.3	0.5	0.5	0.6	0.4	0.4	0.7	–	–	11.0	12.9	16.2	2.8	3.4	6.3
Denmark	0.9	0.8	0.8	1.2	0.7	0.9	0.7	–	–	19.2	14.3	16.3	2.1	11.1	14.2
Finland	0.9	0.8	0.6	0.2	0.7	1.8	6.0	–	–	0.1	5.4	6.9	1.4	1.4	2.3
France	1.2	0.8	0.7	0.8	0.8	0.8	0.1	–	–	11.4	9.9	11.0	1.6	3.7	4.1
West Germany	0.5	0.4	0.4	1.1	1.2	1.2	0.5	–	–	15.2	13.1	13.0	2.2	4.6	5.3
Greece	1.5	0.5	0.8	0.5	0.7	1.0	–	–	–	2.9	8.9	17.0	–	–	–
Italy	1.6	1.3	1.2	0.8	0.8	0.8	–	–	–	9.2	8.8	8.2	0.3	4.0	5.5
Netherlands	0.7	0.3	0.3	0.8	0.7	0.9	0.7	–	–	12.8	10.0	9.9	3.6	1.8	3.6
Norway	1.2	1.2	1.0	1.0	1.4	1.1	–	–	–	7.9	3.7	3.5	2.3	1.8	3.4
Portugal	6.4	5.8	4.5	0.3	0.5	0.5	0.1	–	–	3.4	5.1	5.8	1.0	2.8	4.6
Spain	2.4	1.2	0.4	0.6	0.5	0.3	0.6	–	–	5.6	8.3	11.2	0.3	3.3	8.8
Sweden	0.5	0.3	0.4	0.8	0.7	0.7	0.7	–	–	6.0	5.5	9.1	2.1	2.7	6.0
Switzerland	0.3	0.2	0.2	1.0	0.8	0.8	–	–	–	13.9	13.0	13.3	2.3	2.7	3.2
United Kingdom	0.5	0.2	0.1	1.6	2.3	2.3	2.2	0.9	–	5.9	7.0	9.1	2.2	7.1	7.2
Oceania															
Australia	1.2	0.7	0.7	2.0	2.1	2.1	1.3	0.5	0.4	5.2	5.4	6.3	5.6	5.4	8.3
New Zealand	1.0	0.7	0.7	1.7	1.8	2.1	3.5	2.3	–	–	3.8	4.9	–	3.5	8.5

Table 2.6 – Approximate per capita productions of various product types in different countries kg/capita/annum

Country	Household soaps			Toilet soaps			Soap powders and flake			Syndets (NSDs) Solid			Syndets (NSDs) Liquid		
	1972	1979	1982	1972	1979	1982	1972	1979	1982	1972	1979	1982	1972	1979	1982
Latin America and the Caribbean															
Argentina	5.0	3.2	2.9	1.4	1.2	1.1	1.6	1.0	0.9	–	1.9	3.0	–	0.4	0.2
Brazil	2.6	3.8	4.5	0.4	0.7	1.1	–	–	–	1.3	1.5	2.1	–	0.6	0.8
Chile	2.0	0.6	0.5	–	–	–	1.1	0.4	–	0.4	3.3	3.4	–	–	1.0
Columbia	3.9	2.9	3.6	–	0.6	0.4	–	0.7	–	0.6	2.3	1.7	–	0.3	0.1
Cuba	3.9	3.6	3.4	2.0	1.7	1.5	–	–	–	2.3	2.0	2.4	–	–	–
Dominican Republic	3.2	3.6	4.2	0.2	0.3	0.5	–	–	–	0.5	1.0	1.1	–	–	–
Ecuador	1.1	2.5	4.5	–	–	0.4	–	–	–	–	0.4	0.7	–	–	0.2
Jamacia	3.4	2.9	3.1	–	–	–	–	–	–	2.7	2.9	3.1	–	–	–
Mexico	2.6	2.1	2.3	0.7	0.9	1.3	0.2	–	–	5.6	6.8	9.3	0.1	0.2	0.3
Peru	1.6	3.5	3.0	–	0.8	0.4	–	–	–	1.0	3.0	3.2	–	–	–
Trinidad and Tobago	3.6	2.4	2.2	1.8	1.7	1.4	–	–	–	–	–	–	–	–	–
Uruguay	5.2	4.5	4.8	–	1.4	1.6	–	–	–	1.4	2.1	2.4	–	–	–
Venezuela	2.8	3.1	0.6	–	0.6	1.2	–	–	–	5.5	6.6	8.8	–	0.1	0.3
Asia															
China	0.3	0.7	0.9	–	–	0.1	–	–	–	–	0.4	0.6	–	–	–
India	0.5	1.3	1.2	–	–	0.1	–	–	–	–	0.2	0.2	–	–	–
Iran	1.0	1.0	0.3	–	–	0.3	–	–	–	3.0	3.0	3.6	–	–	–
Iraq	1.0	1.9	1.3	0.8	–	0.7	–	–	–	2.0	1.5	2.2	–	–	–
Israel	1.5	1.5	0.9	1.4	1.1	0.6	1.1	–	1.0	9.7	9.9	6.9	3.8	1.7	–
Japan	1.4	0.1	0.1	–	0.9	0.8	–	0.2	0.3	7.5	4.5	4.8	–	2.5	2.7
Pakistan	1.2	–	2.5	–	–	0.5	–	–	–	–	–	0.1	–	–	–
Philippines	6.3	1.3	0.9	–	0.4	0.8	–	–	–	3.3	1.4	3.1	–	–	–
Syrian Arab Republic	4.0	3.3	2.9	–	–	0.5	0.6	–	–	–	0.7	1.2	–	–	–
Thailand	–	0.2	–	–	–	0.3	–	–	–	1.1	1.4	1.3	–	–	–
Turkey	2.1	1.1	0.8	–	–	0.2	–	–	–	0.1	1.3	1.7	–	–	–

Table 2.6 – Approximate per capita productions of various product types in different countries kg/capita/annum

										Syndets (NSDs)					
	Household soaps			Toilet soaps			Soap powders and flake			Solid			Liquid		
Country	1972	1979	1982	1972	1979	1982	1972	1979	1982	1972	1979	1982	1972	1979	1982
Africa															
Algeria	2.0	4.1	3.8	–	–	0.1	–	–	–	–	1.8	2.2	–	–	0.5
Egypt	–	4.0	6.7	–	0.6	1.1	–	–	–	–	0.5	1.2	–	–	–
Ghana	3.4	2.1	0.1	–	–	–	–	–	–	–	–	–	–	–	–
Kenya	2.3	1.3	0.9	–	–	0.9	–	–	–	–	1.2	1.3	–	–	–
Morocco	1.7	1.6	1.3	–	–	0.2	–	–	–	–	0.7	2.5	–	–	–
Mozambique	2.6	2.4	3.4	0.1	–	0.3	–	–	–	–	0.4	–	–	–	0.8
Nigeria	1.0	0.9	0.8	–	0.3	0.3	–	–	–	–	0.8	1.3	–	–	–
SouthAfrica	1.1	0.3	0.3	0.8	1.4	1.5	2.0	–	–	–	4.7	5.9	0.6	1.3	2.0
Sudan	1.5	2.5	2.7	–	0.1	–	–	–	–	–	–	–	–	–	–
Tanzania	1.4	1.0	0.8	–	–	0.2	–	–	–	–	–	0.3	–	–	–
Tunisia	3.0	3.8	2.9	–	–	–	–	–	2.3	–	–	–	–	–	–
Zaire	1.4	1.7	1.4	–	–	0.1	–	–	–	–	–	0.1	–	–	–

NOTES:

1. NSD solids includes heavy duty detergents, 60°C detergents, light duty and special detergents, dishwashing agents and household cleaners. Liquids include dishwashing agents and household cleaners; in the USA also heavy duty and light duty detergents.
2. Henkel provide special notes against some figures for which the original documents should be consulted.
3. Recent figures for Europe, apart from the UK, include any soap powders with NSD solids.
4. Blanks indicate either negligible production, or that no figures are available. In some cases household soaps includes toilet soaps, and solid NSDs some liquids.

Small quantities of well-known brands in tablet form continue to be sold. As in the USA and countries fed by exports from the USA, soaps which are intermediate between toilet and traditional household soaps are being sold in the UK, primarily for personal washing.

2.5.3.2 Toilet soaps

For some years, the UK toilet soap market has been nearly static. Estimates by AGB/TCA – Lever Bros. show about 40 000 – 45 000 tonnes/annum, but from the way the data are collected for this estimate it is understood that the total market has been somewhat higher. In addition, substantial quantities of toilet soap have been produced in the UK for export. During the last 20 – 30 years considerable efforts have been made to establish brands of toilet tablets containing substantial proportions of non-soapy surfactants. These can be made with a slightly acidic pH, thought beneficial for some delicate skins, and with the properties of not forming scums on water, or deposits on the bath, and of producing an immediate foam. Market success has been confined almost entirely to the USA; housewives in other countries do not appear to have considered the advantages sufficient to justify the extra cost. The tablets also often feel a little different from ordinary soaps, and are hygroscopic, so that they may tend to become soggy in use.

As discussed in this Chapter and in Chapter 6, toilet soaps made with a high proportion of coconut, or palm kernel, oil and superfatted with 5 – 10% free fatty acids have enjoyed a considerable vogue because of the copious, creamy lather they give. These soaps belong to Type S in BS 1914:1984.

Bath additives, particularly foaming NSD products, are depressing the toilet soap market to some extent in USA and Europe.

2.5.3.3 Heavy duty fabric washing products

As already explained, soap powders and then high lathering NSD powders dominated the heavy duty fabric washing market for many years. The UK was unusual in that soap powders continued to have major sales after they had largely been replaced by NSD powders in most European countries. High lathering powders, both soap and NSD, now continue to be replaced by low lathering powders suitable for front-loading automatic washing machines, as shown in the following table for the UK.

Year	Tonnes high lathering powder ('000s)	Tonnes low lathering powder ('000s)	% households with automatic washing machines
1978	224	97	33
1979	213	134	35
1980	187	178	37
1981	160	209	39
1982	142	257	42
1983	128	294	48

Despite considerable publicity, it is not thought that cold water washing has been, or is likely to be, widely employed for fabrics in the UK, but there is now a strong tendency for cool (40 – 60°C) washes to be used instead of very hot ones. As mentioned earlier, over 40% of washes in the UK are estimated to have been carried out at 40°C or below in 1983. This means that nonionic surfactants, which are relatively good at low temperatures and with synthetic fibres (as well as low foaming), have become increasingly important. Soap is being replaced as the foam control agent, because it tends to impair the solubility of a powder. The trend to lower wash temperatures also favours the incorporation of enzymes and of sodium perborate activators.

An impregnated sheet which can be put into a washing machine to activate the perborate in any powder has recently been introduced.

In addition to factors concerned with detergency and the performance of sodium perborate bleach, an important point in relation to washing temperatures is sanitization of articles, such as tea-towels, handkerchiefs, underwear and bed linen, for which purpose high temperatures are used traditionally. In the USA, sodium hypochlorite, which has strong germ killing properties when used correctly, is much more commonly employed in the domestic wash than in the UK and other European countries. Somewhat, although not necessarily, associated with washing temperature is the sale of liquid products for the heavy duty wash. The development of properly built heavy duty liquids is not easy and the products tend to be expensive. Inclusion of bleaches is difficult, if not impracticable, but modern enzymes can probably be introduced into certain types of liquid. Liquid heavy duty products have proved successful in certain parts of the USA, probably because:

- success has occurred mainly in areas where powder quality has been degraded because of anti-phosphate legislation.
- a separate active chlorine bleach is commonly used.
- washing temperatures (40 – 60°C) in machines without heaters are lower than have been usual in Europe, so that a wide range of nonionic surfactants can be included without upper cloud point problems. Liquid products avoid certain difficulties in the incorporation of some nonionic surfactants which can arise with powders.
- the vertical spindle machines common in the USA are often dosed manually and do not need devices for the automatic dosing of liquid products.

In European countries, several companies are trying to establish heavy duty liquids, with results which will be awaited with interest. Trials some 25 years ago were abandoned.

NSD bars have been very successful in some countries, particularly the Philippines, as an alternative to bar soaps for low temperatures hand washing of fabrics. They usually contain ABS, a high level of sodium tripolyphosphate, high silicate, sodium sulphate and some plasticizer to minimize cracking and distortion. One published formula is:

	%
LABS (granules)	48
Stearic acid	2
Sodium tripolyphosphate	39
Sodium metasilicate	10
SCMC	1
	100

2.4.3.4 Light duty products

Only 15 – 20% of households in the UK buy specialist light duty products; the rest prefer to use a heavy duty powder. There is now some small growth in volume as shown by the following figures:

Year	'000 tonnes in the UK
1979	5.6
1980	6.0
1981	6.2
1982	6.1
1983	6.3

This is thought to be because some owners of front loading automatic machines do not like to use a low foaming product in the hand wash. The split between different types of light duty product is 40 – 50% powder, 25 – 30% soap flakes, and about 25% liquid.

2.4.3.5 Dishwashing liquids

Dishwashing liquids, which replaced washing soda and dry soap powders, are now used almost universally in the UK and most of the other more developed countries. The tonnage, which includes a very substantial number of retailers' own-label and other relatively small brands, continued to increase slowly during the period under review as shown in the following table, but the market may have reached a near plateau:

Year	'000 tonnes dishwashing liquid sold in the UK
1978	147
1979	149
1980	153
1981	158
1982	163
1983	162

The use of dishwashing machines in the UK continues to increase and to require specialized products, which must tend to reduce demand for hand dishwashing liquids.

2.5.3.6 Fabric conditioners

When first launched in test market in the UK a cationic fabric conditioner was not successful, but a few years later a similar product did very well, and the market has grown rapidly and continues to expand.

Year	'000 tonnes in the UK
1978	57
1979	71
1980	86
1981	107
1982	125
1983	136

A recent development is the sale of much more concentrated versions of several major brands; this will complicate tonnage comparisons in future years. Fabric conditioners have been particularly successful in some other countries, especially West Germany where the tonnage sold is very high.

In the UK, one company has now launched a low sudsing detergent powder which includes its own built-in fabric conditioner. An impregnated sheet which can be used for fabric conditioning in tumble driers is also now available in the UK.

2.5.3.7 Scouring products

Following a number of solid products containing abrasive, a scouring powder was launched in the UK in 1904, comprising ground silica (quartz), sodium carbonate and soap. With various fomulations, discussed in Chapter 11, scouring powders were used in large quantities for very many years, but sales have now declined and are continuing to fall fast. This decline is associated with:

- replacement of wooden draining boards and stoneware sinks by plastic working surfaces, stainless steel, plastic and enamel sinks, and also the increasing installation of plastic baths. The newer surfaces are more readily scratched and more easily cleaned than the older ones, and housewives have become more conscious of possible damage to baths and other surfaces.
- the rise of milder products, such as creams containing a mild abrasive and cleaning agents without any abrasive.

Recent figures for the UK market are:

Year	Total scourers '000 tones	Scouring powders '000 tonnes	Mild abrasive creams or liquids '000 tonnes
1978	32	23	9
1979	31	19	12
1980	30	16	14
1981	28	13	15
1982	29	10	19
1983	28	9	19

2.5.3.8 Bleaches and lavatory cleaners

Apart from its use for the bleaching of fabrics, sodium hypochlorite is used in the cleaning of lavatories and drains, primarily to kill germs. An important development some 20 years ago was the discovery that sodium hypochlorite solution can be thickened with a suitable surfactant system and that certain perfume ingredients can then be solubilized to modify the bleach odour. Because of the strongly alkaline and oxidizing nature of concentrated sodium hypochlorite solutions, the types of organic compound which are sufficiently stable when incorporated into them is limited. These thickened bleaches are particularly useful in WC bowls because they cling to the surface and do not drain away as quickly as unthickened bleaches.

Apart from sodium hypochlorite, lavatory cleaners have traditionally been acids to remove deposits of lime scale produced by the hardness in the flushing water. Janitors commonly use hydrochloric acid for this purpose, but the acid in domestic lavatory powders is normally sodium bisulphate. Some 20 years ago, improved powders were developed which included sodium bisulphate (acid), trichlorcyanuric acid (source of available chlorine), surfactant (wetting agent and detergent), and mineral abrasive powder (filler and scourer). A proportion of sodium bicarbonate (less than equivalent to the bisulphate) was also sometimes used to disperse the available chlorine more uniformly in the water in the seal (because of the effervescence produced by its reaction with some of the acid). This type of powder can be dispensed from the plastic container to wherever required, including under the rim of the WC bowl.

In 1982, a completely different type of lavatory cleaner was launched in the UK with great success and was followed by similar products from competitors. These products are essentially liquid detergents with a strong, pleasant, fragrance. The success of this type of product, which has doubled the specialist lavatory cleaning market and achieved some 75 – 80% of that market in the first year, is probably due to the consumers' need for a relatively mild, pleasant, cleaner and re-odorizer. Those who buy specialist lavatory cleaners do not normally allow their WC bowls to get into a state which requires a strong acid to remove lime scale, or other deposits. This development is an interesting example of how a radical re-assessment of true requirements can sometimes lead to an important new type of product.

Recent figures for the UK bleach and specialist lavatory cleaner market are ('000 tonnes):

Year	Bleaches	Powdered lavatory cleaners	New type liquid cleaners
1978	98	6	–
1979	101	6	–
1980	99	7	–
1981	100	7	–
1982	99	3.5	10.5
1983	107	3	12

2.5.3.9 General purpose cleaners

Fabric washing powders and dishwashing liquids can be used for floors, walls and other hard surface cleaning, but various types of specialist products, both powders and liquids, are available which are low foaming and more convenient in use. Sales in the UK were:

Year	'000 tonnes hard surface cleaners
1978	17
1979	17
1980	15
1981	14
1982	13
1983	13

Many countries, such as France, Germany and Italy, use much more general purpose hard surface cleaner, but less liquid scourer.

2.6 PRODUCTS FOR USE IN COMMERCIAL AND INSTITUTIONAL LAUNDRIES

The principles of fabric washing in laundries are similar to those in the home, but several factors cause different types of product to be required:

- the cost of the detergents is more important and is much more closely assessed
- laundries are able to make up stock solutions of soap and other materials, and to dispense them to the washing machines at the required stages in the process. When required for high temperature washes, relatively insoluble tallow soaps, which are now cheaper than other types as well as better detergents, can be used without difficulty
- laundries almost always use very soft, normally softened, water. This means that the main functions of the builders are to provide alkalinity and anti-redeposition properties, rather than to deal with water hardness. However, it may sometimes be necessary to remove calcium deposits which have built up in fabrics. Sodium metasilicate is widely used as the alkaline builder. Phosphates are used to a limited extent for special purposes, such as the removal of calcium deposits from articles previously washed in hard water, or if articles have to be rinsed in slightly hard water. Acids, such as weak hydrochloric or acetic, are sometimes used to remove calcium deposits. Oxalic and hydrofluoric acids are used, with due precautions, to remove metal stains, particularly iron.
- base exchange (rather than lime soda) softening is now usual and, as explained earlier, this can produce soft water containing sodium bicarbonate which affects certain wash processes. In particular, sodium sesquicarbonate, which was used with soap for some relatively delicate articles, may not be sufficiently alkaline if the softened water has a high sodium bicarbonate content.

The booklet, 'An Introduction to Laundry Chemistry', published by The Guild of Cleaners and Launderers, provides useful information.

3

The oils and fats; their nature, storage and pretreatment

3.1 THE FATTY ACIDS

As outlined in section 1.3, the soapmaking oils and fats are essentially glycerides, that is fatty acid esters of the trihydric alcohol glycerol. Fairly recent investigations, described in Bailey (1979), have shown that natural oils and fats may contain a wide variety of fatty acids, including some with odd numbers of carbon atoms, branched chains, ketonic groups, or ring structures. Apart from ricinoleic acid, an hydroxy acid which is the major constituent of castor oil, it can for the purposes of this book be assumed that the important fatty acids all have a straight, unsubstituted, hydrocarbon chain with an even number of carbon atoms, including the one in the carboxyl group. That is, the alkyl groups contain an odd number of carbon atoms. The fatty acids can be divided into several groups. The first series comprises the saturated acids $C_nH_{2n+1}-COOH$; and members with various chain lengths are listed in Table 3.1. The other series are unsaturated; that is, they contain less hydrogen than the saturated series and what are known as double bonds. The unsaturated acids have lower melting points than the saturated acids with the same numbers of carbon atoms. The main groups of unsaturated fatty acids are:

- the mono-unsaturated series, $C_nH_{2n-1}-COOH$, with one double bond
- the di-unsaturated series, $C_nH_{2n-3}-COOH$, with two double bonds
- the tri-unsaturated series, $C_nH_{2n-5}-COOH$, with three double bonds
- acids with four, or more, double bonds.

Table 3.1 – The saturated fatty acids $C_nH_{2n+1}-COOH$

Number of carbon atoms	Popular name	Melting point, °C
4 (i.e., $n = 3$)	butyric	–8
6	caproic	–3
8	caprylic	17
10	capric	32
12	lauric	44
14	myristic	54
16	palmitic	63
18	stearic	70
20	arachidic	75
22	behenic	80
24	lignoceric	84

Two types of isomers (that is, compounds with the same empirical formula, such as $C_xH_yO_z$, but with different structures and hence different properties) exist amongst the unsaturated fatty acids:

- positional isomers, in which the difference is in the position of the double bond, or bonds, in the carbon chain
- geometric isomers.

The two carbon atoms, which form the double bonds, each have attached to them a hydrogen atom and a carbon chain. The double bond prevents free rotation; and if the like groups are attached on the same side of the molecule the compound is called the *cis*-isomer; if they are on the opposite sides it is the *trans*-isomer. For example, with the very important $C_{17}H_{33}COOH$ acids, possible isomers include:

$$\begin{array}{r} CH_3-(CH_2)_7-CH \\ \| \\ HOOC-(CH_2)_7-CH \end{array} \qquad \begin{array}{l} CH_3-(CH_2)_7-CH \\ \quad\quad\quad\quad\quad\;\; \| \\ \quad\quad\quad\quad\quad\; HC-(CH_2)_7-COOH \end{array}$$

oleic acid (*cis* form)
mp 15°C

iso-oleic, or elaidic acid
(*trans* form), mp 44°C

Many other isomers are possible with the double bond in other positions in the chain.

Natural acids are usually, but not invariably, *cis*. The higher melting *trans* forms are produced when fats are hydrogenated, or otherwise heated, particularly in the presence of certain catalysts.

It will be appreciated that with variations in chain length, the numbers and positions of the double bonds, and *cis*/*trans*-isomerism at each double bond, there are a very large number of possible unsaturated fatty acids. The Geneva system of nomenclature, generally adopted by academic workers, provides a systematic way of identifying a fatty acid:

- the number of carbon atoms is indicated by a Greek prefix, for example dodeca = 12, tetradeca = 14, hexadeca = 16, and octadeca = 18
- the name of a saturated acid ends in -anoic, so that stearic acid is octadecanoic acid
- the number of double bonds is indicated by a suffix, so that the name of a fatty acid with one double bond ends in -enoic, one with two double bonds in -dienoic, one with three double bonds in -trienoic, and so on
- the position of a double bond is given by the number of the first carbon in the double bond, starting from the carbon in the carboxyl group (or, alternatively, by the numbers of the carbon atoms between which the double bond occurs).

Thus, oleic acid is,

$$\underset{}{CH_3(CH_2)_7}-\underset{10}{CH}=\underset{9}{CH}-\underset{8}{CH_2}-\underset{7}{CH_2}-\underset{6}{CH_2}-\underset{5}{CH_2}-\underset{4}{CH_2}-\underset{3}{CH_2}-\underset{2}{CH_2}-\underset{1}{COOH}$$

cis 9-octadecenoic acid, or *cis* -9, 10-octadecenoic acid. The position of a substituent group is numbered similarly.

In the Geneva system, the names of alcohols end in -ol, so that ethyl alcohol is ethanol and glycerol is 1,2,3-propanetriol.

A few of the more important unsaturated fatty acids are listed in Table 3.2.

The oils from fishes and marine mammals, such as whales and seals, contain fatty acids with a wide range of chain lengths, some of which include three to six double bonds.

In polyunsaturated fatty acids, a special arrangement of the double bonds, known as a conjugated system, –CH=CH–CH=CH–, is possible. This gives oils such as tung, which contains eleostearic acid, 9,11,13-octadecatrienoic acid (an isomer of linolenic acid), specially strong drying properties; that is they can form hard paint, or varnish, films. Such oils are unsuitable for soap.

3.2 THE FATTY ACID COMPOSITIONS OF OILS AND FATS

One fat (or oil) can differ from another in several ways, but the most fundamental is in fatty acid composition. Oils and fats are obtained from a wide range of animal and vegetable sources; and it is to be noted that even those from nominally the same kind of source can vary considerably. In the case of animal fats, this can be related to the part of the animal from which the fat came, the food which the beast ate, and the particular breed of sheep or cattle. Similarly, vegetable oils vary with factors such as plant variety and climate. A striking example is the oil known as rape, or colza, obtained from the seeds of certain varieties of brassica. Traditionally, this oil contained some 50% of erucic (including also eicosenoic) acid, but varieties of seed giving what is sometimes called canbra oil (from Canadian brassica) or canola oil, has now been

Table 3.2 – Some important unsaturated fatty acids

Formula	Scientific name	Popular name	Melting Point °C
One double bond			
$C_{15}H_{29}COOH$	*cis*-9-hexadecenoic	palmitoleic	–
$C_{17}H_{33}COOH$	*cis*-9-octadecenoic	oleic	15
$C_{17}H_{33}COOH$	*trans*-9-octadecenoic	elaidic or iso-oleic	44
$C_{19}H_{37}COOH$	*cis*-eicosenoic	–	–
$C_{21}H_{41}COOH$	*cis*-13-docosenoic	erucic	34
$CH_3–(CH_2)_5–CH(OH)–CH_2–CH{=}CH–(CH_2)_7COOH$	12-hydroxy-*cis*-9-octadenoic	ricinoleic	–
Two double bonds			
$C_{17}H_{31}COOH$	*cis, cis*-9,12-octadecadienoic	linoleic	–5
Three double bonds			
$C_{17}H_{29}COOH$	*cis, cis, cis*-9,12,15-octadecatrienoic	linolenic	–15

Table 3.3 – Approximate fatty acid compositions of some oils and fats

	Saturated acids						Unsaturated acids			
	C_{10} and below	C_{12}	C_{14}	C_{16}	C_{18}	C_{20} and above	oleic	C_{18} linoleic	linolenic	Other
Hard fats										
Beef tallow	–	1	5	27	16	1	42	3	–	5 palmitoleic
Mutton tallow	–	1	4	24	27	–	37	4	–	3 palmitoleic
Palm oil	–	–	1	44	5	–	40	10	–	–
Chinese Vegetable Tallow[b]	–	1	4	61	6	–	28	–	–	–
Lard	–	–	2	28	14	–	44	8	–	3 palmitoleic
Bone grease[a]	–	–	–	–	–	–	–	–	–	1 C_{20} unsaturated
Goat tallow	–	4	2	26	28	2	38	–	–	–
Lauric or nut oils										
Coconut oil	13	48	19	8	3	–	8	1	–	–
Palm kernel oil	7	52	15	7	2	–	16	1	–	–
Babassu oil	11	45	16	7	5	–	14	2	–	–
Soft oils										
Olive oil	–	–	–	10	2	–	82	6	–	–
Groundnut oil	–	–	–	10	3	3	50	34	–	–
Cotton seed oil	–	–	–	25	2	–	21	51	1	–
Maize (corn) oil	–	–	–	13	3	–	36	48	–	–

Soyabean oil	–	–	–	13	4	–	23	52	7	1
Sunflower oil	–	–	–	7	3	1	23	65	1	–
Sesame oil	–	–	–	9	4	–	45	42	–	–
Rapeseed I	–	–	–	3	1	1	15	15	10	55 erucic and similar
Rapeseed II	–	–	–	5	2	1	52	28	12	–
Soft/drying oils										
Linseed oil	–	–	–	6	4	–	22	16	52	
Tung oil	–	–	–	4	1	–	8	4	3	80 eleostearic
Whale (baleen) oil[c]	–	–	8	17	2	–	(73% unsaturated C_{14}–C_{22}, 1–6 double bonds)			
Herring oil	–	–	5	18	2	–	(75% various unsaturated)			
Menhaden oil	–	–	12	22	3	–	(63% various unsaturated)			
Miscellaneous										
Castor oil	–	–	–	1	2	1	4	5	1	86 ricinoleic

[a]See Section 3.5 for characteristics of various greases.

[b]The fruit of the stillingia tree (like the oil palm) yields two commercial fatty products (1) Chinese vegetable tallow from the fruit, and (2) stillingia oil, a drying oil, from the kernels.

[c]Sperm whale oil is quite different. It contains about 76% of wax esters (compounds of fatty acids and long chain alcohols), 23% of glycerides, and 1% minor ingredients.

developed which contains only 0 – 5% erucic acid. Rape oil is widely used for edible purposes, but doubts have arisen regarding the advisability of ingesting much erucic acid; and the new types of rape oil meet this requirement. High erucic oils are still valuable for some industrial purposes.

Much information has been published concerning the fatty acid compositions of oils and fats from many sources; see, for example, Bailey (1979) and Patterson (1983). Table 3.3 gives some approximate compositions of important oils and fats, grouped into categories often used by soapmakers. The *hard fats* contain some 45 – 55% of palmitic plus stearic acids and about 35 – 50% oleic plus palmitoleic acids. They give firm sodium soaps with excellent detergent properties, particularly at temperatures approaching the boil. Their sodium soaps are not very readily soluble, so that lather is obtained only slowly when a piece of the soap is rubbed in cool water.

The *lauric,* or *nut oils* are characterized by a very high proportion, about 70 – 80%, of $C_{10}-C_{14}$ saturated acids, hence the name lauric oils. They give hard, freely soluble, sodium soaps.

The *soft oils* (liquid vegetable oils) contain low proportions of saturated acids and high proportions of unsaturated acids, particularly oleic and linoleic. They give sodium soaps which are softer and more freely soluble than those from the hard fats. It will be noted that there are appreciable variations in the compositions of the oils included in this class.

The oils classed here as *soft/drying oils* are more unsaturated than those in the soft class. They include appreciable proportions of acids containing three, or more, double bonds; and are not generally suitable for use in solid sodium soaps, unless the unsaturation is reduced by hydrogenation or, sometimes, polymerization. Some kinds are used directly in soft soaps.

Rosin, or *colophony,* is not a glyceride, but it was at one time used extensively as a fat in the manufacture of soap. Various grades were available which differed in colour and odour. The better grades were gum rosins, the residue after turpentine has been distilled from the exudate from pine trees. The cheaper ones, wood rosins, were extracted from pine tree stumps. Crude *tall oil,* a by-product from the sulphate process for the production of papermaking pulp from pinewood, contains roughly equal parts of rosin and fatty acids. It is usually distilled to give tall oil rosin and tall oil fatty acids which may contain a few percent rosin. The fatty acids contain about 50% oleic, 7% conjugated linoleic, 45% unconjugated linoleic, and 2% saturated fatty acids, plus a few percent other compounds. Some tall oil fatty acids are used in soap, but, so far as is known, little rosin is now used, partly because it is no longer cheap.

3.3 OTHER DIFFERENCES BETWEEN FATS

Fats can differ from one another in ways other than differences in fatty acid composition:

- in glyceride structure, that is the way the various fatty acids are combined with the three hydroxyl groups in the molecules of glycerol. This is of great importance in some uses of fats, but it is irrelevant when the fats are to be saponified, or hydrolysed, to soaps or fatty acids.

- in the nature and proportions of the non-glyceride substances they contain. These ingredients can loosely be classified into three groups:

  - inherent constitutents, such as vitamins A, D and E (tocopherol); pigments such as carotene, gossypol and chlorophyll; and various sterols. Some of these which are significant to the soapmaker are pigments which need to be removed and natural antioxidants, including tocopherol.
  - free fatty acids which develop when fats are stored under poor conditions, or extracted from materials which have been allowed to deteriorate. Examples are palm oil and rice bran oil. Palm oil produced under good conditions from plantation fruit should have an FFA (free fatty acid) content of 2 – 5%. Oil recovered from the fruit of wild palms by villagers may containers 20 – 30% of FFA, or even more. The oil in rice bran deteriorates seriously in even a few hours if steps are not taken to inactivate enzymes present in the bran.
  - miscellaneous impurities which are found in all soapmaking fats, but particularly in lower grade materials, such as those recovered from wastes. Proteinaceous materials, and coloured and odorous constituents of many kinds, can be present as well as particulate dirt. Deterioration leading to development of a high FFA % commonly also results in the presence in the fat of brownish colouring matter and other undesirable impurities. A relatively modern problem results from the presence in tallows and greases of traces of polythene, which come from plastic film included in fatty materials collected from butchers and other sources.

Additional points in relation to the general quality of fats are discussed later.

3.4 SOME ANALYTICAL CHARACTERISTICS OF OILS AND FATS

3.4.1 General

Well equipped modern laboratories now usually characterize mixtures of fatty acids by GLC (gas liquid chromatography), or similar techniques, using methyl esters formed from the fat or soap. This gives composition in terms of the percentages of individual fatty acids, from which, if required, deductions can be made as to the blend of fats from which the fatty acids probably came. The older tests outlined in the following subsections are still used for various purposes, including specifications. Before the common availability of GLC, analysts deduced approximate fat compositions from these figures and from certain tests for specific oils or fats, but in a routine laboratory it was often only possible to reach fairly generalized conclusions. Table 3.4 gives some values for various oils and fats, and for the fatty acids derived from them; but it must again be emphasized that most fatty materials can have a considerable range of compositions, and hence of values for the various 'constants'.

Table 3.4 – Approximate constants for some fats and the fatty acids derived from them

Oil or fat	Figures on fats						Figures on fatty acids		
	SV		IV						
	Range	Selected	Range	Selected	Unsaponi-fiable	Special tests	Fatty acid ratio	IV	Titre °C
Beef tallow[a]	190–200	197	34–47	45	0.2–1.0		9.1	47	40–46
Mutton tallow[a]	192–198	197	35–46	40	0.2–1.0		9.1	42	43–48
Greases[a]	186–198	–	48–58	–	0.5–2.0		9.4	50	36–40
Palm oil	195–205	200	45–60	53	1.0		8.9	55	38–47
Lard	192–203	196	53–59	59	0.6		9.2	62	32–43
Coconut oil	255–263	257	7–10	9	0.5–1.0		6.8	10	20–24
Palm kernel oil	242–255	246	14–21	17	0.5–1.0		7.2	18	20–28
Olive oil	190–196	192	79–90	84	1.0–1.8		9.2	87	17–26
Groundnut oil	188–196	193	82–99	90	1.0	Arachidic acid test	9.2	95	26–32
Cottonseed oil	189–198	196	100–115	109	1.5	Halphen test	9.2	115	30–38
Maize or corn oil	187–196	192	103–128	122	2.0–2.5		9.2	128	14–20
Soyabean oil	189–195	192	128–143	132	1.0–1.5		9.2	139	20–21
Sunflower oil	188–194	192	122–136	132	Trace–1.5		9.2	139	16–20
Sesame oil	188–193	192	105–118	109	1.5–2.0	Baudouin test	9.2	115	20–25
Rapeseed I oil	168–178	174	97–110	101	1.5	Erucic test	10.4	106	11–15
Rapeseed II oil	–	–	100–105	–	–		–	–	–
Whale oil	185–202	192	97–140	–	Max 2.0	Hexabromides test	9.3	126	22–24
Herring oil	179–192	186	110–155	–	1.0–3.0	Hexabromides test	9.7	147	25
Menhaden oil	188–199	–	150–190	–	1.0–2.0	Hexabromides test	9.2	165	32
Castor oil	177–187	181	81–91	85	1.0	High hydroxyl value	10.0	90	–

[a]See also Section 3.5.
Figures given by different authorities vary considerably; those quoted should be taken only as approximate indications.

3.4.2 Moisture and dirt

For routine purposes, moisture can be determined by loss of weight at 105°C. This can conveniently be estimated by heating a sample in a flat porcelain dish over a small flame, using a very short thermometer to stir and to indicate the temperature. This gives total volatile matter; and moisture can be determined more accurately by the Karl Fischer method. Dirt is determined by filtration of a hot dry sample of the oil through a dry filter paper and weighing the residue after fatty matter has been removed with a solvent such as light petroleum (40 – 60°C boiling range).

3.4.3 Titre

Dry fatty acids are first prepared from the fat or soap. The liquid fatty acids are cooled under standardized conditions; and the titre of the sample is the highest temperature reached when a rise occurs due to the latent heat of solidification. It is, thus, a measure of the melting point of the mixture of fatty acids.

3.4.4 FFA (free fatty acid) content, acid value, saponification value, and ester value

Estimation of the free fatty acid content of a fat is described in section 1.4.5. It can be expressed in terms of Na_2O%; or it can be converted to percentage of a notional fatty acid. For many fats, FFA is quoted in terms of oleic acid which has an equivalent mass of 282, so that percent FFA as oleic equals FFA as $Na_2O \times 282/31$, or $Na_2O \times 9.10$. In the case of palm oil it is sometimes expressed in terms of palmitic acid (factor 8.26); and for lauric oils in terms of lauric acid (factor 6.45). FFA content can also be expressed in terms of mg KOH required to neutralize the FFA in one gram of fat; this figure is known as the acid value (AV), or acid number. If n = FFA% as Na_2O, the FFA in lg fat is equivalent to $n \times 1000/100 = 10\,n$ mg Na_2O, or $10 \times 56.1/31 = n \times 18.1$ mg KOH.

Saponification value (SV), or saponification number, is similarly expressed as the number of milligrams of KOH required to saponify completely one gram of fat. The test is carried out by heating under reflux a suitable weight of fat with a known volume of standard alcoholic potash and back titration of the excess potash with standard acid to the phenolphthalein end point. The net quantity of KOH used has reacted with both FFA and glyceride and gives the SV.

Ester value (EV), or ester number, equals SV–AV; that is it a measure of the glyceride present in the sample.

Saponification value (or, more strictly, the acid value of the fatty acids), and the fatty acid ratio mentioned in section 1.7, are both measures of the average equivalent mass, and hence average chain length, of the mixture of fatty acids in the fat. Relationships between these figures are discussed in section 3.8, where they are also used to estimate the glycerol content of a fat, the quantity of caustic soda required for its saponification, and the quantity of soap produced.

3.4.5 Content of unsaponifiable matter

The percentage of *unsaponifiable matter* in a fat is determined by extraction with ether after complete saponification with caustic potash. A soap may also contain a proportion of free fat, which should normally be very small except when saponifiable superfatting materials are added deliberately. A sample of soap is extracted and, after the extract has been weighed, it is boiled under reflux with caustic potash to complete saponification, and the unsaponifiable is extracted. The first extract gives total unsaponified and the second unsaponifiable. Unsaponified fat equals total unsaponified minus unsaponifiable.

3.4.6 Iodine value

The average degree of unsaturation of a fatty material is measured by determination of its *iodine value,* or iodine number. Several methods are available, but the Wijs method, using iodine monochloride in glacial acetic acid as the reagent is usual. The weighed sample of fat is dissolved in carbon tetrachloride and a measured volume of reagent added. After the appropriate time for addition of iodine to the double bonds has elapsed, water and potassium iodide are added and the excess iodine is titrated with standard sodium thiosulphate. The difference between the titration of a blank without fat and that of the sample represents the iodine absorbed, from which the iodine value is calculated as the number of grams of iodine absorbed by 100g of fat. The weight of fat used in the test must give an excess within narrow limits and a repeat test may be needed after an approximate value has been found. Care is needed in interpreting the results with highly unsaturated materials; in particular, oils containing fatty acids with conjugated double bonds do not take up the stoichiometric quantity of iodine, but these do not normally concern the soapmaker. Iodine values can be determined on dry fats, or on dry fatty acids (normally plus the unsaponifiable matter, that is on total fatty matter), but, with a given fatty acid composition, the value will obviously be lower with the fat than with the fatty acids, because of the diluting effect of the combined glycerol.

3.4.7 Reichert-Meissl, Polenské and Kirschner values

The *Reichert-Meissl, Polenské* and *Kirschner values* of a mixture of fatty acids (derived from a fat or a soap) are determined by distillation under standardized conditions; and are measured in terms of the ml $0.1N$ NaOH needed to neutralize the fatty acids in the distillates. The Reichert-Meissl value is determined on the water soluble fraction, and is a measure of the milk, or butter, fats present. The Kirschner value measures butyric acid more specifically. The Polenské value, of particular interest to soapmakers, determines the water insoluble volatile acids; and is a measure of the lauric oil included, with some indication whether it is coconut or palm kernel.

3.4.8 Specific gravity or relative density

The *specific gravity* of the fatty acids is also useful, particularly when rosin may be present.

3.4.9 Rosin content and tests for specific oils

Tests are available to show the presence of *rosin* (Leibermann-Storch) and to determine its percentage (McNichol or Wolff). There are also tests to indicate specific fats; for example, the Halphen test for cottonseed, or kapok, oils, although a positive may also be obtained from the fats of animals fed on cotton-seed cake.

3.5 SPECIFICATIONS. QUALITIES OF TALLOWS AND GREASES

Many specifications and trading standards have been established, particularly in the USA; see Bailey (1979). The most important for soapmakers are those concerned with tallows and greases, because these fats are the major constituents of most soaps; and because they can vary widely in quality. Years ago, South American beef tallow was favoured for the best soaps, and mutton tallow was somewhat frowned upon. Mutton tallow is usually slightly harder than beef (higher titre and lower iodine value), but it was thought more liable to lead to rancidity after long storage of the soap. Now tallows are often of mixed origin, and are commonly sold on the basis of specification related to analytical characteristics, including colour and/or colour after a standard bleach. Soapmakers often find that a colour of soap, rather than a colour of fat, test is particularly relevant, although it does not appear in the usual specifications.

A convenient colour of soap test for routine purposes is to take a sample of the tallow, or other fat, bleached if appropriate, in a white enamel cup, add a slight excess of alcoholic caustic soda and boil with constant stirring. The fat is saponified and the alcohol boiled off; this leaves granules of soap which are assessed by eye by an experienced observer. Other colour of soap tests use instrumental measurements of the colour of a solution of the soap.

One important specification is BS 3919:1976. Technical tallow and grease. Both tallow and grease are defined as fats derived from the carcases of oxen and sheep, but some admixture of pig and poultry fat is permitted; in practice, inclusion of some horse fat is not unknown. Tallows are defined as fats with a titre of 40°C, or above; and an iodine value below 60 calculated on the fatty acids from a determination on the fat (Wijs method). Greases are defined as having a titre between 36°C and 40°C; and an iodine value below 63 calculated on the fatty acids. Tallows and greases produced by solvent extraction must be so described. Table 3.5 gives gradings suggested in BS 3919:1976 as a basis for the agreement of trading standards.

The SDIA (the British Soap and Detergent Industry Association) uses slightly different trading standards.

American gradings are based upon similar principles, but use different terms. Tables 3.6 and 3.7† give respectively USA standards for tallows and greases. Standard laboratory bleaching tests are included in some specifications; in these nitrogen, rather than vacuum, is often used to produce an inert atmosphere.

3.6 THE PRE-TREATMENT OF OILS AND FATS; AND THE PRODUCTION OF BY-PRODUCTS USED IN SOAPS

3.6.1 General

Most fats are pre-treated prior to use in edible products, as are many intended

†Reprinted from Bailey's 'Industrial oil and fat products', fourth edition, edited D. Swern, Vol. 1, 1979, by permission of the publishers John Wiley & Sons.

Table 3.5 – Tallow gradings from BS 3919:1976

Grade	FFA% (max)	Crude colour (red)	Bleached oil colour lovibond (max. red)	Moisture + dirt %	Unsaponifiable % (max)	Titre °C	Max[a] content of plastic
Tallow 1	3.0	–	0.5 in $5\frac{1}{4}''$[b] cell	0.5	0.5	40 (min)	200
Tallow 2	5.0	–	1.0 in $5\frac{1}{4}''$ cell	1.0	1.0	40 (min)	200
Tallow 3	8.0	–	3.0 in $5\frac{1}{4}''$ cell	1.0	1.0	40 (min)	200
Tallow 4	12.0	–	4.0 in 1″ cell	1.0	1.5	40 (min)	200
Tallow 5	15.0	–	12.0 in 1″ cell	1.0	1.5	40 (min)	200
Tallow 6	20.0	–	No limit	1.0	2.0	40 (min)	200
Grease	20.0	No limit	–	2.0	2.0	36–40	200

[a]Plastic content expressed as mg polythene per kg fat.
Dirt is defined in the specification.

[b]The sign ″ equals 1 inch.

Table 3.6 – Common grades of commercial tallow (Rules of the New York produce exchange (May 1, 1963)

Grade	Titre °C (min.)	FFA% as oleic (max.)	FAC colour (max.)	Colour of refined and bleached tallow FAC (max.)	MIU[a] % (max.)
Edible	41.5	1	5	–	1
Extra fancy	42.5	–[b]	5	–	1
Fancy	41.5	4	7	–	1
Bleachable fancy	41.5	4	–	2.0	1
Choice	41.0	5	9	–	1
Prime or extra	40.5	6	13 or 11B	–	1
Special	40.5	10	19 or 11C	–	2
No 1	40.0	15	33	–	2
No 2	40.0	20	39	–	2
No 3	40.0	35	None	–	2
Naphtha extracted bone	40.0	50	None	–	2

[a]MIU is moisture, impurities and unsaponifiable.

[b]Figure is to be agreed in a contract.

The FAC colour system (devised by the Fat Analysis Committee of the AOCS) makes visual comparision with solutions of inorganic salts provided in sealed vials. The complete set of standards comprises 26 vials, numbered 1 – 45 in odd numbers and divided into five series.

1 – 9 for light coloured fats
11, 11A, 11B and 11C for very yellow fats
13 – 19 for dark, reddish fats
21 – 29 for greenish fats
31 – 45 for very dark fats

The system is somewhat confusing and some fats do not match any standards.

Fat colours can also be measured and defined in terms of absorption spectra, see Bailey (1979) or various AOCS publications.

Table 3.7 – USA standard grades for greases

Grade	Titre °C (min.)	FFA % as oleic (max.)	FAC colour (max.)	MIU % (max.)	Notes
Choice white	37.0	4	11	1	
White A	37.0	8	15	1	
White B	36.0	10	19 or 11C	2	
Yellow	36.0	15	37	2	
House	37.5	20	39	2	
Brown	38.0	50	None specified	2	
No 1 Glue or fleshing	37.0	15	15	2	By-products of
No 2 Glue or fleshing	37.0	40	21	2	glue manufacture.

White greases are similar in characteristics to lard, but generally have not been produced from materials and under conditions required, at least in the USA, for edible grades. Under USA government regulations de-naturants are added to inedible greases to prevent them from being diverted to edible use.

The lowest grades of greases are normally split and distilled to produce fatty acids for use in soaps, or for other purposes.

to be used in soaps, although large quantities of the better grade fats are made into soap without any pre-treatment. Some of the pre-treatments mentioned below are important to the soapmaker mainly because they produce by-products which can be used in soaps.

The main types of pre-treatment are:

- *de-gumming* or *de-sliming*. This is usually used with soyabean and other oils with a high content of phosphatides and other unsaponifiables. Warm water, sometimes with about 0.1% of phosphoric acid, is added to hydrate some of the impurities which are then separated in a centrifuge to leave a relatively dry, clean oil. Commercial lecithin is recovered from the other fraction, particularly in the case of soyabean oil. These gums, known generally as 'break' material, can also be separated by heat.
- *acid refining,* is carried out with concentrated sulphuric acid in the pre-treatment of linseed, and other drying oils, for use in paints. The strong acid chars the phosphatides and related materials. Treatment with sulphuric acid is also used to clean up low grade materials prior to splitting with water (particularly by the Twitchell process).
- *alkali,* or *soda, refining* is normally included in the preparation of fats for edible purposes. It is not commonly employed in the pre-treatment of soapmaking fats, but it can be useful to upgrade certain tallows for use in high quality soap bases. An excess of caustic soda (or sometimes sodium carbonate) solution is used to neutralize the FFA in the fat and to carry away in the lower aqueous soapstock layer other impurities, plus some neutral fat. Careful attention is necessary to avoid an undue loss to the soapstock of good neutral fat. The use in soapmaking of the by-product soapstock is discussed in section 3.6.2.
- *steam distillation* can be used to reduce the free fatty acid content of a fat; a process which has been employed particularly with fairly high FFA palm oil. The fatty acids separated are used in soap. The relatively high temperatures involved bleach some fats, particularly palm oil; but some colouring matter tends to be 'set', and so more difficult to remove after the treatment.
- *deodorization* is a steam distillation process carried out under fairly high vacuum to remove odiferous and taste producing impurities from fats. Deodorization of edible fats is normally carried out on materials which have been soda refined and bleached; otherwise FFA would be removed in this process with other volatile constituents and this is better done by soda refining. The process is sometimes used to remove highly odorous constituents from coconut and palm kernel oils to be included in high proportions in certain toilet soaps. Metal scavengers, such as citric, tartaric and phosphoric acids are often added to fats prior to deodorization.
- *bleaching,* widely used for soapmaking as well as edible fats, can be by adsorption or by chemical means. Both are discussed in section 3.6.3.
- *hydrolysis,* or *splitting,* often followed by distillation of the fatty acids, is, with the manufacture of soaps from fatty acids, the subject of Chapter 5.
- *hydrogenation,* or *hardening,* in which the unsaturation of a fat is reduced

by catalytic hydrogenation. This process is used extensively in the edible fat industry. At one time, particularly when whale oil was plentiful and cheap, it was also used to produce soapmaking hard fats from oils which were too unsaturated for direct use in most soap products. With current prices it is now little used in the soapmaking field, but it may be useful in countries which produce large quantities of unsaturated oils and, for politico-economic reasons, need to use these materials for the production of soap. The hydrogenation process decolorizes oils such as palm by converting carotinoid pigments into colourless compounds.

The chemistry of hydrogenation is complex and the product obtained depends upon processing conditions as well as upon the starting material. This is because in the production of fats which retain a substantial degree of unsaturation, as is normally required, the hydrogenation can be more, or less, selective; and isomerization can occur. These matters are discussed in Bailey (1982) and Patterson (1983).

- *solvent separation processes,* such as the Solexol process using propane under pressure, have been developed to separate impurities from fats; but they are not generally important. Other solvent processes using furfural are for separation of one group of glycerides from another, rather than for removal of impurities.

3.6.2 Soapstocks and acid oils, or acidulated soapstocks

Soapstocks, the by-products from the soda refining of fats, are an important source of fatty matter for soapmaking. They can be used in several ways.

- directly as the soapstocks
- after concentration with additional fat
- after acidification with sulphuric acid to produce acid oils, sometimes known as acidulated soapstocks.

Advantages of using soapstocks directly in soapmaking are:

- the alkali used in the soda refining process is transferred to soapmaking so that caustic soda is saved.
- sulphuric acid does not have to be used to convert the soapstock to acid oil.
- glycerol liberated by saponification of glycerides in the soda refining process is transferred to soapmaking and not lost as when acid oil is produced.

Advantages of transfer to soapmaking via an acid oil are:

- transport costs are saved, unless the refinery producing the soapstock is adjacent to the soap factory. Soapstocks should usually contain 30 – 50% TFM, but may be weaker, particularly when washings are included.
- the water content of the soapstock can have an adverse effect on the efficiency of the soapmaking process which is avoided by the use of acid oil.
- the acidification, and disposal to waste of the aqueous layer, followed by washing of the acid oil, removes various impurities which are not then transferred to soapmaking. If required, the quality of the acid oil can be

improved by bleaching, or if necessary, by splitting and distillation. Alternatively, certain chemical bleaching processes can be used with soapstocks. Soapstocks can also be purified by completion of saponification, followed by graining with salt and the discard of the lye. This will result in the loss of some of the free glycerol in the soapstock.

Concentration of soapstock is achieved by mixing the soapstock with a carefully controlled quantity of additional fat at a temperature above 70°C. The quantity of fat needed should be determined by experiment, but it is often found that about 3.8 parts of new fat should be added to soapstock containing one part of fatty matter. Circulation by a pump, with return to the vessel through a fish-tail jet, as described in Chapter 4 for soapmaking, is convenient. On settling for 3 – 4 hours, an aqueous layer should separate, containing perhaps 0.5 – 1.0% glycerol, which can be introduced in an appropriate manner into the soapmaking system. The upper fatty layer, containing the soap, can be handled as an oil. If conditions, particularly temperatures, are not right an emulsion may be formed and no proper separation achieved. Such emulsions can often be broken by addition of brine, or treated soap lye. Benefits which can accrue from the concentration of soapstock are mainly improvements in soapmaking efficiency, for reasons discussed in Chapter 4. Transport costs may also be reduced if only the fatty matter needs to be moved, but this saving will be negatived if the aqueous layer also has to be transported.

Acidification of soapstocks is usually carried out with sulphuric acid in wooden, plastic, or lead-lined vats. After sufficient acid has been added to leave free mineral acid, the mixture is boiled with steam until the soap is split and the emulsion broken. After the mass has been settled, the aqueous layer containing various impurities is run to waste; and the acid oil is washed with hot water as many times as necessary.

Crude cottonseed oil is often soda refined twice. The first treatment, plus alkali and water washes, gives washed cottonseed oil and a very dark soapstock, sometimes known as black cottonseed grease. This black grease, which contains substantial proportions of gossypol and phospholipids, is not normally suitable for incorporation in soaps directly, or as an acid oil. Usable, but rather dark, fatty acids can be recovered from it by splitting and distillation.† The washed cottonseed oil is soda refined again to yield a soapstock, or acid oil, of reasonable quality for soapmaking.

The fatty matter in soapstocks, and acid oils, has a very similar fatty acid composition to the parent oil, or fat, but the colour and general quality are inferior.

3.6.3 Bleaching of oils and fats

3.6.3.1

Methods used to bleach oils and fats can be divided into adsorption bleaching, used for both edible and soapmaking fats, and chemical bleaching used only for soapmaking fats.

†The Alpha-Laval Company, *Soap, Perfumary and Cosmetics* (1960), claims a process in which black cottonseed grease can be used directly in soap, the dirt being extracted and removed in part of the soap nigre which is rejected.

3.6.3.2

In *adsorption bleaching,* the molten fat, or oil, is brought into contact with a powdered solid adsorbent, which can be natural fullers earth, or activated earth or clay, or, to a limited extent, activated carbon used in conjunction with activated earth, particularly in the USA and with edible oils. Activated earths are made by the action of mineral acids on suitable mineral substances, such as bentonite or montmorillonite. The adsorbent removes colouring matter from the oil which is then separated by filtration. Sulphuric acid markedly improves the performance of natural fullers earth and may also be useful with activated earths. An addition to the bleaching vessel of about 10% of the weight of natural earth used is often found suitable for soapmaking oils. Davidsohn, Better and Davidsohn (1953) refers to the use of 2 – 3% sulphuric acid, followed by 3 – 5% earth, in the bleaching of certain dark liver oils.

Only about half the ratio of earth to oil is required with a good activated earth to give the same result as that obtained with natural earth. Various grades of activated earth are available, some acidic and some more nearly neutral. Acidic earths are usually more effective than neutral types and are commonly used with soapmaking oils. They may cause a very slight increase in the FFA of a soda refined oil and, for this reason, they may not be desirable with some edible oils. The proportion of earth needed to achieve the desired bleach should be determined by experiment. It varies widely from one oil to another, and with the colour of bleached oil required. 2 – 6% of activated earth, or 4 – 10% of natural earth, is common for soapmaking oils; the highest proportions often being needed with palm oil.

Batchwise bleaching in single vessels is usual for soapmaking oils, but continuous bleaching processes are described in Bailey (1979), and are sometimes used in edible fat refineries operating on a large scale. Batchwise bleaching can be carried out in open vessels, fitted with closed steam coils and with an agitator to keep the earth in suspension in the oil and to promote heat transfer without local overheating; but closed vessels working under vacuum (50 – 70 mmHg absolute pressure) are usually used and are better because:

- moisture is removed more effectively.
- there is less risk of undesirable oxidation of the oil by air, particularly if steps are taken to deaerate the oil as much as possible before it reaches bleaching temperature.

Vacuum bleachers can be of various designs, but vertical, cylindrical, vessels with dished base and top are common. Various types of agitator, including paddles or tubines are possible. Heating is usually by internal steam coils, but in the case of very large bleaching vessels, external circulation through a tubular or similar heater is sometimes used, and is preferable. Coolers may be fitted to cool the oil before it is filtered. In the case of edible oils, this is to minimize risk of oxidation whilst the oil is exposed to air; with soapmaking tallows and greases it is primarily to remove polythene, for which purpose the oil should be filtered at below about 70°C. Bleaching temperature is not very critical; typically, good quality tallows and other oils and fats are bleached at 90 – 95°C, poorer qualities at about 110°C, and palm oil at 120 – 130°C. About 15 minutes is said to be sufficient for equilibrium to be reached, but the complete bleaching cycle takes several hours.

The role of water is important. Complete drying of the bleaching earth will destroy its efficiency; and water in the oil charged to the bleacher does no harm as regards final quality. It has been reported that an addition of water can improve the performance of activated earths when bleaching tallow; other experience shows that sufficient water is important when palm oil is being bleached with activated earth, or with natural earth plus 10% acid; and, if necessary, water is added to bring the level to about 2%. On the other hand, it is usually found necessary to adjust temperature, pressure and time, so that the oil is substantially dry before the bleaching stage is completed in order to achieve proper adsorption of colouring matter.

Mild steel, or cast iron, is normally used for the vessels and coils of bleachers, even when sulphuric acid is added, but a splash plate near the acid inlet is desirable to direct the acid to the oil rather than the vessel wall. The earth is commonly sucked in from a hopper by the vacuum.

It is necessary to separate the earth from the oil completely and in such a way that the quantity of oil left in the filter cake is minimized. This requires a substantial pressure, so that vacuum filters are not suitable, and facilities are needed to permit the cake to be air blow and/or steamed. Filter presses, manual or, more recently, mechanized, are traditional; but some types of leaf-filter can be convenient and require little labour. The oil discharge arrangements should minimize the risk of fume being dispersed in the work area, particularly during steaming or blowing. Filters are normally cast iron. Various filter cloths, of different weights, weaves and materials are now available; and the following table, courtesy of Johnson-Progress Limited, provides useful information concerning the main fabric materials.

Material	pH range	Maximum temperature °C
Cotton	6 – 10	100[a]
Polyamide	7 – 12	110
Polypropylene	2 – 12	100
Polyester	2 – 7	160
Polyvinylidene chloride	2 – 11	80

[a]higher temperatures sometimes used successfully.

Cotton cloths were traditional for bleaching earths and are still widely used; but synthetics, particularly polyester, withstand acidic conditions better and may have several times the life of cotton, which can outweigh the higher initial cost.

Cotton cloths seal the joints between the plates well, but synthetic cloths do not; and special treatment of the areas of the cloths which form the seal, or special mechanical devices, may be needed. Fine weave metal gauzes are used in some mechanized filters. As in other filtrations, the cloth, or other primary medium, must retain most of the solids, but the initial layer of earth residue on the cloths forms the true filter medium, and it is usually necessary to return the first runnings for re-filtration.

A second filtration through a polishing filter is sometimes used to remove traces of fines which pass through the first filter. Paper may be used over the cloths, particularly in polishing filters, to improve the clarity of the filtered oil. When filter-presses are to be used, there is a choice between the plate and frame type and the recessed plate type. Either can be satisfactory for the main filtration; but, contrary to what might be expected, a plate and frame press should be chosen for polishing. This is because in a polishing filter a thin layer of very fine earth builds up, and becomes impervious, before the chamber is nearly full. It is then necessary to drain the oil and to remove the paper and residue. This can be done with a plate and frame press, but not satisfactorily with a recessed plate press having a central feed inlet.

The proportion of oil left in the filter cake after it has been steamed and/or blown with air as effectively as possible, varies considerably with the type of earth used. With natural earth it is often possible to achieve about 30% on the weight of residue, but with activated earths it is usually higher and 40 – 50% is common; up to 70% has been mentioned with edible oils. The loss of oil per tonne of crude fat treated is controlled by the percentage of bleaching earth used, as well as by the percentage of oil left in the residue, so that it is frequently less with an activated than with a natural earth, despite the higher retention of activated types. This loss of oil is a major item in the cost of bleaching, often exceeding the actual cost of the earth. Even a small percentage of activated carbon can greatly increase the oil retention of an earth.

One reason why plant bleaches may be more effective than laboratory bleaches lies in the bleaching which occurs whilst the mix is being passed through the cake in the filter. Multi-stage counter-current operation has theoretical advantages, but is not often practised. It is, however, sometimes useful to clean up, and bleach slightly, an oil for a relatively low grade soap by pumping it through full presses. The oily residue discharged from the filters can sometimes be sold for use in animal feedstuffs (in which substantial percentages of some kind of fat are often included). Processes for recovery of oil from these residues include:

- wet extraction with a solution of alkali, which is not very efficient.
- extraction with a hydrocarbon solvent. This gives a rather low yield of oil of moderate quality, but requires all the precautions for a process with a flammable solvent.
- extraction using a mix and settle process with a chlorinated solvent, such as trichloroethylene. This gives a higher yield of a poorer quality fat, because the chlorinated solvent extracts more coloured and oxidized material than the hydrocarbon solvent. Better settling, and an even higher yield, but of a very dark brown oil, can be achieved by steaming the solvent residue mixture.

It is often found that it is not worthwhile to recover oil from bleaching earth residues. Disposal of earth residues, extracted or not extracted, can be difficult; particularly if residues from the bleaching of highly unsaturated oils, which can be subject to spontaneous combustion, are involved.

3.6.3.3

Chemical bleaching normally involves the oxidation of the colouring matter in the oil or fat. The result varies greatly with the kind of fat; and palm oil, and by-products containing palm oil residues, are particularly responsive to oxidation bleaching. In some cases, an undesirable odour develops; and chemical bleaching is not used with edible fats, or, usually, with fats to be made into the better quality toilet soaps.

The oxidation can be carried out with air (or oxygen), with alkali dichromate (plus acid), with sources of available chlorine, or of chlorine dioxide, or with other reagents. Air can be blown through hot palm oil to produce air bleached palm oil. Soap made from this oil has a slight brownish shade and a characteristic odour. These qualities limit its use, but it has been employed in orange/red carbolic toilet soaps. Bleaching of palm oil with chromic acid has been described by Thomssen and Kemp (1937). The oil is agitated with air for several hours after addition of sodium dichromate, salt, and hydrochloric acid. After the mix has settled, the aqueous layer is run to waste and the oil is washed several times with water. More recently, most chemical bleaching processes have used chlorine compounds.

Oils and fats, particularly acid oils, can be bleached under acid conditions. Sodium chlorite is usually employed, activated with acid; and the active species is thought to be chlorine dioxide (ClO_2). Tuttle and Woodward (1946) describe the bleaching of tallows and greases using sodium chlorite plus sufficient sulphuric acid to reduce the pH to 4, or below. Other operators have found that no additional activator is necessary if the fatty matter contains sufficient FFA (usually above about 20%); and when the fat is not sufficiently acidic, the preferred activator is phosphoric acid. About 0.5 – 1.0% of sodium chlorite is used. The process can be carried out in a simple vessel in which agitation is provided by compressed air from a perforated pipe. The vessel can be constructed of mild steel, but stainless steel is preferred. Alternatively, continuous bleaching plants can be designed for use if the scale of operation is sufficiently large. When phosphoric, or other mineral, acid is used, particular care must be taken to avoid addition of reagents when no fat is present; otherwise an explosion is liable to occur. The bleached fat should be saponified as soon as possible. Soaps made from chemically bleached fats tend to deteriorate more than those made from adsorption bleached materials.

Chlorine compounds can also be used under alkaline conditions to bleach soaps, including soapstocks. Soapstocks should be completely saponified, and soaps should preferably be washed to remove the glycerol before the bleaching operation. Sodium hypochlorite has long been used to bleach soap in the pan, but, particularly for soaps containing some palm material, there are advantages in using:

- sodium hypochlorite activated with hydrogen peroxide.
- sodium chlorite activated with sodium hypochlorite, although various other activators are possible.

The sodium hypochlorite/hydrogen peroxide process can conveniently be carried out in a continuous reactor in the form of a vertical cylinder with two turbo impellers; one near the base and the other part way up. The soap flows upwards, is mixed with the sodium hypochlorite in the lower impeller, and the hydrogen peroxide is fed to the eye of the upper impeller. Sufficient hot water must be introduced to avoid the grained condition in the reactor, it being remembered that sodium hypochlorite with its associated salt and caustic soda is a strong electrolyte. Hypochlorite to give about 2% available chlorine plus 0.1% hydrogen peroxide, both expressed on TFM, may be found suitable.

As mentioned in section 1.6.6, sodium hydrosulphite has been used in the pan as a reduction bleach to improve the colour of good coloured soaps; it is not very effective and is now seldom employed. The hydrosulphite powder is sprinkled onto the surface of the soap when in a 'closed' condition, and the pan contents are boiled for about ten minutes.

3.7 POINTS IN THE SELECTION OF FAT BLENDS FOR VARIOUS SOAPS

Soaps can be made from a wide range of fats; and unusual fats, or blends, are used in some countries for politico-economic reasons. For example, soaps high in coconut oil have been made in the Philippines and high in cottonseed oil in the Sudan. In such cases, it might be better, if possible, to buy tallow and to sell the replaced vegetable oils abroad. As well as the technical advantages which would accrue, coconut and good soft oils now have higher world market prices than technical tallows, but relative prices have varied considerably over the years. The great reduction in household soap consumption in the USA, and other developed countries, has provided much technical tallow for export from the USA, which is being used in very many countries.

For many purposes, including conventional toilet soaps and household soaps made by the modern vacuum chilling/extrusion processes, a blend of about 15 – 25% lauric oil (coconut or palm kernel) and 75 – 85% hard fat (usually tallow or palm oil) gives a soap base with an excellent combination of properties. The hard fat gives a firm texture and good detergency at high washing temperatures; the lauric oil also gives a firm soap, but improves the rate of solution and freedom of lather. Particularly for household soaps, but also for cheaper toilet soaps, less than 15% lauric oil is sometimes used because of cost. Too high a content of lauric oil tends to give a soap which is rather harsh (irritant) to the skin, unless superfatting is employed. High lauric oil soaps are sometimes used for certain textile operations and when the soap is to be used in sea, or brackish, water; but tablets containing non-soapy detergents may now provide a better answer to the salt water problem.

Assuming that titre and iodine value for blends are additive, values for a mixture of 80% tallow, of titre 41°C and IV of fatty acids 45, with 20% coconut oil, titre 22°C and IV 10, can easily be calculated:

		Titre °C	Iodine value
80% tallow	0.8 x 41 =	32.8	0.8 x 45 = 36.0
20% coconut oil	0.2 x 22 =	4.4	0.2 x 10 = 2.0
blend		37.2	38.0

A titre of 36 – 38°C and an iodine value of about 37 are often considered ideal for a conventional toilet soap. If a rather hard, high titre, tallow is to be used, a proportion of good quality soft material, such as lard (choice white grease) or groundnut oil, will adjust the titre and iodine value of the blend; and produce a soap with an improved balance of plastic properties. Oils with significant proportions of highly unsaturated acids are liable to give soaps with tendencies to deteriorate through oxidation on long storage, particularly at relatively elevated temperatures. Care must, therefore, be exercised in the choice of soft material for high quality soaps and minor ingredients in the fats can be important. Some years ago an example occurred in which inclusion of a proportion of distilled bone grease fatty acids to soften the blend resulted in abnormal deterioration of the colour of a white toilet soap after some months storage. Olive oil, which has a very high content of oleic acid and little polyunsaturated acids, is a particularly good soft oil. It was a traditional soapmaking material in Mediterranean countries, but, apart possibly from some exceptionally poor, inedible, grades it is now much too expensive for use in soap. The use of preservatives, primarily but not exclusively, to minimize the adverse affects of heavy metal catalysts in producing deterioration on storage, is discussed later.

Tallow and palm oil are the main hard fats. Apart from some difficulty in the production of very white soap, palm oil is an excellent soapmaking material, but it is now normally too expensive compared with tallow. Some 30 – 70 years ago, whale oil was relatively cheap and much was hardened (hydrogenated) for use in soap. Special attention needed to be paid to the hardening conditions to produce an acceptable tallow substitute, but hardened whale oil was generally less good than tallow. Little whale oil is now available and the soft vegetable oils are normally more expensive than tallow. Hardening may be useful in countries where there are pressures to use soft oils in soapmaking; and, although not very good technically, hardened fish oils may be economically attractive for low grade soaps under some circumstances.

For household soaps made by the old frame, or water-cooling, processes, a substantial proportion of soft oil is advisable to produce a good lathering soap. Although soft oils can be, and are, used with the modern household soap processes, they are not generally desirable, as explained in Chapter 6.

Rosin was used extensively in soaps at one time, primarily because it was cheap; but it does also have useful properties. To some extent, it is an alternative to lauric oils as a means by which the solubility and lather of a frame-cooled household soap made from hard fats can be improved; and tallow/rosin soaps were popular. A good quality rosin also gives the soap a characteristic odour which was generally thought pleasant and preferable to that of a tallow soap. At one time, 2 – 3% of good quality rosin was included in toilet, and other, soap blends as a preservative. Soaps with a high rosin content tend to be sticky, which makes them unsuitable for some purposes, including soap powders.

Particularly for household soaps, the fat is the predominant item in the cost of the product, so that it is very important commercially to use as effectively as possible any relatively cheap materials, such as by-products, which can be obtained. These by-products are often of somewhat uncertain composition, although described as 'soft acid oil', etc. A preliminary assessment of their suitability can be made from an analysis giving fatty acid composition by GLC; or, less precisely, from fatty acid constants, such as titre and iodine value. Colour, odour and stability on storage must also be considered.

3.8 FATTY ACID AND GLYCEROL CONTENTS OF FATS, AND RELATED MATTERS. GLYCERINE CREDIT

Initially, it will be assumed that the fat contains only triglycerides and fatty acids; complications due to the possible presence of di- and mono-glycerides will be considered subsequently. Also, the calculations will be based on 100% fatty matter; but the small corrections for the 1 – 2% of moisture and dirt commonly found in crude fats can readily be made.

As explained in section 3.4.4, the saponfication value, acid value and ester value of a fat are all expressed in terms of mg KOH per g of fat.

The quantity of caustic soda needed theoretically to saponify the fat is thus equivalent to SV mg of KOH per g fat. This is equal to

$$\frac{SV}{1000} \times \frac{31}{56.1} = SV \times 0.000553 \text{ g } Na_2O\text{/g fat (or t/t fat)}$$

The glycerol content of the fat is proportional to the ester value; and EV = SV – AV.

Glyceride + 3 KOH $\rightarrow$ $C_3H_5(OH)_3$ + 3 potassium soap.

$$\begin{matrix} 3 \times 56.1 & \\ = 168.3 & 92 \end{matrix}$$

EV = mg KOH equivalent to the glyceride in lg fat

Hence, percentage glycerol in fat

$$= \frac{EV}{1000} \times \frac{92}{168.3} \times 100$$

$$= EV \times 0.0547$$

Estimation of the fatty acid content of a fat is more complicated, but the formula in terms of SV, EV and AV can be derived as follows:

Consider first a pure triglyceride with no FFA and let M = the average equivalent (= molecular) mass of the *fatty acids* it contains.

$$C_3H_5(OOC{-}R)_3 + 3H_2O \rightarrow C_3H_5(OH)_3 + 3R.COOH$$

$$\begin{matrix} 3 \times 18 & & \\ = 54 & 92 & 3M \end{matrix}$$

To make the equation balance, the molecular mass of the glyceride = $3M + 92 - 54 = 3M + 38$, that is the equivalent mass of the glyceride is

$$\frac{3M + 38}{3} = M + 12.67$$

lg of fat is equivalent to $(56.1 \times 1000)/(M + 12.67)$ mg KOH, and, by definition, to SV mg KOH. Hence,

$$\frac{56.1 \times 1000}{(M + 12.67)} = \text{SV, or } M = \frac{56\,100 - 12.67\ \text{SV}}{\text{SV}}$$

Considering now pure fatty acids, the equivalent mass is M, instead of $(M + 12.67)$, and SV = AV; so that $M = 56\,100/\text{SV}$, or $56100/\text{AV}$. In actual fat containing both triglycerides and free fatty acids, the ratio of combined to total fatty acids is EV/SV (assuming, as is usually approximately the case, that the composition of the free is similar to that of the combined fatty acids). The addition of 12.67 to the equivalent mass of the fatty acids applies only to the combined portion, so that, from the equation for pure glycerides,

$$M = \frac{56\,100 - 12.67\ \text{SV} \times \dfrac{\text{EV}}{\text{SV}}}{\text{SV}}$$

or

$$M = \frac{56\,100 - 12.67\ \text{EV}}{\text{SV}}$$

The total mass of fatty acids in 1000 mg of fat equals the quantity of KOH required to saponify it $\times M/56.1 = \text{SV} \times M/56.1$. Hence, from the expression just derived for M, the total fatty acids in 1000 mg fat

$$= \frac{\text{SV}}{56.1} \times \frac{56\,100 - 12.67\ \text{EV}}{\text{SV}}$$

$$= 1000 - \frac{12.67\ \text{EV}}{56.1} = (1000 - 0.226\ \text{EV})\ \text{mg}$$

Hence, 100 parts of fat contain (100 − 0.0226 EV) parts total FA.

$$\text{FFA\%} = \frac{\text{AV}}{\text{SV}} \times \text{total FAs in 100 parts fat}$$

$$= \frac{\text{AV}}{\text{SV}} \times (100 - 0.0226\ \text{EV})$$

If it is required to use a determination of FFA% rather than of AV, the figure in terms of Na_2O% should be converted to FFA% using an estimated molecular mass for the actual fatty acid blend, rather than a notional one, such as that for oleic acid. In many cases the difference is not great.

It may also be helpful to note that

$$M = \text{fatty acid ratio} \times 31$$

As an example, consider a tallow with SV = 195 and AV = 15. EV = 195 − 15 = 180.

$$\%\text{FFA} = \frac{\text{AV}}{\text{SV}}(100 - 0.0226\ \text{EV}) = \frac{15}{195}\{100 - (0.0226 \times 180)\}$$

$$= \frac{15}{195}(100 - 4.1) = 7.4\%.$$

Theoretical requirement of caustic soda for saponification

$$= \text{SV} \times 0.000553$$

$$= 195 \times 0.000553 = 0.108\text{g } Na_2O\text{/g fat}$$

or 0.139 NaOH/g fat.

glycerol content = EV × 0.0547

= 180 × 0.0547 = 9.85%

fatty acid content = (100 − 0.0226 EV)

= 100 − (0.0226 × 180) = 95.9%.

As mentioned previously, all these figures are calculated for pure fatty matter in the fat, and strictly they should be corrected for its moisture and dirt content. This correction is usually only significant for the fatty acid content.

Table 3.8 shows figures calculated in this way for a number of typical soapmaking fats. It will be seen that:

- the theoretical caustic soda requirements fall into two groups, one for coconut (and other lauric) oils, and one for other fats.
- the glycerol content varies widely with both the equivalent, or molecular, mass of the fatty acids and the percentage of free fatty acids.
- the total fatty acid content does not vary much, although it is a little lower for the lauric oils, and a little higher when the FFA is high. After allowance for moisture and dirt, fatty acid content is about 94 – 96% for most fats.

As fats contain about 95% TFM, 1 tonne of fat yields about 95/63 = 1.51 tonnes of 63% TFM soap. This supports the common approximation that 1 tonne of fat should yield around $1\frac{1}{2}$ tonnes of soap base at 62 – 64% TFM.

Saponification almost certainly occurs in stages; but mono- and di-glycerides are not normally found in partly saponified fat, probably because they react with caustic soda more rapidly than triglycerides. In the case of hydrolysis of fats with water, substantial proportions of mono- and di-glycerides are found during the process; and some are likely to be present in fats which have deteriorated with development of FFA, although presumably not in acid oils, unless they have deteriorated after production. The following calculations show that

Table 3.8 – Contents of glycerol and fatty acids, and caustic soda requirements for saponification, of some typical fats (no allowance for possible presence of mono- and di-glycerides)

Fat	SV	AV	EV	FFA% (calc)	Theoretical caustic soda requirement Na_2O % on fat	Glycerol content %	FA content %
Tallow	195	5	190	2.5	10.8	10.4	95.7
Tallow	195	15	180	7.4	10.8	9.9	95.9
Grease	192	90	102	45.8	10.6	5.6	97.7
Coconut oil	257	5	252	1.8	14.2	13.8	94.3
Coconut oil	257	12	245	4.4	14.2	13.4	94.5
Coconut acid oil	257	150	107	57.0	14.2	5.9	97.6
Cottonseed oil	194	4	190	2.0	10.7	10.4	95.7

this can be important in the estimation of the glycerol content of a fat from its SV and AV.

Consider a low grade tallow, or grease, having an equivalent mass of its fatty acids of 279 (fatty acid ratio 9.0), so that the group (O·CO·R) is equal to 279−1 = 278. The molecular mass of the triglyceride, $C_3H_5(O{\cdot}CO{\cdot}R)_3$ is $(3 \times 12) + 5 + 3\,(278) = 875$; and it yields $(92 \times 100)/875 = 10.5\%$ glycerol and $(3 \times 279 \times 100)/875 = 95.4\%$ FA. Similarly, the diglyceride, $C_3H_5(OH)(O{\cdot}CO{\cdot}R)_2$, has a molecular mass of 614; and yields 15.0% glycerol and 90.9% FA.

Again, the monoglyceride, $C_3H_5(OH)_2(O{\cdot}CO{\cdot}R)$, has a molecular mass of 353; and yields 26.1% glycerol and 79.0% FA.

Table 3.9 shows the glycerol and FA contents of various hypothetical fats build up from these components. It will be seen that glycerol content varies considerably between three compositions, all with 10% FFA, but that, as before, the fatty acid contents do not vary very much.

Table 3.9 – Compositions of hypothetical fats with varying proportions of the different glycerides and free fatty acid

Composition of fat	Glycerol %	Fatty acid %
100% triglyceride	10.5	95.4
90% triglyceride + 10% FA	9.5	95.9
80% triglyceride + 5% diglyceride + 5% monoglyceride + 10% FA	10.5	94.8
70% triglyceride + 10% diglyceride + 10% monoglyceride + 10% FA	11.5	93.8
100% FA	–	100

Glycerol and fatty acid contents of the fats used in a soap factory are required for process control and for costing. For many fats the methods described in this section give satisfactory figures, but if substantial quantities of fats likely to contain mono- or di-glycerides are used, it is better to have actual analytical estimations of glycerol and fatty acid contents. A useful compromise is to have a sufficient number of analyses of low grade materials made to permit the compilation of tables, or graphs, relating glycerol and fatty acid contents to AV, or FFA%, for the various materials. Values can then be read off when similar materials are used again.

When a quantity of fat is used to make soap it contains both fatty acids and glycerol; and there is no absolute way in which to decide how much has been paid for each. The lyes from lauric oils and from low grade fats are more difficult to purify than those from a good grade tallow; and, if the lyes were worked-up separately, they would tend to produce a poorer quality crude glycerine. In practice, lyes from the various types of fat are usually treated and evaporated together. On this basis the glycerol in one fat has the same value as that in another; and crude glycerine normally has a reasonably well-defined price. It is, therefore, convenient to allow a glycerine credit for the value of the glycerol and to estimate the cost of the fatty acids by difference. This can be done as follows:

Let £G = value per tonne of crude glycerine at 80% glycerol
£F = cost per tonne of the fat
g = percentage of glycerol in fat
f = percentage of fatty acids in fat
z = percentage yield of glycerol recovered in crude glycerine
y = percentage yield of fatty acids recovered in soap
£E = cost of recovering one tonne of 80% crude glycerine (treatment and evaporation).

Consider 1 tonne of fat, that is $g/100$ tonnes of glycerol and $f/100$ tonnes of fatty acids. Crude glycerine, at 80%, recovered $= g \times z/100(100 \times 0.8) = gz/8000$ tonnes, worth £$Ggz/8000$. The cost of recovery (disregarding costs in the soap-making department) = £ $(Egz/8000)$, so that the net value of the glycerol in the fat, the glycerine credit, is £ $(G-E)$ $(gz/8000)$.

The fatty acids recovered in soap equal $fy/(100 \times 100)$ tonnes, and the price deemed to be paid for one tonne of fatty acids in soap equals

$$F - \frac{[G-E)gz]}{8000} \text{ divided by } \frac{fy}{10\,000}, \text{ or } \frac{[8000F-(G-E)gz]}{0.8\,fy}$$

As an arbitrary example, say $G = 400$, $F = 200$, $E = 50$, $g = 10$, $f = 95$, $z = 90$ and $y = 97$. Substitution in the formulae gives the net cost of the fatty acids in this fat as £174 per tonne, and the glycerine credit as £39.4. When a fat is to be bleached, the calculation should start with the bleached fat; the cost of bleaching, including losses, being allowed in F. This procedure is useful:

- in the costing of the various soap bases made.
- to permit a proper comparison of the cost to soapmaking of various fat blends. For example, to allow for the fact that the higher glycerine credit for an oil than for its acid oil offsets to some extent the higher cost of the genuine oil. Similarly in a comparision of the cost of a lauric oil and a tallow or soft oil.

3.9 HANDLING AND STORAGE OF OILS AND FATS. DETERIORATION OF FATTY MATERIALS

3.9.1

The handling, storage, and processing of oil seeds, animal tissues, and other fat-bearing materials is outside the scope of this book; and the interested reader is referred to Bailey (1979). As mentioned earlier, the quality of oils such as palm and rice bran can be poor unless the source materials are treated carefully and promptly to minimise the action of air, water, enzymes and micro-organisms.

Oils and fats, particularly those containing the more highly unsaturated fatty acids, are subject to oxidation and development of rancidity. This, and the associated topic of flavour deterioration, is of major importance in relation to fatty foods. Care is also important with soapmaking fats, but the problem is somewhat less critical. The chemistry of rancidity development is complex, and only a few broad generalizations will be discussed here:

- initially oxidation occurs at a double bond with the formation of peroxides. As the process proceeds, the peroxides break down to aldehydes, ketones, hydroxy-acids and other compounds; in consequence the content of peroxides (often determined as a measure of rancidity) passes through a maximum and then tends to fall. The Kreis test for rancidity depends upon the red colour which develops when the fat is reacted with phloroglucinol in ether solution; and is some measure of the aldehyde and ketone content.
- the oxidation starts slowly but, after an induction period of varying length, it accelerates markedly. Pro-oxidants and antioxidants can be present. Pro-oxidants are mainly heavy metal catalysts, such as copper and iron; and one type of antioxidant sequesters, and so inactivates, these metallic catalysts. Other types of antioxidant interfere with the oxidation mechanisms. They can be natural, such as the tocopheraols (vitamin E), or artificial. Sequestrant antioxidants are widely used in soaps, as discussed later, but also with fats, as in the addition of citric acid before deodorization. Natural antioxidants are more general in vegetable than in animal fats, so that the former tend to keep better than the latter with similar unsaturation. Antioxidants are usually partially removed when an oil is refined; and crude oils often keep better than refined. Beef tallow also tends to deteriorate less than some other animal fats. Various organic compounds can be added to a fat to improve its keeping properties, but some, such as hydroquinone, can lead to the development of undesirable colour if the fat is made into soap. The most important artificial fat antioxidants, which have all been cleared by the USA authorities for use in edible fats at not more than certain specified concentrations, are:

  - BHA (butylated hydroxyanisole)
  - BHT (butylated hydroxytoluene), and
  - THBP (2,4,5-trihydroxybutyrophenone)

  BHT, sold under various trade names such as Stavox and Ionol, is commonly used in soapmaking fats and in soaps. These compounds can also have some adverse affects on the colour of the soap after storage.
- oxidation bleaching naturally affects unsaturated fatty acids, but such chemically bleached fats are, or at least have been, widely used in soaps, although generally not in the higher grades. Care should be taken not to continue air bleaching of palm oil beyond the optimum point, after which colour deteriorates and undesirable reactions may occur. The high peroxide value of an air bleached palm oil can be reduced by soda refining, adsorption bleaching, or deodorization, but, apart perhaps from passage through a

filter press loaded with spent bleaching earth, these additional processes are unlikely to be commercially attractive.

- complete saponification is generally considered important to ensure good stability of a soap. It is also commonly considered desirable to leave a small percentage of free alkali in a soap; but superfatting with fatty acids, which eliminates the free alkali in the base, is not unusual.

3.9.2

At one time oils and fats were normally handled in barrels, casks, or metal (usually mild steel) drums, but they are now commonly transported in bulk, both on land and over the sea. Tallow, the world's most important soapmaking fat, is handled in liquid form in very large quantities. At least for long journeys heating coils are required in the tanks.

3.9.3

When solid fats are received at a factory in drums, or barrels, a melting out stage is required; this normally comprises a tank, or tanks, covered with a staging onto which the drums can be put. Steam jets discharge into the bung holes of the drums, which face downwards, and the molten fat and condensed water drip into the tank below. The mixture in the tank is settled, the water run off, and the molten fat pumped away.

3.9.4

Conventional tank parks, with pumps and pipelines are used; and after a quantity of oil has been pumped the line is usually blown with steam. It is also desirable to blow through a line before starting to pump, to ensure that the cocks (or valves) are set correctly, and to warm the system. A 'go-devil' (a plug which fits loosely in the pipe and is blown along with air or steam) is sometimes used to clear long, relatively straight, pipelines of large diameter.

3.9.5

At least in cold and temperate countries, storage tanks for oils and fats usually need heating coils. In vertical, cylindrical tanks (the most usual kind), flat helical coils placed near the bottom are common. These coils often use low pressure steam, with discharge of condensate controlled by a manually adjusted valve. More effective heating can be achieved with hairpin coils installed with an adequate slope from the centre to the outside of the tank base. Steam traps can then be used; and a higher steam condensing pressure, and temperature, maintained in the coils. The better use of coil surface must be balanced against the greater loss of heat in the higher pressure condensate discharged, as compared with the near atmospheric pressure condensate from the older type of coil with manually adjusted valve discharge.

Tanks containing hot oil lose heat from all surfaces, but a surface pyrometer, or the hand, shows that most of the loss is from the 'wetted' part of the vertical walls. The surface temperature of this part of the tank is markedly higher than that of the 'unwetted' part of the vertical wall, and the roof. It is also usually higher than that of the under surface of the tank base, because there is normally a layer of water and dirt which does not mix much with the oil, and so cools more quickly and insulates the tank base. Need to conserve fuel raises the question of economic justification for the insulation (lagging) of part, or all, of the tank surface. Lagging is most likely to be worthwhile on tanks which are situated out of doors, and which need to contain hot oil continuously; for example, tanks which service a bleaching, or soapmaking, plant. It is unlikely to be justified for tanks holding strategic stocks which can be allowed to become cold for extended periods.

Storage tanks contain varying quantities of oil at differing temperatures at different times, and are subject to a range of atmospheric conditions. These, and other, factors make difficult the assessment of the heat saving likely to accrue from lagging. The most satisfactory estimates are made from direct measurements of the rate of cooling of lagged and unlagged tanks. Those who do not have access to relevant lagged tanks may find useful the following notes on some experiments carried out by the author many years ago. The tests were conducted with large (500 tonne) tanks which contained soapmaking palm oil, but similar results are likely with other fats, such as tallow, which separate solid constituents at temperatures well above atmospheric. When a tank was allowed to cool undisturbed and with no supply of steam to the coils, it was found that:

- whilst the bulk of the oil was fully molten, natural convection kept it at a uniform temperature, as determined from dip samples taken from various positions.
- a probe thermocouple inserted through the vertical wall of the tank, and arranged so that the sensitive element could be positioned at known distances (down to zero) from the inner surface, confirmed the existence of what is often called a 'self-lagging' layer of viscous or semi-solid, material. With the bulk of the oil at about 45 – 50°C, there was a gradient of some 5°C over the two, or three, inches nearest to the tank wall. This is distinct from the thin stagnant layer usually considered in heat transfer calculations.
- as a first approximation, it was found that over the limited range of temperatures at which a tank contents is likely to be kept if it is in regular use to feed a process, straight line cooling curves could be drawn of the form,

$$Q/\theta = A + By\,\Delta t, \text{ where}$$

Q/θ = heat loss (CHu/hr); which was easily calculated from the rate of temperature fall × mass of oil × specific heat of oil (about 0.5).
Δt = temperature difference between oil and atmosphere (°C).
y = fullness of the tank (for example, if the tank was half-full, $y = 0.5$).

A and B were constants found from the cooling curves. B was a measure of the heat loss from the wetted wall, and A of the loss from unwetted wall, roof and base.

If the tank had a diameter of a ft and a height of b ft, the area of the part of the wall wetted by oil was πaby ft^2; and the overall heat transfer coefficient for the loss of heat from the wetted part of the wall was

$$\frac{By\,\Delta t}{\pi aby\,\Delta t} = \frac{B}{\pi ab}\ \text{CHu/ft}^2\text{/hr/}^\circ\text{C.}$$

- the temperature of the outer surface of the wetted wall was determined with a surface pyrometer, from which the coefficient of heat loss from the outer surface of the wall was calculated. The values found for both overall coefficient of heat loss from the oil and for heat loss from the outer wall surface were lower than might have been expected from the literature. Values may vary widely with climatic conditions and other factors; and should be determined for the tanks under consideration by determination of rates of cooling with varying fullness fractions.

Various types of lagging, including cork plus bitumen, can be applied to tanks; but adequate weatherproofing is important if the tanks are situated outdoors. Suppose a layer of a good insulating material z inches thick, with a thermal conductivity of 0.04 CHu/ft·hr·°C per foot, that is CHu·ft/ft^{2}hr°C, is applied to the cylindrical wall of the tank. Resistances to heat flow are additive, so that if H_i is the overall coefficient for heat loss from the lagged wetted wall, $1/H_i = 1/(B/\pi ab) + (z/12)/0.04$, from which H_i can be calculated and substituted in the heat flow equation $Q/\theta = A + H_i\,\pi aby\cdot\Delta t$. The rate of temperature drop, and the saving of heat due to the presence of the lagging, can then be estimated approximately for various conditions; and the economic case for lagging assessed.

The heat loss could be reduced still more by lagging the top and bottom of the tank, but this is probably not worthwhile. Lagging only the lower part of the vertical wall can also be considered, because the upper part is relatively seldom in contact with fat.

3.9.6

Groups of tanks are best surrounded by a low bund wall to trap any leakages of fat. Water from tank coils, melting out stages, or other sources which may carry fat, should be discarded through a fat trap to avoid loss of valuable material and to minimize risk of pollution of sewers or rivers, or the blocking of pipes.

4

The manufacture of soap base by direct saponification of oils and fats

4.1 GENERAL

Some soapmaking terms are defined, and the processes for the manufacture of soap are outlined, in section 1.3.

This chapter is concerned with the production of soap base from fats using the fully boiled process, and more modern processes derived from it; leaving the production of fatty acids, their purification, and the manufacture of soap base from them to Chapter 5. Topics will be discussed in the following order:

- some principles.
- the traditional process in single vessels.
- developments including the use of groups of vessels.
- an additional principle.
- the components of modern continuous soapmaking plants:
    - equipment for saponification.
    - equipment for washing.
    - equipment for fitting and final separation.
- complete continuous soapmaking plants.
- plant performance.
- preliminary treatment of lyes in the soapmaking department.

4.2 SOME PRINCIPLES

4.2.1 Phase diagram

Fig. 4.1 shows diagrammatically a simplified phase diagram for a typical soap (say 80% tallow and 20% coconut oil) at about 90 – 100°C, indicating only

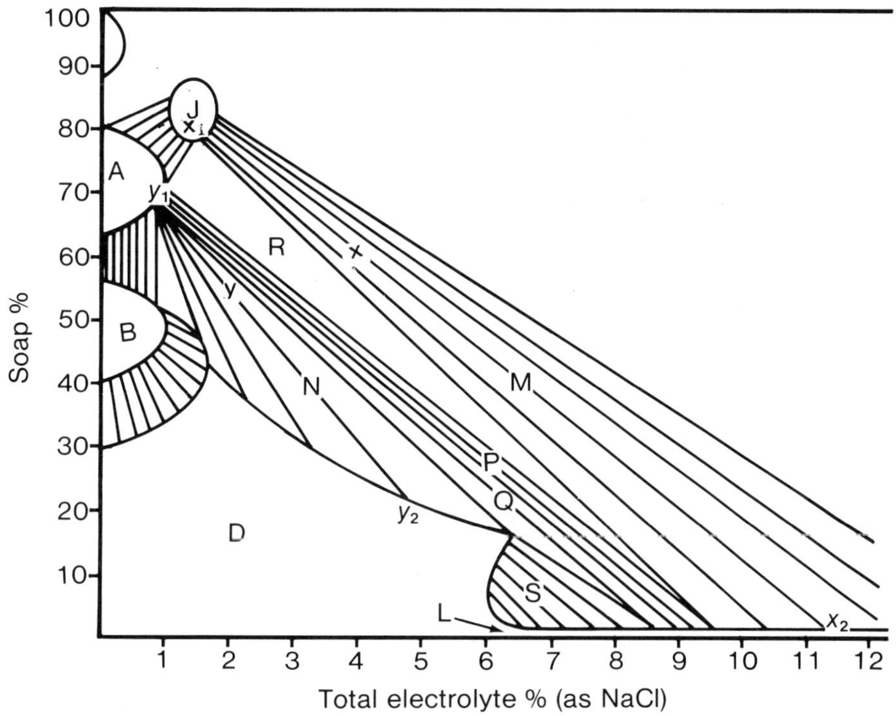

Fig. 4.1 – Simplified diagram for phases significant in the production of soap base by the fully boiled process.

phases which are relevant to the soapmaking process. The shapes of the various areas are influenced by the fat blend, particularly its content of lauric oils. The main electrolytes are salt (sodium chloride) and caustic soda (sodium hydroxide); but some sodium carbonate and sodium sulphate (derived from salt returned from the glycerine recovery process) may also be present. Figures quoted by various authorities for the relative salting out powers of these electrolytes vary, but roughly one part of salt is equivalent to 0.85 parts of caustic soda or 2.0 parts of sodium carbonate. No figures are available for sodium sulphate, but it is probably similar to sodium carbonate. The effects of the various electrolytes are additive, and are commonly expressed in terms of salt.

Area A on the diagram represents neat soap, or soap base, the product of the soapmaking operation, from which most products are made. It contains about 67 – 70% anhydrous soap, or 61 – 64% TFM.

Area J represents curd, or grained-out soap, sometimes called kettle wax. As discussed later, the high anhydrous soap (or TFM) content indicated on the diagram for this phase is somewhat misleading in that the curd soap produced by settling (or even centrifugal separation) contains a variable amount of enmeshed lye.

Area B represents 'middle soap'. This phase appears as 'water lumps' (possibly augmented by acid soap) if the electrolyte content is allowed to fall too low in any part of the vessel. The area is avoided by experienced soapmakers, as, once formed, water lumps are difficult to eliminate.

Area D is nigre, an impure solution of soap in water.

Area L, an extension of D, represents lye. As the electrolyte content increases, this phase changes from a thin, low soap, nigre, or nigre lye, to an aqueous solution which contains hardly any soap, known as lye.

Area M is a two-phase area which, theoretically at least, can separate into curd and lye. A mixture represented by point x, comprises curd x_1 and lye x_2. The proportions of the two phases are given by the relationship

$$\frac{\text{quantity of curd}}{\text{quantity of lye}} = \frac{xx_2}{xx_1}$$

that is the quantities of the phases are inversely proportional to the lengths of the tie lines from the mixture point x to the phase points x_1 and x_2.

Similarly, area N is another two phase area representing mixtures of neat soap and nigre. Mixture of composition y can separate into neat soap y_1 and nigre y_2, in the ratio

$$\frac{\text{quantity of neat soap}}{\text{quantity of nigre}} = \frac{yy_2}{yy_1} .$$

Area P represents the neat soap-lye zone. The so-called 'bay area', S, shows the equilibrium between nigre and lye.

The triangular area Q represents mixtures which separate into three phases of fixed compositions represented by the points of the triangle, neat soap, nigre and lye. Similarly, area R.

Fig. 4.1 is used later to illustrate various points in the discussion of soapmaking processes.

4.2.2 Saponification

Many studies have been carried out on the soapmaking, or saponification, reaction; and some are discussed in Davidsohn, Better and Davidsohn (1953). Important practical points concerning saponification are:

- Fats and aqueous caustic soda are not miscible at temperatures which can be achieved without pressures well above atmospheric; and the reaction with neutral fats, particularly those other than the lauric oils, does not start readily. It is autocatalytic, that is catalysed by the product of the reaction, soap. Hence, the reaction rate accelerates greatly until most of the fat has reacted, when it slows down again. The soap formed/time curve is, therefore, somewhat S shaped as shown diagrammatically in Fig. 4.2. The slow initial phase is a heterogeneous, surface, reaction helped by vigorous agitation and the formation of an emulsion stabilized by soap. The main part of the reaction is homogeneous, with the fat probably 'dissolved' in the soap micelles. The final stage is slow because the con-

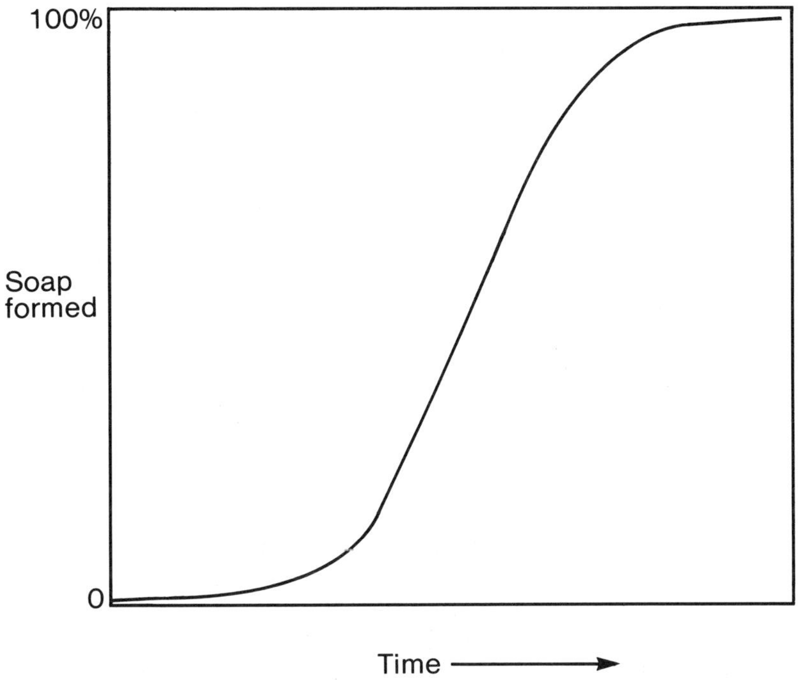

Fig. 4.2 – Rate of saponification curve for neutral fat.

centration of residual fat becomes low. In factory practice, the slow initial stage rarely causes much difficulty because fatty acids present in most soapmaking fats react readily to provide some soap; and because saponification is commonly carried out in a way which adds the reactants to a mass of soap. As with other chemical reactions, an excess of caustic soda helps to minimize the residual unreacted fat, but steps have to be taken to avoid an unnecessary excess of alkali in the lyes which are transferred from soapmaking to glycerine recovery. One way is to carry out the saponification in two stages; in the first an excess of fat is allowed so as to reduce the free alkali in the lye, and in the second the saponification is completed, using an excess of alkali, and the lye containing this excess is passed to the first stage. At one time, soapmakers used drastic treatments to minimize free fat in which the soap mass was boiled for long periods with a substantial excess of caustic soda, using either open, or closed, steam coils. This process certainly produced soaps with excellent keeping properties. The author once had a piece of soap which was perfectly sweet when over sixty years old; but much soap fully adequate for all normal needs, at least when supported by preservatives, has been made without this special treatment with alkali. Prolonged heating with caustic soda possibly breaks down some components in what is normally regarded as the unsaponifiable portion of the fat.

- Like other reactions, saponification is accelerated by increases in temperature; and some processes use temperatures which can only be achieved in an aqueous system by the use of elevated pressures.
- When some soap is present, too much electrolyte will salt, or grain, out the soap and greatly slow down the reaction. Too little will cause 'bunching', or thickening, so that it is important to ensure that the electrolyte content of the saponifying mass is within the correct range at all times. The actual requirement depends upon the composition of the fatty acids in the vessel, that is upon the fat blend. Caustic soda is a powerful electrolyte, but allowance must be made for the fact that it is continuously being removed by the reaction.

It was mentioned earlier that mono- and di-glycerides are not found in partly saponified masses from the fully-boiled process (although Davidsohn, Better and Davidsohn (1953) reports some found in soaps being made by the cold process).

4.2.3 The '66%' rule of Wigner; and related matters

Various authors have reported studies to quantify soapmaking operations; see Davidsohn, Better and Davidsohn (1953). In this field, work by J. H. Wigner, carried out at the Gossage factory at Widness in England during the 1930s and reported in his book *Soap Manufacture: The Chemical Processes* (E. and F. N. Spon Limited, 1940), is particularly interesting. This work is based upon examination of the results of large numbers of full scale boils of soap, as well as upon laboratory and pilot scale experiments, and, by kind permission of the publishers, figures and deductions from Wigner are quoted at various points in this chapter. Much of Wigner's work is based upon his, so-called, '66% rule', which says 'a soap curd may be regarded as consisting of soap hydrate containing 66% of fatty acids (called TFM in this book), and of lye identical in composition with that from which it has separated'. This does not necessarily imply the physical existence of a soap hydrate not containing any salt; and the nearest approach achieved by Wigner in removing additional lye from a curd by settling and centrifuging in the laboratory contained about 64% TFM and 0.3% NaCl. Wigner deduced his rule as an empirical observation from a study of large numbers of boils, which showed that when a pan contents has been grained and settled,

$$\frac{\text{Percentage salt in soap curd}}{\text{Percentage salt in lye}} = 1 - K_s \frac{\text{Percentage TFM in curd}}{100}$$

K_s is nearly constant for almost all soaps, with a range of about 1.50 – 1.53, average 1.515. If electrolytes other than salt, particularly caustic soda, are present in significant quantities, equivalent salting out figures must be added to the salt concentrations. By substituting salt in soap curd = 0, it follows that for a salt-free curd, the percentage TFM in curd equals $100/K_s = 100/1.515 = 66\%$. Table 4.1 gives some analyatical data for full scale boils of different types of soap presented by Wigner in support of his rule.

Table 4.1 – Data in support of the 66% rule for soap curds from Wigner (1940)

TFM% in soap	NaCl % in lye	NaCl% in soap found	NaCl% in soap calculated	K_s†	TFM% in soap hydrate
53.7	9.5	1.76	1.76	1.518	66.0
52.3	9.1	1.87	1.89	1.520	65.9
54.6	9.0	1.44	1.56	1.538	65.0
57.8	8.8	1.10	1.10	1.514	66.0
56.0	8.7	1.33	1.32	1.513	66.1
53.5	9.1	1.69	1.71	1.522	65.7
57.1	8.2	1.17	1.10	1.501	66.6
57.8	9.4	1.17	1.17	1.516	66.0
58.0	9.1	1.12	1.10	1.512	66.2
56.7	9.2	1.34	1.30	1.506	66.3
54.7	9.6	1.68	1.65	1.508	66.3

†The column for K_s has been added by the present author.

Wigner also found that the 66% rule could be applied to fitted pans using 'free solution' in place of lye. Free solution is defined as water which is not combined in the 66% TFM hydrate. For the neat soap which separates when a fitted pan is allowed to settle, Wigner says that 'neat soap is a soap hydrate of a composition which varies according to the concentration of salt in the free solution, always containing more water than the hypothetical 66% hydrate, and containing a small amount of salt which increases with the increase in the degree of hydration'. Figures in support are given in Table 4.2. It will be seen that as the concentration of electrolyte in free solution falls and the fit becomes 'finer', or 'closer':

Table 4.2 – Data in support of the 66% rule in fitted soaps from Wigner (1940)

Sample	NaCl in free solution %	TFM%	Neat soap NaCl% on sample	NaCl% at 63% TFM†	Nigre TFM%
1	11.0	63.2	0.47	0.47	<1
2	10.9	63.35	0.435	0.43	1.2
3	10.7	63.3	0.425	0.42	3.5
4	10.3	63.2	0.435	0.43	9.6
5	10.25	63.2	0.44	0.44	8.6
6	9.3	62.8	0.445	0.45	21.0
7	8.25	62.35	0.45	0.45	32.6
8	7.85	61.85	0.48	0.49	36.2
9	7.55	61.3	0.53	0.54	37.2
10	7.25	60.85	0.55	0.57	39.3

†The figures for salt content of neat soap at 63.0% TFM have been calculated by the present author.

- the TFM content of the nigre increases very substantially.
- the TFM content of the neat soap falls a little.
- the salt content of the neat soap adjusted to the figures at 63% TFM, or any other fixed level such as 79% for a dried toilet soap, is nearly constant

for samples 2 – 7, but tends to rise as the fit becomes finer. It is also a little higher for sample 1 separated from a nigre lye rather than a nigre.

It is also found that soap given a coarse, or open, fit settles much more rapidly than when the fit is close, or fine; this is due to the greater difference in density between the phases, as well as to changes in viscosity.

These results conflict slightly with what might be expected from the phase diagram as usually drawn, for example in Fig. 4.1, in that the electrolyte content of the neat soap, as well as its TFM content, should fall as the fit becomes finer. As shown in Table 4.3, this latter trend is found in results reported by Palmqvist (1961) for centrifugal separation of neat soap and nigre in a De Laval continuous soapmaking plant. Phase equilibrium is not necessarily reached with either gravity or centrifugal separation; and degree of separation can also affect the compositions obtained. Some practical implications are:

- rapid separation and the low TFM content of the nigre/nigre lye, make open fits attractive whenever they can be used.
- household soaps made by the traditional frame process require fine fits with long settling times. On the other hand, household soaps made by the modern spray chill/vacuum plodding method usually benefit from an open fit and relatively incomplete separation, which leaves a high salt content in the soap. This type of soap is also generally acceptable for soap powder manufacture.
- traditionally, bases for toilet soaps and soap flakes were given rather more open fits than those for framed household soaps. This was understood to give lower electrolyte contents, as stated by Wigner.

Table 4.3 – Centrifugal separation of fitted soap in a De Laval continuous soapmaking plant

Recorder		Fitting column (that is soap before separation)				From separator Neat soap	Nigre†	
Index	Distance from minimum	Composition by volume: Neat soap %	Nigre %	Lye %	NaOH + NaCl %	NaOH + NaCl %	NaOH + NaCl %	TFM %
37	0	84	16	–	1.01	–	–	–
41	4	85	15	–	1.12	0.36	4.55	31.6
43	6	85	15	–	1.15	0.43	5.11	29.9
45	8	86	14	–	1.17	0.47	5.54	28.0
50	13	86	14	–	1.23	0.49	6.06	–
53	16	87	13	–	1.25	0.53	6.29	21.2
57	20	87.5	12	0.5	1.29	–	6.70	18.1
62	25	88.5	10	1.5	1.33	0.57	7.23	14.5

†including lye when present.

The De Laval process and equipment will be discussed later. The figures for the recorder indicate the setting of the controller for addition of electrolyte solution; and, as can be seen from the analyses, the higher the index the larger the addition of electrolyte and the more open the fit.

- modern continuous soapmaking plants using centrifuges for the final separation tend to use quite open fits, often with the separation of a very thin nigre, or nigre lye.

Some of these points are elaborated in subsequent sections. Wigner himself mentions that there are certain exceptions to the 66% rule; for example, when a groundnut oil soap is boiled for long periods with a lye having a high salt content in order to produce a curd soap which is as hard as is possible from GNO. The rule is, however, commonly accepted as applicable in all normal soapmaking operations; and some applications are mentioned later.

4.3 PRODUCTION OF SOAP BASE BY THE TRADITIONAL METHOD IN SINGLE VESSELS

4.3.1 Plant

Pans, sometimes called kettles (particularly if they are covered), for the manufacture of soap base can have a range of shapes and sizes. Cylindrical pans are common, but one major company standardized on 14 ft (4.27 m) cube vessels which fit conveniently together in groups of four, with alleyways between each group. Whatever the shape, they should have a slightly dished base so that lyes (or nigres) can be run-off completely from the bottom discharge valve, or cock. Webb (1927) and Davidsohn, Better and Davidsohn (1953) discuss optimum pan dimensions. Points are that the heat loss should be as low as possible and the settling distance not too great; and the cube is probably a good compromise. Neat soap and nigre do not separate properly at temperatures below about 70 – 75°C, so that pans should not be too small and should be lagged for this reason, as well as to minimize steam consumption. Very small pans used for experimental boils should usually be steam jacketed, so that a little steam can be used to prevent unduly rapid cooling while neat soap and nigre are separating. Pan covers are also desirable to reduce heat loss from the soap surface, and to reduce risk of contamination with dirt from the atmosphere; but they are by no means always fitted. Pans are usually made of mild steel, but it has been found useful to line the upper parts with stainless steel, or other corrosion-resistant material, to reduce contamination of the soap with iron.

The usual fittings, in addition to a bottom discharge facility are:

- an open steam coil to provide both heat and agitation. Closed steam coils are no longer necessary, or usual.
- a skimmer pipe, which can be raised or lowered, and through which the neat soap layer can be withdrawn when the settling is complete and the pan is 'cleansed'; that is the neat soap removed.
- pipes arranged to discharge into the pan above its contents for fat, caustic soda solution, water, and other materials. Salt can be shovelled over the side of the pan, but it is now commonly added in the form of saturated brine.

  Particularly in large panrooms making a variety of soap bases, a fairly comprehensive pipe system to permit transfer from pan to pan, and to other departments, is desirable.

4.3.2 Saponification

Traditionally, saponification is carried out by pumping the fat into the pan whilst the soapboiler adjusts the additions of caustic soda, salt, water and steam, so as to maintain the conditions necessary for steady reaction. This is a skilled operation; too much electrolyte can cause the soap mass to grain, and too little to bunch or thicken. If the pan contents are inadvertently grained with too much caustic soda, the reaction slows and great care is necessary when conditions are adjusted to re-establish it; otherwise the heat of reaction may cause the pan contents to boil over.

Nigre and other soap from the previous boil is often left in the pan to provide the soap needed to aid saponification; although it may be found desirable to remove some, or all, the nigre from time to time, for transfer to a lower quality base, or for it to be cleaned by graining, or fitting.

Rosin, which contains no glycerol, is preferably added to the rest of the soap after as much as possible of the glycerol has been removed. Solid rosin can be tipped into the pan, but, particularly when substantial proportions are to be included in the fat blend, it is better to partially saponify it separately. Rosin contains complex acids related to the terpenes, such as abietic and pimaric, which react readily with sodium carbonate solution. When rosin was widely used in soaps, it was common practice to saponify it partially with sodium carbonate in a separate vessel, grain, and discard the dirty lye (which rises to the top instead of sinking to the bottom as with ordinary soap lyes). The partially saponified rosin soap was added to the rest of the soap in the pan at a convenient stage, preferably after most of the glycerol had been removed; and some caustic soda was then needed to complete the saponification.

Saponification in a pan can be made easier by the use of a saponification jet, which can be made quite simply. One design which has been used widely is shown in Fig. 4.3. A slightly more complex design, which is better, is shown diagrammatically in Fig. 4.7c. The partially saponified emulsion of fat, soap and caustic soda solution is discharged through a flexible pipe, which can often be arranged to serve more than one pan; the metal hose being securely clamped to the pan in use. When the jet is used as an aid to saponification in a pan it is only necessary to regulate the flow of caustic soda solution roughly in proportion to the flow of fat. Steam is used in the jet as required, depending upon the temperature of the fat and the reaction rate. The mixture should flow into the pan as an emulsion in which saponification is taking place vigorously. The reaction is completed by boiling in the pan, with additions of caustic soda and salt as necessary.

4.3.3 Graining or washing

After saponification has been completed, the pan content is grained with salt. When in the grained state, clear lye can be seen between the curds as the pan content boils gently, or in a sample lifted on a paddle. The pan is allowed to settle and the saponification lye is run off. The soap is 'closed' by addition of water and grained again with salt. During this operation sufficient water must be added to produce a wash lye of the desired size, say 0.3 tonnes of lye per

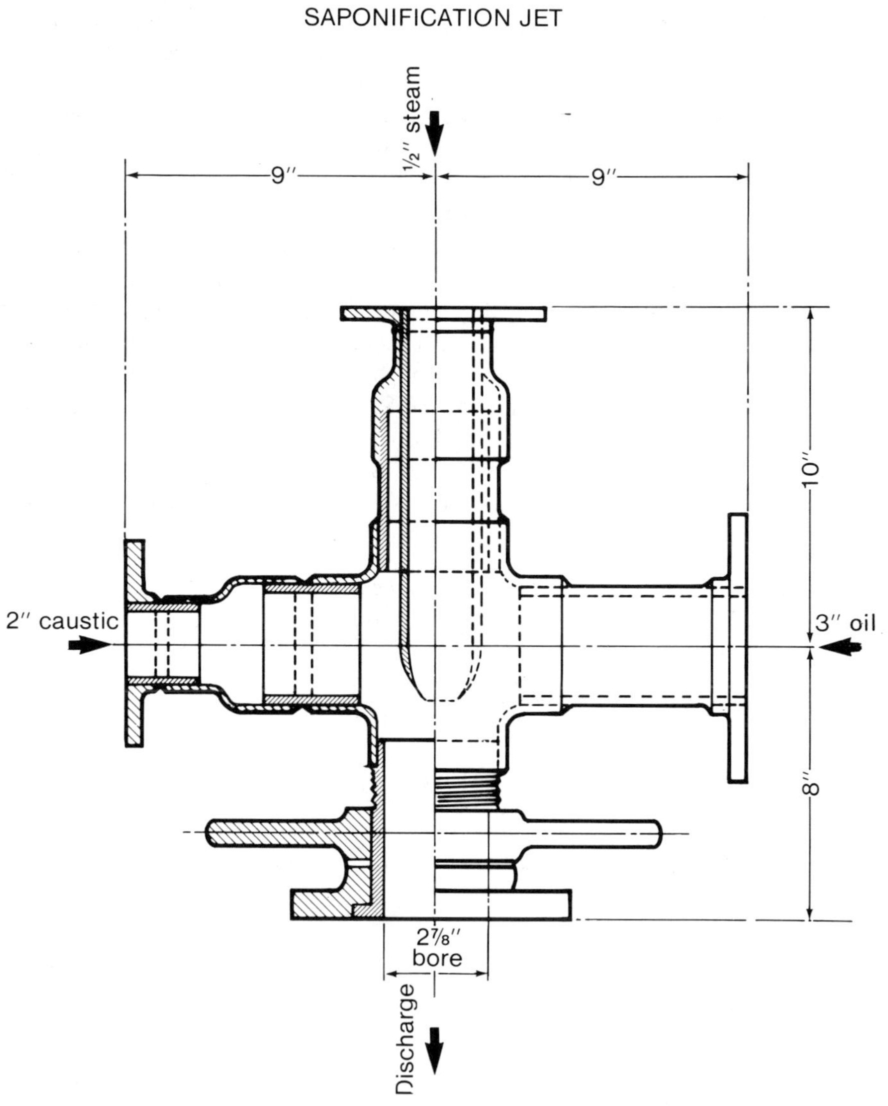

Fig. 4.3 – Saponification jet.

tonne of soap curd; care being taken to avoid the formation of water lumps. The process is repeated as many times as is worthwhile, it being appreciated that the glycerol content of each wash is lower than that of the previous wash, so that, using this process, it is uneconomical to wash to reduce the glycerol in soap to a very low level. The capacity of the glycerine recovery plant may also be a limiting factor.

Dirt is removed from the soap mass in the washes, but the primary objective is to separate the glycerol. As indicated in the previous paragraph, there is an economic balance between the yield (percentage of the glycerol in the fats which is recovered) and the average concentration of glycerol in the lye produced, because the cost of lye treatment and evaporation is related to the lye concentration; that is to the quantity of lye which must be purified and the quantity of water which must be evaporated to produce a tonne of crude glycerine.

It is found that the glycerol is distributed between curd and lye approximately in proportion to the total water in each layer†; and K_G is defined as concentration of glycerol in curd/concentration of glycerol in lye. The lower K_G the more effective the wash.

Alternatively, it can be said that the drier the curd, that is the higher its TFM%, and the less enmeshed lye it retains, the better. This is analogous to the improved removal of dirty detergent solution from domestic washing which results if the articles are wrung, or spun, more effectively between rinses. This important point can be illustrated by some simple calculations. Assume that the lye contains 11.5% salt and 88.5% water plus glycerol. Consider first a very well settled curd at 60% TFM. Its anhydrous soap content is about 60.0 × 1.085 = 65.1% and, if it contains 1% salt (substantially in accord with the 66% rule), its content of water plus glycerol is 100.0 − 65.1 − 1.0 = 33.9%. As water and glycerol occur in constant proportion in the layers,

$$K_G = \frac{\text{concentration of water in curd}}{\text{concentration of water in lye}}$$

$$= \frac{\text{concentration of glycerol plus water in curd}}{\text{concentration of glycerol plus water in lye}}$$

$$= 33.9/88.5 = 0.383.$$

Less well settled curds are equivalent to mixtures of the 60% TFM curd and additional lye. For a mixture of 1 part 60% TFM curd and 0.05 parts lye:

	Soap	Salt	Water	K_G
from 1 part 60% curd	0.651	0.010	0.339	0.383
from 0.05 parts lye	—	0.006	0.044	—
in 1.05 parts new curd	0.651	0.016	0.383	
in 1 part new curd	0.620	0.015	0.365	$\frac{0.365}{0.885} = 0.412$
	= 57.1% TFM			

Similar calculations and graphical interpolation provide the data for Table 4.4.

A carefully grained and well settled pan should result in a curd at about 58% TFM, giving a K_G of about 0.4, but it is easy to produce less dry curds. It is particularly important to use the minimum concentration of electrolyte which

†See also Govan (1950) where it is stated that the ratio of the concentrations of glycerol in the water in the two layers can vary with the type of grain used. This would alter some of the figures quoted from Wigner, but not the general conclusions.

Table 4.4 – Variation in K_G with dryness of soap curd

TMF% in curd	Soap %	Salt %	Water + glycerol %	Partition coefficient K_G
60	65.1	1.0	33.9	0.383
58	62.9	1.3	35.8	0.405
56	60.7	1.7	37.6	0.425
54	58.5	2.0	39.5	0.447
52	56.3	2.3	41.4	0.468
50	54.2	2.6	43.2	0.488

is sufficient to produce lye substantially free from soap, as coarse, overgrained, curds occlude more lye than those which are only just grained. Wigner quotes the following figures for the graining of a 'washer' type soap.

NaCl in free solution %	TFM%
10.5	4.15 (a thin nigre)
10.84	0.18
11.4	0.14
11.7	0.17
12.0	0.17
12.5	0.12

It will be seen that nothing is gained by using more salt than about 1.0% over that needed for a very rough fit giving a thin nigre. The actual concentration of salt required depends upon the fat blend. That used in Wigner's 'washer' soap is not quoted, but it may well have been something like 30% hard fat, 30% lauric oil, 30% soft oil and 10% rosin. Other factors equal, the quantity of lye needed to remove a given percentage of glycerol from the soap varies with the number of washes used. Wigner calculates the figures given in Table 4.5.

Table 4.5 – Quantity of lye needed to obtain different percentage recoveries of glycerol with various numbers of washes (Wigner)

Number of washes	Parts of lye per part of curd for yield of	
	80%	85%
1	2.47	3.92
2	1.53	2.12
3	1.31	1.83
4	1.24	1.69
5	1.18	1.60
6	1.14	1.54
7	1.12	1.48

As might be expected:

- the quantity of lye which has to be produced with the traditional system of washing for a given percentage recovery is reduced very substantially by increasing the number of washes from one to three or four; but the benefit which accrues from more washes becomes increasingly small (also very large individual washes might well be impractical because of pan size).
- with this system of washing yields of glycerol above about 80% become increasingly difficult to achieve.

Settling time is also very important, so that for a given total quantity of lye produced, it may be counter-productive to increase the number of washes further if this leads to an inadequate time to settle each wash.

A useful, but very simple arithmetical calculation shows that if each wash removes, say, 0.4 parts of the glycerol present in a curd when it is washed, so that 0.6 parts are left in the washed curd, n washes will leave 0.6^n in the final washed curd.

If the weakest lye which it is economical to process contains, say, 2% of glycerol, and K_G equal to 0.4 is achieved, the curd from which the last lye came contains $2 \times 0.4 = 0.8\%$ glycerol. It follows that soaps made from fat blends with a high glycerol content and those from fats with a low content, should both be washed down to 0.8% glycerol; so that more lye should be produced from the glycerol rich soaps.

When, as is usual, salt recovered in the process of glycerine lye evaporation is re-used in soapmaking its glycerol content has a considerable adverse effect upon the efficiency of glycerol recovery, or on the quantity of water which has to be evaporated to achieve a given recovery, particularly when the level of recovery is high. With regard to glycerol recovery, it is best to use in the final washes whatever new salt is needed to make up for losses. From the angle of soap quality the new salt is better used in the manufacture of the highest quality bases; but, for convenience, new and recovered salt are often used indiscriminately.

4.3.4 Fitting

The washed curd should be well settled, preferably overnight, before it is fitted, because the quantity of water in the fitted soap prior to settling affects the yield of neat soap, as can be shown by a simple materials balance. Let the total mass of material at the completion of the fit equal 1 and the mass of nigre produced equal x, so that the mass of neat soap equals $(1 - x)$. Then if

a = TFM% of the total pan contents
b = TFM% of the nigre, and
c = TFM% of the neat soap,

a TFM balance can be written

$$1 \times a/100 = bx/100 + c(1 - x)/100, \text{ or } x = \frac{c-a}{c-b}$$

In words this can be written, nigre produced (as a fraction of the whole mass)

$$= \frac{\text{TFM\% of neat soap} - \text{TFM\% of pan contents}}{\text{TFM\% of neat soap} - \text{TFM\% of the nigre}}$$

Even more significant than the dryness of the curd, which affects a, is the nature of the fit used, which governs the TFM content of the nigre, b, as discussed earlier; but this has to be decided in relation to quality requirements, including removal of dirt. Table 4.6, from Wigner, shows examples of the influence of different pan drynesses with two types of fit. The significance of type of fit is illustrated further in Table 4.7, again from Wigner with some additional columns from Davidsohn, Better and Davidsohn (1953). Recalculation suggests minor errors in some of the figures, but these do not affect the conclusions. The curd is assumed to contain 58% TFM and 1.45% salt, equivalent to 12% in lye. Columns 1 – 3 are taken from smoothed curves drawn from the data in Table 4.2. Column 4, the TFM% in the total mass after fitting, assumes that only water is added to the curd; and is calculated for the first line by saying that free solution (occluded lye) in 100 parts of curd is (1.45 x 100)/12 = 12.08. This is diluted to (12.08 x 12)/9.65 = 15.02 to reduce its salt concentration from 12.0% to 9.65%. Hence water added = 15.02 – 12.08 = 2.94; and the TFM% in fitted soap is

Table 4.6 – Influence of dryness of pan contents and type of fit on loss of usable soap to nigre

Average TFM% in pan	Nigre, percentage of pan contents		Percentage of TFM in nigre	
	Coarse fit yielding neat soap at 63% TFM and nigre at 18.5% TFM	Fine fit yielding neat soap at 61% TFM and nigre at 38.5% TFM	Coarse fit	Fine fit
56	15.7	22.2	5.18	15.3
54	20.3	31.2	6.98	22.2
52	24.6	35.6	8.80	26.4
50	29.3	49.0	10.84	37.7
48	33.8	57.8	13.02	46.2

Table 4.7 – Additional information concerning the fitting operation (the curd is assumed to contain 58% TFM and 1.45 salt, equivalent to 12% in lye)

Line	Neat soap TFM% (1)	Nigre TFM% (2)	NaCl % in free solution (3)	TFM % on total mass (4)	% Nigre on total mass (5)	% of total TFM in nigre (6)	NaCl% in total mass (7)	% Free solution in total mass (8)
a	63.0	18.5	9.65	56.35	15.2	4.8	1.41	15.0
b	62.5	28.5	8.75	55.52	20.6	9.9	1.39	16.6
c	62.0	33.5	8.15	54.88	25.2	14.6	1.37	17.8
d	61.5	36.5	7.70	54.34	28.8	18.1	1.35	18.85
e	61.0	38.5	7.35	53.89	32.0	21.3	1.34	19.75
f	60.5	40.3	7.10	53.65	34.4	24.0	1.33	20.4
g	60.0	41.5	6.85	53.18	36.8	26.4	1.33	21.2

As classified by Davidsohn, Better and Davidsohn, lines *a* – *c* represent 'coarse fits', *d* – *f* represent 'medium fits', and line *g* represents a 'fine fit'.

(58.0 x 100)/102.94 = 56.35%. Column 5 is obtained by application of the materials balance equation. Percentage nigre in total mass equals (63.0 − 56.35)/(63.0 − 18.5) x 100 = 14.9% (compare 15.2% reported, as shown in the table). Column 6 equals (column 2 x column 5)/column 4 = (18.5 x 15.2)/56.35 = 5.0% (4.8% reported). Column 7 equals salt in curd recalculated as a percentage of the new total mass = 1.45 x 100/102.94 = 1.41%. Column 8 equals new free solution as a percentage of new total mass = (15.02 x 100)/102.94 = 14.6% (15.0% reported).

The composition of the pan contents needed for any selected type of fit can be obtained either by adding water when the soap mass has too high an electrolyte content, or by adding electrolyte solution when it is too low. In addition, it is necessary to adjust the caustic soda content so that, after the electrolyte has been partitioned as indicated in Table 4.3, the neat soap will have the required alkalinity. Because of the large 'loss' (to reprocessing) of soap when a rather wet curd is given a fine fit, it is often advisable to use a rough fit prior to the finishing fine fit. This removes water and electrolyte, together with a relatively small mass of soap because of the low TFM content of the rough fit nigre. Hence, the quantity of the nigre from the final fit is small and its removal does not involve the rejection of very much soap despite its high TFM.

Fitting, as carried out in the traditional way, requires skill and experience; and is best learned by personal tuition by an experience soapboiler. The soap mass is boiled extensively; and the end point is judged by the look of the boiling soap and by the way it leaves a heated trowel. The soapboiler's trowel test is carried out using a bricklayer's trowel. The trowel is heated by immersion in the boiling soap, after which the adhering soap is scraped off. The trowel is then again immersed, drawn out edgewise and tilted to about 60° to see how the soap runs off. A soap which is too close for a typical medium/fine fit travels slowly, is apt to stick to the trowel, and the trailing edge tends to leave streaks of soap adhering to the blade. A soap which is too open slides off the trowel fairly rapidly and breaks into a number of small pieces. In the ideal condition the soap film should be transparent; and should slide slowly and steadily from the trowel, leaving the surface of the trowel clean and bright. It should break up into two, or three, large broad flakes. The trailing edge of the moving film should thin out gradually to an almost imperceptible transparent wafer of soap which may extend to $\frac{1}{4}$ inch behind the main body of soap. Readers who require more information concerning details of practical soap boiling by the traditional method are referred to the older books on soapmaking, particularly Webb (1927).

The actual yield of usable neat soap is less than that which might be expected from the loss of nigre, because crusts of solid soap form on the soap surface and on the sides of a pan which has cooled during settling; and so cannot be cleansed with the liquid neat soap.

As mentioned earlier, soap given an open fit settles relatively rapidly, and neat soap suitable for soap powders, and for household soaps made using the modern chilling and extrustion processes, can commonly be cleansed after 1 – 2 nights settling, instead of the 3 – 4 nights needed after a medium/fine fit.

4.3.5 Wigner system of quantitative soapboiling

An important feature of the book by Wigner (1940) is the disclosure of the quantitative method of soapboiling which he had developed. It has been seen that data can be collected from experimental boils for the phase equilibria for the various fat blends used in a factory. If the composition of the pan contents is known, it is then possible to calculate what additions should be made to achieve a desired wash, or fit. A problem in the application of this idea is that steam condenses as the soap is boiled with open steam, so that it is difficult to know how much water is present. Also, the density of the pan contents is influenced by the bubbles of steam and total mass cannot be determined by staffing to measure volume. Wigner overcame these difficulties by the use of a pneumercator type of gauge shown diagrammatically in Fig. 4.4. When air has bubbled gently through the soap and the compressed air supply is shut off, the column in the manometer balances the soap above the air orifice; and the manometer can be calibrated directly in terms of the mass of material in the pan, regardless of its density. Allowance must be made for the material in the pan dish and Wigner describes some refinements. Water content is determined by difference between the total mass and the known additions. Wigner's book explains in detail how the quantitative method of control of the traditional process can be achieved, together with the use of nomographs to minimize calculations. Possible extended use of this system has been overtaken by other developments to be described, which have made the traditional process largely obsolete, although it is still used by many small, and/or, less technically developed, companies.

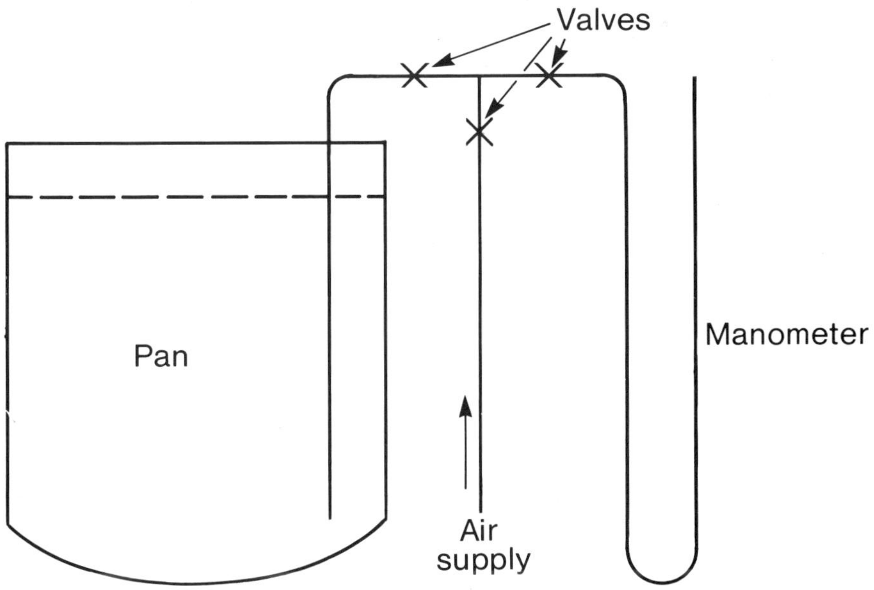

Fig. 4.4 – Pneumercator gauge for measurement of pan contents.

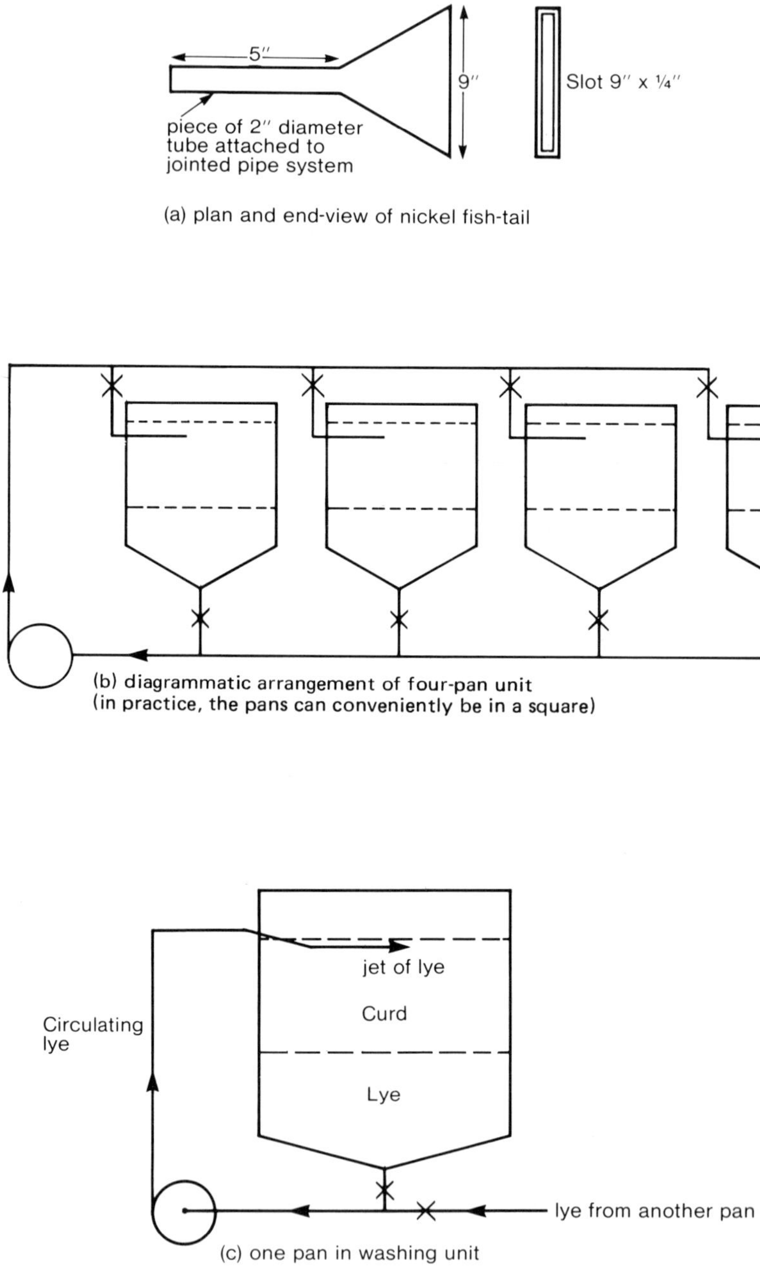

Fig. 4.5 – Lye diffuser system of washing soap in pans.

4.4 DEVELOPMENTS OF THE TRADITIONAL PROCESS INCLUDING THE USE OF GROUPS OF VESSELS

One important weakness of the traditional process is the low glycerol content of the later washes. This results in the production of large quantities of weak lyes; and seriously limits the glycerol recovery which is economical, or even practical. Some weak liquors can be used to dilute the caustic soda, or to dissolve salt, but it has long been appreciated that counter current operation of the washing system is needed, as used in other extraction processes. Govan (1946), quoted extensively by Davidsohn, Better and Davidsohn (1953), describes, and assesses quantitatively, systems of counter-current washing in comparison with a sequence of simple washes. Counter-current washing is complicated by the addition of water to close the soap and then of salt, or brine, to grain it in each washing stage; and a major advance, some forty years ago, was the discovery that soap can be washed quite satisfactorily by just mixing curd and lye, and allowing the mix to settle. All modern washing processes wash the soap in the grained condition in this way.

So far as is known, this principle was first applied in pans. Steam jets can be used to mix and transfer the pan contents; or a mechanical system can be used. Fig. 4.5 shows diagrammatically a system which was used satisfactorily for a number of years. Fig. 4.5(a) shows a fish-tail shaped nozzle, sometimes called a lye diffuser, through which lye is pumped into the upper part of the layer of soap curd in a pan. The nozzle is installed at the end of a jointed pipe system arranged so that the discharge is horizontal, which is also able to be adjusted to the level required just below the soap surface. Fig. 4.5(b) shows a group of four pans (diagrammatically in line, but, in practice, conveniently arranged in a square), with the essential pipe connections; and Fig. 4.5(c) shows the operation of one pan. Lye is transferred from one pan to the next by the pump through the lye diffuser; and is then circulated from the lye layer at the base of the pan into the curd layer. The action of the jet of liquid from the lye diffuser causes circulation and mixing of lye with curd, so that near equilibrium is achieved after a period of circulation. The pan is then allowed to settle before its lye is transferred and the curd is ready for another wash, or to be fitted. With an eight-pan unit (two sets of four as just described), the complete process, including several nights settling after fitting, is possible, but this is only suitable for a large production of a single type, or similar types, of base. For smaller productions a set of four pans fitted with lye diffusers can be used as a washing unit, and the washed curd transferred to other pans to be fitted and settled.

4.5 AN ADDITIONAL PRINCIPLE

The influence of quantity of water added, and the number of washing stages used, on the efficiency of glycerol recovery have been considered. The way in which the water is added is also important. The principle is simple, it being that fresh water should, as far as possible, be added at the last washing stage. This is because water added at the saponification stage is used only once for the removal of glycerol from the soap; whereas water added at the end of a multi-stage process is used several times. It is also to be noted that although processes can be divided into saponification, washing, and fitting stages, glycerol is transferred from the soap in each stage. Various liquors containing some glycerol, such as those from nigres, or from the concentration of soapstock, often need to be introduced into the soapmaking process. The principle for the most effective return of such liquors is that they should, as far as possible, be introduced into the soapmaking system at the point where the concentration of glycerol in the system is the same as that in the liquor. This discussion explains the point made earlier that it is advantageous as regards recovery to concentrate a soapstock, so that a portion containing the fatty matter, but relatively little water, can be added at the saponification stage, and a liquor containing a small percentage of glycerol can be introduced at a later stage.

The saponification of 1 tonne of a typical 80% tallow 20% coconut oil fat blend can be shown by the methods used in section 3.8 to require about 0.118 tonnes Na_2O, allowing 2% excess. This is equivalent to 0.317 tonnes caustic soda liquor at 102°Tw, containing 48% NaOH or 37.2% Na_2O. The 1.317 tonnes reaction mixture which is produced from 1 tonne of fat plus 0.317 tonnes 102°Tw caustic soda contains about 0.95 tonnes fatty matter, or 72% TFM†. When mixed with liquors and grained, this will produce lye plus a curd containing, say, 58% TFM. To convert the 72% TFM soap to 58% TFM soap the total mass has to be increased to 0.95/0.58 = 1.64 tonnes; that is 1.64 − 1.32 = 0.32 tonnes of material must be added. This could be, and sometimes is, salt plus water; but water introduced in this way is added at the wrong end of the process and is used inefficiently. It is better to use clean spent lye; a procedure sometimes known as use of the 'lye absorption principle'. It may seem peculiar to return to the fresh soap mass glycerol which has already been separated, but the procedure is analogous to what is done in a stage of a counter-current washing process; and it can be shown to be beneficial both by calculation and in practice. The best way to return lye to the saponfication may vary with the soapmaking process in use and various arrangements are discussed later. Assume first that the lye is used to dilute the approximately 100°Tw caustic soda liquor. If 0.317 tonnes of caustic soda liquor containing $0.48 \times 0.317 = 0.152$ tonnes of NaOH and 0.165 tonnes of water is diluted with 0.32 tonnes of spent lye containing, say, 11% NaCl, 11% glycerol and 78% water, the mixture will contain:

NaOH	0.152	+ —	= 0.152	tonnes
NaCl	—	+ 0.035	= 0.035	tonnes
Glycerol	—	+ 0.035	= 0.035	tonnes
Water	0.165	+ 0.250	= 0.415	tonnes
			0.637	tonnes,

so that the diluted caustic soda solution contains 23.9% NaOH and 5.5% NaCl. In practice, it may be advisable to add rather less spent lye, to give a caustic

†Any condensation of steam during saponification is disregarded in this calculation.

soda solution for use in saponificaton containing about 30% NaOH and 4% NaCl. A small percentage of salt, of the order of 4%, is useful in the caustic soda solution to ensure that an adequate concentration of electrolyte is maintained. If sufficient final lye is re-cycled to the 48% NaOH liquor to give a saponified mass containing about 58% TFM and the lye from the next stage is added in the usual way, the saponification lye will be about the same size as the lye separated in the other washing stages. In some processes, the objective of not using fresh water to dilute the 48% caustic soda is achieved by adding lye from the second stage (that is the first wash if the saponification wash is not counted) instead of final lye. In this case, the volume of the saponification lye is less than that from subsequent washes by the volume needed to reduce the soap mass from 72% to 58% TFM.

4.6 THE COMPONENTS USED IN MODERN SOAPMAKING PROCESSES

4.6.1 Equipment for saponification

Modern continuous saponification units use either:

- devices to provide a very intimate mixture of fat and caustic soda solution, followed by completion of the reaction in a mass of soap.
- reaction in a mass of soap, which is usually recirculated, to which the reagents are added, and from which nearly completely saponified material is bled off.

Reaction in a circulating mass of soap is an application of what is called the dominant bath procedure. The soap may be circulated under atmospheric pressure, which limits the temperature to just over 100°C; or the system may be closed to that higher pressure and temperature can be used.

The *Monsavon* saponification unit, shown diagrammatically in Fig. 4.6, uses a colloid mill to bring the fat and caustic soda solution into intimate contact followed by flow through a reaction tube and mixer.

BP 605 995 to J. Crosfield and Sons Limited (part of *Unilever*) and R. V. Owen (1946) uses a saponification jet in which a stable emulsion is formed, and a baffled tank through which the soap mass flows quietly whilst saponification proceeds almost to completion. Fig. 4.7(a) shows diagrammatically the design of the jet and 4.7(b) that of the saponification tank disclosed in the patent. Fig. 4.7(c) shows a more complex, but superior, design of jet; and it is possible to use several designs of saponification tank with various numbers of baffles.

According to the patent, solutions containing up to 15% of electrolyte favour the formation of an oil in water emulsion, 15 – 20% no emulsion, and above 20% a water in oil emulsion; a water in oil emulsion is preferred. Conditions, including design of jet, temperatures and electrolyte content (including caustic soda) must be adjusted to give a good emulsion. If the fat blend is nearly neutral, some free fatty acid, or soap, can be added if thought necessary. The patent also discloses the following example: a jet with $1\frac{1}{2}$ inch (38 mm) feed pipes and a $\frac{1}{4}$-inch (6.4 mm) steam jet fed with 70 psig steam is capable of emulsifying

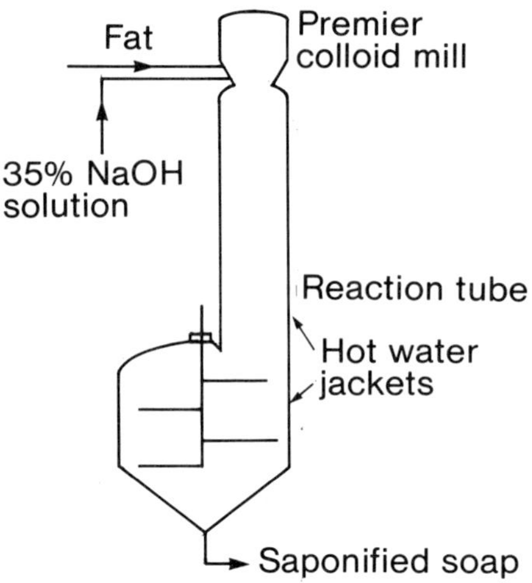

Fig. 4.6 – Monsavon saponification unit.

7 – 8 tonnes of mixed fats containing 10 – 15% FFA per hour using a caustic soda solution containing 27 – 30% NaOH and 3% NaCl. Temperatures should be arranged so that the emulsion only heats to just about 100°C when left in a small vessel to saponify. Table 4.8 from the patent gives results obtained with various fats containing a range of FFA contents. It will be noted that fats with a low FFA% required more time for self-saponification than others with a higher FFA%. To avoid risk of jet blockage the flow of fat should always be started before that of caustic soda. The principles of BP 605 995 can be combined with the idea of saponfication in a circulating mass of soap. Two possible arrangements are shown in Fig. 4.8. The soap is recirculated at a high rate and the emulsion from the jet falls, or overflows, into it. The pressure is substantially atmospheric. The jet and tank system, preferably with recirculation, provides a sound, but inexpensive, saponification unit.

Table 4.8 – Data from BP 605 995. Continuous saponification unit.

Type of fat	FFA%	Temperature of of oil, °C	Temperature of emulsion, °C	Time for self-saponfication 98–100% (mins)
Neutral palm oil	0.6	54	61	23
Palm oil	46.2	52	85	2
Palm kernal oil	6.4	52	62	5
Groundnut oil	17.4	54	74	4
Mixed oils	1.7	46	59	3.5
Mixed oils	20.0	43	70	1

The final temperature of the self saponified soap mass is approximately 100°C

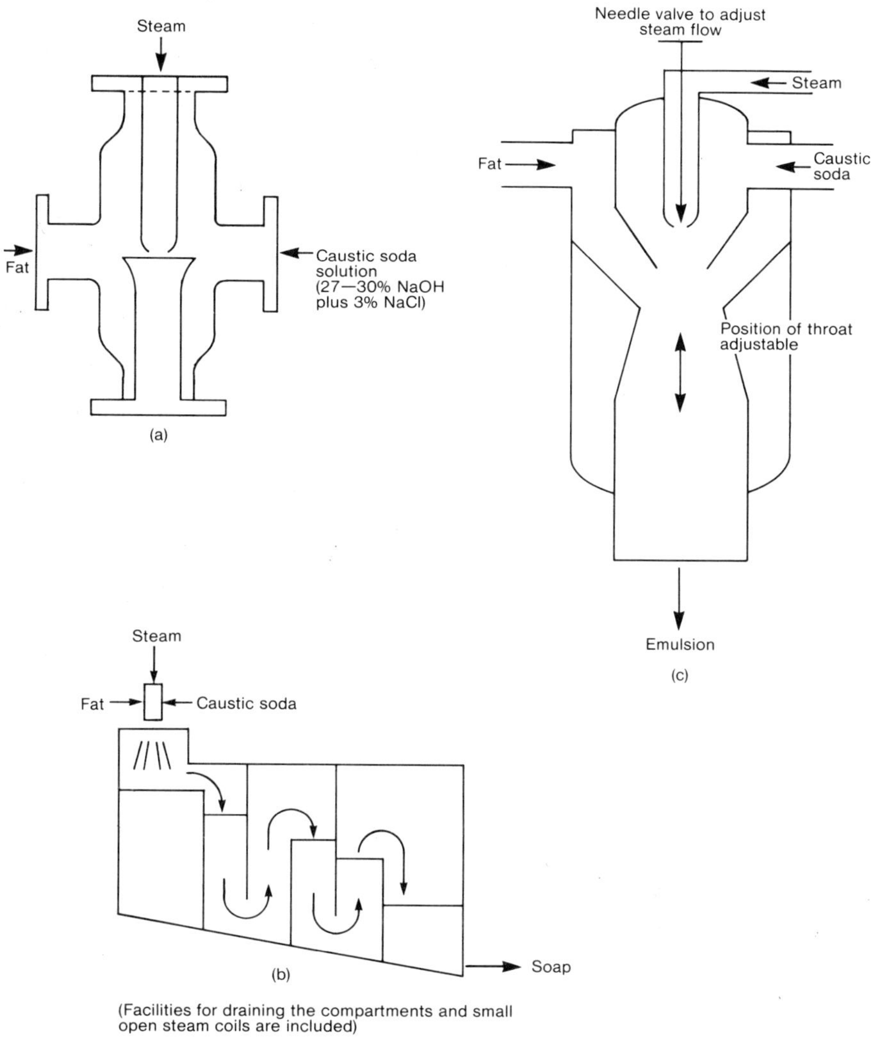

Fig. 4.7 – Jet and tank system of saponification.

As part of the *De Laval Centripure* process the saponification equipment shown diagrammatically in Fig. 4.9 has been developed. This unit, and other parts of the process considered later, are described in articles by Palmqvist (1959, 1960a, 1960b and 1961). An article by Palmqvist and Harris (1964) describes the Centripure plant installed in the Gerrard factory, Nottingham, England; this includes some developments not mentioned in the earlier articles by Palmqvist. Another article, Platt (1970), describes a Centripure plant installed in the Unilever factory at Port Sunlight, England. As shown in Fig. 4.9, the saponification unit comprises a lower tower divided into two sections *a* and *b*, and an upper tower *c*. Section *b* is further divided by a baffle which restricts flow from its lower to its upper part. Pump 1 circulates material from the

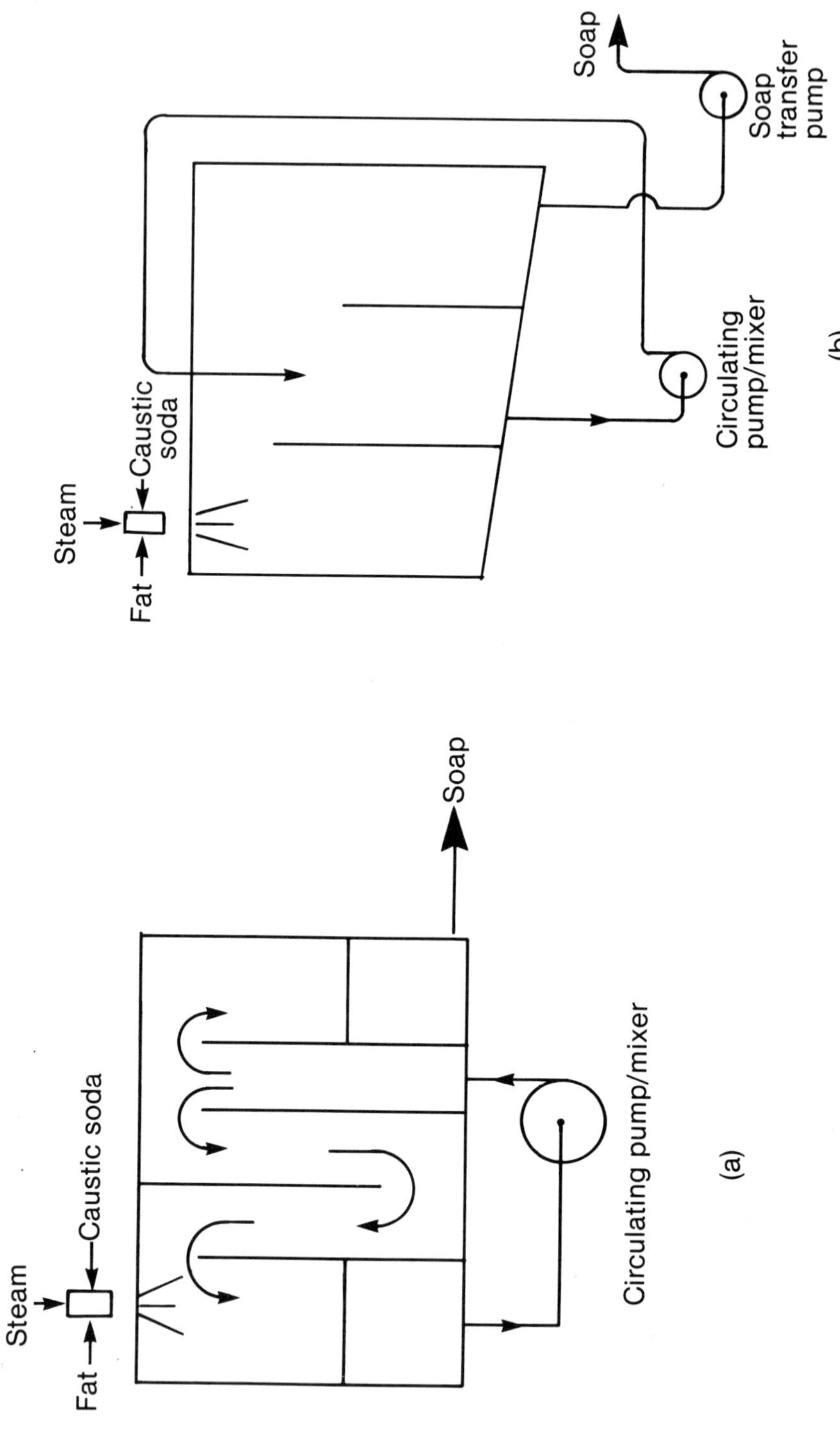

Fig. 4.8 – Jet saponification using tank with recirculation.

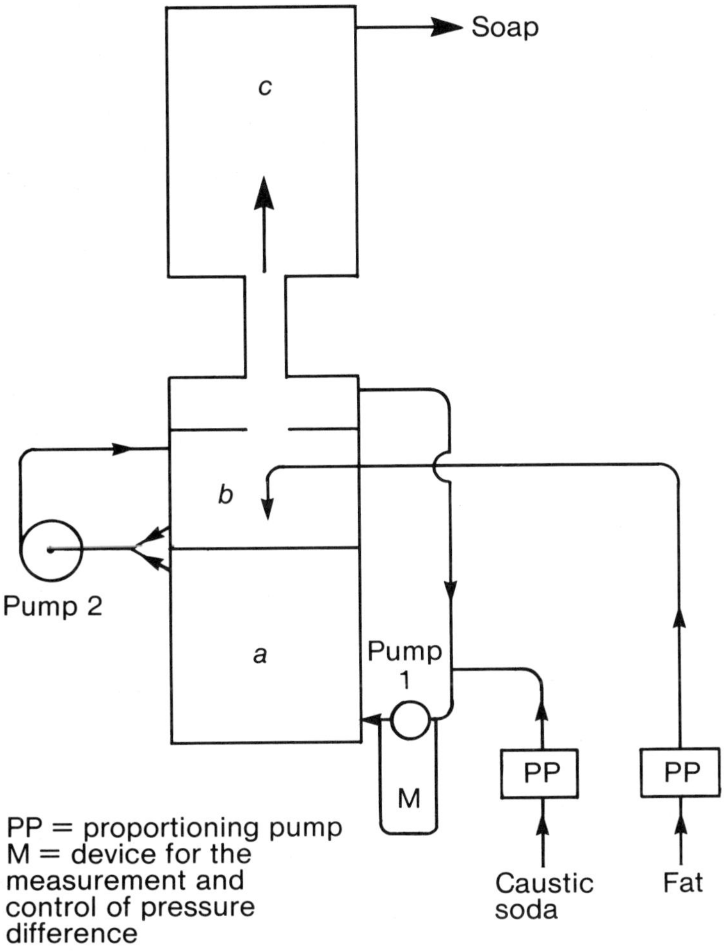

Fig. 4.9 – De Laval saponification unit.

top of section *b* back to the bottom of section *a*; and the caustic soda solution is fed into its circulating line. Pump 2, which has several times the capacity of pump 1, draws material from both section *a* and section *b*; and returns it to near the top of the lower part of section *b*. The fat is introduced into section *b*. The whole Centripure plant is hermetically sealed; and the saponifier is understood to operate at up to about 125°C under sufficient pressure to keep most of the water liquid. Contact parts are of stainless steel. After only a few minutes in section *b* the soap passes to the upper tower *c*, which is a holding vessel. Saponification is said to be about 99.8% complete on transfer from *b* to *c*, and over 99.95% complete when the soap leaves *c*, with only about 0.1% free alkali. Earlier designs used caustic soda solution at about 28% NaOH, giving a saponified mass at about 62% TFM, and with this design lye absorption can be practised by dilution of the strong caustic soda with spent lye. The 1964 article describes a more robust unit which can produce soap at 72 – 73% TFM by the use of 48% NaOH caustic soda. Spent lye is then added to the mass leaving the saponifier to achieve both lye absorption and cooling. An important feature of the De Laval saponifier is a control system based upon changes in viscosity which occur when the quantity of caustic soda added varies slightly from the stoichiometric quantity (that is the theoretical requirement for complete reaction). Pump 1 is of the positive rotary, gear, type which transfers soap at a constant rate, so that the pressure rise induced by the pump increases as viscosity increases. This is measured and controlled by a pneumatic controller, indicated diagrammatically in Fig. 4.9 as a U tube manometer. Fig. 4.10 shows roughly how viscosity, measured as a differential pressure, varies with excess, or deficit, of caustic soda at three different TFM levels. The minimum, or maximum, point can easily be determined by simple manipulation of the flows. Operation is normally on the right-hand side of the point of inflexion; and the device is said to be capable of control to ± 0.01% NaOH. From Fig. 4.9 it will be appreciated that the caustic soda content at the point of measurement is not the same as that in the soap leaving the saponifier. The De Laval saponifier is excellent, but very costly.

The saponifier and part of the washing section of the *Sharples* plant, made and sold by the Pennwalt Company, are shown diagrammatically in Fig. 4.11. The first wash section is shown here because it also forms part of the saponification process. Sharples number the units in their plant 1, 2, 3 and 4, starting at the saponification end. They also call the liquor leaving a separator lye *n*; but when this liquor is used in the previous section, after addition of caustic soda, or other incoming liquors, as necessary, it becomes reagent (*n*–1). The fat and the first reagent, comprising second lye plus 48% caustic soda, are introduced into saponification vessel 1. The soap is circulated externally by pump and a portion is bled-off into the holding vessel, saponification vessel II, from which it passes to the centrifuge S_1. Saponification is not complete at this stage and the spent lye separated is low in free alkali. Literature from Pennwalt indicates 98% completion, but Bailey (1979) says that 10–25% of the fat remains unreacted which seems excessive. The soap from S_1 is mixed with reagent 2 in mixer 2 where saponification is completed with liquor containing an adequate excess of caustic soda. Additional 48% caustic soda can be added to R_2 as indicated, but sufficient usually remains from additions later in the process and the provision for addition to R_2 is not normally included. It will be noted that the Sharples process uses two-stage saponification for the reasons discussed in section 4.2.2. Actually provision is made for the completion of saponification in the washing stages of most processes, even though free fat may not be left deliberately after the main saponification. The Sharples process also includes the version of the lye absorption principle which uses second stage lye.

Meccaniche Moderne and *Mazzoni* continuous soapmaking plants use very similar saponification systems shown diagrammatically in Fig. 4.12. Saponification takes place in a four compartment autoclave operating at about 120°C under two atmospheres pressure; and soap is re-cycled. The soap then moves to the mixer/cooler, which has four compartments with axial mixers. Saponification is substantially completed in this vessel. The first two compartments have water

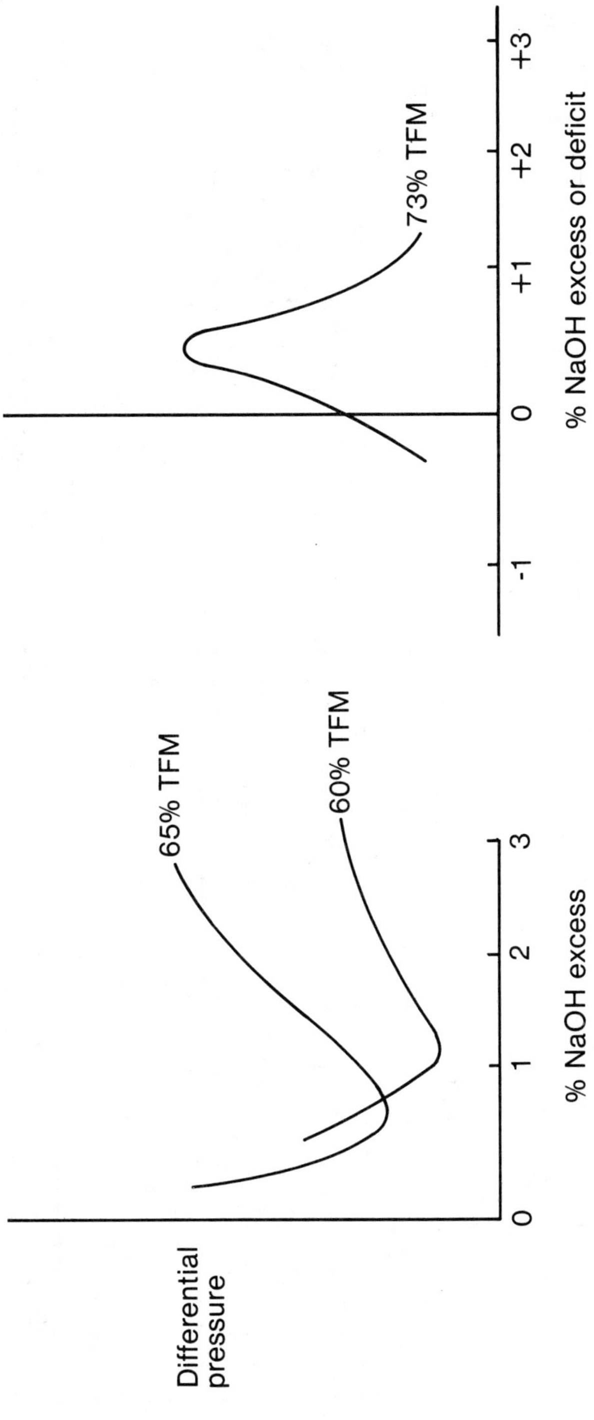

Fig. 4.10 – Variation in viscosity, or differential pressure, with excess, or deficit, of caustic soda – De Laval saponifier.

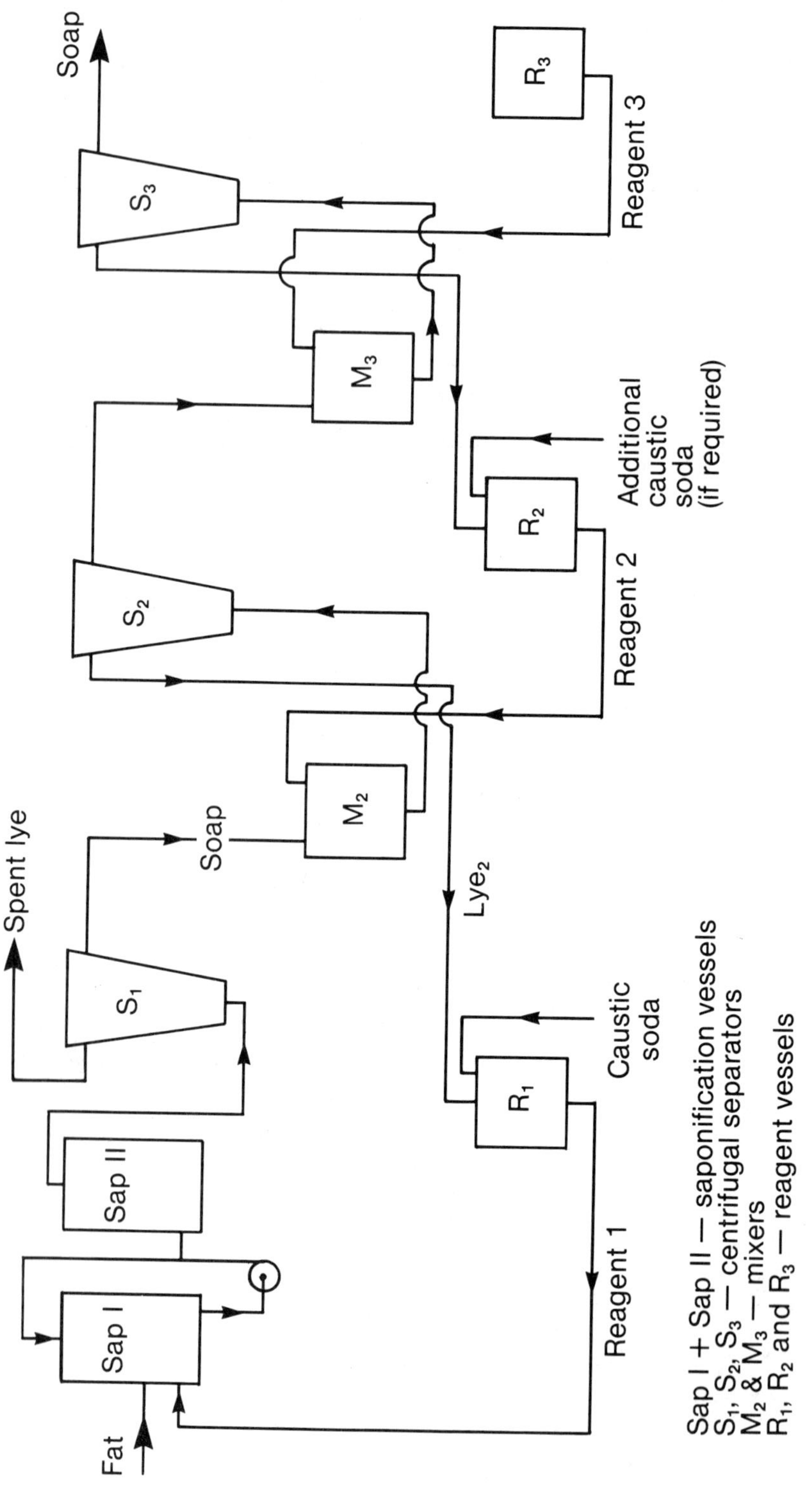

Fig. 4.11 – Sharples soapmaking plant. Saponification including part of washing system.

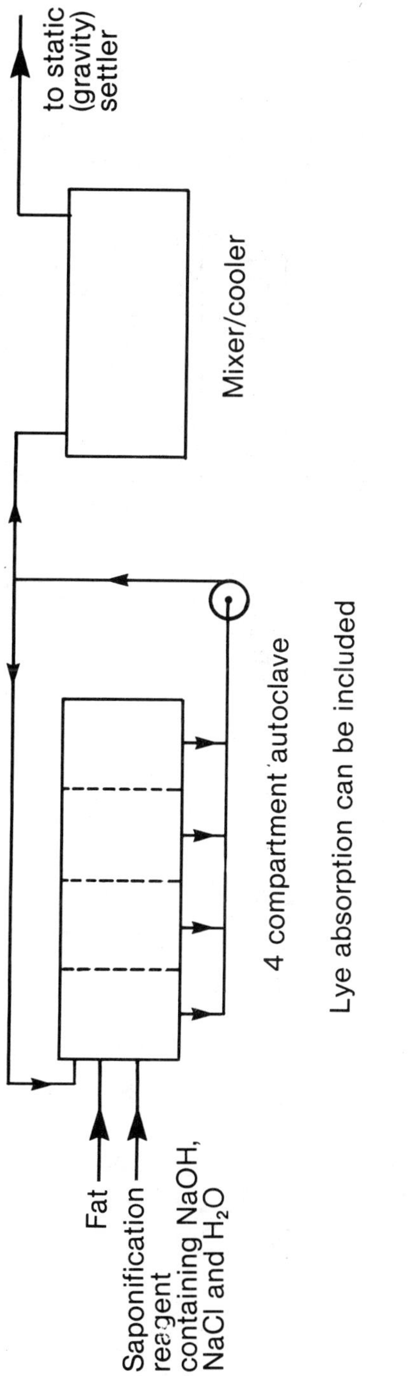

Fig. 4.12 – Saponification system for Meccaniche Moderne and Mazzoni SCN plants.

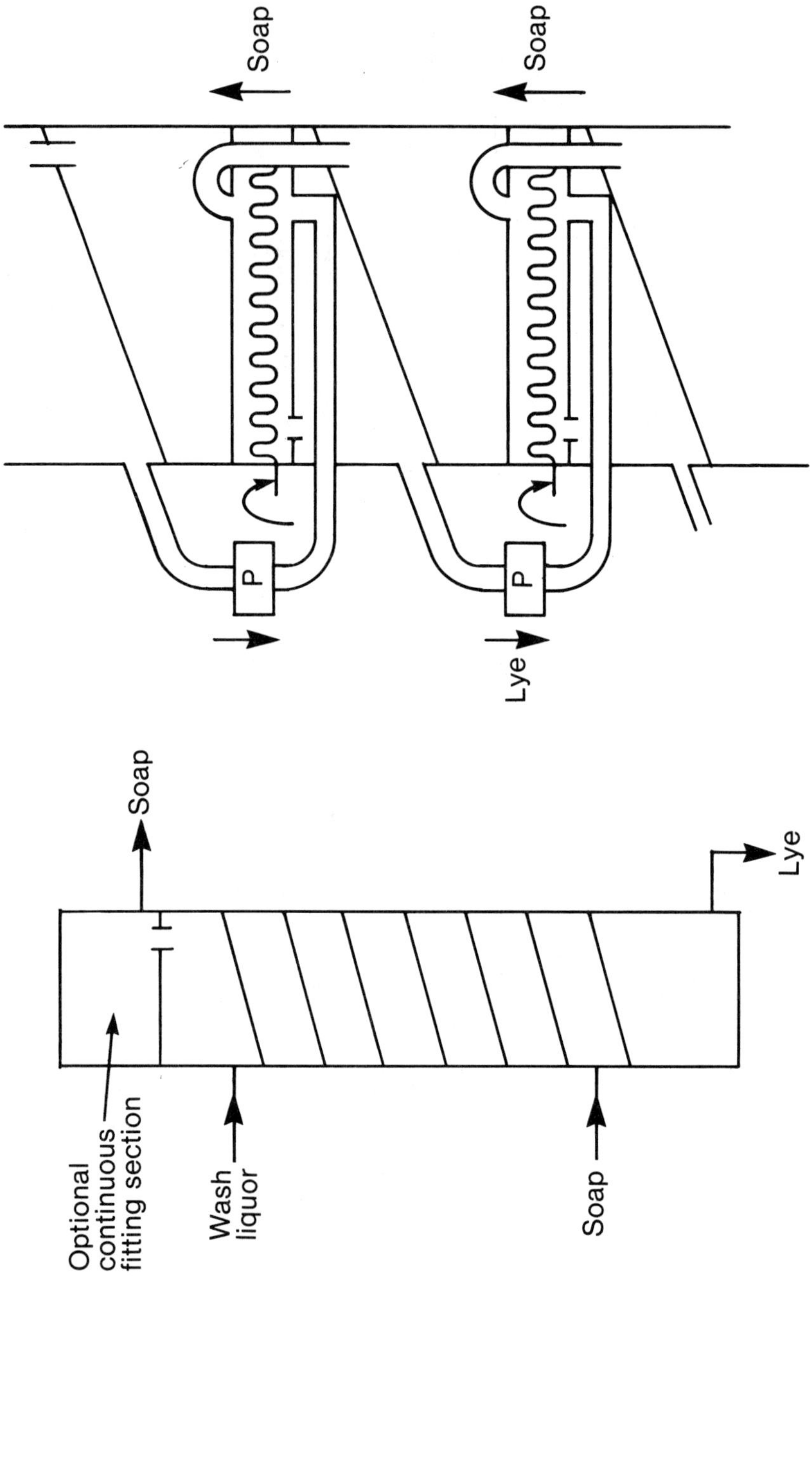

Fig. 4.13 – Monsavon washing tower.

cooling coils and the temperature is reduced to about 90°C; this reduces solubility of soap in lye, and the two phases separate so that they can be parted in the static, gravity, separator to which the mixture is passed.

4.6.2 Equipment for washing

As mentioned earlier, all modern soapmaking plants wash the glycerol out of the soap by counter-current transfer of glycerol from soap in the grained condition to the lye. Many types of equipment can be used, but all fall into one of the following kinds:

- mixers and static, or gravity, settlers.
- mixers and centrifugal separators.
- truly continuous counter-current contactors.

In the 1940s *R. V. Owen* in the Crosfields, Warrington, factory (part of Unilever) developed systems in which lye was circulated and transferred from stage to stage using steam-jet injectors in each pan of a group. He also devised a tray-type multi-stage washer, somewhat similar to the washing tower subsequently patented and made by the French company *Monsavon.* This is shown diagrammatically in Fig. 4.13. Six stages of mixer/settler washing are included in a single washing tower. Each stage comprises a settling section with inclined top and bottom, through which is inserted a tube with a mechanically rotated mixer. The soap rises through a tube from the top of a settler section into the next mixer; and lye joins it from the section above at a rate controlled by a positive displacement (Mono) pump. Monsavon towers were also made by Baker Perkins Limited, of Peterborough, England; but they are no longer available.

At Port Sunlight, England, another part of the Unilever organization, a continuous washing unit, known as a *DPU (divided pan unit)* was evolved from the mechanical circulation system of washing in pans discussed in section 4.4. The main features are shown in Fig. 4.14. In the original unit a 14 foot (4.27 m) cube pan was divided into six sections, but a tank divided into any reasonable number of sections and arranged to suit a required layout can be used. The flows are those standard for a mixer/settler system. Open inpeller centrifugal pumps are used to mix lye and soap curd, and to return the mixture to the middle level in the next stage. Recirculation round the pumps to increase mixing was not found to improve performamce. Large floats designed to position themselves at the curd/lye interface control the outflow of lye by actuating gate, or treacle, type valves. To minimize the force needed to move the valve disc, a small clearance is left between the stationary and the moving plate; in consequence these valves will not stop the flow completely, but do all that is necessary for control. The float rods extend above the top of the unit, so that the operator can push them down momentarily from time to time to ensure that the valves are working properly. A simple basket-type filter is installed to protect the crude soap feed pump. This pump is preferably of a variable delivery positive type, such as the Plenty pump, so that it can be used to control the rate at which the soap is fed. The unit made from a 14 foot cube pan was found capable of washing 4 – 15 tonnes/hour of 58% curd soap. Simple, flap-type, non-return valves are fitted in the lye lines near to the centrifugal pumps to prevent back-flow of soap. The unit should be well lagged; and fitted with a stainless steel cover, provided with inspection doors.

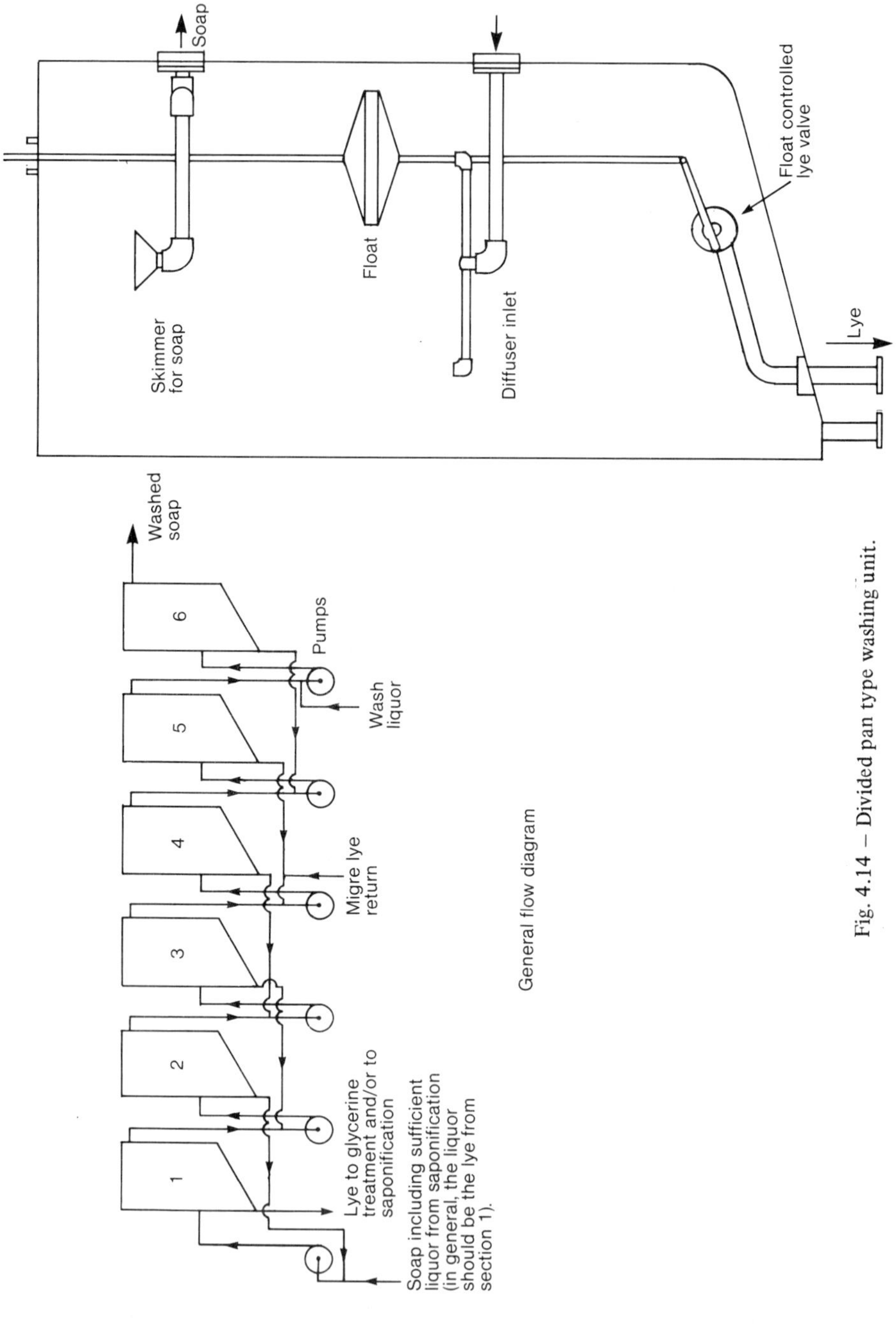

Fig. 4.14 – Divided pan type washing unit.

Meccaniche Moderne and *Mazzoni* use four stage mixer/settler washing units similar in principle to the DPU, but they differ in design from each other and from the DPU. Fig. 4.15 shows the main features of a Mazzoni washing stage. The general arrangement of the Meccaniche Moderne washer is shown later in Fig. 4.19 for the complete plant. In this unit, the four washing stages are placed together in the compartments of a single tank. Soap curd overflows from the tops of the stages and is mixed with lye in the pumps.

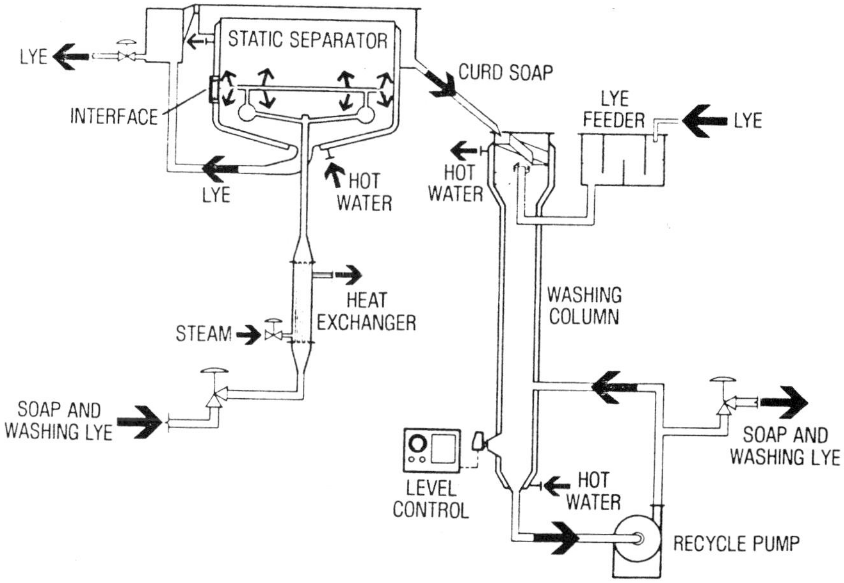

Fig. 4.15 – Main features of a Mazzoni washing stage.

Plants which use mixers and centrifugal separators use essentially the same flow sheet as those with gravity settlers, as shown in Fig. 4.11 for the Sharples plant. Some De Laval plants use a similar system, with a different type of centrifuge, as shown in Fig. 4.16. A viscosity recorder, is used to aid the control of the grain; the proportions of water and concentrated brine being varied manually. Because centrifuges normally separate more efficiently than gravity settlers, less washing stages are required; and often only two are included, apart from the centrifuges used in the saponification and fitting stages. Centrifuges also collect most of the particulate dirt in the material processed; and the bowls need to be cleaned when they become as full as is permissible.

So far as is known only two types of continuous contactor have been used commercially to wash soap, but at least one other type, the pulsed column, has been tried experimentally. De Laval once used a rotating helical contactor, similar in principle to the Podbielniak contactor employed in other extraction processes. This is illustrated by Fig. 4.17 and is described in more detail in the article by Palmqvist and Harris (1964), but it is no longer available. In brief,

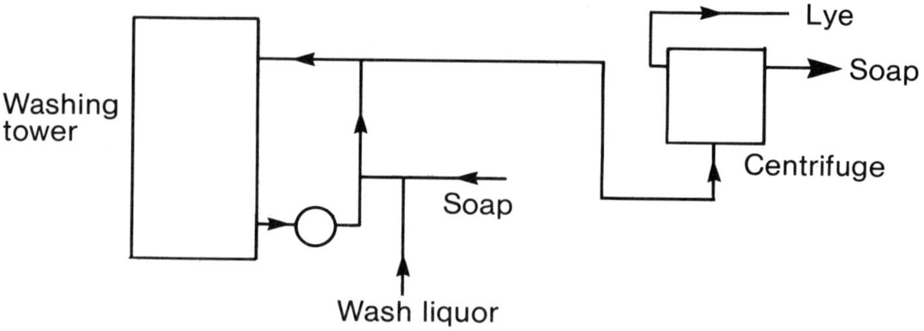

Fig. 4.16 – Unit in De Laval washing system.

it comprised a rotating helix in a casing; the soap was led to the outer end and the wash liquor to the inner, so that, because of the difference in density, the soap phase moved inwards and the aqueous phase outwards. Some mixing holes were also provided.

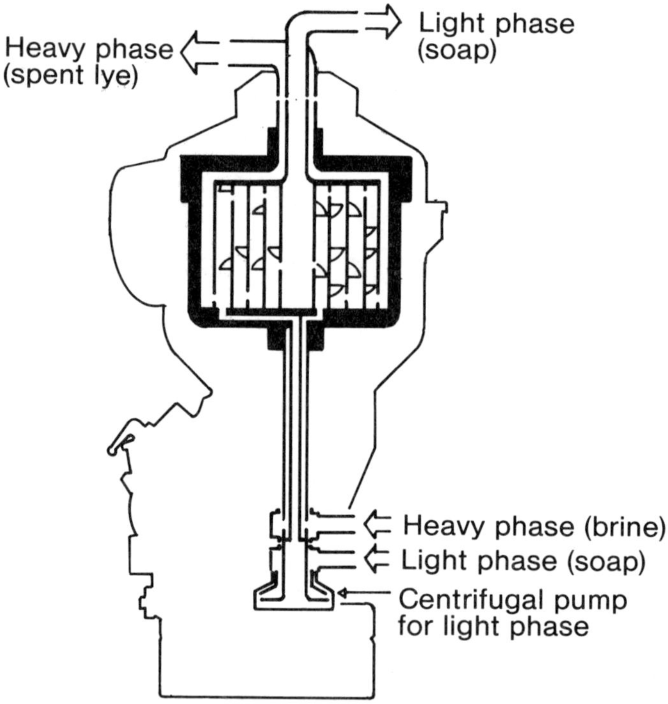

Fig. 4.17 – De Laval helical extractor.

Unilever companies have found that the *rotating disc contactor,* shown diagrammatically in Fig. 4.18, can provide a particularly effective unit for the washing of soap. The shaft rotates at about 80 rpm; and the diameter of the

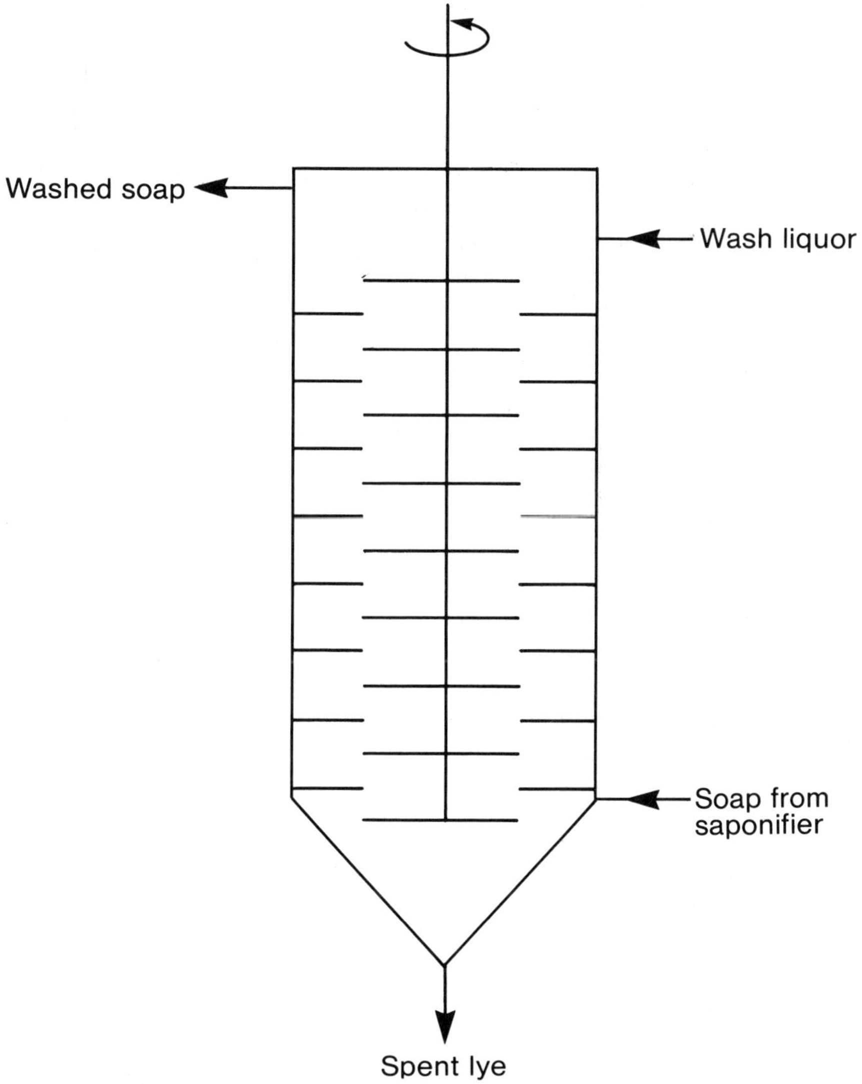

Fig. 4.18 – Rotating disc contactor.

rotating discs is slightly smaller than that of the holes in the fixed discs, so that the shaft can be withdrawn. Lye recirculation to use the lye absorption principle can be arranged in various ways to suit the saponification system employed.

It must be re-emphasized that, whichever system of washing is used, it is particularly important to ensure that the soap is only just grained sufficiently; overgrained soap occludes excessive amounts of lye in the curd and hence reduces efficiency. Concentrated brine and water in the proportions required for the wash liquor can be blended in various ways; some systems use an automatic density controller, others blend manually with the aid of flow measurers, such as rotameters.

4.6.3 Equipment for fitting and final separation

In the traditional process fitting was a complex operation involving much boiling, but it has been found that it is only necessary to mix the washed curd with the water and, if required, electrolyte needed to give the correct composition for the desired type of fit. An early example of mechanical fitting in a pan used the lye diffuser described in section 4.4, when it was found that soap fitted mechanically settled more quickly than soap which was boiled in the conventional way, probably because it was not filled with bubbles of steam.

Some modern continuous processes use only a simple line mixer, or even just injection of fitting liquor into the pipeline through which the soap flows. Others use a holding and mixing column into which the fitting liquor is injected. with recirculation of the soap. The De Laval process uses such a column with circulation and the addition of fitting liquor is again controlled by differential pressure measurement directly related to viscosity. The mixture passes to a second column intended to ensure full establishment of equilibrium.

Obviously, long settling under gravity is not practicable in a continuous process; and centrifuges are always used for the final separation. Open fits giving a thin nigre, or nigre lye, are also common; and arrangements are usually included for the continuous return of the nigre lye to the washing system.

Mazzoni normally uses a neat soap/lye separation, but offers an additional unit in which this is followed by a fit giving a neat soap/nigre separation. The use of this additional stage is recommended when poor quality fats are used and extra cleaning of the neat soap is required. The nigre is then degraded to a lower quality soap, cleaned prior to return to the system, or partially re-circulated. With this system, the initial neat soap/lye separation is the equivalent of a rough fit. The additional stage can, if thought desirable, also be used for the production of toilet soap base, but Mazzoni does not normally recommend it for this purpose.

The effective concentration of electrolyte controls the type of fit, but it is possible to vary the proportions of caustic soda and salt. This affects the free alkali and salt contents of the neat soap produced. If desired, the free alkali in the neat soap can be reduced by neutralization with fatty acid or coconut oil; and this provides a means by which soaps with a very low salt content can be made.

4.7 COMPLETE SOAPMAKING PLANTS

It is, in general, possible to combine any method of saponification with any of washing, and any of fitting, to form a complete soapmaking plant. It is now usual for the whole plant to be continuous, with the elements integrated so that the plant operates substantially automatically. However, it is also possible to use a combination of batch and continuous elements, for example:

- batch saponification, followed by continuous washing, and batch fitting and final settling.
- continuous saponification and washing, followed by batch fitting and final settling in pans.

Such combined systems provide some, although not all, of the advantages of a proprietory continuous plant, as compared with the traditional pan process. They might possibly appeal to some small manufacturers who do not feel able to justify the initial cost of a continuous plant; or to any who require the quality of neat soap produced by gravity settling after a fine fit. Additional advantages from the purchase of a proprietary plant are:

- the elements are balanced and connected with the controls needed for simple automatic operation.
- the plant manufacturer is responsible for design and commissioning.
- some aspects of efficiency are likely to be greater.
- less space is required than if the soap is settled in pans.

Some flow diagrams for complete soapmaking plants are rather complex, but consideration of the elements and principles already discussed should help in interpretation. Plants offered by the main manufacturers are now discussed briefly.

Fig. 4.19 shows the *Meccaniche Moderne* type SAES–N/AG–Bu–To plant. It includes a saponification autoclave and cooler, and a four-stage mixer/settler washing unit. The composition of the final washed curd is adjusted and neat soap is separated in a centrifuge. The nigre, or nigre lye, is returned to the third stage of the washer. This may sometimes lead to problems with foam. The figure also shows a mixer in which the alkalinity of the soap can be reduced by a small addition of fatty acid, or other neutralizing agent, such as coconut oil, boric acid or phosphoric acid.

Fig. 4.20 shows the very similar *Mazzoni* SCN–N plant proposed for toilet soaps. Type SCN, intended for household soaps, is the same as the SCN–N, except that the facility for neutralization of the neat soap is omitted. A pH type controller is used in conjunction with the neutralizer.

Fig. 4.11 discussed in the section on saponification equipment also shows the arrangement of a Sharples washing stage. In a complete *Sharples* plant, the soap curd from centrifuge S_3 is mixed with fourth stage reagent (water, brine and 48% caustic soda) in a fitting mixer, equipped for recirculation by external pump, and the mixture is centrifuged to give neat soap and thin nigre which is returned to R_3 where it forms part of the third reagent. Two centrifuges in parallel may be used for the final stage, because a lower rate of throughput is needed for proper separation of neat soap than for curd.

The elements of the *De Laval* plant have already been discussed in some detail and they are connected essentially as in the other plants just considered. The plant described in Palmqvist (1960) included two mixer/separator washing stages, whereas the plant at the Gerrard factory, Palmqvist and Harris (1964), used two rotating helix type washing units in series; other arrangements of the washing system are possible.

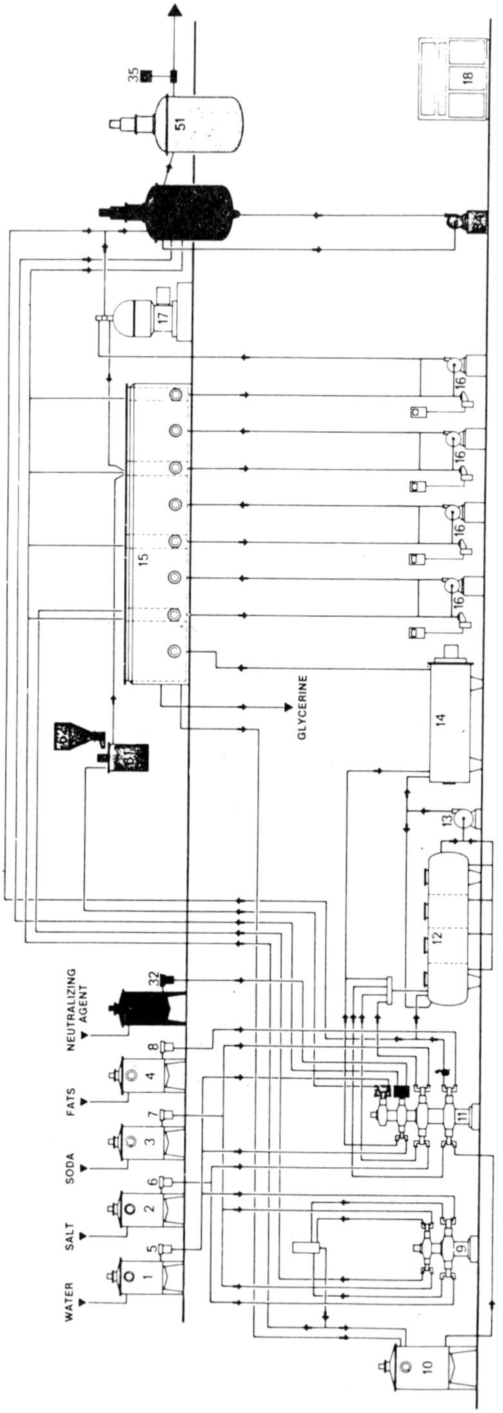

Fig. 4.19 – Meccaniche Moderne continuous saponification plant type SAES.N/AG–BU–TO.

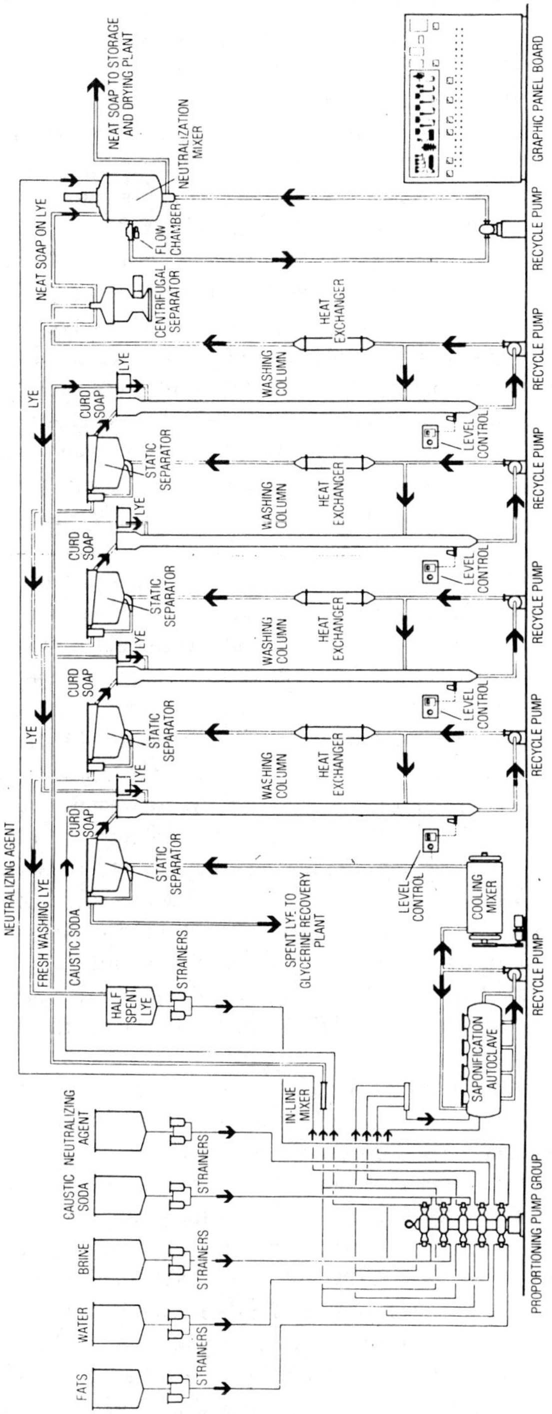

Fig. 4.20 – Mazzoni type SCN–N continuous soapmaking plant.

4.8 PLANT PERFORMANCE

Apart from matters of quality and capacity, the performance of a soapmaking plant is judged by the efficiency with which it separates the glycerol using a given quantity of water. Both these important criteria can be expressed in various ways. The most direct way in which to express recovery is in terms of percentage yield, that is glycerol in spent lye as a percentage of glycerol in fats. It is also often reported in terms of the glycerol content of the soap produced (usually calculated at 63% TFM, or at 100% TFM). The relationship between the two measures depends upon the glycerol content of the fat blend used; and Table 4.9 gives figures calculated by simple arithmetic for (a) a high glycerol fat blend containing 10.5% glycerol, and (b) an acid oil type of blend containing 6% glycerol. The quantity of water used can be expressed as lye/fat ratio, lye/63% TFM soap ratio, or as the concentration of glycerol in the final spent lye. Some figures showing the relationships between these criteria for the two fat blends are given in Table 4.10. The relationship between percentage recovery and lye/fat ratio depends upon the process used and is a measure of its efficiency. The glycerol content of the wash liquor used, which depends upon the glycerol added in recovered salt and in any weak liquors introduced at the final wash stage, affects recovery and becomes increasingly significant as the level of recovery increases and the glycerol in soap decreases. In general, it is possible to improve recovery at the expense of lye/fat ratio (or concentration of glycerol in lye) and vice versa. Just as with simple washing in pans considered in section 4.3.3, the efficiency of a mixer/separator washing stage is directly influenced by the dryness of the separated curd, as can be measured by determination of K_G. With gravity settlers, values of K_G around 0.4 have been found which indicates that the mixing of lye and curd is as effective as washing by the traditional method of closing the soap and re-graining at each stage. Centrifuges usually produce curds containing less enmeshed lye than gravity settlers; and so provide more efficient washing stages with lower values of K_G. Operators can check the performance of a plant by estimation of K_G for the various stages, or even by just determination of the TFMs of the various curds from time to time. Inadequate settling time (that is too great a throughput for the particular plant) can result in wet curds; but, as indicated earlier, poor performance is often due to over-graining.

Glycerol cannot be separated until it has been liberated by the chemical reaction, so that it is important to ensure that saponification is as complete as possible at the earliest possible stage.

Most of the performance figures available for modern continuous plants are reports, or estimates, by plant manufacturers and are not necessarily strictly comparable. Broadly, it can be said that around 95% recovery is common with a lye/fat ratio of, say 0.7; which, from Table 4.10, is equivalent to a lye containing about 14% glycerol from a fat blend containing 10.5% glycerol. Better figures have been reported for some plants, but others have been found to be less efficient. For comparision, Govan (1946) estimates that with three traditional washes a yield of 72% is obtained with a lye/fat ratio of 1.0, and 87% with a ratio of 2.0. Rather better results stem from the figures in Table 4.5 calculated by Wigner for optimized operation. A factory producing a range of soap bases from fats of varying quality by the traditional process commonly sent lyes to the glycerine department with an average glycerol content of about 4 – 5%.

Table 4.9 – Relationship between glycerol percentage in soap and percentage recovery from (a) fat containing 10.5% glycerol and (b) fat containing 6% glycerol

Percentage glycerol in 63% TFM soap	(a) Fat blend containing 10.5% glycerol		(b) Fat blend containing 6% glycerol	
	Percentage of the glycerol in the fats left in the soap	Percentage transferred to lyes	Percentage of the glycerol in the fats left in the soap	Percentage transferred to lyes
0.1	1.4	98.6	2.5	97.5
0.2	2.9	97.1	5.0	95.0
0.3	4.3	95.7	7.5	92.5
0.4	5.7	94.3	10.0	90.0
0.5	7.1	92.9	12.5	87.5
0.75	10.7	89.3	18.8	81.2
1.0	14.3	85.7	25.0	75.0
1.5	21.4	78.6	37.5	62.5

No allowance for any losses is made in this table.
It is assumed that 1 tonne of fat produces 1.5 tonnes of 63% TFM soap.

Table 4.10 – Relationships between criteria for quantity of water used in soap-making under various circumstances

Ratio lye/fat	Ratio lye/63% TFM soap	Concentration of glycerol in lye			
		Fat containing 10.5% glycerol		Fat containing 6% glycerol	
		Recovery = 95%	Recovery = 85%	Recovery = 95%	Recovery = 85%
0.4	0.27	25.0	22.3	14.3	12.8
0.5	0.33	20.0	17.9	11.4	10.2
0.6	0.40	16.6	14.9	9.5	8.5
0.7	0.47	14.3	12.8	8.1	7.3
0.8	0.53	12.5	11.2	7.1	6.4
0.9	0.60	11.1	9.9	6.3	5.7
1.0	0.67	10.0	8.9	5.7	5.1
1.1	0.73	9.1	8.1	5.2	4.6
1.2	0.80	8.3	7.4	4.8	4.3

Other points in a comparison between the modern and the traditional pan processes are:

- the old process took about 5 – 8 days, so that much space was required. Material passes through the modern processes in an hour or so, or even less.

- the continuous processes normally re-cycle the nigre, or nigre-lye, and are completely self-contained.
- in addition to the large saving in steam, and in capacity needed for lye treatment and evaporation, which results from the higher concentration of glycerol in lyes, the modern processes use much less steam than the pan process itself.
- the modern processes use more power than the old, but the cost per tonne of soap is not high.
- comparison of quality is less clear. Advantages for the modern processes are that:
  - there is little opportunity for airborne dirt to contaminate the soap; and centrifuges remove particulate dirt effectively.
  - there may be less risk of damage by oxidation and prolonged heating, but it seems doubtful whether this difference is substantial.

On the other hand:

- very thorough saponification is possible with the pan process.
- although soap from properly operated continuous plants is entirely satisfactory for modern finishing processes, the author considers that there may be some doubt if good quality framed household soap is required. This doubt is based upon experiments many years ago in which soap fitted normally in pans was separated in a centrifuge. The neat soap produced was satisfactory for many purposes, but it showed nigre spots when frame cooled unless the centrifuge throughput was very low. Further work might show that any such problem can be avoided by adjustment of conditions.

4.9 PRELIMINARY TREATMENT OF SPENT LYES IN THE SOAPMAKING DEPARTMENT

Spent lyes contain variable quantities of free alkali and soap (particularly low molecular weight soaps), as well as colouring matter, particulate dirt and other impurities. The system of two-stage saponification, in which the final spent lye is separated from soap containing some free fat, produces lye with a low free alkali content; and in modern counter-current processes this is the only lye sent to the glycerine department. Some soapmakers prefer to saponify as completely as possible in the first stage, which, as well as having possible quality advantages, ensures that the glycerol is all liberated as early as possible. With the old process most lyes are transferred for recovery of crude glycerine and the traditional procedure was to leave the lyes in tanks below the pans where they cooled. Soap separated on the surface, was skimmed, and then returned to a suitable boil of probably low grade soap. The cooled and skimmed lyes were transferred to the glycerine department where they were acidified to a pH of about 4.6 with mineral acid, preferably hydrochloric to avoid unnecessary introduction of sulphate into the system. The tank contents was agitated with compressed air and allowed to settle. The lower layer of acidified lye was transferred to the first treatment process described in Chapter 8. When a sufficient layer of fatty material had collected from several acidifications, it was skimmed off and either discarded or returned to a low grade soap, possibly after filtration. The value of this scum fatty acid was doubtful. It was a poor soapmaking material and contained low molecular weight and hydroxy acids, both of which produce rather soluble soaps liable to be rejected to the next lyes; and so to build up in the system. The acidified lye was well suited to the rest of the lye treatment process, but, although it continued to be recommended by authors such as Govan, the process had important disadvantages:

- it involved a loss of alkali and the need to purchase additional mineral acid to neutralize this alkali.
- it led to a loss of heat which had to be replaced later, although some of this loss is liable to occur anyway, if a substantial buffer stock of lyes is held in the factory.
- it was a messy procedure which occupied considerable space and labour.
- unless particular care was taken to keep the alkaline part of the system clean, and mud was generally deposited in the lye tanks, it could provide conditions favourable to the growth of the micro-organisms which destroy glycerol and produce trimethylene glycol (TMG, 1,3-propane diol), a substance which is troublesome in the glycerine refinery, as well as an indication that glycerol has been destroyed.

By the late 1930s it had been found that lyes could with advantage be neutralized with fatty acids, or with fats containing a substantial proportion of free fatty acid. The operation, known as lye neutralization, can be carried out in various ways, but it is necessary to ensure intimate contact between fatty acid and lye. Wigner describes batch neutralization in a pan, using a steam

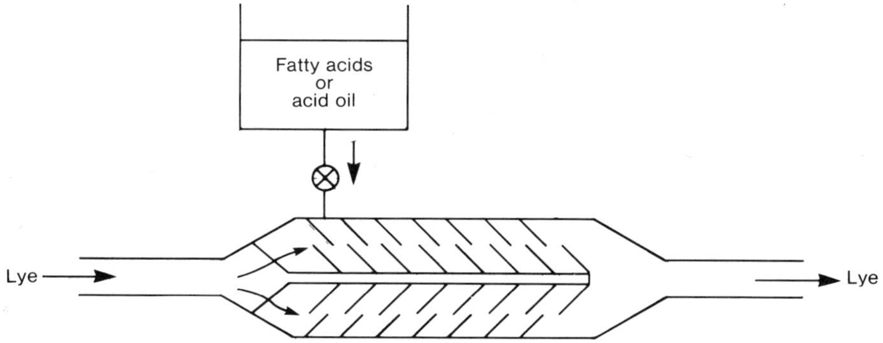

Fig. 4.21 – Korting type jet for lye neutralization.

injector near to the open steam coil to introduce the fat. The jets of steam from the coil perforations then disperse the fat in the lye. Another system is to use a Korting, or similar, jet, as shown diagrammatically in Fig. 4.21. The lye is pumped through the jet which contains annular and central baffles; and the fatty acid, or acid oil, is introduced by gravity, or otherwise, at a roughly controlled rate. The mixture is boiled and settled. The neutralized lye has a reduced soap, as well as free alkali, content and is transferred to glycerine treatment. The upper layer contains fat, soap, occluded lye and dirt. It is often saponified and grained to remove dirt before the soap is returned to a low grade base. Wigner states that if fatty acids are used, a small addition of glyceride is advisable as its presence makes soap insoluble in lye which contains several percent less salt than would otherwise be the case.

In a factory making a wide range of soap bases, and using some low grade materials with a high FFA content, it is often convenient to collect the lyes from all the soaps and to neutralize them in the way just described. If these conditions do not apply, it may be difficult to find suitable material for neutralization and to dispose of the low grade soap produced. In order to neutralize the sodium carbonate normally present in the lye with glyceride rather than fatty acid, it is necessary to use somewhat elevated temperature and pressure, say 140°C and 40 psig; continuous processes with circulation loops, heat exchangers and continuous gravity settlers have been devised to operate in this way.

5

Manufacture and purification of fatty acids and the production of soap base from them. Miscellaneous methods of soap manufacture

5.1 GENERAL BACKGROUND

Fatty acids have been produced, purified and separated into fractions for very many years. The solid fraction, known as stearin (which actually usually contains a substantial proportion of palmitic as well as stearic acid), was used particularly in candles; and the liquid fraction, known as red oil or olein, in the processing of wool, although there were, and are, other uses for both materials. Relatively minor quantities of various fatty acids were used in soap; and high quality stearin was, and is, used in shaving soaps. In the late 1930s and 1940s, major American companies developed much improved processes for the manufacture of soap by the fatty acid route, which achieved the more efficient recovery of glycerol and the much shorter processing time obtained later by other companies using the developments of the direct saponification process discussed in Chapter 4. Points related to a comparison between the two methods are considered in section 5.7. Despite the great reduction in soap production in the USA, the quantity of fatty acids used in soap in that country rose from 100 million pounds in the late 1930s to 956 million pounds in 1978 (excluding tall oil); and these figures apparently do not include the major combined fat-splitting and soapmaking plants. There were said to be fifteen producers of fatty acids in the USA in 1978, mostly divisions of large companies. A development in many countries is the purchase by medium-sized soap companies of refined fatty acids tailored to close specifications. Provided a reliable source, or sources, can be found from which the fatty acids can be transported economically and quickly, this is satisfactory and may be commercially attractive. It is important to avoid undue delay between production and use of most fatty acids; and to store and transport them carefully, not in mild steel. The purchase of fatty acids simplifies the soapmaking process and obviates the need for lye, or sweet water, treatment and evaporation equipment. It also transfers to the fatty acid maker the problems of economical purchase and pre-treatment of crude fats. Many small producers of toilet soaps, the so-called remillers, have for many years gone even further and have bought dried soap base to which they add perfume and other ingredients, and just complete the processing and packaging.

Fatty acids can be produced for special purposes by the acidification of soap with mineral acid, usually sulphuric; and this process can be used to recover fatty acids from soapstocks after saponification has been completed and, if required, the soap has been grained and/or fitted to purify it. This procedure is used in the laboratory, but, on the plant scale, it is expensive and obviously does not provide an alternative method for the manufacture of soap. All important methods for the production of fatty acids from natural fats involve hydrolysis, or splitting, with water, with, or without, a catalyst. Processes can be divided into those carried out at atmospheric pressure and those which use high pressures and temperatures. The processes at atmospheric pressure are:

- acid splitting,
- Twitchell type splitting,
- splitting with lipases, the fat splitting enzymes.

The high pressure processes use water alone, or water aided by a basic catalyst, usually zinc oxide. Higher temperatures, and/or longer times, are needed when no catalyst is employed, but processes without catalyst avoid the expense of the catalyst and of removing it from the hydrolysis products, and are now usual in modern installations. When a zinc oxide catalyst is used, most of it reaches the fatty acid still residues. High temperature and pressure processes can be batch or continuous. As explained later, the normal high temperature processes require the presence of a liquid aqueous phase and this makes high pressures also necessary to avoid vaporization of the water. USP 3 253 007 (May 1966) to K. E. Lundi and the Carad Corporation claims that tallow can be split with highly superheated steam without high pressure, but, so far as it is known, this method is not in commerical use.

5.2 FAT SPLITTING

5.2.1 Principles of fat splitting

The hydrolysis of fats takes place in stages and the reactions are reversible:

$$\underset{\text{triglyceride}}{C_3H_8(OOCR)_3} + \underset{\text{water}}{H_2O} \rightleftharpoons \underset{\text{diglyceride}}{C_3H_8(OH)(OOCR)_2} + \underset{\text{fatty acid}}{R{-}COOH}$$

$$C_3H_8(OH)(OOCR)_2 + H_2O \rightleftharpoons C_3H_8(OH)_2(OOCR) + R\text{--}COOH$$

diglyceride monoglyceride fatty acid

$$C_3H_8(OH)_2(OOCR) + H_2O \rightleftharpoons C_3H_8(OH)_3 + R\text{--}COOH$$

monoglyceride glycerol fatty acid

By the usual principles of mass-action, the degree of split can be increased by the removal of a product from the reaction zone; and in practice this is the glycerol. Batch processes use two or more stages, in which the sweet water is removed and replaced by fresh water, or by a sweet water with a concentration of glycerol lower than that in the sweet water removed. This is analogous to the way in which soap is washed in a number of stages; and, for the same reasons as in soapmaking, counter-current contact is desirable in order to reach as a high a degree of split as possible with the minimum quantity of water. In the case of splitting, however, the need to recover the maximum yield of glycerol in as concentrated a form as possible is augmented by the need to liberate the maximum percentage of fatty acid from the fat.

The solubility of water in fatty acids and fats increases with rise in temperature. At temperatures up to 100°C it is slight; between 100°C and 200°C it increases slowly, and then, at temperatures above 200°C the rise is rapid. Ittner and Mills, in patents concerned with continuous splitting, report that at 235°C coconut fatty acids dissolve about 17% of their weight of water when excess water is present, and the pressure is sufficiently high to keep it liquid. At 245°C coconut fatty acids dissolve about 20% of water and tallow fatty acids about 11%. At 287°C (with a pressure in excess of about 1050 psig) coconut fatty acids blend freely with more than their own weight of water to form a single clear liquid phase. Fatty acids of low molecular mass dissolve more water than those of higher molecular mass, and fatty acids more than the corresponding fats, possibly greatly so. In the high pressure processes, the main reactions occur almost entirely in the fatty phase. Free glycerol is partitioned between the fatty and aqueous phases, and, as already explained, two or more changes of water are normally used to achieve a high degree of split in batchwise processes. Each replacement of sweet water by fresh water (or by a sweet water low in glycerol content) causes a transfer of glycerol from the fatty to the aqueous phase and so disturbs the hydrolysis reaction in the fatty phase in the required direction. In a counter-current splitting tower, a similar effect is achieved by a transfer of glycerol from the fatty to the aqueous phase, as discussed in more detail later.

In the Eisenlohr process, fat and water, without catalyst, are pumped through Inconel tubes at 3,500 psig and heated to 330°C. These conditions with concurrent contact give rapid reaction, presumably in a single phase, but the equilibrium is not very favourable, because there is no removal of liberated glycerol from the reaction zone. So far as is known, this process is not used commercially.

Sonntag (1979a) quotes data for rate of splitting at various temperatures from Sturzenegger and Sturm (1951) as shown in Table 5.1. Temperature and catalyst have very substantial effects, but there is little difference between the

Table 5.1 – Influence of temperature on splitting time

Temp. °C	Time to reach equilibrium (mins.)			Half-life time (mins)		
	Beef tallow	Coconut oil	Groundnut oil	Beef tallow	Coconut oil	Groundnut oil
225	156	158	156	62	58	50
240	82	85	85	34	33	25
260	47	46	53	18	23	16
280	34	33	33	8	10	10
260 (with 0.2% ZnO catalyst)	21	–	–	4	–	–

various fats tested. Fig. 5.1, courtesy of Fratelli Gianazza SpA, shows that the reaction between a neutral fat and water at 240°C starts slowly, accelerates and then slows down as equilibrium is approached; the curve being very similar to that for the saponification of a fat with caustic soda. The shape of the curve, and the fact that the hydrolysis starts more rapidly if the fat contains more than

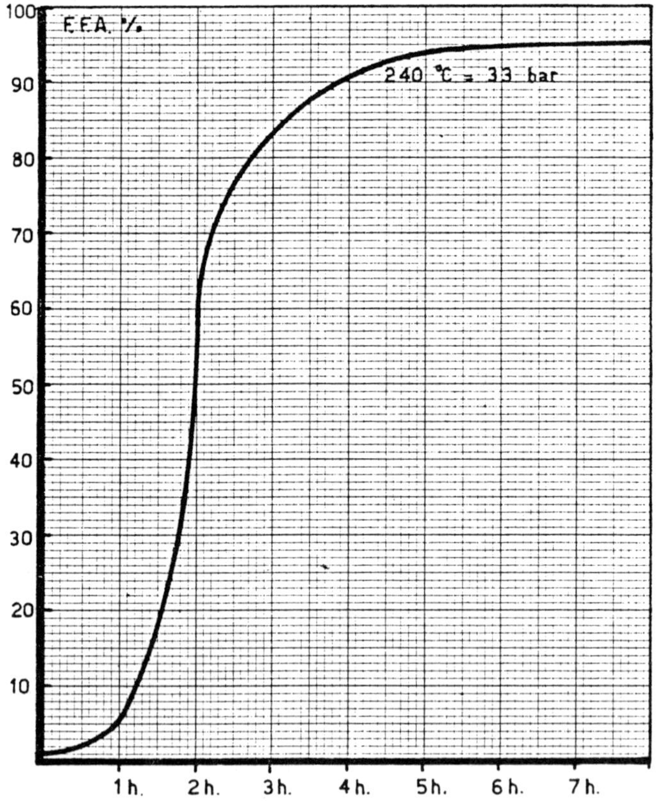

Fig. 5.1 – Rate of hydrolysis curve for neutral fat.

about 4% FFA, confirms that the reactions are autocatalytic and suggest that the slow early stage with a neutral fat is because the reaction is then mainly at the fat/water interface. Fig. 5.2, also from Gianazza, shows how the degree of

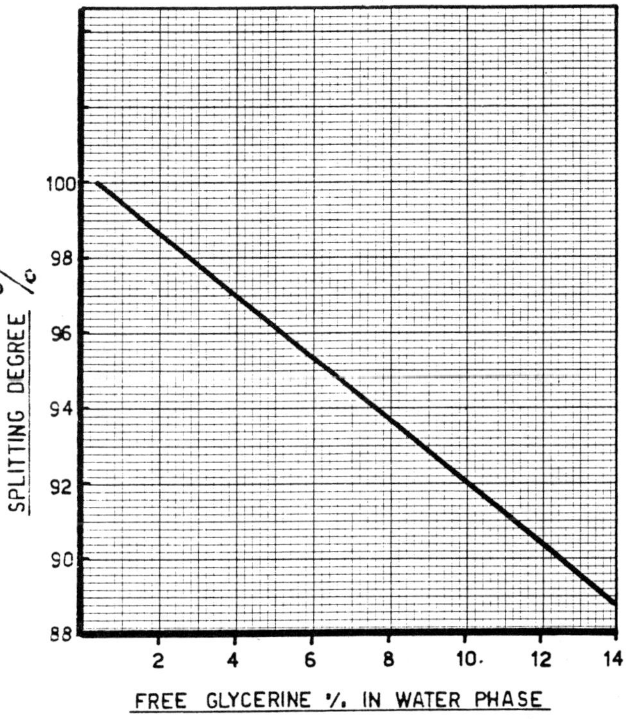

Fig. 5.2 – Degree of split v. glycerol content of water phase.

split at which the reaction rate becomes negligible (that is, equilibrium is nearly reached) varies with the percentage of glycerol in the water phase. These facts show that a fat which contains more FFA than about 4% starts to react more rapidly than one with less. Also, that to maintain an adequate reaction rate, whilst producing a reasonably concentrated sweet water, the water must be changed at least before the split is 85 – 90% complete. The actual figures vary with the glycerol content of the fats being split and the ratio of fat to water which is used.

Splitting processes carried out at atmospheric pressure use acidic conditions. With the fat-splitting enzyme from castor beans, a pH of about 5 is maintained with acetic acid; and a small proportion of manganese sulphate is said to be added. The other atmospheric pressure processes are carried out with addition of sulphuric acid to give strongly acidic conditions. The earliest process, dating from the 1820s, used concentrated sulphuric acid (around 80% H_2SO_4) with fat at 110 – 120°C, which previously had been washed and dried. Severe discoloration occurred; and organic sulphuric or sulphonic acids formed may have helped the acid catalyst by aiding emulsification. In 1898, Twitchell patented a modification in which a sulphonic acid was used in conjunction with water and sulphuric acid; and the mixture was boiled for considerable periods. The original Twitchell reagent, or catalyst, was made by reacting a fatty acid (usually oleic), and naphthalene, and/or benzene, with concentrated sulphuric acid. Most reagents used more recently, such as the 'Kontakt' type, are complex sulphonates, which are by-products from the petroleum industry.

Medium-pressure processes use water with a basic catalyst, normally zinc oxide, ZnO. A wide range of concentrations of ZnO are reported, from 0.2%, or even less, to 4.0% on weight of fat. Zinc soap is fat soluble and probably functions as an emulsifier. It behaves as a true catalyst in that it increases the rate of reaction under given conditions, but does not affect the equilibrium composition significantly. Zinc powder is sometimes also added and is said to improve the colour of the fatty acids. Other catalysts are magnesium oxide, calcium oxide, caustic soda, and other alkalis; but zinc oxide has been found to be far the most effective. Despite the catalytic effect of calcium and magnesium oxides, it is usually considered advisable to use soft, or even demineralized, water in the pressure splitting, as well as the Twitchell, process. Water hardness can affect emulsification in the Twitchell process, and cause problems with insoluble soap scums in the high pressure processes.

5.2.2 Splitting plant and processes

Unless the fats to be split are of high quality, pre-treatment as described in section 3.6.1 is usually necessary, or at least desirable. Break material can cause emulsification in towers which may interfere with the counter-current operation. Polythene in tallow is converted into black plastic masses in towers, or other autoclaves, which can block sparge pipes and other plant items. As mentioned in Chapter 3, polythene can be removed by the use of bleaching earth with filtration at below about 70°C. If a dilute sulphuric acid treatment is used prior to high pressure splitting, it is very important to remove the mineral acid by thorough washing to avoid risk of corrosion damage to the high pressure plant.

The enzyme process operates best at about 35°C and is therefore most suited to the splitting of liquid oils. It does little damage to sensitive materials and gives fatty acids of good colour, but it is currently of little, or no, commercial importance, although it may be used to produce highly unsaturated fatty acids in the laboratory. Enzyme developments may possibly change this situation.

Splitting in batch autoclaves was patented in the 1850s, but declined after the development of the Twitchell process because of a lack of suitable materials of construction.

The Twitchell process is carried out in vats which are lead lined, or otherwise resistant to aqueous sulphuric acid. Open steam is used to heat and agitate the mix; and two, or three, 'boils', with change of water, are normally given. With water equal to 40% of the weight of the fat, 1 – 5% of catalyst, and sufficient sulphuric acid (about 0.3 – 3%), a split of 80 – 85% can be obtained after the first boil. This is followed by a second boil which should raise the total split to at least 95%. Alternatively, three stages can be used giving, say, 60%, 80–85% and over 98% total splits. More details can be found in some of the older books, and Cox (1949) demonstrated from full-scale factory experience the merits of counter-current use of the sweet waters. The Twitchell process is simple and does not require complex plant, but it gives dark coloured fatty acids. It has now largely been replaced by medium and high-pressure systems.

The high pressure processes received a great impetus from the development of continuous splitting columns by major American companies, particularly Colgate Palmolive Peet (now Colgate Palmolive), Procter and Gamble, Armour, and Emery Industries around and since the late 1930s. Early patents, such as USP 2 139 589 (1938) to M. H. Ittner (Colgate), and USP 2 159 397 (1939) to V. Mills (Procter and Gamble), are still of interest. This process is often known as the Colgate-Emery process. Its core is the splitting column shown diagrammatically in Fig. 5.3. Designs from different plant manufacturers may vary, but the main features are:

- the fat is pumped in near the base of the column, through the perforations in a sparge pipe. The droplets rise through the sweet water zone; and the sweet water, substantially free from fat, leaves from the bottom of the settling section.
- similarly, the water is introduced near the top of the column and is suitably distributed into droplets which fall through the fatty matter phase. The fatty acids leave after passage through a settling zone.
- fairly low down in the column there is an interface above which the continuous phase is fatty, and below which it is aqueous.
- internal heat exchange, from fatty acid to incoming water near the top and from sweet water to incoming fat near the bottom, is important, but the Mills patent USP 2 159 397 refers to the discharge of the fatty acids at high temperature to a neutralization mixer for the production of soap powders. At least in some designs, special heat transfer devices are included in appropriate parts of the column.
- high pressure steam is introduced at suitable levels to establish and maintain the required reaction temperature. As an alternative to a sufficiently high pressure boiler it is possible to use a steam compressor to increase the pressure of the available steam.
- the fatty acids are further cooled and dried by a reduction in pressure which allows dissolved (and suspended, if any) water to 'flash' off. Similarly, the sweet water is partially concentrated by allowing it to flash. In some installations, the flashed vapour is used in multiple effect to continue the concentration of the sweet water (see Chapter 8); and other heat exchange may be possible to improve thermal efficiency. Condensate is collected for re-use as process water, made up with demineralized water.

A sufficiently high operating temperature is selected to give adequate solubility of water in the fatty phase, but not so high as to eliminate the aqeuous phase when the desired ratio of water to fat is supplied. The pressure is substantially above that just necessary to avoid vaporization.

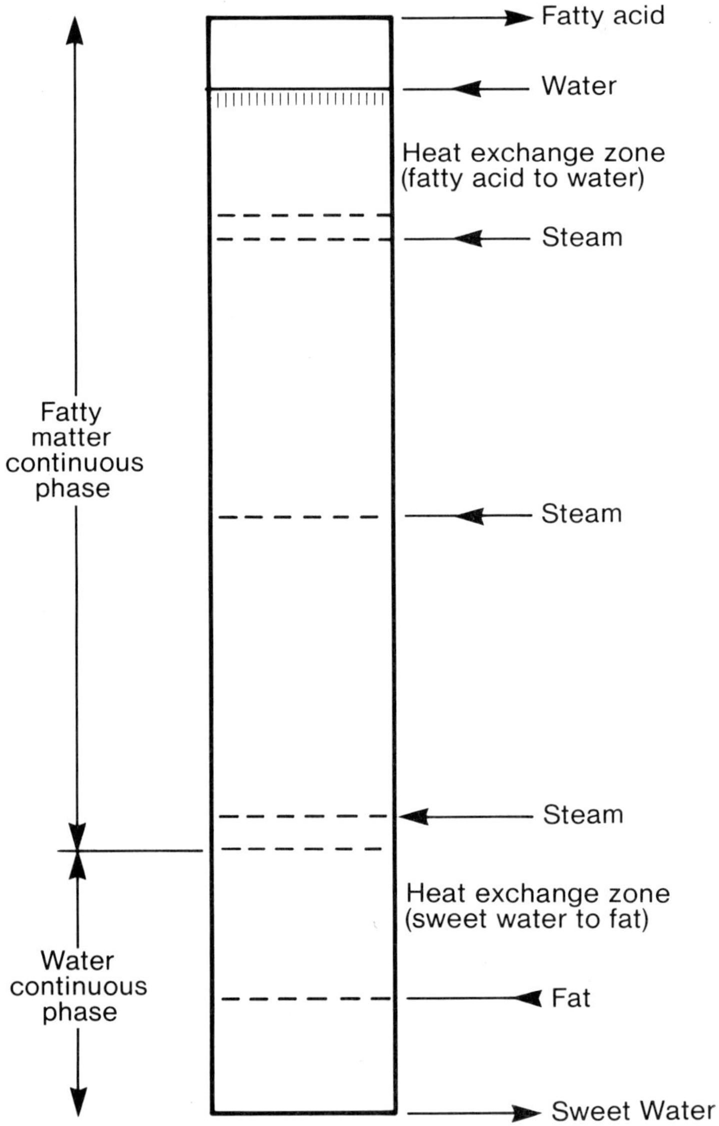

Fig. 5.3 – High pressure splitting column.

An adequate height of tower, and proper dispersion of the phases, is necessary to enable the water to extract the glycerol effectively from the fatty phase. This is needed to enable the reaction to proceed substantially to completion, but also to prevent removal of some free glycerol in the fatty phase instead of in the sweet water. Mills (USP 2 159 397) quotes for his type of tower:

height of tower, ft:	10	20	50	(as used)
percentage glycerol content of water dissolved in the fatty acid leaving the tower:	5	1	0.1	

The quantity of water required for the chemical reaction is directly proportional to the ester value, EV, of the fat. One molecule of water (HOH) is equivalent to one molecule of KOH in the test, so that the water required for the reaction equals EV/1000 × 18/56.1 × 100 = 0.032 EV% of the mass of fat. For example, using figures from Table 3.8, a blend of 80% good tallow and 20% good coconut oil has an EV of about 200, and requires about 6.4% water. The grease in Table 3.8 has an EV of 102, and requires about 3.3% water. In USP 2 139 589 Ittner states that for rapid hydrolysis it is necessary to use conditions in which the fatty material dissolves more water than that needed for the chemical reaction. This is surprising in that water used in the reaction is replaced by transfer from the aqueous phase. Perhaps it should be said that the rate of hydrolysis increases with temperature in the usual way and also with the concentration of the reactant water in the fatty phase, rather than that it is directly related to the presence in the fatty phase of more than the stoichiometric concentration of water. In consequence, conditions which meet the Ittner criterion are required but for different reasons.

The total quantity of water, including condensed steam used in the process, must be sufficient to provide the amount needed for the reaction, plus enough to dissolve in the fatty phase and to leave sufficient to form an aqueous phase to wash the glycerol out of the fatty phase. Table 5.2, obtained by simple arithmetic, illustrates these points. The fatty acid and glycerol contents of the

Table 5.2 – Water requirements of fat-splitting columns and glycerol concentrations in sweet waters produced

(Basis 1 tonne fat)

	80% tallow plus 20% coconut oil. 50% water on fat	Grease 50% water on fat	Grease 40% water on fat
Fatty acid content	0.954t	0.977t	0.977t
Glycerol content	0.111t	0.056t	0.056t
Water required for reaction	0.064t	0.033t	0.033t
Water dissolved in the fatty phase at 245°C (from solubilities given earlier and mass of FA)	0.122t	0.107t	0.107t
Total water	0.500t	0.500t	0.400t
Water in aqueous phase in reaction zone (by difference)	0.314t	0.360t	0.260t
Percentage glycerol in sweet water, if water dissolved in fatty acids leaves with FA at reaction temperature	35	15.5	21.5
Percentage glycerol in sweet water if FA is cooled before leaving the tower to a temperature at which solubility of water in it is negligible	25.5	12	15

fats are again taken from Table 3.8. The first figures for the glycerol content of the sweet water assume that the fatty acids leave the column saturated with water at the reaction temperature, but that the glycerol is all transferred to the aqueous phase. The second set assume that the fatty acids are cooled in the heat exchange zone to a temperature at which they contain a negligible percentage of dissolved water and that the separated water joins the new water descending the column.

In practice, the glycerol content of the sweet water leaving a tower seems commonly to be about 12 – 18%, depending upon the type of fat and the conditions used. Provided the sweet water leaves the tower at a temperature sufficiently high for it to flash when the pressure is reduced, further concentration then occurs, and the flashed vapours may be used to produce additional evaporation.

Jeffreys, Jenson and Miles (1961) provide a mathematical analysis of the performance of a continuous counter-current splitting tower, and similar information is given in the book by Jenson and Jeffreys (1963).

The main manufacturers of fat-splitting plants, Lurgi, Gianazza, Mazzoni, and Wurster and Sanger, each make several types of plant. Plants based upon counter-current splitting columns are now usual for large-scale operations, but batch, or multi-stage, types may be more suitable when the production required is small. High temperature processes lead to a reduction in iodine value when highly unsaturated oils are handled, and milder methods may need to be selected if this is important. Figs. 5.4, 5.5, 5.6, 5.7 and 5.8 show flow sheets for different plants from the various manufacturers; and Fig. 5.9, courtesy of Gianazza, is a photograph of a continuous splitting column. The choice of diagram is somewhat arbitrary; and it should be noted that plants similar in flow sheet, although doubtless different in engineering detail, are often offered by the other makers.

Fig. 5.4 shows the *Lurgi* batch-splitting plant. A feature is the continuous circulation of sweet water by external pump from the base to the top of the autoclave. Standard operating conditions are 230°C and 28 bar (406 psi); and it is stated that 95 – 96% split should be achieved with two changes of water (the split will also be influenced by the quantities of water and fat used, and the glycerol content of the fat). In the past, Lurgi has made continuous plants comprising three autoclaves of this general type linked in series for counter-current flows of fatty matter and water, but this equipment is no longer offered.

Fig. 5.5 shows the *Gianazza 'F.G.A.P.'* batch-splitting plant. Two autoclaves in parallel are included. As with some other designs, open steam injected near the base of the autoclaves provides agitation as well as heat. This plant is recommended for up to 36t of fat per 24 hour day; and the working conditions are 230°C and 30 bar (435 psi). The first sweet water is removed when the split reaches 85 – 90% and a second charge of water is used to raise the split to 94 – 95% (again, results will depend to some extent upon other factors). If required, Gianazza supply the F.G.A.P. unit with the facilities needed to permit conversion to the *'Contipercol'* continuous plant shown in Fig. 5.6 by addition of the extra plant items needed. This plant uses the same working conditions as the F.G.A.P. type. A split of 96 – 98% is claimed when purified fats are being processed. The contipercol plant is recommended for 36 – 85t fat per 24 hour day.

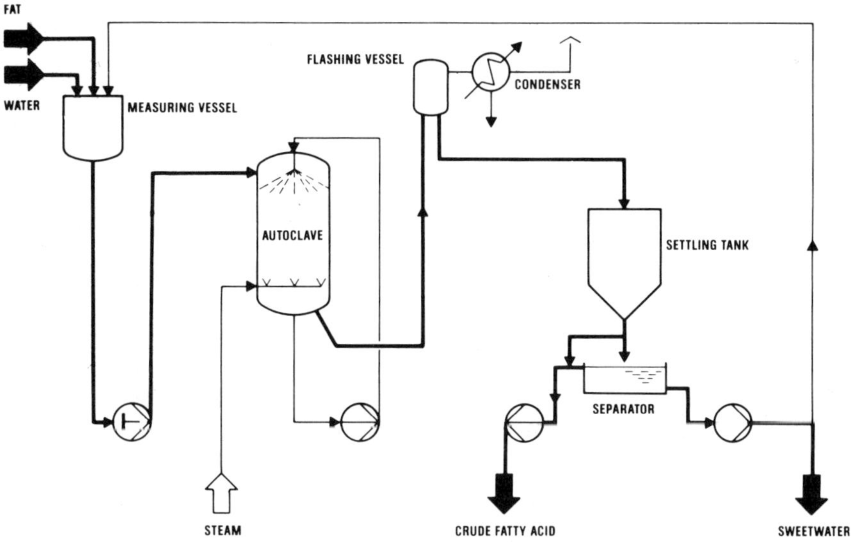

Fig. 5.4 – Lurgi batch-splitting plant.

The flow diagrams 5.5 and 5.6 are largely self-explanatory. In 5.5, CO3 is a flash vessel from which the water vapour passes to condenser EO1 and the liquid to settlers STO3 and STO4.

In 5.6, EO1 and EO2 are heaters using steam at 35 bar (507 psi), or other high temperature heat transfer medium, to bring the ingoing fat and water to operating temperature. CO3 is a sweet water flash pre-evaporator which also takes heat from the fatty acids passed through the coil on the way to the fatty acid flash vessel CO4. The flashed fatty acids are settled further in STO3 and STO4, where a little more water separates. DCO1 is a continuous sweet water decanter in which a little fatty matter separates.

Fig. 5.7 shows the *Colgate-Emery type* continuous splitting plant from *Wurster and Sanger.* As mentioned earlier, plants of this general type are offered by all the manufacturers. Typical working conditions for this type of plant are 250 – 260°C and a pressure of about 50 – 55 bar (725 – 800 psi), but some variation is possible (it will be appreciated from the earlier discussion that, for a given split, a higher temperature, which gives a higher reaction rate, permits a lower holding time, that is a smaller plant for a given throughput).

Mazzoni makes Colgate-Emery type plants with one tower (SAC1), two towers (SAC2) and three towers (SAC3). Fig. 5.8 shows an SAC2 plant with two towers. The use of two, or more towers in series, is equivalent to one very tall tower; and Mazzoni uses the multiple tower system as a convenient way to provide larger throughputs than that of their SAC1 type plant. The Mazzoni

SAC units are designed to use direct steam at 35 bar (507 psi), so that they presumably work at slightly lower temperatures than some other Colgate-Emery type plants.

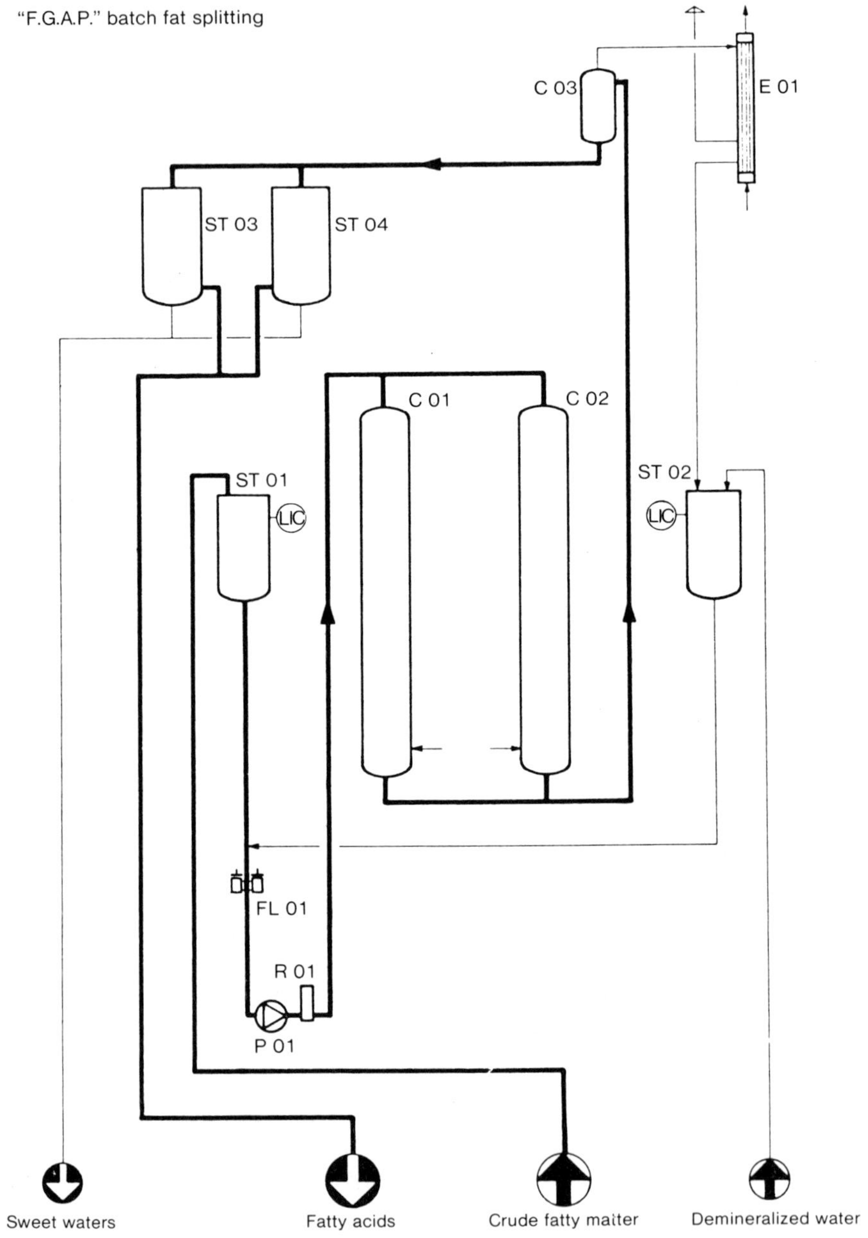

Fig. 5.5 – Gianazza batch-splitting plant.

5.3 MATERIALS OF CONSTRUCTION FOR SPLITTING AND OTHER FATTY ACID PLANTS

Materials of construction for use in the fatty acid industry are reviewed by Rice (1979). Austenitic stainless steels are widely used, particularly types:

304: 8 – 12% Ni and 18 – 20% Cr

316: 10 – 14% Ni, 16 – 18% Cr and 2 – 3% Mo.

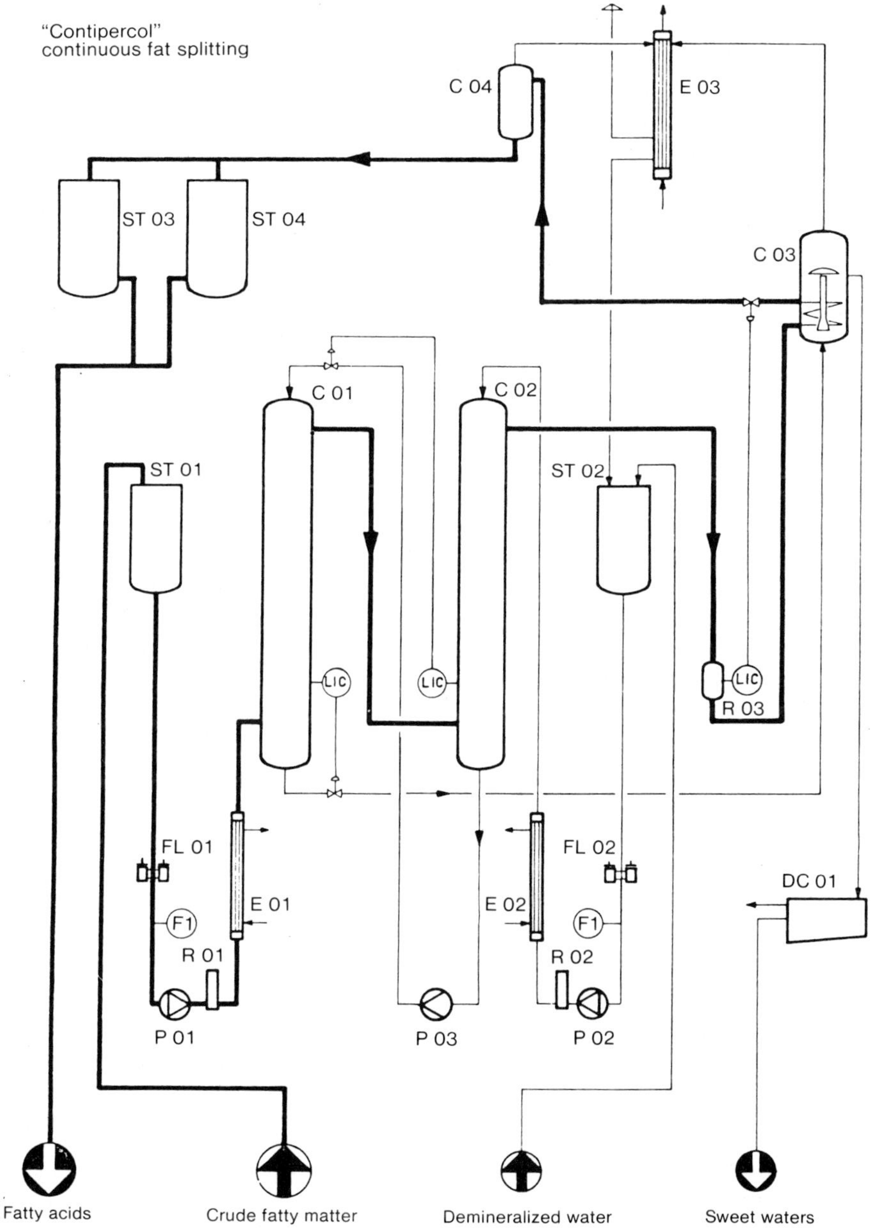

Fig. 5.6 – Gianazza continuous splitting plant developed from batch plant.

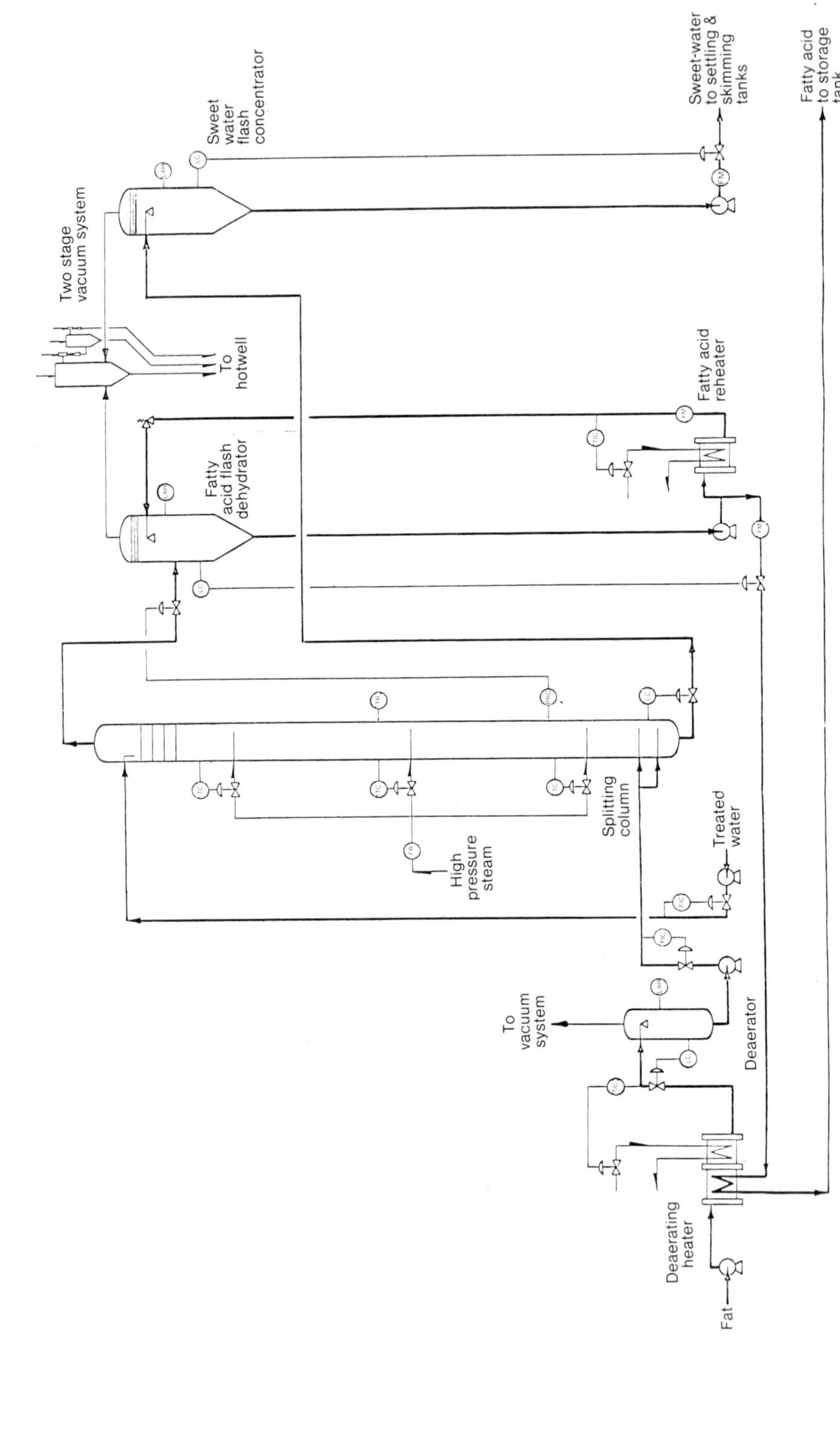

Fig. 5.7 – Wurster and Sanger continuous splitting plant with counter-current column.

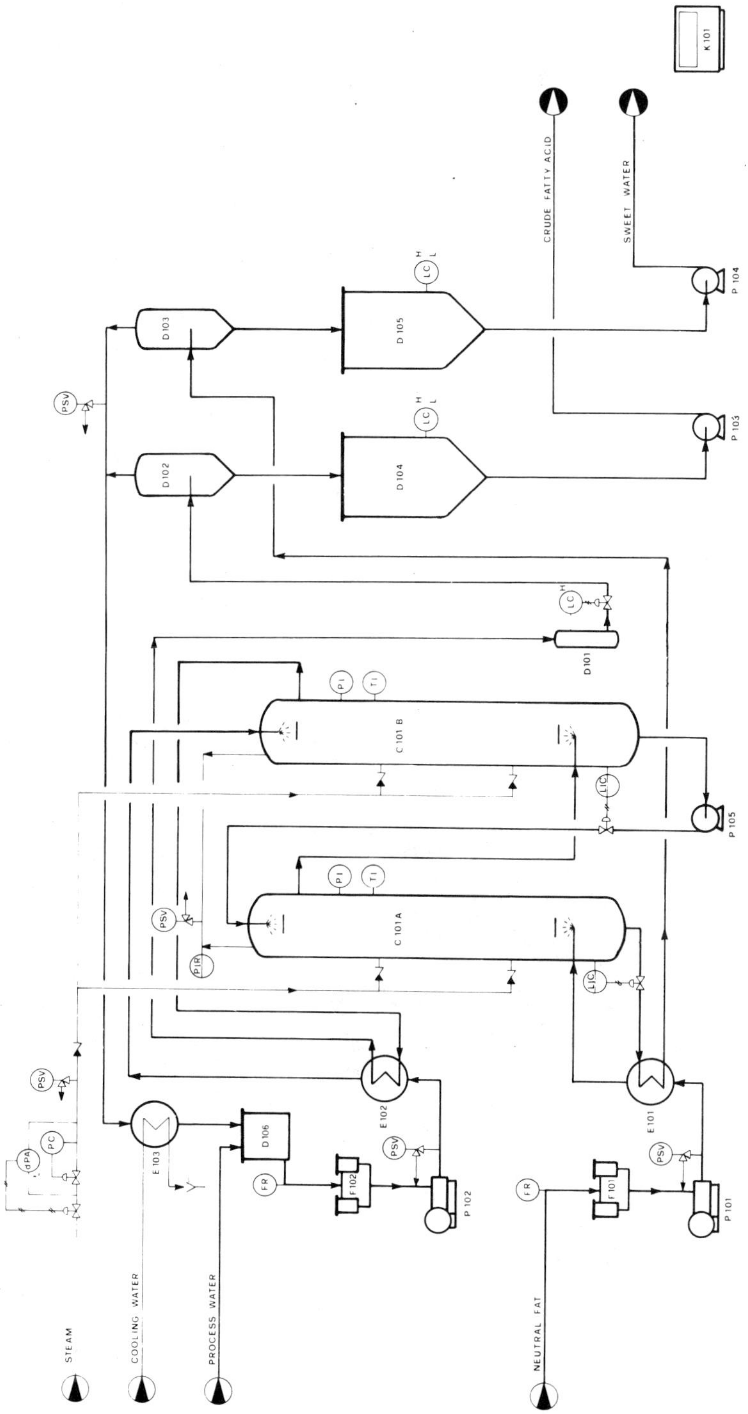

Fig. 5.8 — Mazzoni continuous splitting plant with two counter-current towers.

Fig. 5.9 – Photograph of Gianazza continuous fat splitting tower.

Type 304 is suggested for temperatures up to 150°C, but the more resistant type 316 is needed for higher temperatures. The special low carbon versions, or steels stabilized with titanium or columium, are necessary for many plant components to avoid risk of weld decay. Higher nickel alloys, such as Inconel (about 77% Ni, 7% Fe and 16% Cr), are also used. Carbon steel clad with corrosion resistant stainless steel, or high nickel alloy, is used for autoclaves and similar vessels for which a considerable thickness of metal is needed for mechanical reasons.

Aluminium is sometimes used for storage tanks, or other low temperature, low pressure, duties; but it is only suitable if the fatty acids are completely dry. Non-metallic materials, such as fibre glass–polyester for tanks, and phenolic or epoxy resin coatings, have limited applications. Fatty acids can be stored in mild steel tanks with little damage to the equipment, but iron pick-up can cause deterioration of the fatty acids and of the soaps made from them so that more resistant materials are usually necessary.

5.4 DISTILLATION OF FATTY ACIDS

5.4.1 General

Fatty acids directly from the splitting plant can be, and sometimes are, made into soap, but it is often necessary to improve their quality by distillation. Apart from the requirements of high grade soap bases, this is because one of the merits of the fatty acid route to soap is the possibility of using lower quality, and hence cheaper, raw materials than would otherwise be possible, by taking advantage of the upgrading which can be achieved by distillation of the fatty acids. Simple distillation of fatty acids using direct fired stills initially operated at atmospheric pressure, but later under vacuum, with considerable open steam, has been used for over a hundred years. These stills caused considerable damage to the fatty acids and are largely obsolete.

Distillation separates the volatile fatty acids from non-volatile glycerides and other compounds. If justified, the residue can be re-split with the liberation of more fatty acids, from which a low grade distillate may be obtained (or the re-split material may be returned to the main stream of crude fatty acids, although a residue of non-volatile, non-hydrolysable material must eventually be discarded). A high degree of split in the main hydrolysis process minimizes the amount of residue left on distillation and is important. The ultimate pitch left after as much fatty acid has been vaporized as is economically possible is a low value material which may have to be burnt. It is also usually desirable to separate the very low molecular mass. acids (C_8 or C_{10} and below), together with other very volatile constituents, to improve odour; and to remove acids from lauric oils which give somewhat irritant sodium salts. Mixtures of fatty acids, as produced from the natural oils and fats, can be separated into fractions useful for specialized applications, but this is not normally necessary for use in soaps.

The glycerides are not significantly volatile under the conditions used for fatty acid distillation, but they can be distilled in 'molecular', or 'short-path' stills; and this process is sometimes used to produce mono-glycerides and similar products.

5.4.2 Distillation principles

Fatty acids, particularly hydroxy and highly unsaturated types, are heat sensitive and are liable to be dehydrated, polymerized and/or isomerized. These, usually, but not invariably, undesirable, reactions are minimized by the use of the lowest possible temperature for the shortest possible time. Time and temperature are interdependent, so that a very short time at a relatively high temperature may be better than a longer time at a lower temperature; and vice versa.

Vapour pressure/boiling temperature data for some important fatty acids are given in Table 5.3 and for the region of greatest interest in Fig. 5.10. Little information appears to have been published for unsaturated acids other than oleic. Markley (1960) reviews available data on this and related topics.

Table 5.3 – Vapour pressure/boiling temperatures for some fatty acids

Fatty acid	Vapour Pressure mmHg						
	0.01	0.1	1.0	2.0	5.0	10.0	100.0
	Boiling temperatures °C						
Saturated C_4	–	–	26.0	35.5	50.0	61.5	107.5
Saturated C_6	11.5	34.0	62.0	72.5	86.5	98.5	146.5
Saturated C_8	33.5	57.5	87.5	98.5	113.5	126.5	177.0
Saturated C_{10}	55.0	79.0	110.0	121.5	137.0	150.5	202.5
Saturated C_{12}	73.0	98.5	131.0	142.0	158.0	172.5	226.5
Saturated C_{14}	89.0	116.0	149.5	161.0	177.5	192.5	249.0
Saturated C_{16}	105.0	132.5	167.0	179.0	197.0	211.5	–
Saturated C_{18}	119.5	147.5	183.5	196.0	213.0	–	–
Oleic	–	–	198	204	215	225	

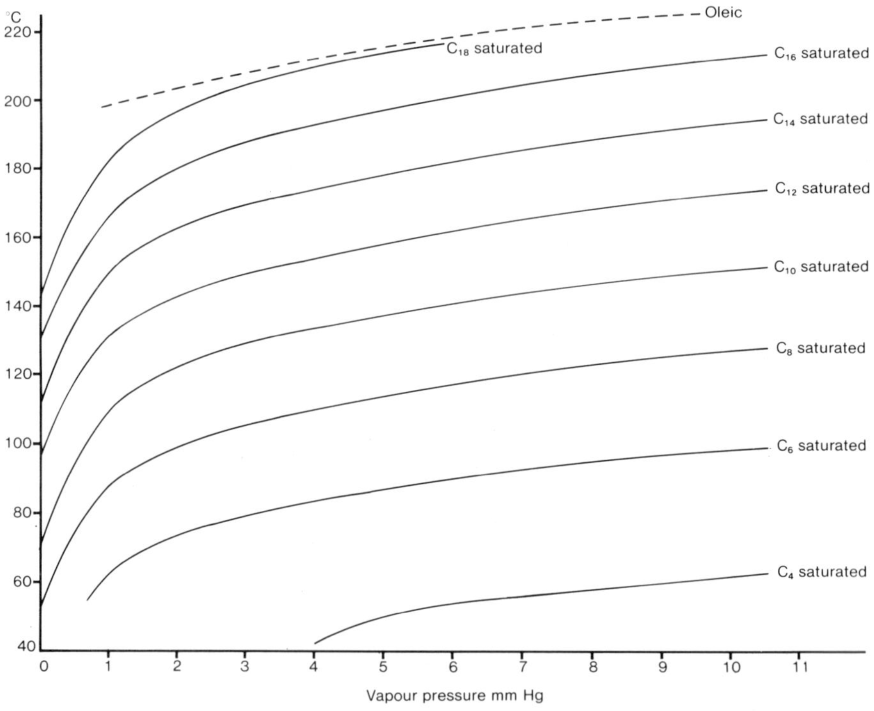

Fig. 5.10 – Vapour pressure-boiling temperature curves for some fatty acids.

The compositions of vapour and liquid in equilibrium at various temperatures for a binary (two component) mixture are shown diagrammatically in Fig. 5.11,

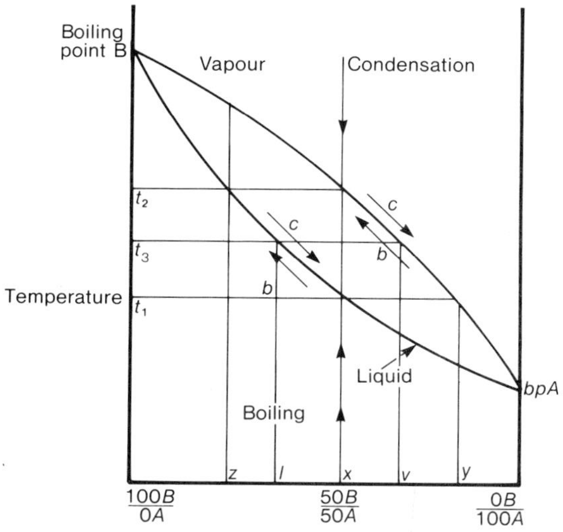

Fig. 5.11 – Vapour–liquid equilibria for a binary mixture at a given pressure.

drawn for a fixed total pressure. If a liquid of composition x is heated, boiling commences at temperature t_1 and the composition of the first vapour given off is y, that is it is richer in the lower boiling component A than the liquid from which it comes. This means that the liquid residue left is slightly richer in B, the higher boiling component. As vaporization proceeds, the boiling temperature rises and compositions change along the liquid and vapour curves as indicated by the arrows marked b (boiling). At temperature t_2 the whole of the liquid is vaporized and the mixed vapours have the original composition x. At intermediate temperatures, such as t_3, a vapour of composition v is in equilibrium with liquid of composition l; and the quantities are inversely proportional to the distances from composition x. This is the basis upon which fractional distillation is founded. The separation in one stage cannot be more than that for equilibrium at the relevant temperature; and a number of stages are required to effect more than a limited degree of separation. This is usually carried out in a continuous fractionating still shown diagrammatically in Fig. 5.12. The column comprises a number of plates on each of which rising vapour and descending liquid are brought into contact, so that some of component A is transferred from the liquid to the vapour in an approach to the new equilibrium relevant to the temperature of the plate in question. Instead of plates, which give a stepwise, stage by stage, separation, various types of packed columns are often used in which counter-current contact is continuous. Reverting to Fig. 5.11, if vapour of composition x is cooled, condensation will start at temperature t_2 and the first drop of liquid will have composition z. As cooling is continued, the composition of the vapour moves along the vapour curve and that of the liquid along the liquid curve, as indicated by the arrows marked c (condensation). At temperature t_3 the last drop of liquid condensed under equilibrium conditions has composition *l* and leaves a residual vapour of composition *v*, to be condensed separately; this is the basis of fractionation by partial condensation.

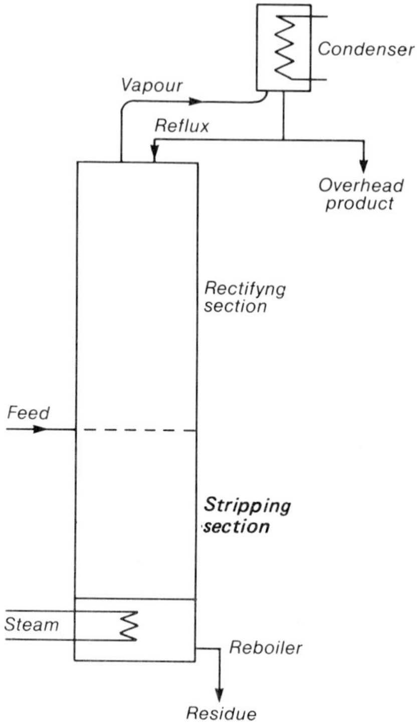

Fig. 5.12 – Fractionating still for a binary mixture.

Theoretical analysis of the fractionation of multi-component mixtures is complex, or even impossible, but some useful generalizations can be deduced from information concerning binary systems:

- the function of the part of the column above the feed point, called the rectifying section, is to remove less volatile components from the rising vapours which become the distillate.
- the function of the part of the column below the feed point, called the stripping section, is to strip more volatile components from the descending liquid which becomes the residue.
- a fractionating column can only work if liquid reflux is returned to the top, and a reboiler provides heat and vapour to the bottom.
- in general, only one substantially pure component can be separated from a single column. Hence an *n*-component mixture requires $(n-1)$ columns to separate it into its *n* individual compounds. More than one complete unit, each with column, reflux and reboiler, can sometimes be included in one shell. In the case of fatty acids, it is seldom required to separate the mixture into individual compounds.

Fractionating columns can be fed with liquid or vapour, or a mixture of liquid and vapour. Liquid feed increases the flow down the stripping section of the column, whereas vapour feed increases the flow of vapour up the rectifying section.

From Table 5.3 it will be appreciated that very high temperatures are required to vaporize the fatty acids from a fat under atmospheric pressure; and this would result in major damage. It is, therefore, necessary to reduce the boiling temperature by the use of reduced pressure (that is vacuum) and/or a carrier fluid, normally steam. The open steam used as a carrier fluid commonly also provides the agitation needed for good heat transfer and the avoidance of local overheating. Consider first the distillation of a pure liquid A which is immiscible with liquid water (W). The vapour pressure of A is plotted as shown in Fig. 5.13 and its boiling point under total pressure P_T is t_A. The lower P_T the lower t_A in the usual way; and as illustrated in Table 5.3. The vapour pressure of water is plotted downwards from the total pressure point P_T, so that the

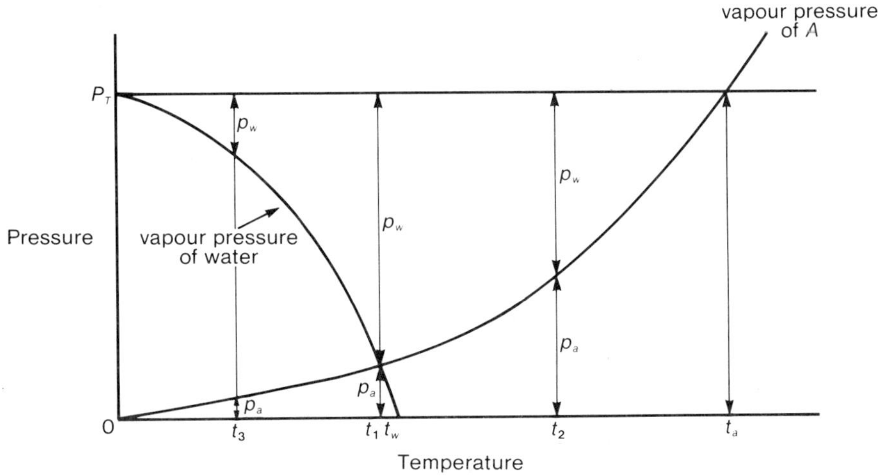

Fig. 5.13 – Steam distillation of a pure liquid.

boiling point of water under pressure P_T is t_w. p_A represents the partial pressure of A in the vapour at any temperature and p_w that of water. At $t_1, p_A + p_w = P_T$; and this is the only condition under which steam distillation can occur with liquid water as well as liquid A in the still. At t_3, $p_A + p_w$ is less than P_T, so that the liquids do not boil. At t_2, $p_A + p_w = P_T$ and liquid A boils, but there is no liquid water in the still. From Dalton's law of partial pressures

$$\frac{p_w}{p_A} = \frac{\text{moles } W}{\text{moles} A}, \text{ so that } \frac{\text{mass } W}{\text{mass } A} = \frac{p_w \times M_w}{p_A \times M_A}$$

where

$$M_w = \text{molecular mass of water} = 18$$

and

$$M_A = \text{molecular mass of } A.$$

Since $p_w = (P_T - p_A)$, this equation can be rewritten, mass in lb of open steam/lb of A distilled

$$= \frac{(P_T - p_A)\, 18}{p_A \times M_A} \, .$$

When A is substantially a single compound, as in the case of glycerine distillation, useful quantitative results can be obtained from this equation (see section 8.6.3). For complex mixtures such as fatty acids, the equation provides qualitative indications of the significance of changes in total pressure, use of open steam and supply of indirect heat. The indirect heat provides the latent heat of vaporization plus other heat requirements, such as sensible heat to raise the temperature of the feed to its boiling point and to replace heat lost by radiation and convection from the still. The actual distillation conditions establish themselves so that both the vaporization conditions just discussed and the condition of heat balance between heat input and requirements are met. Consider a still in which heat transfer conditions and a certain flow of open steam are set. A boiling temperature will be established so that (a) $p_A + p_w = P_T$, and (b) the heat input balances the requirements. This also determines the rate of vaporization and the ratio of open steam to fatty acids vaporized.

In the discussion so far, it is assumed that the partial pressure of A in the vapours equals the vapour pressure at the distillation temperature; that is the vaporization efficiency, defined as p_A actual/p_A theoretical, is 1. If contact between open steam and boiling liquid is good, so that the steam is nearly saturated with organic vapour, this is often approximately true (see section 8.6.3). In *The Chemical Engineers Handbook* (edited by Perry), J. S. Carey from his own work states that vaporization efficiency, E, varies with the substance being distilled as well as with physical conditions in the still. With the steam dispensed through small holes at an adequate depth below the surface of the liquid, he quotes 0.9 – 0.95 for organic compounds of molecular mass below 100, but only about 0.5 for the steam distillation of lubricating oils of molecular mass greater than 250. Data for fatty acid distillation are not available. If a figure for E is known, or assumed, Ep_A should be substituted for p_A in the expressions in this section.

It will be noted that the usage of open steam per lb fatty acids distilled is reduced by:

- reduction in the operating pressure at the point of vaporization. This depends upon the pressure drop through the system as well as the absolute pressure induced by the vacuum maintenance system (see Appendix 2). In the limit, no open steam is used and the mix boils at t_A.
- an increase in the boiling temperature which increases p_A. This results from higher indirect heat input relative to the flow of open steam.
- an improvement in the contact between open steam and boiling liquid if E is initially low.

Conversely, boiling temperature can be reduced by an increased usage of open steam, as well as by a reduction in total pressure. If open steam usage is increased, care must be taken to avoid an undue rise in P_T due to the increased flow of open steam through the system to the vacuum equipment, and consequent increase in pressure drop, (see Appendix 2 again).

As well as volatile material, mainly fatty acids, the crude fatty acids fed to a still contain non-volatile components, such as glycerides, unsaponifiables and proteinaceous matter. For convenience, the volatile part of the mixture is called V and the non-volatile R (residue), but both include compounds of varying volatility. By the laws of physical chemistry concerning solutions, the vapour pressure of V in equilibrium with a mixture of V and R is given by

$$p_{V\,\text{mix}} = p_V \times \frac{M_V}{M_V + M_R}\text{ , where}$$

$$M_V = \text{moles of } V \text{ present in the mixture}$$

and

$$M_R = \text{moles of } R \text{ present in the mixture}$$

Hence, as V is removed from the liquid by vaporization, $p_{V\,\text{mix}}$ at a given temperature falls. In a batch, or semi-batch, still the more volatile components are vaporized first; and the residue in the pot becomes increasingly rich in the higher boiling and non-volatile components. The boiling temperature and the ratio of open steam to material vaporized rise until it is not economical to proceed further. In a continuous still, as now normally used, it is necessary to design the vaporizer so that the residue discharged is as free as possible from volatile matter. Methods used are discussed in section 5.4.3.1.

The direct heat can be supplied in two ways:

- by transfer through a surface to the liquid at its boiling point under still conditions; or approximately so.
- in an external heater in which the material is heated under a relatively high pressure so that the temperature rises, after which the hot liquid is discharged into the still where the lower pressure results in flash evaporation by conversion of sensible into latent heat. Under still conditions, the specific heat of fatty acids appears to be about 0.7 (0.5 at 25 – 100°C) and latent heat of vaporization about 100 CHu/lb. Hence, roughly, $1 \times \Delta t \times 0.7 = x \times 100$ where Δt°C is the fall in temperature when the mixture is flashed and x is the fraction vaporized.

  From this:

Δt°C	x
50	0.35
100	0.70
143	1.0

5.4.3 Distillation equipment

5.4.3.1 Vaporizers

In addition to points just elaborated, the requirements for a good vaporizer can be summarized as follows:

- the best combination of time and temperature, with both as low as possible, should be achieved. This means that any reservoir of boiling liquid should be as small as possible.
- the pressure under which the liquid boils should be as low as possible. This is primarily controlled by the vacuum equipment, and by the pressure drop through the condensing system and connecting ducts, but it can also be affected by the vaporizer. In particular, the hydrostatic head under which the liquid is heated is important. One foot of liquid of specific gravity 0.8 produces an additional pressure of about 19 mmHg which is very significant when the pressure in the vapour space is only, say, 4 mmHg.
- if open steam is used, the contact between steam and liquid should be intimate, so that E is as near 1 as possible.
- velocity over the heating surface should be sufficient to avoid local overheating and to promote good heat transfer.

Semi-batch stills commonly use *simple pot vaporizers,* with open steam to agitate the boiling liquid and to act as carrier fluid; and direct heat supplied from a furnace, or via a carrier such as Dowtherm. A simple still of this type is shown in Fig. 5.14. As the initial charge of crude fatty acids is distilled more feed is

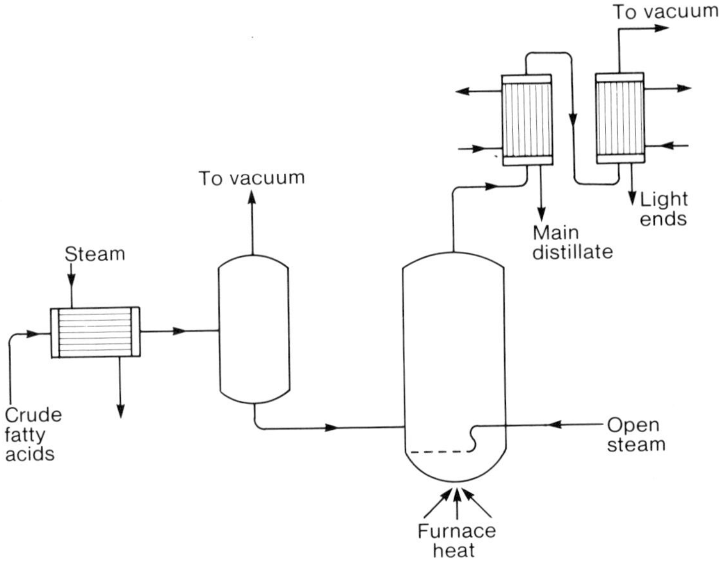

Fig. 5.14 – Simple semi-batch still.

introduced. The concentration of non-volatile matter in the still liquid increases until it is necessary to stop the feed and to work down the residue as completely as is economically possible. This type of still fails to meet the criterion regarding time and temperature. It is possible to collect the distillate in separate fractions, and an impure first fraction (and possible also last fraction) may be separated.

Otherwise, the only fractionation is by slight partial condensation; and distillate quality and yield tend to be poor because of thermal damage. The design is obsolete, but may still be used in some small capacity plants.

The *Lurgi vaporizer,* shown diagrammatically in Fig. 5.15, uses a number of compartments to give a continuous vaporization system. The crude fatty acid feed is partially vaporized in each compartment; and the residual liquid flows

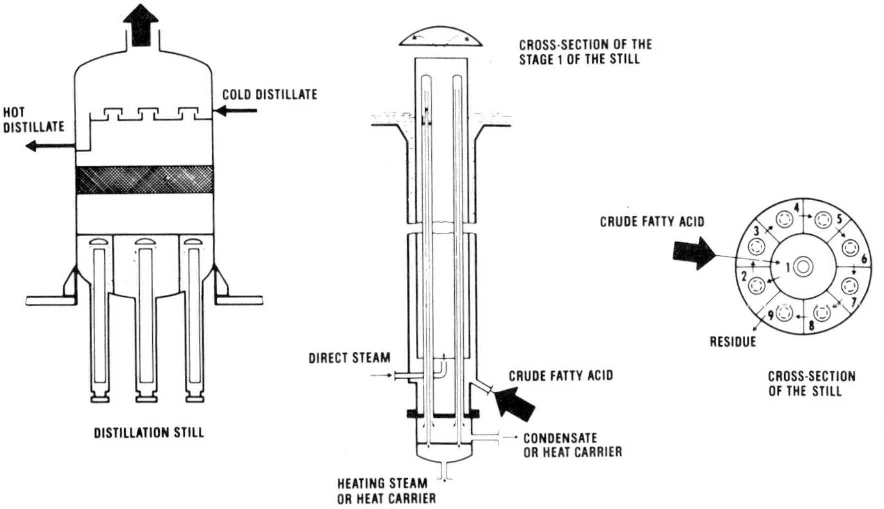

Fig. 5.15 – Lurgi vaporizer.

from compartment to compartment. Volatile components are removed progressively, so that the residue leaving the last compartment is substantially free of volatiles. In addition, the boiling temperature rises gradually from compartment to compartment as the concentration of non-volatiles increases, so that the bulk of the vaporization occurs at the lowest possible temperature. The annular heating elements ensure a positive flow of high pressure steam, or other heating medium. The open steam produces a 'vapour-lift' effect, similar to that in an air-lift pump, which causes a good flow of the fatty matter over the heating elements and minimizes the adverse influence of hydrostatic head. As shown, the mixture of steam and fatty matter impinges on a baffle, the vapours separate, and the liquid drops back to be recirculated, or to pass on to the next compartment. 5–10 compartments are used, depending upon the capacity of the plant.

Some other plant manufacturers use a broadly similar type of vaporizer.

The still shown diagrammatically in Fig. 5.16, was patented by *Mills,* and assigned to Procter and Gamble around 1940. External heaters are used to raise the temperature of the fatty matter so that it flashes when discharged into the still body which is maintained at about 3 mmHg pressure. As shown, the small volume of liquid in the still is circulated by a vapour-lift effect. The liquid which remains after the first stage is reheated in the secondary vaporization unit. The boiling point of a mixture of stearic, oleic, and palmitic acids, as from a tallow

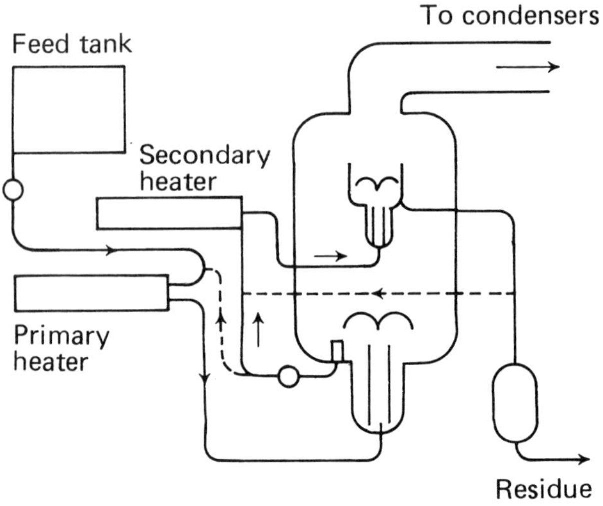

Fig. 5.16 – Victor Mills still.

or grease, is about 200°C at 3 mmHg pressure, disregarding the influence of the non-volatile components which, in practice, seriously impede the later stages of vaporization. The temperature to which the mixture is heated in the primary heater is said to be about 450°F (232°C), so that if no vaporization occurs in the heater, the equation in section 5.4.2 indicates that the fraction vaporized when the material is flashed equals $(32 \times 0.7)/100 = 0.22$. Similar treatment of the liquid fraction (0.78) in the secondary stage would vaporize another 0.17, making a total of 0.39 (39% of the feed). This may have been increased by substantial vaporization in the heater, but facilities are also provided for recirculation round each stage, as shown dotted. As discussed earlier, the vapour pressure falls to a very low level as the residue is stripped nearly free from volatiles; and this makes flash distillation alone difficult.

Wurster and Sanger use a vaporizing column comprising bubble cap trays heated by Dowtherm vapour as shown in Fig. 5.17. The liquid flows across the trays and the unvaporized portion passes through the downcomers to the next tray below. Just as in a fractionating column, the vapour through the slots in the bubble caps produces violent agitation and good contact between liquid and vapour. A separate vapour outlet is provided above each pair of trays to minimize pressure drop. Open steam is injected into the liquid in the reservoir at the base of the column, and below the lower tray in each higher pair.

Falling film vaporizers are used by some plant manufacturers when the fatty acids to be distilled are highly unsaturated and very heat sensitive. In some

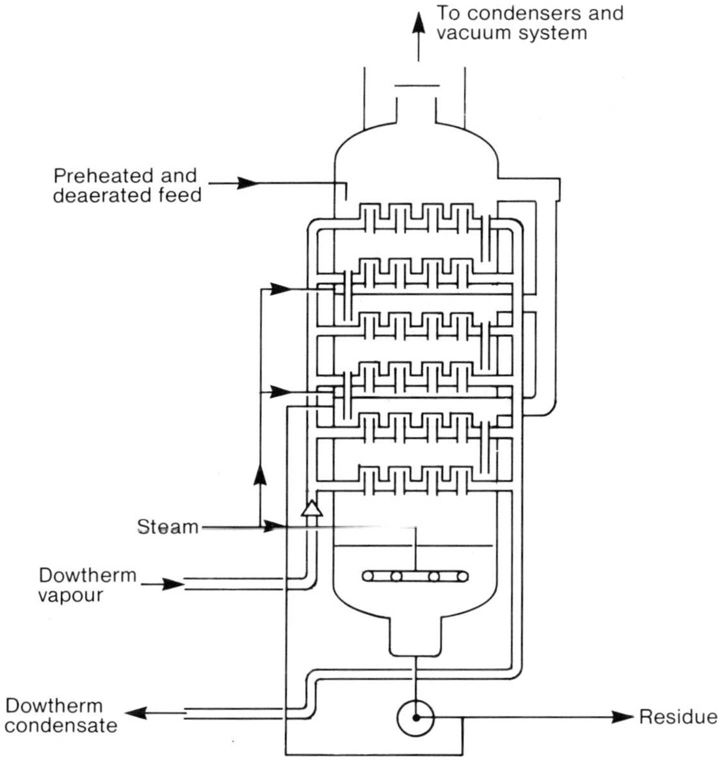

Fig. 5.17 – Wurster and Sanger vaporizer.

designs, the vaporizers are tubular heat exchangers with devices to cause the liquid to flow down as annular films inside the tubes. It might also be possible to use scraped wall vaporizers such as the type made by *Luwa*, as discussed later (section 8.6.4) for glycerine distillation, provided the residue is sufficiently liquid to emerge satisfactorily from the base of the unit.

5.4.3.2 Fractionators, condensers and strippers

If the vapours leaving the vaporizer are liable to carry droplets of liquid, it is desirable to remove these droplets as completely as possible to avoid contamination of the distillate; baffles, packed beds, or cyclones are used, but, as for all parts of the system, pressure drop is important.

To achieve a major separation of the components in a mixed vapour, as in the production of nearly pure lauric acid from coconut, or palm kernel, fatty acids, fractionating columns with substantial numbers of plates (or equivalent heights of packing), and using reflux and reboilers, are necessary. This exposes the liquid to increased time and temperature despite efforts which are usually made to hold as little as possible in the reboiler. Lauric oil fatty acids can withstand more drastic thermal conditions than the more unsaturated types (and they also have lower boiling points), so that more efficient fractionating columns can be used.

Simple columns may be used to separate good quality distillate from a relatively small low boiling fraction containing low molecular mass acids and volatile impurities; but these are usually stripping devices rather than columns with reboilers. This separation of an impure, low boiling, fraction is often achieved by the use of two separate condensers, arranged so that the vapours leave the first at a temperature suitable to provide the required partial condensation. Condensers can be surface or direct contact. With surface condensers the coolant should be at least 5°C above the titre of the fatty acids to avoid risk of tube blockage by solidification. In a direct contact condenser, cooled condensed fatty acids are re-circulated to the top of the column as shown diagrammatically in Fig. 5.18. The heat extracted in the liquid cooler balances that

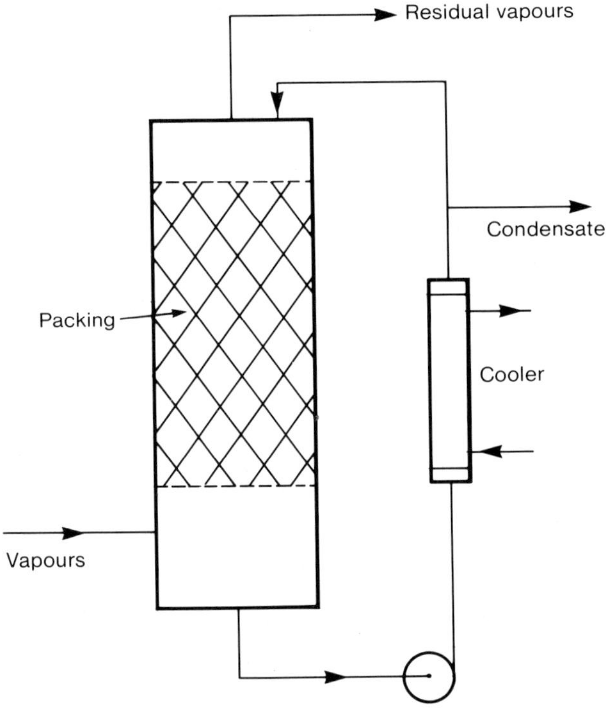

Fig. 5.18 – Direct contact condenser/stripping column.

needed to condense the required quantity of fatty acid vapour in the packed column. This piece of equipment can also be called a stripper, or stripping column, because the intimate counter-current contact between rising vapour and descending liquid permits a transfer of components. Some of the more volatile components of the liquid fed to the top of the column are vaporized and replaced by less volatile components from the vapour. A similar effect can be achieved in other devices, as explained in the next section.

Because of the ease with which some fatty acids can be oxidized, distillates should be adequately cooled before they are exposed to atmospheric air.

5.4.3.3 Complete stills

Many different types of fatty acid still have been, or still are, made and used. In this section a number of designs taken from recent brochures of the major manufacturers are illustrated and described briefly. Most of the specialized items which they use have been discussed in the previous sections.

Fig. 5.19 shows the *Lurgi simple distillation plant.* This uses the multi-compartment vaporizer (Fig. 5.15), a surface condenser for the main distillate, and a direct contact tower to strip the most volatile fraction from the steam. A special feature is the bubble cap tray in the top of the vaporizer on which the more volatile components are stripped from the main distillate by the rising vapours.

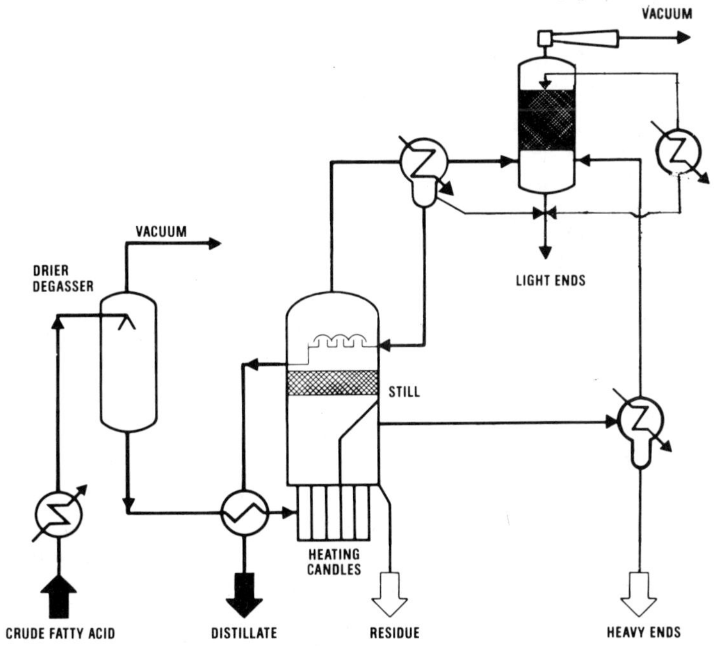

Fig. 5.19 – Lurgi simple distillation unit.

The *Lurgi fractionating fatty acid still* is represented in Fig. 5.20. This includes two fractionating columns each with reflux and reboil; and the reboilers are said to have low liquid volumes to minimize thermal deterioration of the fatty acids being processed. The main condensation is in surface condensers,

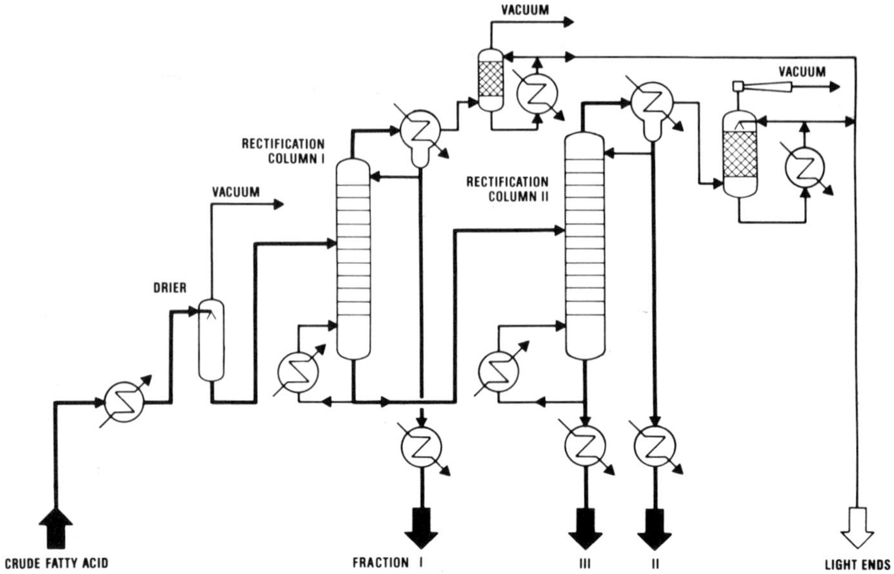

Fig. 5.20 – Lurgi fractionating still.

but direct contact columns are again used to condense the most volatile components. Open steam is used as required to aid vaporization, depending upon the material being distilled. The plant shown produces three fractions, plus the light ends, in fully continuous operation. For smaller plants, other arrangements are possible in which fractions are produced semi-batchwise from a single column.

Fig. 5.21 gives a flow sheet for the *Gianazza type 'V70' fatty acid distillation plant* intended for normal materials; and Fig. 5.22 is a photograph of the vaporizer. It uses a multi-compartment vapour-lift vaporizer similar in principle to the Lurgi. The first compartment operates at a relatively low temperature which vaporizes the lowest boiling fraction and these vapours are condensed separately. Similarly, the last compartment operates at a higher temperature than the rest of the still to vaporize the least volatile components as completely as possible. The condensate from this compartment is returned to the main still for redistillation. Condensation is in surface condensers, followed by a direct contact condensing tower to collect the most volatile fraction. The surface condensers which are not used to heat the process liquids are cooled with demineralized water at an appropriate temperature in order to avoid scaling; and this water is cooled with ordinary water in EPO3 shown in the detached section of the diagram.

Gianazza also offer a *type 'FF80'* still which uses two falling film vaporizers as indicated in Fig. 5.23. The crude fatty acids are preheated, and then deodorized and dehydrated in EO2/CO1. They pass to the first falling film unit EO4. The unvaporized portion is partly re-circulated and partly transferred to the second stage falling film section (EO5).

Fig. 5.24 shows the *Wurster and Sanger still.* The vaporizer has already been described and the rest of the system is evident from the diagram.

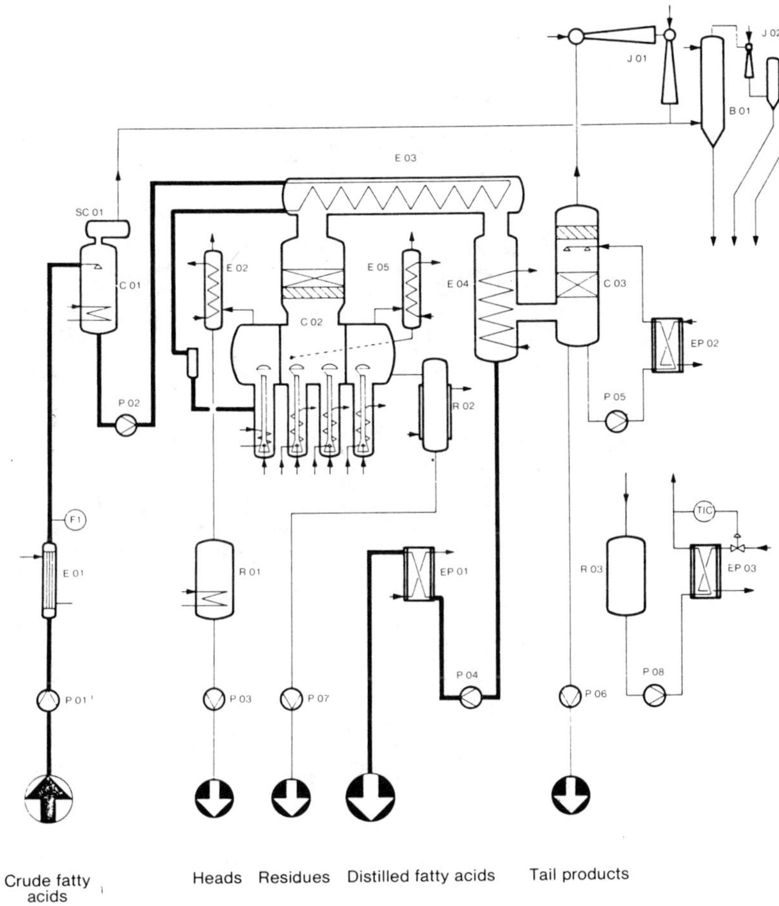

Fig. 5.21 – Gianazza type V70 fatty acid distillation plant.

The *Mazzoni type DAG simple continuous still* and their *type FAG/06 fractionating still* are shown in Figs. 5.25 and 5.26 respectively. The vaporizer 7 in Fig. 5.25 is of the multi-compartment type with a direct heating element and a supply of open steam in each section. In Fig. 5.26 TO/EO is a dryer/deaerator and E1.2 a falling film vaporizer. Because the material is fed to the column T with a substantial portion in vapour form the upper part of the column has a larger diameter than the lower section in which volatiles are stripped from the residue.

5.5 SEPARATION OF SOLID AND LIQUID FATTY ACIDS

The separation of fatty acids from fats such as tallow into a solid fraction, stearin, and a liquid fraction, red oil or olein, is the concern of fatty acid manufacturers rather than soapmakers (although stearin is used in shaving soaps); and the processes will only be mentioned briefly. The traditional method is to

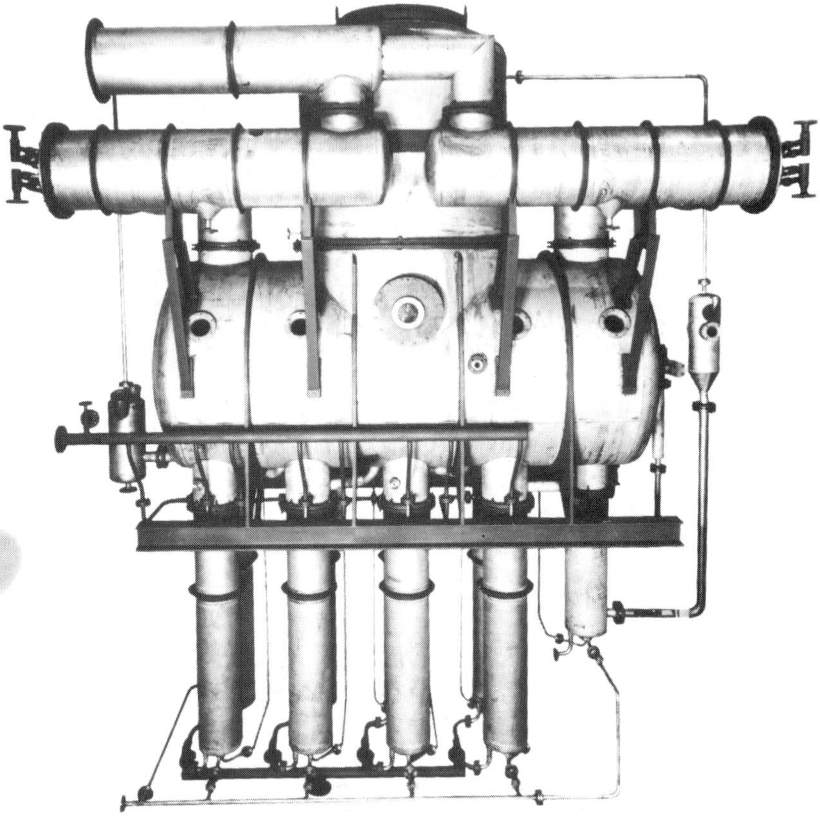

Fig. 5.22 – Photograph of vaporizer in V70 type still.

crystallize the fatty acids in trays, wrap the blocks in cloth and press. This gives single pressed stearin, and reprocessing with heated presses gives double and triple pressed stearin, which contain less unsaturated material than the single pressed grade. A more recent method involves crystallization from a solvent; which is methanol in the Emersol process. Lurgi has a third type of process in which stearin crystals are wetted with a solution of wetting agent to give an aqueous dispersion which can be separated centrifugally from the non-dispersed olein phase.

5.6 PRODUCTION OF SOAP BASE FROM FATTY ACIDS

Points to be considered in the selection of a process for the production of soap base from fatty acids include:

- as there is no glycerol to be recovered and, at least if distilled, the fatty acids should be clean, 'all-in' processes are attractive and are normally used. Alternatively, it is possible to follow neutralization of the fatty acids by

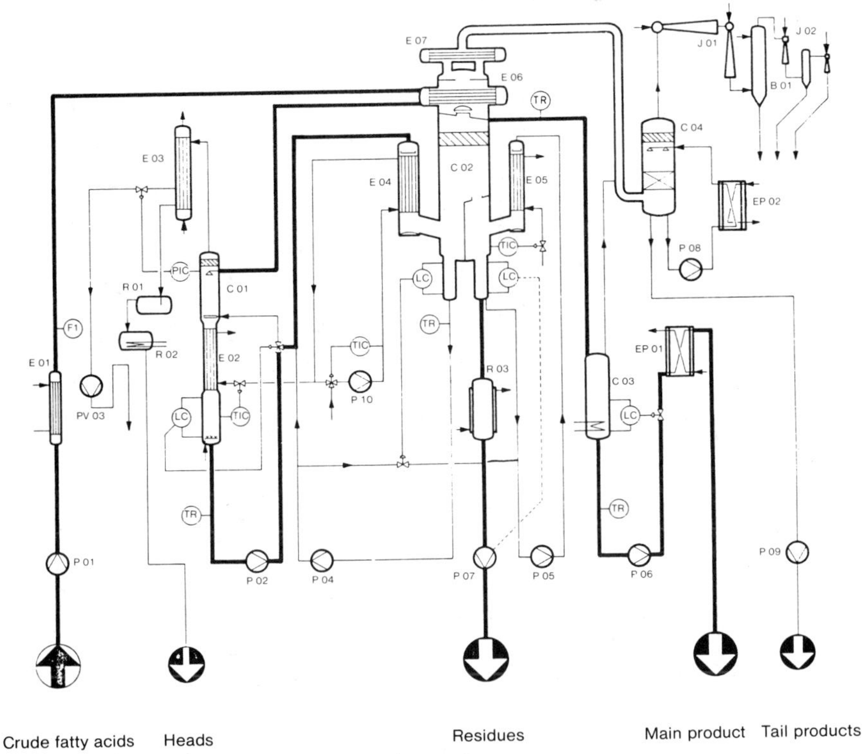

Fig. 5.23 – Gianazza type FF80 fatty acid distillation plant.

soapmaking operations including removal of a lye and/or a nigre; this makes the correct proportionation of fatty acids and alkali somewhat less critical, and removes dirt if the fatty acids are crude.

- sodium carbonate (soda ash) may be cheaper than caustic soda per unit of Na_2O and it can be used to neutralize fatty acids, although the operation is normally completed with caustic soda. Problems arise with the use of sodium carbonate because of the evolution of carbon dioxide gas, which needs to be separated and removed, and because of the formation of sodium bicarbonate. When neutralization with sodium carbonate (to, say, 80%) is completed with caustic soda, any sodium bicarbonate reverts to sodium carbonate. The caustic soda can also saponify any glyceride in the fatty acids. Relevant reactions are:

$$2R\cdot COOH + Na_2CO_3 \rightarrow 2R\cdot COONa + H_2O + CO_2$$
$$R\cdot COOH + Na_2CO_3 \rightarrow R\cdot COONa + NaHCO_3$$
$$NaHCO_3 + NaOH \rightarrow Na_2CO_3 + H_2O.$$

- neutralization of fatty acids with concentrated (48% NaOH) caustic soda gives a soap at well above 63% TFM unless additional water is used and, if the soap is to be dried for the production of toilet soaps, it appears attractive

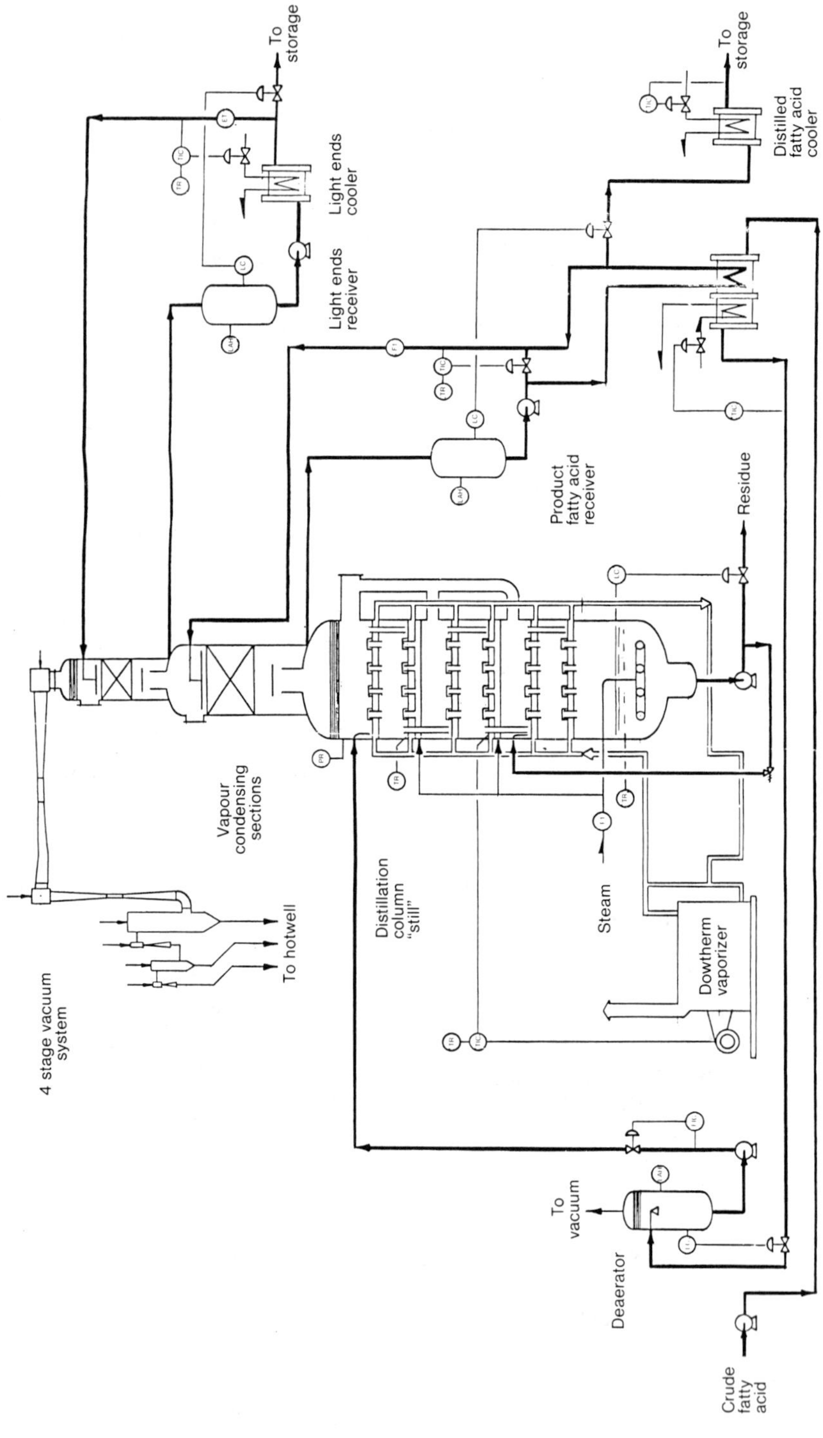

Fig. 5.24 – Wurster and Sanger continuous fatty acid distillation plant.

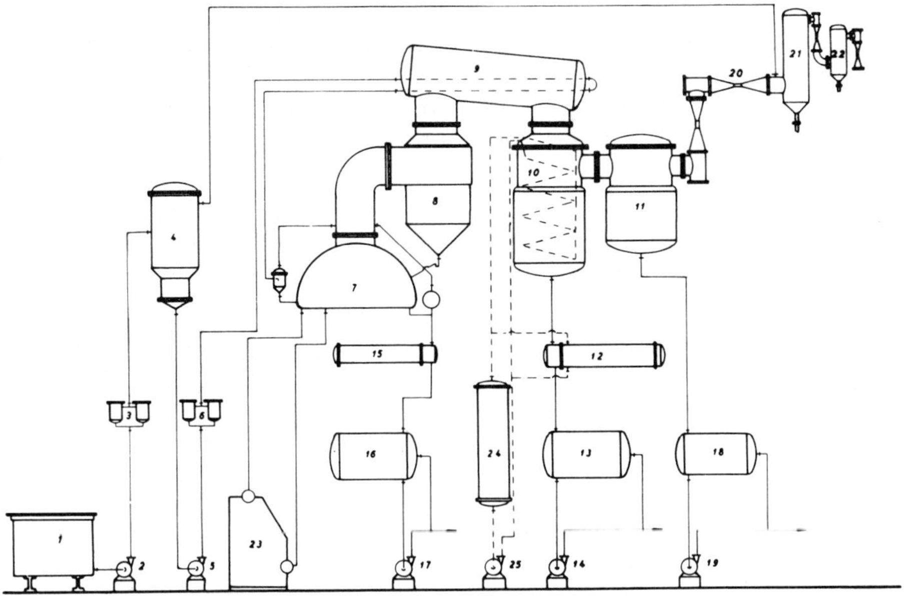

1. Raw fatty acid feed tank.
2. Power-driven feed pump for crude fatty acids.
3. Filters.
4. De-aerator dehydrator for crude fatty acids.
5. Power-driven feed pump for dried crude fatty acids.
6. Filters.
7. Still.
8. Cyclone separator.
9. Heat exchanger.
10. Surface condenser.
11. Mist separator.
12. Distillate cooler.
13. Distillate receiver.
14. Power-driven pump for withdrawal of distillate.
15. Residue cooler.
16. Residue receiver.
17. Power-driven pump for the withdrawal of the residue.
18. Light end fraction receiver.
19. Power-driven pump for withdrawing the light end fraction.
20. Vacuum steam-jet.
21. Barometric condenser.
22. Steam ejectors group.
23. Steam boiler plant.
24. Distilled water cooler.
25. Electric distilled water circulation pump.

Fig. 5.25 – Mazzoni type DAG fatty acid still.

to save heat in the drying process by neutralization at a high TFM. Consider a blend of 80% tallow and 20% coconut oil (fatty acid ratio of blend about 8.55) neutralized with caustic soda solution containing 48% NaOH, or 37.2% Na_2O. Assume also an addition of 20% brine sufficient to give a salt content of 0.5% NaCl and of refined glycerine to give 0.4%, both calculated on TFM. Salt is normally added, and glycerine sometimes, to give a composition similar to that of neat soap.

1 tonne fatty acid requires 1/8.55 tonnes Na_2O, or $1/8.55 \times 100/37.2 = 0.314$ tonnes of 48% caustic soda solution. Add 0.4/100 tonnes glycerine and $0.5/100 \times 100/20 = 0.025$ tonnes of brine, to give a total mass of $1.000 + 0.314 + 0.004 + 0.025 = 1.343$ tonnes of soap with a TFM of $100/1.343 = 74.5\%$. If it is desired to produce soap at 63% TFM, the total mass must be increased to $1.343 \times 74.5/63 = 1.588$ tonnes which requires additional water equal to 0.245 tonnes, or a dilution of the caustic soda solution to about $48 \times 0.314/0.559 = 27\%$.

A plant has been described in which concentrated caustic soda is used; but at least one plant manufacturer has decided to cease experimental work on this project because of difficulties, particularly with automatic control of the free alkali using an electrode potential method which is analogous to pH. It is thought that dilution of the caustic soda to give a soap at about 63% TFM is usual.

The simplest way to neutralize fatty acids with sodium carbonate is in a pan. An approximately saturated sodium carbonate solution is brought nearly to the boil and the fatty acids are added gradually. Sufficient time must be allowed between additions to permit the carbon dioxide to escape without the pan contents boiling or effervescing over. The saponification is completed with caustic soda.

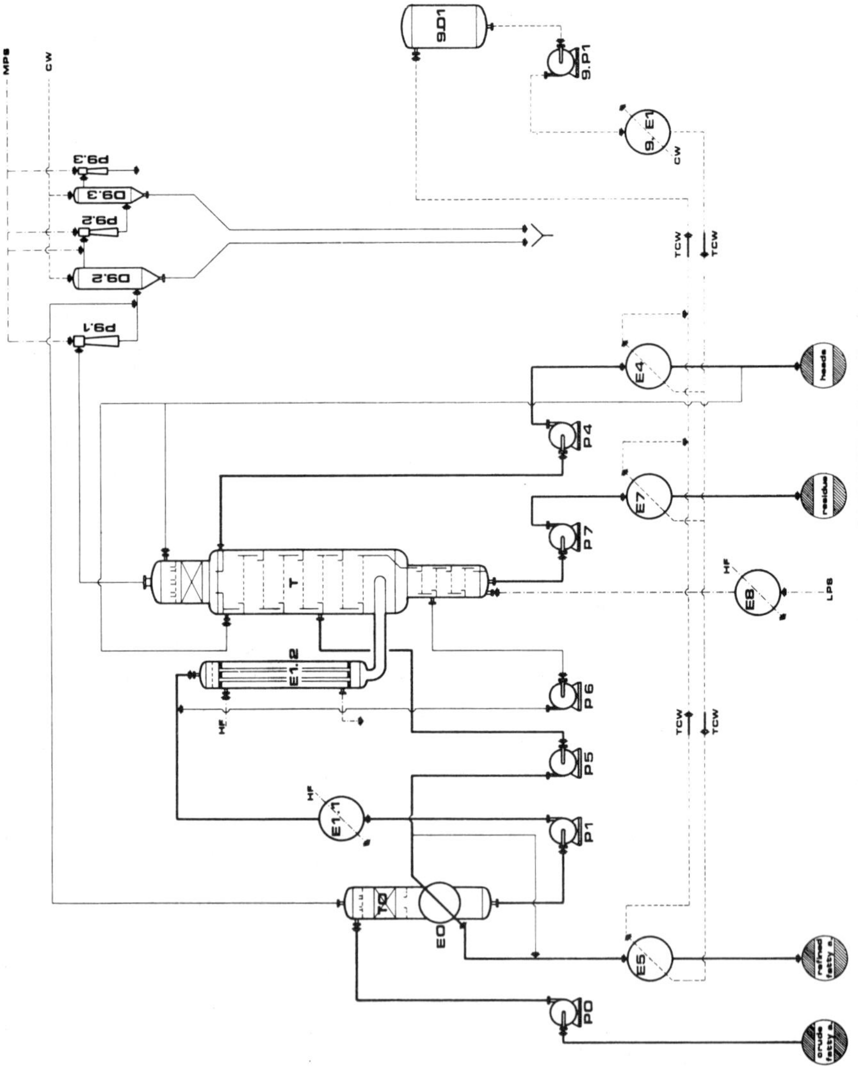

Fig. 5.26 – Mazzoni type FAG/06 fractional distillation plant.

Mazzoni offers two continuous plants for the production of soap base from fatty acids, the SC using caustic soda, and the SCC using both sodium carbonate and caustic soda. *Meccaniche Moderne* also offers plants. Fig. 5.27 shows the *Mazzoni SC plant*. The reagents, in proper proportion, are added to a circulating

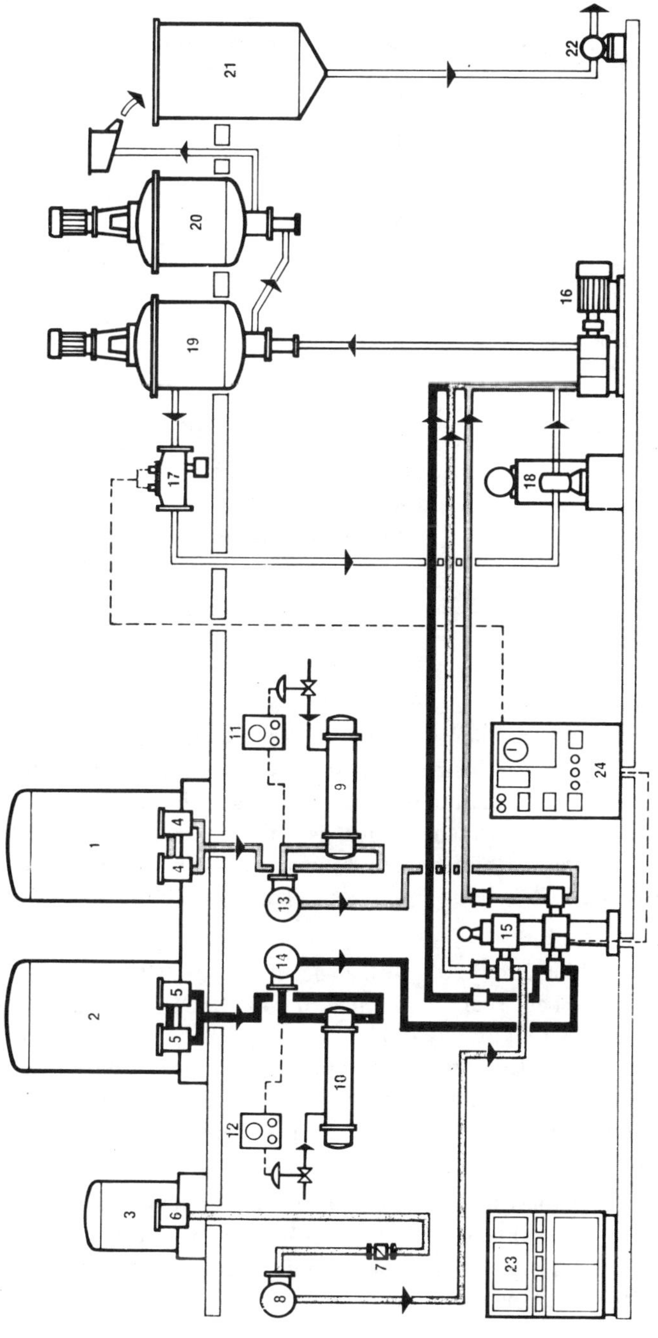

Fig. 5.27 – Mazzoni SC plant for neutralization of fatty acids with caustic soda.

stream of soap in saponifier 16. The mix is circulated from 16 through the first stage mixer 19 and control device 17 by pump 18. A proportion is bled off to the second stage mixer 20, where the reaction is completed.

An electrode assembly in 17 measures the free alkali in the circulating soap, as shown in Fig. 5.28, and controls the addition of caustic soda solution.

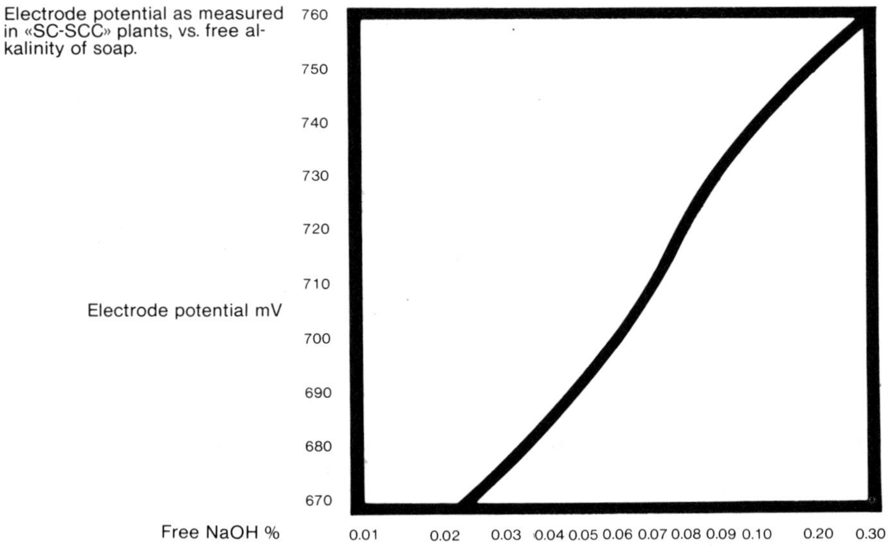

Fig. 5.28 – Mazzoni free alkali indicator.

The *Mazzoni SCC plant* is shown in Fig. 5.29. About 80% of the neutralization is carried out with sodium carbonate and the process is completed with caustic soda using equipment identical with the corresponding part of the SC process, as shown in Fig. 5.27. The major neutralization with sodium carbonate is carried out in mixer 13 and column 14. Column 14 is a film reactor-separator supplied with open stripping stream which is said to complete decomposition of sodium carbonate and bicarbonate, remove the carbon dioxide, and also remove any volatile odiferous bodies. Mazzoni claims the following results from these plants:

- a free alkali of 0 to 0.20% with an accuracy of 0.01%.
- with a finished soap containing 0.04% NaOH an unsaponified content below 0.08%, and usually 0.03 – 0.05%.
- a sodium chloride content at any set value between 0 and over 1%.
- with the SCC process using 80% neutralization with sodium carbonate, the production of a clear, transparent, gas free, liquid soap containing about 0.05% Na_2CO_3.

Other companies have devised plants for the neutralization of fatty acids, usually using a dominant bath of circulating soap, broadly as just described. At least one uses adapted De Laval equipment, including the viscosity device to control the addition of the caustic soda solution.

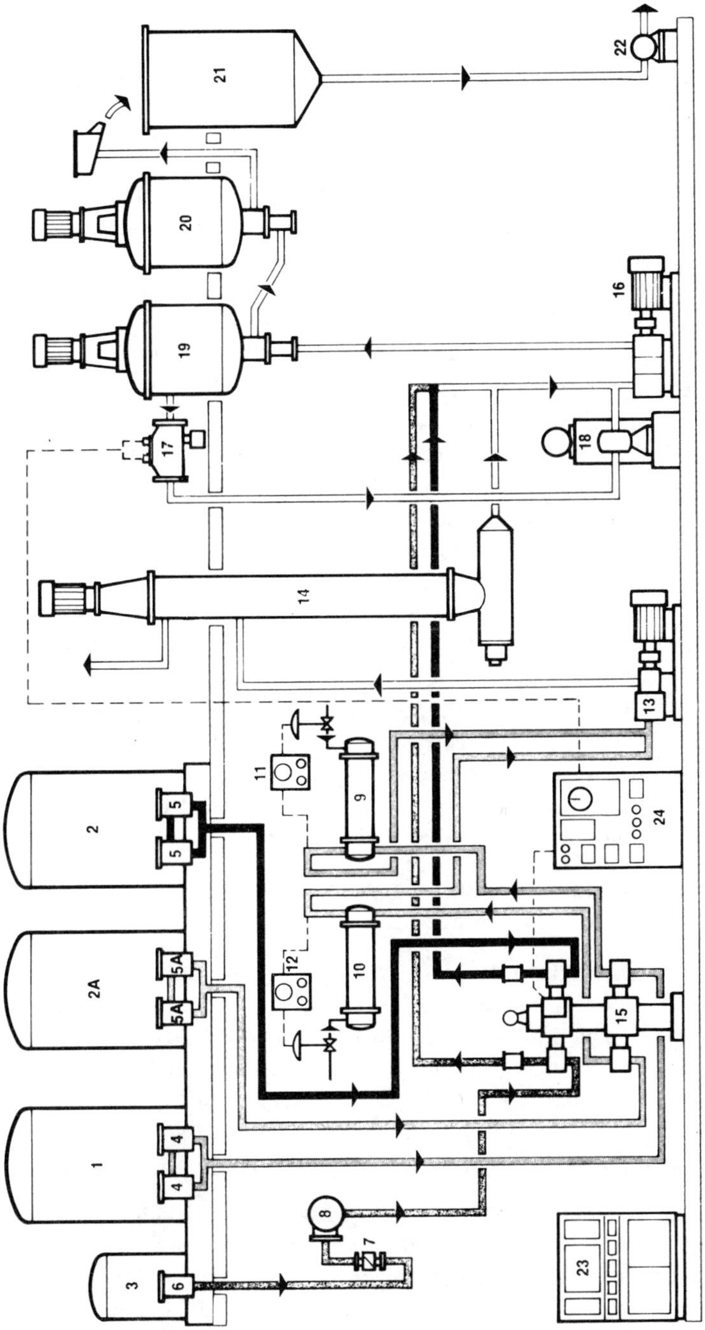

Fig. 5.29 – Mazzoni SCC plant for neutralization of fatty acids with sodium carbonate and caustic soda. 1. Fatty acid tank: 2. Caustic soda tank:2A. Soda ash solution tank: 3. Brine feed tank.

5.7 COMPARISION BETWEEN SOAP MANUFACTURE BY DIRECT SAPONIFICATION AND BY THE FATTY ACID ROUTE

The major companies do not publish the data required for a proper comparison between the costs of the direct saponification and the fatty acid route to soap. The comparison depends upon a number of circumstances which vary from factory to factory; and some important points are:

- the cost of the fatty raw materials. Splitting and distillation often permits the use of lower grade materials than are needed to produce soap of the required quality by the direct saponification process, which may lead to substantial savings. The availability of suitable low cost by-products from edible oil plants may be important.
- possible use of sodium carbonate with the fatty acid route.
- the yields of fatty acid converted into soap and of glycerol, taking into account material downgraded into lower quality soaps and to pitch discharged from fatty acid stills. Losses in pre-treatments, such as earth bleaching.
- initial capital costs; and capital and maintenance charges. High pressure splitting plants and fatty acid stills are expensive. Some continuous direct soapmaking plants are costly, but efficient direct soapmaking plants can probably cost substantially less than modern splitting and distillation equipment.
- cost of glycerine recovery. One of the factors which led to the development of the fatty acid route was the low concentration of glycerol in soap lyes then common; but, as explained in Chapter 4, this difference has now been eliminated or greatly reduced. The fact that sweet waters do not contain salt is significant, as discussed in Chapter 8.
- other operating costs, including that for the high pressure steam, or other heating medium, used in the fatty acid route.

In addition to the cost comparision between the two routes as the major method of soap manufacture:

- fatty acids may be needed for some products, such as certain modern soap powders (see section 7.3).
- companies which produce low grade by-products, such as black cotton grease, may find it worthwhile to split and distil such materials, even if their main soapmaking is by direct saponification.

5.8 MISCELLANEOUS METHODS OF SOAP PRODUCTION

5.8.1 Sodium carbonate saponification of oils and fats

Although glycerides cannot be saponified with sodium carbonate at atmospheric pressure, it is possible to make soap by this reaction at elevated temperatures under sufficient pressure to maintain an aqueous phase. Batch autoclaves were used, at least experimentally, for this purpose in Czechoslovakia around 1938. Satisfactory soap was made but, so far as is known, the process has not been developed further.

5.8.2 Saponification with vaporization of the glycerol

Processes have been patented in which fats are saponified with caustic soda at high temperatures and pressures in a tubular reactor; and the glycerol is separated by vaporization when the pressure is released. So far as is known such processes have not been commercially successful.

5.8.3 Manufacture of soap from sperm whale oil and low grade fish oils

The oil from the sperm whale, including the spermiceti from the head, is almost unique in that it contains a large proportion of waxes, the fatty acid esters of long chain fatty alcohols, such as cetyl oleate. Bailey (1979) quotes as an analysis of sperm whale oil:

	%
wax esters	76
triglycerides	22.6
hydrocarbons	0.18
more polar compounds	1.2

The fatty alcohols and the fatty acids range from C_{10} to C_{20}, and the acids are mainly unsaturated.

Sperm whale oil has been used for various purposes including the production of fatty alcohols, such as stearyl, cetyl and oleyl. This involves hydrolysis of the wax esters with caustic soda, a reaction which requires fairly drastic conditions. The alcohols also react with caustic soda to produce soap with liberation of hydrogen $R \cdot CH_2OH + NaOH = R\text{-}COONa + 2H_2$, and soap is produced by ordinary saponification of the triglycerides in the oil. When the objective is the production of fatty alcohols, the hydrolysis is carried out as gently as possible and the alcohols are distilled off. The residual soap is dark; and fatty acids are usually liberated with sulphuric acid and distilled for use as an alternative to soft oil in soaps. In the late 1940s and early 1950s there was a glut of sperm whale oil which then only commanded a low price (a situation which no longer exists with the great reduction in the number of whales caught). Efforts were, therefore, made to find additional uses for sperm whale oil; and a company in Chile developed a process by which it could be made more completely into soap. The oil is reacted continuously with fused caustic soda in a vessel containing molten anhydrous soap at 300–325°C. Under these very drastic conditions, the waxes are converted into soap; and hydrogen with some fatty alcohols, hydrocarbons and other compounds is vaporized. The liquid condensate from these vapours is not suitable for the commercial recovery of fatty alcohols because it contains too much other material. In molten anhydrous soap plus caustic soda, the soaps of unsaturated fatty acids react to produce less unsaturated soaps of slightly shorter chain length. This reaction was described by Varrentrap in 1853 when it was found that tallow could be ‘hardened’ by reaction with fused caustic soda to convert the oleate to saturated soaps, from which hard fatty acids could be produced for use in candles. With the Chilean process it was found that polyunsaturated acids react selectively in preference to oleic acid, so that the soap contains oleic, or other mono-unsaturated acids, but little more highly unsaturated material. This reduction in unsaturation was also used in Chile to convert low grade fish oils, or fish acid oils, to a usable, although not very good quality, soap. In the original plant in Chile the soap was discharged continuously onto a steel band where it cooled and solidified. In a similar plant built in England under licence, the anhydrous soap was converted directly into a soap solution at about 30% TFM in a mixer equipped with a turbo-disperser. An adjustable gate allowed some of the solution to overflow into another vessel; and the level in the turbo-disperser vessel was set so that the streams of molten anhydrous soap and of water dropped into a vortex depression and were drawn immediately into the impeller where they were dispersed in the circulating soap. Other methods of dissolving the soap failed because it either solidified into difficulty soluble masses, or burst into flames if brought into contact with hot metal and air. The soap solution was grained, the lye which contained a range of compounds was discarded, and the soap was introduced into a main soapmaking system. It was only of moderate quality. Soap made from the best quality sperm oil was, however, used in toilet soaps in Chile.

5.9 SOAPS MADE FROM SYNTHETIC FATTY ACIDS

Fatty acids were produced in Germany before and during the Second World War by oxidation of suitable paraffins with air. Some were used in soaps and some were esterified with glycerine to give eatable fats. The paraffins were obtained from coal by the Fischer-Tropsch process. Synthetic fatty acids are said to have been produced also in Russia, where naphthenic acids, by-products from the petroleum industry, have also been used in soap. These materials are discussed in Davidsohn, Better and Davidsohn (1953), where references to reports and articles are given.

According to Sonntag (1981) synthetic fatty acids have been widely used in soap in Russia and other Eastern European countries during the last 30 years; so far as is known they are not now used in the West.

6

Production of solid soaps from soap base. Household soaps and toilet soaps

6.1 GENERAL

Some definitions of terms used in this book for various types of solid soap are given in section 1.3.3; and the main requirements of BS 1910 (carbolic soaps), BS 1911 (genuine hard soaps) and BS 1914 (toilet soaps) are quoted in section 1.8.

Genuine household type soaps can be made from soap base in various ways; that is by allowing the soap to cool in frames or water-coolers, or by extrusion, which is now almost universal, apart from some small traditional manufacturers who still use the old processes. Similar processes are used for filled household soaps, but consideration of these soaps is deferred to near the end of this chapter.

Toilet soaps can also be made in several ways, but for very many years extrusion processes have been dominant. Extrusion processes for household soaps and for toilet soaps have features in common; and it is convenient to discuss some general aspects of the two processes after the older household soap processes have been considered, but before extrusion processes for household and toilet soaps are discussed more specifically.

6.2 THE OLDER PROCESSES FOR HOUSEHOLD SOAPS

Genuine soaps are produced by solidifying soap base without major additions; but minor additions of colour, perfume, preservative, etc., and of cresols in the case of carbolic soaps, are frequently made in simple mixing pots called crutchers. The oldest and simplest process is then to run the soap into moulds in which it is left to cool. Almost any containers can be used, but rectangular boxes with detachable sides, known as frames, are usual. Frames can be static, arranged in rows and fed with liquid soap by an arrangement of troughs and gates, or mobile, that is fitted with wheels, so that they can be put under the outlet from the crutcher and then moved elsewhere to cool. Solidification in production sized frames often takes about six days. Two operations should usually be carried out during the solidification of soap in a frame:

- hand crutching whilst the soap is still substantially liquid. This is mixing by a man who normally stands on the sides of the frame using a hand crutcher, which is a perforated steel plate with a shaft and handle at right angles to the plate, pushed up and down in the soap in the frame.
- knocking out, in which the bolts which hold the sides of the frame are loosened and the still plastic, or partly liquid, soap is pressed down with a pummel. In this way the soap which has contracted on cooling is pressed out to fill the frame. Otherwise some bars from zones which have contracted will be short.

The frame joints are liable to leak; and gaps are caulked with paper pressed into the joints as necessary.

When the frame contents are completely solid, the sides of the frames are removed and the blocks are cut into slabs with a mobile cutter using a frame of wires drawn through the soap. The slabs are then cut into bars with another wire cutter. The bars are usually kept for some time in an 'open pile', that is, each layer of bars is placed with spaces between the bars, and each layer is laid at right angles to the one below. This allows air to penetrate the stack; and a slightly dried skin is formed on most of each bar. The bars are trimmed and, usually, given a simple imprint with a stamper struck with a mallet (or a mechanical equivalent) before being packed in cases as bar soap. Alternatively, the bars are cut into billets and stamped into tablets using dies and shells as described later for toilet soaps.

Mechanical, or hand, stampers can be used; and the tablets may be wrapped and/or cartonned. A considerable quantity of unusable soap is produced in the frame process. In particular, the top and bottom slices and other off-cuts have to be re-boiled, or re-melted, before being returned to a crutching pot with new soap.

To reduce the time (and hence space) needed to convert liquid soap into the solid form, machines were devised in which cooling by water was used. Fig. 6.1 shows a slab cooler, sometimes known as a Jacobi cooler, although types were made by various firms. The liquid soap is run from the pan, or from a crutcher if additions are made, into a pressure vessel from which it is forced by compressed air into the cooler. The cooler is something like a filter-press, in which water-cooled plates alternate with frames in which the soap is solidified. The plates are faced with stainless steel, or other corrosion-resistant metal; and baffles are provided as shown inside the plates to cause the water to flow at a good velocity over the whole of the cooling area. Solidification occurs in a few hours, and the slabs of solid soap are handled like slabs from the frame cooling process.

Other water coolers produce bars directly. Tubes of non-rusting metal, with the cross-sections required to give bars of the desired dimensions, are arranged

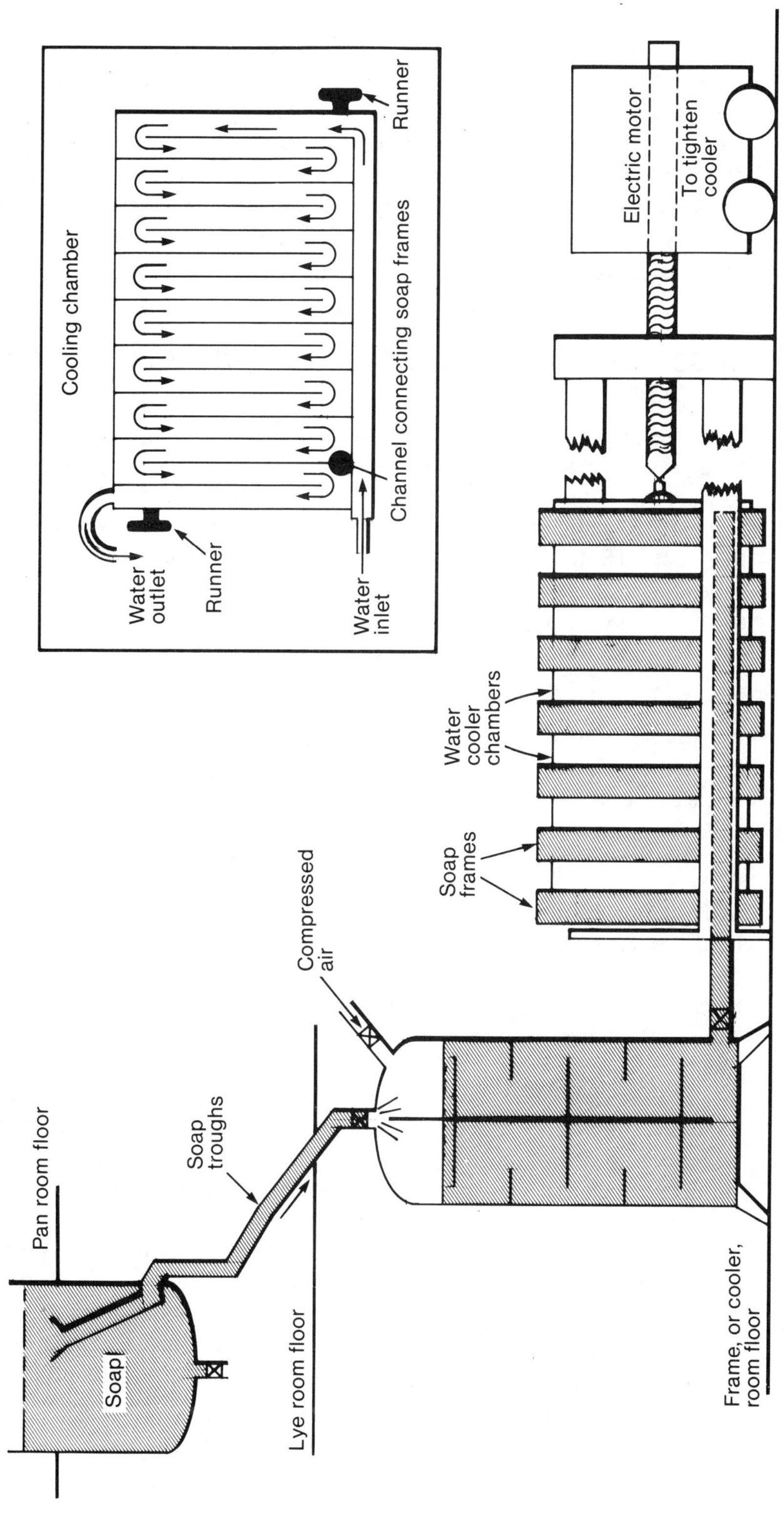

Fig. 6.1 — Water cooler. Slab type.

in a water-jacket (the unit being rather like a calandria). Each tube is fitted with a plunger which is moved downwards to draw in liquid soap from the header tank situated at the top of the unit immediately above the upper tube plate and the tube mouths. When the tube contents have been solidified by the water flowing through the jacket, they are pushed out in the form of bars by upward movement of the plungers. Experiments have been carried out on the moulding of individual billets of the required shape and dimensions, but, so far as is known, this procedure has never been used commercially.†

6.3 SOME PRINCIPLES RELATED TO THE PRODUCTION OF EXTRUDED SOAPS

6.3.1 General

Soap base produced using a conventional fit normally leaves the settling pan, or final centrifuge, at about 62 – 64% TFM, as discussed in Chapter 4. British Standard Specifications require a genuine household soap to contain not less than 62% TFM, and a toilet soap not less than 76.5, 78.0, or 79.0% TFM, depending upon type. Other specifications are possible, but the soap base needs to be dried in the production of all normal toilet soaps.

The operations considered in the following sections are:

- solidification
- drying, now often combined with solidification
- homogenization of soap and ingredients, and plastic working
- extrusion.

Additional operations are discussed later in connection with specific processes.

6.3.2 Solidification and drying

In the traditional toilet soap process, the 63% TFM soap base is solidified and dried in separate operations. Frame cooled, or water cooled, soap is cut into chips, which are then dried on trays with mesh bases in a warm air cupboard. This method has long been obsolete except for experimental batches and very small scale production.

The next method, employed for many years, uses a chilling drum and a band drier in which ribbons of soap are conveyed slowly through a chamber in which warm air is circulated to give a continuous drying process. Several bands, through which the air can flow, are normally used, arranged so that the soap falls from one band to the next below. Some moist air must be vented and replaced by fresh air, but internal and/or external recirculation of air is necessary for a good rate of drying and proper thermal efficiency. The soap ribbons are dried from their surface with diffusion of water from the interior to the surface; and they become drier on the outside than in the interior. Some driers include an intermediate mill to homogenize the soap part way through the operation, and so improve the later stages of drying. The fully dried soap is usually milled

†See also the converter process for floating soaps, section 6.10.

to crimped chips, or converted into noodles, (see later), to make it more compact for the next process and/or for transport and storage, as well as more homogeneous. Particles of overdried soap can impair the smooth texture of a soap tablet and make it feel sandy, or gritty. Special care is needed to avoid soap ribbons with irregular protruberances, or undue thinness, which can become overdried; and to remove dust whenever necessary. The removal of dust applies also to conveyors. The chilling drums used to convert liquid soap base to solid ribbons are large diameter water cooled drums on which the soap is spread in a thin layer. This is often done by providing a small roll to work close to the upper part of the drum. A small reservoir of liquid soap is held in the 'nip' between the two rolls and a layer of soap is formed on the cooled drum which is subsequently removed as ribbons by a serrated knife followed by a full-bladed knife. The large drum into which the cooling water is sprayed can operate half-full of water, or a discharge pipe introduced through the trunnion and bent downwards can keep the drum nearly empty.

Band driers have now largely been replaced by driers in which the liquid soap is heated in a heat exchanger, usually tubular or plate type, and then flashed by reduction in pressure. Three types are used which in principle differ only in the pressure to which the soap is flashed:

- tubular, or plate, driers in which the liquid soap is heated and the pressure then reduced to atmospheric by passage of the fluid through a throttle. Vapour is discharged and the concentrated liquid soap is solidified on a chilling drum, or other device.
- vacuum flash driers in which the hot liquid soap is flashed into an evacuated chamber. Flashing to a vacuum also chills the soap and converts it to a solid form.
- driers in which partial drying by flash to atmospheric pressure is followed by a second stage in which the residual liquid soap is reheated and flashed to vacuum.

The *vacuum flash drier/chiller,* shown diagrammatically in Fig. 6.2(a) and in more detail in Fig. 6.2(b), is a key element in modern plants for the production of both household and toilet soaps. This system was patented by Mazzoni in 1945, and the figures show the Mazzoni single stage unit, but equipment similar at least in principle is also offered by Meccaniche Moderne, and by Weber + Seelander.

When used to produce *household soap,* the heater is not needed, except as a pipe. The liquid soap enters the spray chamber at about 80 – 90°C and sufficient evaporation occurs to cool and solidify the soap by conversion of sensible to latent heat of vaporization.

The liquid soap is sprayed onto the walls of the evacuated chamber and is scraped off by blades on the same shaft set at 180° to the nozzles. Water vapour is drawn through dust collectors by the vacuum system. The solidified soap falls to the base of the chamber and is removed continuously by the first worm extruder (refiner). The noodles of soap fall into the vacuum chest from which they are extruded in the form of a continuous bar by the second extruder (plodder). The two extruders are shown in Fig. 6.2(b), but only one in Fig. 6.2(a).

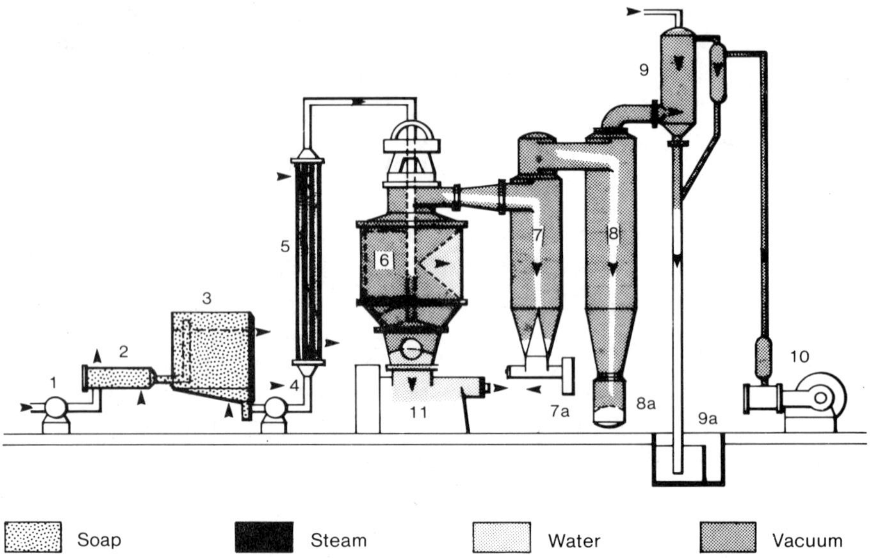

Fig. 6.2(a) – Mazzoni single stage drier/chiller. 1. Filtration pump: 2. Filter: 3. Feed tank: 4. Feed pump: 5. Heat exchanger: 6. Vacuum spray chamber: 7. First fines separator: 7a. Fines extrusion plodder: 8. Second fines separator: 8a. Fines collector: 9. Barometric condenser: 9a. Barometric hotwell: 10. Vacuum pump: 11. Single-stage plodder.

The worm extruders, which are considered in detail later, also serve to seal the chambers so that vacuum can be maintained.

When the unit is used as a drier to produce *toilet soap noodles,* heat is supplied in the heat exchanger to increase the evaporation, and the second extruder is changed from a plodder to a refiner by removal of the cone and substitution of an end with a perforated plate and rotating knives to give noodles of dried soap.

Some rough figures illustrate several important points. Fig. 6.3 shows the vapour pressure/temperature curve for water, which also indicates the temperature to which water is chilled when warm water is sprayed into an evacuated chamber in which equilibrium between liquid and vapour is reached. Because of the elevation of the boiling point (depression of the vapour pressure) due to dissolved substances, the curve for soap of any given water content indicates a somewhat higher temperature than that for water at the same pressure; but the curve for water shows roughly how the temperature to which the soap is cooled varies with the absolute pressure in the evacuated chamber. Mechanical work done on the solid soap in the extruder, or extruders, raises the temperature to an extent which increases as the soap increases in hardness, that is as soap from a given fat blend becomes drier and/or colder. Table 6.1 shows some limited experimental figures for the temperature of the extruded soap with various pressures in the spray chamber, together with the corresponding temperatures for water taken from Fig. 6.3. There are some inconsistencies in these very

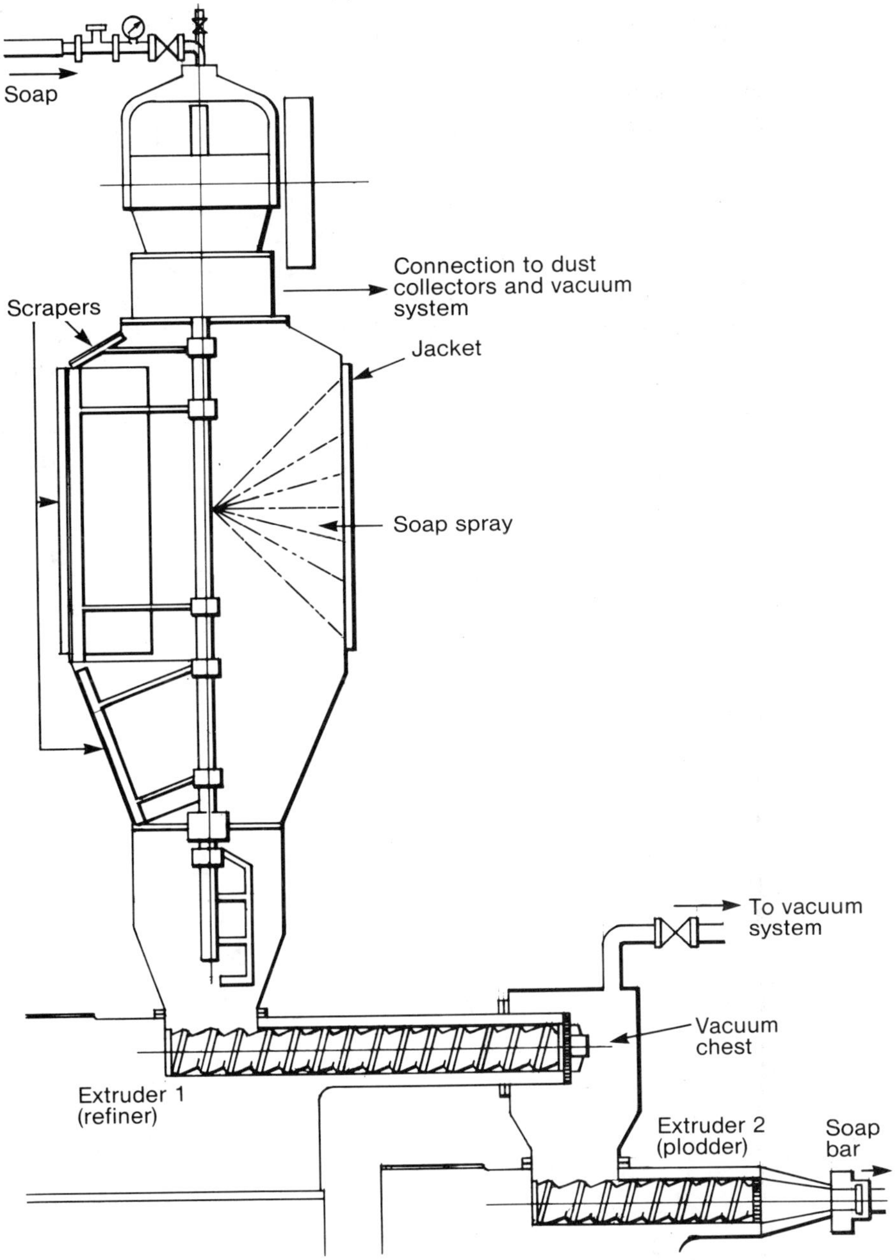

Fig. 6.2(b) – Vacuum drier/chiller unit.

limited observations, but they give a general indication. The low pressure and temperature were used for soap which was inherently soft, whereas the high pressures were for soaps which were exceptionally hard, so that the data in Table 6.1 cannot be used to confirm the expectation that the increase in temperature between evacuated chamber and extruded soap is lower at high soap

Table 6.1 – Temperatures and pressures for soap leaving a vacuum spray drier/chiller

Absolute pressure in spray chamber (mmHg)	Experimentally determined temperatures of extruded soap (°C)	Water temperatures corresponding to pressures (°C)
10	18–24	10
20–25	–	22–26
35	35	31
39	39	33
42	35	35

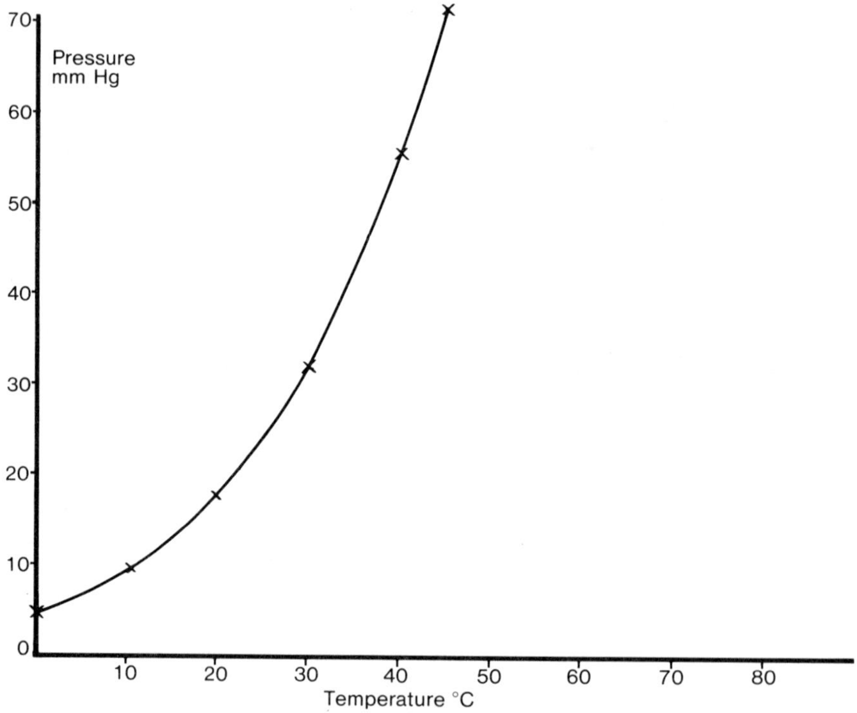

Fig. 6.3 – Vapour pressure/temperature curve for water.

temperatures than at low. Heat transfer between the soap and the water in the extruder jacket is also a factor affecting the temperature of the extruded soap. As explained in Appendix 2, a vacuum system as shown in Fig. 6.2(a), with no booster jets, can only reduce the pressure to the vapour pressure of water at the temperature at which the cooling water leaves the main condenser. If cooling water at 15°C is used and the quantity is such that it leaves at 25°C, the pressure in the soap spray chamber will be 24 mmHg plus a few millimetres for the non-condensible gas present and the pressure drop through the vapour system; say 28 – 30 mmHg. This is sufficient for some reasonably firm soaps at about 63% TFM, but lower pressures of 10 – 20 mmHg are often better for household soaps. To obtain lower pressures than those which can be produced by a condenser and air pump (mechanical or jet) alone, it is necessary to include in the vacuum system a booster jet, or thermocompressor, to compress the vapours before they reach the main condenser. A booster jet is essential when the cooling water is warm, or only available in limited quantity; also when filled, or other soft, soaps are being processed. No steam need be used in the booster jet, if fitted, for the production of toilet soap noodles. If very hard soaps, such as those from 80 – 100% coconut oil, are to be processed it is usually necessary to use a higher pressure in the spray chamber but, often preferably, not in the vacuum chest. This can be achieved by a partially closed valve, or similar throttle, in the relevant vapour line. As well as necessary for the production of extrudable soap, this higher pressure and temperature avoids the inclusion of specks of extra hard soap. These specks can be removed by fine gauzes in the extruders (see also later); but frequent stoppages to clean the gauzes are then needed. The output with 80 – 100% coconut oil soaps is well below that with soaps of more normal composition. Chilled water for use in the jackets round the extruder worms is also sometimes provided by partial evaporation under vacuum, a process known as spray chilling. The use of chilled water is often desirable with soaps that are soft.

An approximate indication of the amount of drying which results from the use of various liquid and solid soap temperatures can be obtained by the same method as that used for flash evaporation in fatty acid distillation. Assume initially that no vaporization occurs in the soap pipe prior to the spray nozzles. The specific heat of soap is about 0.75; and the latent heat of vaporization of water is about 566 CHu/lb (1019 Btu/lb) for the range 90 – 20°C, 545 CHu/lb (981 Btu/lb) for the range 150 – 20°C, and 523 CHu/lb (941 Btu/lb) for the range 150 – 100°C. For 1 lb of liquid soap, let the evaporation be y lb, so that the dried soap produced equals $(1 - y)$ lb. The heat balance for the case in which liquid soap at 90°C is chilled to solid soap at the base of the chamber at 25°C is approximately $1 \times 0.75 \times (90 - 25) = y \times 566$ or $y = (0.75 \times 65)/566 = 0.086$, or 8.6%. If the liquid soap fed contains 63% TFM, the solid produced contains $(63 \times 100)/(100 - 8.6) = 68.9\%$ TFM.

Table 6.2 shows the results of similar calculations for various conditions. In the estimation of the temperature of the extruded bar (or noodles), experience has been used, with allowance for the heating effect of the mechanical work in the extruders and the influence of likely barrel cooling water temperatures. In the production of household soap (with no heat added by the heater) a feed at 63% TFM emerges at about 69% TFM in good agreement with the figures calculated in Table 6.2. As shown in Table 6.2, the feed must be reduced to about 57 – 58% TFM, if a bar at about 63% is to be produced, provided the liquid soap is sprayed at about 90°C. This dilution of the feed with weak electrolyte (brine or sodium silicate) can be carried out in a crutching pot, or, more conveniently, by direct injection into the soap feed pipeline. At least one factory operates continuous soapmaking units so that household soap base leaves the final centrifuges at 57 – 58% TFM. The evaporations, and hence final TFMs, estimated in Table 6.2 for toilet soap base are low; this is doubtless primarily because some vaporization occurs in the heater, so that additional heat is added without a further increase in soap temperature.

Table 6.2 – Evaporation and TFM of chilled soap under various conditions of flash drying (no vaporization prior to nozzle assumed)

TFM	Feed Temperature (°C)	Pressure in spray chamber (mmHg)	Temperature of water equivalent to pressure (°C)	Evaporation (%)	Extruded soap (°C) (est.)	TFM (%)
63.0	90	30	29	8.1	34	68.6
63.0	90	25	26	8.6	32	68.9
63.0	90	20	22	9.0	28	69.2
63.0	90	10	11	10.4	23	70.3
62.0	90	20	22	9.0	28	68.1
57.0	90	25	26	8.6	30	62.4
63.0	125	25	26	13.6	31	72.9
63.0	140	25	26	15.6	31	74.6
63.0	150	25	26	17.0	32	75.9
63.0	160	25	26	18.4	32	77.2
63.0	140	atmospheric	100	5.7	–	66.8
63.0	160	atmospheric	100	8.6	–	68.9

In one series of experiments, the dust collected from the vapours when making soap at about 63% TFM was 0.08% of the throughput, whereas it rose to 1.0 – 1.2% when toilet soap noodles were being produced. Soap lost in the cooling water for the vacuum system was negligible.

Steam and water requirements of flash driers/chillers, can vary considerably. The sensible heat in the liquid soap is utilized in vacuum units, as compared with processes in which the soap is solidified on chilling drums, but steam and/or power is required for vacuum maintenance, particularly when booster jets are needed to achieve the required soap temperature. In their brochure, Weber + Seelander discuss consumptions of steam, power and water for various plant arrangements, including vacuum, atmospheric flash, and combined units. The other manufacturers also provide data for various systems and requirements.

Reasons why flash drier/chiller units have largely replaced other types of plant for both household and toilet soaps are:

- they are versatile, although, as explained later, some different ancillary equipment is needed for toilet and for household soaps.

- they take up much less floor space, and involve much less soap in process, than the frame, or water cooling, processes for household soaps; or band driers plus the necessary finishing equipment for toilet soaps.
- relatively little labour is needed.
- more steam, power and cooling water are needed than for some of the older alternatives; but the overall cost comparison is normally favourable.
- as discussed later, there are commonly quality advantages, although household soap is softer and more readily dissolved than that made from the same fat blend, and at the same TFM%, by the older processes. This may be an advantage with tallow now usually the cheapest good quality fat.

6.3.3 Extrusion, homogenization and plastic working

Worm type extruders are used for two purposes:

- to homogenize and work soap in order to get it into the required plastic condition. Such machines are often called refiners. They push the soap through gauzes backed by perforated plates; and it emerges in the form of rods which are cut into noodles by knives which rotate with the worms. Refiners are also used to convert soap ribbons into noodles for pneumatic, and other, transport.
- to form the soap into bars of the required cross-section which can be cut into bars for sale, or into billets to be stamped into tablets. Machines which carry out this task are called plodders.

The quality of an extruded soap is considerably affected by plodder performance which will now be considered in some detail. Fig. 6.4 shows diagrammatically the main features of the most common type of non-vacuum plodder, with a single worm and single orifice. Small pieces of homogenized and worked soap, in the form of noodles, or crimped chips, are fed into the hopper and then

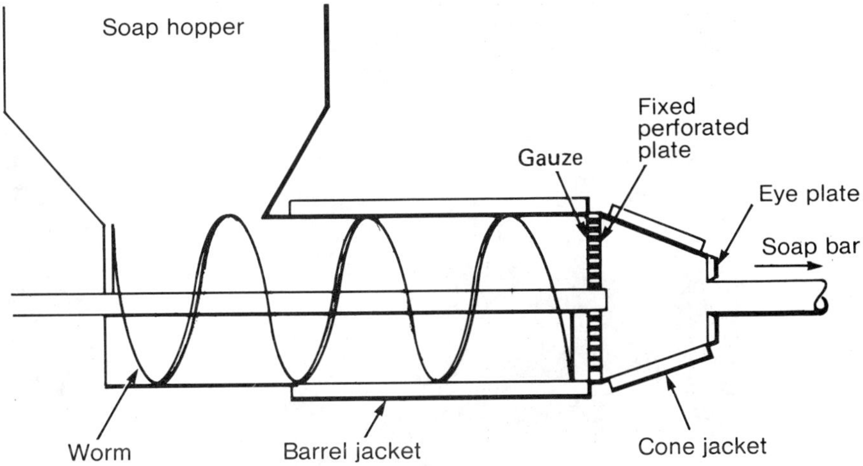

Fig. 6.4 – Plodder.

pushed along the barrel by the worm. Air is squeezed out; and the cylinder of soap is forced along like a nut on a screw, arranged so that the screw can only rotate and the nut only move longitudinally. The soap is pushed through the holes in the perforated plate in the form of rods which are welded together in the cone; and a bar of the required cross-section emerges continuously from the eye plate. A gauze, say 16 mesh BS (0.385 mm holes), in front of the perforated plate is often advantageous, but is not always used, and it does tend to reduce throughput.

An important advance during the last 40 years is the vacuum plodder. The soap hopper which feeds the plodder is enclosed and soap is fed into it by a refiner, roughly as shown in the lower part of Fig. 6.2(b). The hopper, or vacuum chest, is connected to a vacuum system, so that the air between the pieces of soap taken into the plodder worm is at a pressure much below atmospheric and the mass of air is much lower than when an open hopper is used.

Under some circumstances it is possible to use an eye-plate with two holes, so that two bars are extruded simultaneously. Plodders, called duplex or twin-worm, are also available in which two worms and cones are fitted in a single casing, as shown in Fig. 6.5, which is a photograph kindly provided by Weber + Seelander.

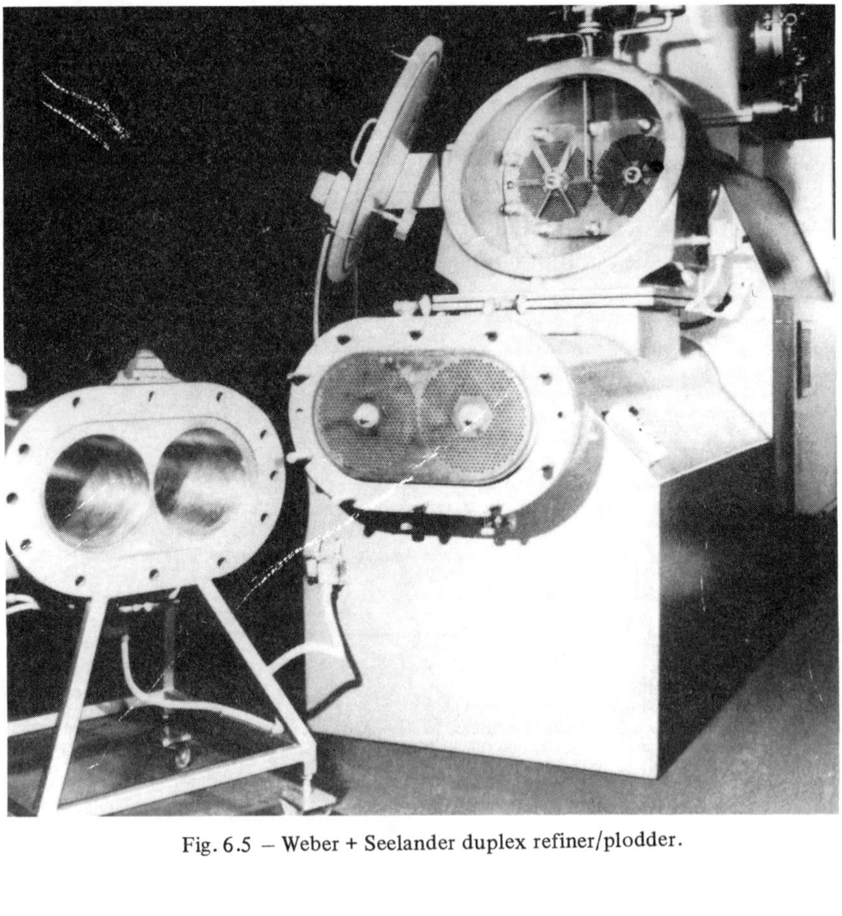

Fig. 6.5 – Weber + Seelander duplex refiner/plodder.

Cold, or cool, water is usually used in the jacket to cool the plodder barrel; and warm water or other heating medium in the jacket on the cone. When collar, or shell, type dies without cutting, or shaping, boxes are used to stamp tablets, the mass (weight) of the tablets is controlled by the cross-section of the billet, the length to which the billets are cut, and the density of the soap. It is often convenient to use an eye plate in which small adjustments can be made to one dimension of the hole with a slide, as shown diagrammatically in Fig. 6.6. In order to obtain smooth surfaced bars and tablets, it is normally necessary to use a heated eye-plate and a small electrical heater, of about 100 watts power, is a convenient way to achieve this. It is important to ensure that the edges of the eye plate are free from defects as lines on extruded bars do not usually stamp out.

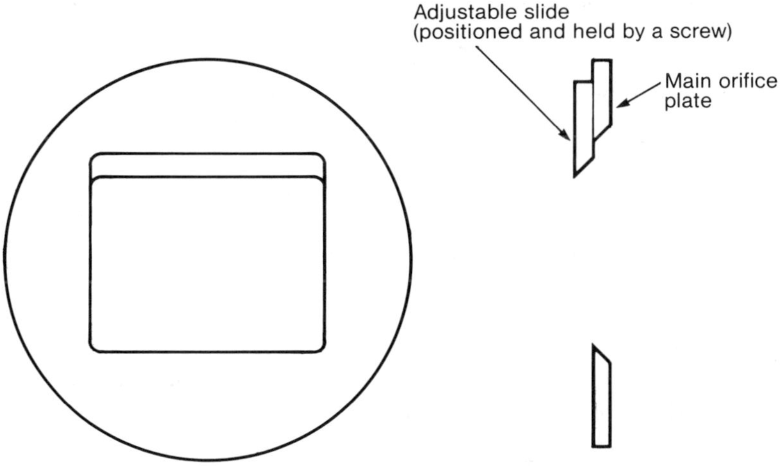

Fig. 6.6 – Eye plate with adjustable slide.

Worm extrusion is an apparently simple operation, but results can be influenced by a number of factors in a rather complex way. One variable tends to interact with others, and it is helpful to consider first some general principles. As already described, the noodles, or crimped chips, are fed into the worm with air between the pieces (much less in mass when the hopper is under vacuum than when it is open to atmosphere). This air should be forced back as the soap is compacted by the worm in the barrel; and the effectiveness with which this is done is influenced by the pressure developed in the plodder. The presence of undue air in the bars, or tablets, shows up in the form of blisters when the pieces of soap are allowed to stand for a few hours. Two convenient tests for this defect, sometimes known as blister, or thumbnail, cracking are (1) to cut tablets longitudinally with a sharp knife, or wire, and to leave the halves cut face upwards for a period, after which thumbnail-shaped cracks develop in faulty soap; and (2) to determine the density of the soap, which will be lower than normal if air is present.

The rods of soap formed by the perforated plate must weld properly in the cone to produce a substantially homogeneous bar. All plodded bars are weaker in the direction of extrusion than in the cross-direction, but the difference should be relatively small. A useful test is to take a length of plodded bar and to twist it backwards and forwards in the hands to ascertain how well the rods have welded together and how brittle the soap is. The aim is to produce a bar which is uniform; and free from streakiness and tendencies to blister and crack. This requires conditions under which sufficient pressure is developed, but the soap is also sufficiently plastic for the rods to weld together well. To some extent these requirements are opposed, in that soap which is hard leads to the development of high pressure, but tends to be lacking in ability to weld. For many years, toilet soaps were made with non-vacuum plodders, but problems arose particularly with types which contained high percentages of perfume, or other softening ingredients such as the cresols in carbolic soaps; and vacuum plodders are now widely used. It was once thought that household soaps were too soft to be plodded; but it has been found they can be made quite satisfactorily with vacuum plodders; and this is now the normal method of production. An interesting experiment which led to the development by Unilever of an early extrusion process for household soaps showed that whilst plodding with air in the hopper of a non-vacuum machine led to a bar which fell apart, replacement of the air by carbon dioxide gas gave a coherent bar. This was because carbon dioxide left in the bar was absorbed chemically, whereas air was not. The use of a vacuum plodder was a practical next step. The author has seen what was nominally household soap, but at about 68 – 69% TFM and made from 100% tallow base, extruded satisfactorily without vacuum, but these conditions are not normal for household soap.

When non-vacuum plodders are used to produce toilet soaps it is particularly important to ensure that the worm is kept well covered with properly compacted soap.

Apart from the additional cost, the only disadvantage of vacuum compared with non-vacuum plodders is that there can be some additional loss of the more volatile components of a perfume.

Plodding is affected by soap factors and by machine factors. *Soap factors* include:

- composition of the fat blend
- moisture, or TFM, content
- salt content
- soap temperature
- nature and quantity of added ingredients. As already mentioned, high contents of perfume, cresols, and some superfatting agents, soften the soap and can cause problems even with toilet soaps at around 79% TFM.

Further points concerning soap composition will be discussed later (section 6.4).

Machine factors include:

- use, or non-use, of vacuum

- worm (and barrel) dimensions
- clearance between worm and barrel, which increases with time due to wear, until the machine has to be reconditioned
- worm speed
- temperature of barrel cooling water
- use of gauzes behind the perforated pressure plate.

It is important to balance various factors so as to achieve the best results. For example, if the soap is of an inherently soft composition, it may be advisable to dry it to a lower moisture content, to cool the noodles to a lower temperature, or, in the case of household soaps, to include a higher content of salt. Conditions in the layer of material adjacent to the plodder barrel wall are important. Plodder performance tends to fall when the clearance between worm and barrel increases, but it may be possible to mitigate this to some extent by the use of very cold water in the barrel jacket which hardens the layer of soap adjacent to the barrel wall.

Before a further discussion of machine factors, it is convenient to present some design and performance data for a number of plodders tested some years ago. Most of these plodders are no longer available; and new machines are now usually obtained from the main producers of soap finishing plants, Mazzoni, Meccaniche Moderne, and Weber + Seelander, although there may be other makers. Table 6.3 gives design details for the plodders tested. Table 6.4 shows outputs from the three Vickers Armstrong plodders at various worm speeds. The first set of figures is experimental; the second set is calculated from the outputs at the lowest speeds *pro rata* to the worm speeds. It can be seen that:

- plodder output can vary substantially under nominally constant conditions.
- within the proper working range output varies roughly in proportion to worm speed. If too high a worm speed is attempted churning of the soap may occur with a decrease in output and unsatisfactory performance.

The maximum permissible worm speed (and output) is higher when the plodder is in good mechanical condition, with only a small clearance between worm and barrel, than when it has become worn. Tests on a $9\frac{1}{8}$ inch diameter Condon plodder showed that one with a worm clearance of 0.055 inch had a 25% lower output than one with a 0.010 inch clearance. When new the Vickers Armstrong plodders had a clearance of about 0.003 inch. Machines were considered to require re-conditioning when the clearance reached 0.030 inch, but typical figures during normal operation were about 0.010 – 0.013 inch. Other operators regard 0.040 inch (1 mm) clearance, that is 0.080 inch difference in diameter, as the maximum permissible. Because of the possible variations in output at a given plodder worm speed, it is most desirable in a continuous production line to be able to vary this speed to keep the output close to that required by subsequent units. Similarly, the speed of the refiner in a vacuum plodder should be adjustable to keep its output in accord with that of the plodder. Otherwise the vacuum chest will fill up, or the plodder will become starved.

Table 6.3 – Mechanical details of non-vacuum plodders tested

	10″ Vickers Armstrong	12″ Vickers Armstrong	14″ Vickers Armstrong	MIAG	Condon
Length of barrel inches	–	30(75.2 cm)	24(61.0 cm)	$25\frac{3}{4}$(65.4 cm)	19(48.3 cm)
Diameter of worm, inches	10(25.4 cm)	12(30.5 cm)	14(35.6 cm)	$11\frac{1}{2}$(29.2 cm)	$9\frac{1}{8}$(23.2 cm)
Pitch of worm, inches	–	$7\frac{1}{4} - 5$ (18.4–12.7 cm)	$11\frac{1}{4} - 7$ (28.6–17.8 cm)	$10\frac{3}{4} - 4\frac{7}{8}$ (27.3–12.4 cm)	$6\frac{3}{4} - 4\frac{1}{8}$ (17.1–10.5 cm)
Perforated plate					
Number of holes – outer	–	24	12	810	556
Diameter of holes – outer inches	–	$1-\frac{7}{8}$(2.5–2.2 cm)	$\frac{3}{4}$(1.9 cm)	$\frac{5}{16}$(0.8 cm)	$\frac{5}{16}$(0.8 cm)
Number of holes – inner	–	37	139	–	–
Diameter of holes – inner inches	–	$1\frac{3}{16}-1\frac{1}{8}$ (3.0–2.9 cm)	$\frac{15}{16}-\frac{7}{8}$ (2.4–2.2 cm)	–	–
Gauze BS mesh	–	16(0.39 mm)	16(0.39 mm)	16(0.39 mm)	16(0.39 mm)
Motor HP	–	20	35	–	–
Motor KW	–	15	26	–	–
Typical worm speed rpm	18	13–15	12	$12\frac{1}{2}$	14

As indicated, the holes in the perforated plates of some Vickers plodders were tapered towards the front. The MIAG and Condon plodders had only one size of hole.

Table 6.4 – Outputs of Vickers-Armstrong plodders at various worm speeds

Orifice size:	Vickers 10″ Toilet (rpm)	Vickers 10″ Toilet (lb/min)	Vickers 12″ Bath (rpm)	Vickers 12″ Bath (lb/min)	Vickers 14″ Bath (rpm)	Vickers 14″ Bath (lb/min)
Experimental						
	15	18–24	7	14–20	7	18–27
	18	24–27	$10\frac{3}{4}$	26–30	$10\frac{1}{3}$	40–45
	20	25–28	15	36–40	$11\frac{3}{4}$	49–55
Outputs at higher worm speeds calculated pro-rata to output at lowest speed						
	15	–	7	–	7	–
	18	22–29	$10\frac{3}{4}$	22–31	$10\frac{1}{3}$	27–40
	20	24–32	15	30–43	$11\frac{3}{4}$	30–45

These tests were done with conventional toilet soap at 77–79% TFM.

Experiments have shown that small diameter plodders are much more sensitive to the temperature of the cooling water in the barrel jackets than those of large diameter. When this temperature is found to be important, it is best to arrange a circulating system to which cold water is added as required to maintain the required temperature and the surplus water is allowed to overflow. The normal pattern is for output to increase as barrel jacket water temperature increases up to near the point at which churning starts with a fall in output. This again emphasizes the significance of the soap layer near to the barrel wall. Even when the required output can be obtained from a small diameter plodder, it is usually found desirable for quality reasons to use a large diameter machine when bars of large cross-section are to be made.

Special perforated pressure plates which rotate with the worm, or are fitted in the final flight of the worm, have been described, but it is doubtful whether they are beneficial.

Plodder worms are often made with the pitch greater at the hopper than at the discharge end in order to allow for the increase in bulk density and consequent decrease in volume as the noodles are compacted and the air squeezed out. This seems logical, but limited tests failed to show an advantage. So far as is known, little systematic work has been done, or at least published, on this and other major design features of plodders, such as ratio of barrel length to barrel diameter and angle of cone.

The plastics industry, which also uses worm extruders, usually with a much higher ratio of barrel length to diameter than is normal for soap plodders, has published some work on theoretical factors in the design of machines to meet various requirements. Attempts along these lines in the soap industry have, again so far as is known, achieved little.

It is possible to gain some insight into the working of a plodder by stopping the machine, removing soap from the hopper, filling up again with similar soap of a different colour, and running the machine for a short time until the second coloured soap has reached a point in the cone. The cone of soap is then removed as a whole and cut longitudinally, or into a series of slices. The boundaries between the two soaps provide interesting information as to the flow of soap through the machine.

When ingredients, such as perfume, pigments and colour solutions, are added in appropriate proportions to dried toilet soap noodles the ingredients are spread on the surface of the soap. It is then necessary to homogenize the mixture, and to work the material; this can be done with either refiners or mills, or a combination of the two.

As already indicated, refiners are worm extruders, basically similar to plodders, except that the cone is omitted and a multibladed knife is fitted to rotate with the worm and to cut off the rods of soap as they emerge from the perforated plate in the form of short pieces called noodles. In continuous production lines units of three refiners in series are often used. The three barrels are usually arranged in a stepped horizontal fashion, or inclined. In either case, this saves space and the soap which emerges from one barrel drops into the hopper of the next stage. Refiners use gauzes behind the perforated plates, to improve the homogenization and working of the soap; and to strain out any particles, such as hard soap, dirt, or fibres. The gauze in one stage may be finer than the gauze in the preceding stage because the soap is warmed and made more plastic as the operation proceeds. Too fine a gauze in the first stage would tend to result in excessive pressure and power consumption; and/or reduced throughput. The percentage of hole area in a gauze depends upon the wire diameter, as well as the number of holes per inch (mesh size), but the wire must be sufficiently strong to withstand the pressure on the unsupported areas over the holes in the perforated plate. A 52 mesh gauze (0.3 mm hole) is about the finest likely to be used. In some cases the fine gauze can be supported by a coarse gauze, say 12 or 16 mesh (1.4 or 1.0 mm hole), between it and the pressure plate. The higher temperature induced by a fine gauze may be compensated by increased barrel cooling, that is by the use of more, or colder, water.

Some mechanical and performance details for one design of triple refiner unit made by Vickers Armstrong are given in Table 6.5. The performance figures vary somewhat with worm clearance, soap quality, and other factors.

It should be noted that vacuum flash driers provide one, or two, stages of refining (before some important ingredients have been added); and a vacuum plodder one additional stage of refining.

Table 6.5 – Vickers-Armstrong 14 inch (35.6 cm) triple refiner unit

The unit comprises three barrels, each of 14 inches internal diameter, inclined at 20° to the horizontal. The perforated plates have 456 holes with diameter 0.495 inch (12.6 mm) on the inlet side tapering to 0.377 inch (9.6 mm) on the discharge side. This taper is intended to improve compaction of the rods. The rotating cutters are eight-bladed.

	Stage 1	Stage 2	Stage 3
Gauze:	22 BS (0.03 inch– 0.7 mm–holes)	30 BS (0.02 inch– 0.5 mm–holes)	44 BS (0.015 inch– 0.35 mm–holes)
Speed, rpm	16	16	16
Motor HP installed	40	40	40
kW installed	30	30	30
Actual power consumption HP	36	30	27

Throughput 2.25 – 2.5 tonnes/h depending upon the build-up of debris on the meshes, which are usually cleaned about every two hours. The total power consumption is about 26 kWh per tonne of soap at 78% TFM. The soap enters the first stage at 30 – 35°C and leaves the last at 38 – 42°C.

The barrels are water cooled in series using 4800 – 6000 lbs/hr. Water temperature rises from 13 – 14°C to 21 – 28°C.

Mills can be used as an alternative to refiners; and they were so used many years before refiners were developed. Mills can have 3, 4 or 5 rolls. For continuous toilet soap lines sets of three three-roll mills in series are often used, but large new installations now more often select refiners. At one time, granite rolls were common, but now cast iron rolls (often with a chilled skin), at least some of which are water cooled, are almost universal. Granite rolls are sometimes considered desirable if medicated soaps containing corrosive ingredients, particularly mercuric iodide, are to be processed. Roll arrangements for 3, 4 and 5 roll mills are shown diagrammatically in Fig. 6.7. Each roll rotates at a substantially higher speed than the proceeding one, so that:

- the soap transfers from roll to roll, and so moves through the mill as indicated in the diagram.
- there is a shearing action at each nip which homogenizes the mix and tends to break down any particles of hard soap.

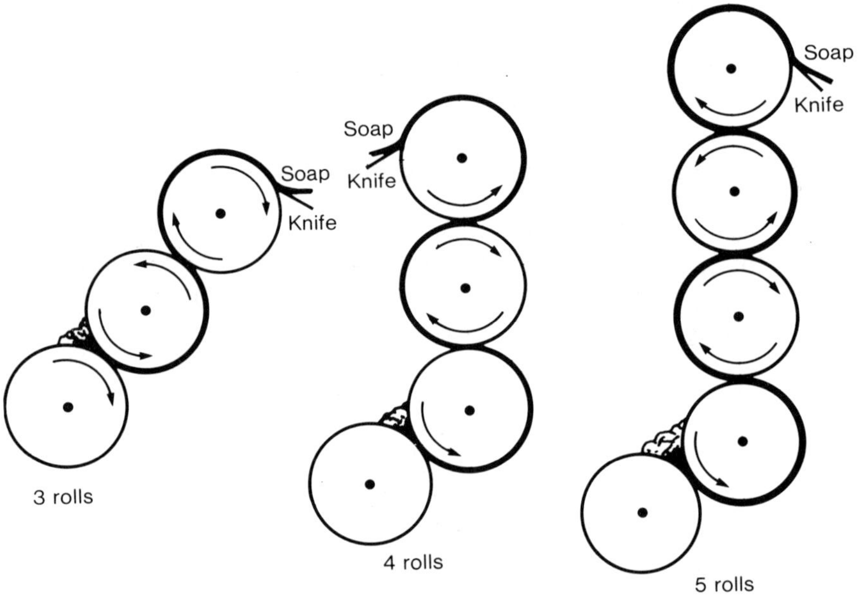

Fig. 6.7 – Mills.

Traditionally, the final roll was fitted with two knives. The first was serrated and removed alternate ribbons of soap; and the second was a complete blade which removed the remaining ribbons. These knives were often followed by crimping bars arranged to impede the flow of the ribbons and so to produce compact, crimped, pieces of soap. The more modern procedure is to use a comb-like knife which produces conveniently crimped pieces of soap directly. With some mills the knife, or knives, is/are positioned mechanically; in others, pressure is maintained hydraulically. Some data for Vickers Armstrong three roll mills are given in Table 6.6. The position of the middle roll is fixed and the top and bottom rolls are adjusted to give the required clearances. In the Vickers Armstrong mill this is done manually with hand wheels; in some more modern mills, such as those sold by Marchant Brothers, the gaps are set and held hydraulically. Because of the increase in roll speed upwards, the gap must decrease from nip to nip in order to maintain a full ribbon on the rolls.

The most satisfactory way to measure the true running clearance between a pair of rolls is to insert a piece of lead wire whilst the mill is running normally and to measure it after it emerges. Neither the thickness of the soap flake produced, or that of a piece of lead wire passed through the mill when it is empty, gives the correct answer. One test showed:

lead wire-rolls empty	0.004 inches (0.1 mm)
lead wire-normal operation	0.011 inches (0.3 mm)
soap flake on the roll	0.020 inches (0.5 mm)

Typically, the lower gap is 0.010 inches (0.25 mm) and the upper gap 0.006 inches (0.15 mm).

Table 6.6 – Vickers-Armstrong three roll mills

Roll length	32 inches (81.3 cm)
Roll diameter	14 inches (35.6 cm)
Rotational speeds	
bottom roll 23 rpm (peripheral speed 84 ft/min or 0.43 m/s)	
middle roll 72 rpm (peripheral speed 264 ft/min or 1.34 m/s)	
top roll 144 rpm (peripheral speed 528 ft/min or 2.68 m/s)	
Motor	35 HP (26 kW)
Castellated knife $\frac{3}{8}$ inch (9.5 mm) teeth at $\frac{3}{8}$ inch (9.5 mm) intervals.	
Water cooling (middle and top rolls only) 3300 lb/h (1500 kg/h) for a set of three mills. Inlet temperature 13 – 14°C, outlet 21 – 24°C.	
Soap fed at 30 – 35°C and leaves final mill of set of three at 38 – 41°C.†	
Running clearance between rolls 0.011 inches (0.3 mm).	
Under these conditions the throughout is 1.75 tonnes/h; and the power consumption with conventional toilet soap at 78 – 79% TFM is about 25 kWh/tonne, very similar to that for a set of three refiners.	

†Some operators like to control cooling and other conditions so that the soap leaves as near as possible to 34 – 36°C. The need for adequate cooling restricts the speed of the top roll to about 290 rpm (much higher than that of the old Vickers Armstrong mill); and this also limits the speeds of the lower rolls.

Whilst sets of three by three roll mills are convenient for continuous large scale production, small productions may require the use of one mill. Methods by which more than one milling operation can be achieved with only one mill are:

- the mill operates in two halves, with chutes arranged so that the soap passes through one half and is then diverted back to pass through the second half of the mill.

- the mill is equipped with two sets of knives, and with upper and lower storage hoppers. On the first pass the soap is removed by the back knives and held in the upper storage hopper. When a batch has been collected, the once milled soap is dropped into the lower hopper and is given a second milling. This process is repeated if a third milling is required. When the soap has been milled sufficiently, it is taken off by the front knives and leaves the machine.

Points in a comparison between milling and refining of toilet soap are:

- with proper operation soap quality is similar; but mills have an advantage if overdried soap has to be incorporated.
- studies some years ago showed little difference in cost.
- mills are more easily cleaned for a change from one soap to another; and so have an advantage when the plant is to be used for short runs of different varieties of soap. Mills are also preferable if the plant is only operated intermittently, as the soap in refiner worms cools and hardens (and hence has to be removed) on standing for several hours.
- refiner systems can be substantially enclosed to minimise loss of perfume by vaporization and pick-up of air-borne dust.
- it is sometimes claimed that the noodles from a refiner pack into the plodder worm better than crimped chips from a mill. This is of little significance if a vacuum plodder is used, because the actual plodder worm is fed with noddles from the refiner part of the machine.

The use of three refiners, plus refining in a flash vacuum chiller/drier and in a vacuum plodder, is typical modern practice, but some plants use less working and a few more. Some operators still consider the inclusion of a milling stage desirable. If mills and refiners are used in series, it seems sensible to refine first, in order to strain out unwanted particles, before any such particles are broken by the shearing action in a mill. The converse arrangement, in which the mill comes first, might impede effective straining in the refiner by breaking down the particles.

So as to avoid damage, mills, refiners and plodders are usually provided with a replaceable link which will break if the machine is overloaded.

6.4 PROPERTIES OF SOLID SOAPS

Important properties of a piece of solid soap are:

- hardness, including sufficient firmness to avoid damage to the tablet whilst it is being handled in the factory, in transit and in use. Soaps dry out and become harder on storage.
- lathering properties, which are closely related to the rate at which soap is removed from the tablet surface when it is rubbed, for example, with a damp cloth or the hand. This rate is more important than the true solubility of the soap.

- freedom from undue tendency to form mush when the tablet is left damp, particularly when it is left on a damp surface. Much formation seems to be related to tendency to swell in water and to disintegrate; that is to crack in use.
- tendency to crack on storage, or, more particularly, in use.
- smoothness to the touch in use; that is freedom from grit, or sandiness.
- freedom from deterioration of odour and of colour, including the development of spots or patches. Also freedom from undue distortion.
- detergent properties. Differences between one soap and another can be significant when bar soaps are used for laundry purposes, but they are largely unnoticed when tablets are used for personal washing, which is now the major application in the more developed countries.

Some of these properties are inter-related and a balance may be needed. In particular, the rate of wear should be sufficient to give an acceptable rate of lathering, but not so great as to cause users to regard the soap as wasteful. When household soap was widely used in the UK for general purposes, prudent housewives often kept soap for a considerable time in a warm place, such as an airing cupboard, before use in order to make it more economical. In addition to the points already made with regard to good plodding, it is important to ensure thorough homogeneity and uniformity of temperature at the correct level during processing.

An experienced person can assess the hardness of a piece of soap by pressing it with a thumb, but an instrument is desirable to provide more reliable and comparative data. One type is a penetrometer, in which the force necessary to push a sphere (or hemisphere) of given size a set distance into a soap surface is measured. The design used by Wigner (1940) determines the quantity of water which has to be run into a can on a lever arm, as shown diagrammatically in Fig. 6.8(a). Another type of instrument sometimes called a sectilometer measures the force needed to cut a piece of soap of known width with a wire of given diameter drawn through it at a steady speed. In one version the wire is held on a structure connected to the frame of the instrument by springs, and the extension of the springs, that is the position taken up by the wire, is a measure of the force being exerted. A trace as shown in Fig. 6.8(b) is obtained, and the height is an arbitrary measure of hardness. The initial peak, due to the extra hardness of the outer skin, can normally be disregarded. Another version which uses the same principle was described by Rayner (1938 and 1939). The cutting wire, attached to a spring balance, was drawn through the soap and the reading of the balance was a measure of hardness.

Apart from the influence of chemical composition, the properties of a piece of soap can be affected by its physical state. There is an extensive literature relating to phase differences, crystal compositions, crystal size, and crystal orientation. It is, however, conflicting; and published information is not readily applied to practical situations, and no further discussion is included in this book. Many operators find an empirical approach more useful, although observations concerning factors such as temperatures and conditions under which the soap is worked may apply only to specific soaps under a limited range of circumstances.

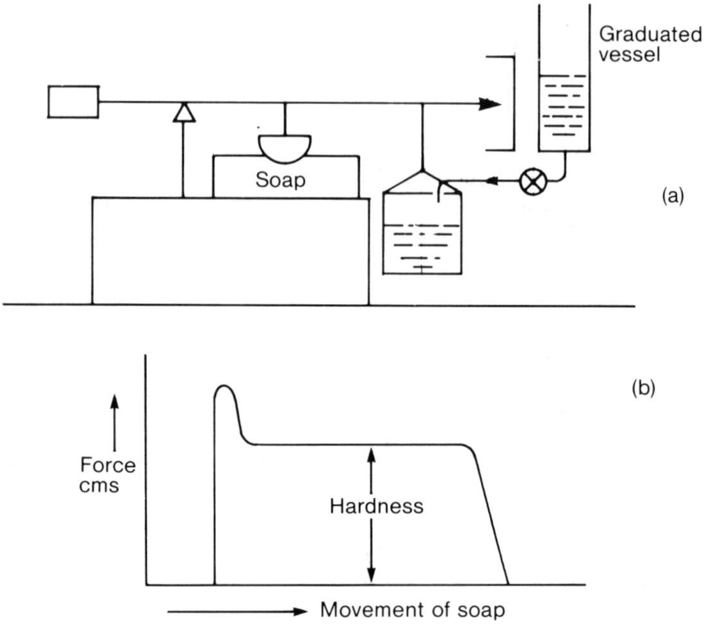

Fig. 6.8 – Measurement of hardness. (a) Penetrometer (an electrical contact in a circuit with flashlamp bulb and battery, can be used to indicate the end-point). (b) Sectilometer trace.

There are major differences between soaps produced by slow cooling in a frame, and to a lesser extent in water coolers, and those made by the modern chill, work and extrude methods. Fractional crystallization, with solidification of sodium stearate and sodium palmitate before that of the unsaturated and lower molecular mass soaps, can occur when soap cools slowly in a frame, to give a very different structure from that in an extruded bar. Framed soaps are much harder and slower lathering than plodded household soaps of similar composition. Unless filled, they also have a rough surface, showing what is called 'feather', and they tend to distort as they dry out. Plodded soaps show a smooth surface, provided a properly shaped and heated eye plate is used. When dried out, plodded soaps may become wedge-shaped if the moisture loss is mainly from one side of the case, but they retain a more pleasing shape than framed tablets. The softer and freer lathering characteristics of plodded soaps make desirable the use of a harder fat blend than in best quality framed soaps. The so-called 'washer' type of framed soap used substantial proportions of lauric and soft oils with the hard fat, whereas hard fat with 10 – 15% lauric oils is often found ideal for a plodded household soap. The hard fat used in soaps is now usually tallow. Palm oil was once an important soapmaking fat and, as discussed later in this section, it was superior to tallow in a plodded soap process which used chilling drums and mills. With the now usual flash vacuum chill processes tallow is perfectly satisfactory. As already explained, it is easy with the modern processes to produce household type soaps with a TFM% higher than about 63%. This can have some technical advantages, but the cost per lb when made is naturally increased, although not generally that per lb of anhydrous soap or TFM. Salt content should normally be kept low in framed soaps and in toilet soaps, in which too much salt is liable to cause brittleness and a tendency to crack in use. In plodded household soaps a somewhat higher level of salt than in framed soaps is often desirable as it hardens and improves the texture, particularly if soft oils have to be used in the fat blend. In one factory using the early chilling drum type of plodded bar process, it was observed that soap which had settled in a pan for a shorter period than normal behaved better than the longer settled equivalent which would have been superior if framed. This was particularly the case when the hard fat was tallow rather than palm oil. It was later found that the same advantage could be achieved by an addition of an appropriate quantity of brine to well settled soap. As already mentioned, at least one factory feeds its flash vacuum chilling units with soap at 57 – 58% TFM directly from the final centrifuges of continuous soapmaking plants, and this soap includes a rather high salt content; British Standard Specification BS 1911:1966, Genuine Hard Soaps, allows up to 1.25% salt in extruded soap, but only up to 0.75% in batch (that is framed or water cooled) soaps. Rayner (1938 and 1939), quoted by Davidsohn, Better and Davidsohn (1953), gives some interesting data for the hardness, solubility, and rate of abrasion of soaps made from individual fats and blends; but, apart from some figures for toilet soaps, this information refers to frame cooled soaps and is not directly relevant to modern extruded household soaps. Ginn, Steinhauer, Liebman and Jungermann (1968) report systematic tests on a series of toilet soaps made by the fatty acid route from blends ranging from 95% tallow and 5% coconut oil to 75 tallow/25 coconut oil. It was found that, within the range of compositions tested, hardness, swelling and cracking after immersion in water did not vary greatly. Mechanical tests on rate of foam generation in hard water with sebum present showed a definite peak at 85/15; but subjective assessments by panels of users carried out under controlled conditions showed that preference increased with increase in coconut oil content up to the maximum tested. The panel members used more soap with the higher coconut oil soaps; 0.5 g per use with the 95/5 soap and 0.8 g with 75/25. Slushing (mush formation) also increased linearly with coconut oil content.

When it is required to estimate the composition when made of competitive, or other stored, soaps it is important to allow for the fact that salt and caustic soda migrate into the core of the tablet as drying takes place from the outside. As explained in section 1.9, an average sample for analysis can be obtained by taking an aliquot portion of a solution of the whole tablet, or of an eighth cut to represent the whole tablet. Sodium carbonate concentration tends first to increase and then to decrease on moving inwards from the outside of a tablet. This may be associated with carbonation of caustic soda by carbon dioxide absorbed from the atmosphere.

6.5 INGREDIENTS ADDED TO HOUSEHOLD AND TOILET SOAPS

6.5.1 General

Many types of ingredient are added to soaps, particularly toilet soaps. The main classes of additives are:

- preservatives.
- perfumes.
- colours, opacifiers and optical brighteners.
- germicides and other medicinal ingredients.
- superfatting agents.
- scum dispersants, leading to NSD tablets and bars.
- alkaline fillers (discussion of filled, or run, soaps is deferred to section 6.12).
- miscellaneous.

6.5.2 Preservatives

Points already discussed which concern the stability of a soap are:

- selection of a fat blend without substantial proportions of highly unsaturated fatty acids. At one time 2 – 3% of rosin was included in toilet soap and other bases to improve keeping properties, but rosin is little used now.
- preservation of fats by natural and artificial antioxidants.
- the need for thorough saponification, although much less drastic treatment with caustic soda than was once common is now usual.
- the need to minimize the presence in the soap base of heavy metals, particularly iron and copper, and perhaps, nickel.

The most important soap preservatives are substances which negate the adverse catalytic effects of heavy metals which often cause brown spots, or patches, as well as more general deterioration:

- organic sequestrants, such as EDTA and, less usually, NTA (see section 2.3.5.2). Too much EDTA can form coloured (yellow) complexes with iron; and it is often better to use EHDP (ethane hydroxydiphosphonic acid

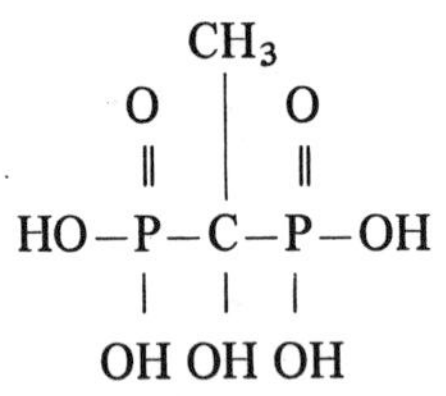

)

to sequester the iron, plus a small concentration of EDTA to deal with the copper. EHDP is sold by the German company Henkel under the name Product 1000 SL, which is a 60% solution. Concentrations of these preservatives should be determined by experiment to suit local conditions, but about 0.03% of Product 1000 SL plus about 0.025% EDTA (expressed as the tetra-sodium salt) may be a good starting point for trials.

- Magnesium silicates which can also be used to inactivate heavy metals in soaps other than those superfatted with fatty acids; presumably they work by adsorption. Magnesium silicate can be made separately and added as such, but it is usual to add the ingredients magnesium sulphate and sodium silicate solutions to the soap so as to form magnesium silicate *in situ.* When modern driers which include tubular or plate heaters are used, it is advisable to inject the silicate after the soap has passed through the heater; otherwise scale is liable to form on the heating surfaces.
  Typical additions are:
  for toilet soaps:

  0.33% of a 50% solution of $MgSO_4 \cdot 7H_2O$ (equals 0.027% of MgO) plus 0.6% of 57°Tw (1.285 specific gravity) neutral sodium silicate (equals about 0.12% SiO_2).

  for household soaps, say, 0.004% magnesium sulphate expressed as MgO plus 0.006% sodium silicate expressed as SiO_2.

  Magnesium silicate preservative is not used in soaps superfatted with fatty acids because it causes grittiness. Some technologists consider that it may suffer this defect with conventional toilet soaps; but it has been widely used for many years and has the merit of cheapness and ready availability of the ingredients.

Antioxidants of the BHT type (see section 3.9.1) are sometimes used in soaps as well as in fats.

6.5.3 Perfumes

Between 0.5% and 2.0% of perfume is usual in toilet soaps although higher concentrations are sometimes used, with about 0.8 – 1.0% common in the mass produced popular brands. Great care is necessary in the selection of a perfume, both as regards appeal to users and technical factors. It is usually advisable to approach one, or more, companies with experience in the field of soap perfumery and to give them as much information as possible regarding requirements. Detailed marketing considerations are outside the scope of this book, but it can be said:

- the perfume should accord with the general image possessed by, or desired for, the brand.
- full market research, or at least the reactions of panels of people typical of the market in which it is hoped to achieve sales should be obtained; the taste of the managing director's wife is not necessarily a good indication of popular appeal. Particular care may be necessary in countries where people of two, or more, races are important potential customers. Positive and negative reactions to a perfume are important. Sometimes it is necessary to decide between a perfume which is well liked by many and disliked by few, and one which produces stronger reactions, some for and some against.

- perfumes are expensive and are an important item in the cost of a toilet soap. It is important to obtain good value for money and to decide how much can reasonably be spent on perfume. This should be assessed on the basis of cost per tonne of soap, rather than per kilogram of perfume, as some blends need to be used at higher concentrations than others.
- some perfumes are more substantive to the skin than others; and it is important to decide whether substantivity is desirable and how far it can be achieved (or avoided).

It is necessary to make up soaps containing the perfumes to be considered at the concentration, or usually series of concentrations, which seems appropriate. Storage tests, usually including accelerated tests at temperatures higher than ambient, should be carried out; and these should include exposure to strong light both naked and in the proposed packaging. Consumer tests should be carried out on soaps which have been matured for an adequate time. Perfumes for toilet soaps are normally blends made from a large number of ingredients many of which contain a large number of different chemical compounds. Perfumers with adequate experience in the soap field should only submit blends which are sufficiently stable to the alkalinity of soap, light, etc; and which will not deteriorate excessively in colour and/or odour on storage. Additives such as EDTA can affect the alkalinity of a soap and proposed perfumes should be checked by the soap manufacturer under conditions relevant to his particular product and market. Vacuum plodders are more liable than simple plodders to remove very volatile components from a perfume. In addition to possible deterioration of the soap, it is important to ensure that the soap containing a perfume under consideration is substantially free from allergic, or irritant, reactions by potential users. This requires extensive tests, but, again, experienced perfume houses have built up knowledge and should be able to provide blends for which they can give adequate assurances. The small soap manufacturer is not normally in a position to carry out adequate tests.

Household soaps are often made without perfume. When perfume is added, it is usually a relatively cheap but powerful type, such as citronella, at a low concentration. Many years ago oil of myrbane, nitro-benzene, was also popular particularly for tropical markets, despite its toxic nature.

6.5.4 Colours, opacifiers and optical brighteners

Colours used in soaps are of two types, dyes soluble in water or in an organic solvent, commonly the perfume, and pigments which are added as suspensions. As for perfumes, various checks are necessary in addition to an assessment of aesthetic appeal before a colour system is accepted as satisfactory:

- adequate stability when the soap is stored and/or exposed to light (naked and packaged). Some colours have been used which are not very stable to alkali and which could only be used with dried soap base exceptionally low in free caustic soda; such dyes are best avoided if possible.
- safety for the user. Some dyestuffs, or impurities they contain, have carcino-

genetic properties; and many countries have strict regulations concerning colouring matters which are used in cosmetics, as well as in foods and drugs. Concentrations of colours used in soaps are low compared with those in products like lipsticks (which may also be ingested), but care in selection is still necessary. Anstead (1973) discusses the legislative position at that date for cosmetics in various countries. The US Food and Drugs Administration has for many years issued detailed regulations; and it is probably wise to use only colours approved by a recognized authority, even when this is not essential to meet the laws of the country concerned.

Opacifying agents are also added to soaps to obtain the desired appearance. Titanium dioxide (of which there are two major types, rutile and anastase) is now the normal material used; and 0.2% is a typical addition to give an opaque toilet soap. Zinc oxide has also been used, but is less effective than titanium dioxide. At one time magnesium carbonate was employed as an opacifier in filled household soaps. Extruded soaps, particularly household types, are liable to show a pattern of relatively translucent soap on the surfaces cut at right angles to the direction of extrusion, known as end marks. If the end marks cannot be eliminated by adjustment of the conditions of plodding, a small addition of titanium dioxide will largely obscure the pattern and make the soap more attractive in appearance. A little opacifier also helps to minimize changes in colour as a soap dries out and becomes more translucent.

As mentioned in section 2.3.7 special types of optical brightener, such as Tinopal GS, are used to improve the colour and brightness of white soaps. It is also possible to use a low concentration of cotton substantive optical brightener to improve slightly the performance of household soaps in laundry work; but care is necessary to avoid risk of patchy effects when the soap is rubbed on extra dirty parts of articles.

6.5.5 Germicides and other medicinal ingredients

Many brands of soap have been sold on the basis of germicidal, or freshness, claims; and a number of different bacteriostatic, or bacteriocidal, ingredients have been and/or are used to justify such claims. The traditional ingredient was a mixture of cresols and related phenols, often known as cresylic acid, which gave 'carbolic' soaps. Carbolic soaps, commonly, but not necessarily, coloured red, usually contain about 2% of phenols when intended for the domestic market. It is necessary to select cresylic acid which is reasonably free from pyridine, and other nitrogenous bodies, which can lead to unpleasant smells when the cresylic acid is included in alkaline soap. Naphthalene is often dissolved in the cresylic acid to improve the odour of the soap. Phenols tend to be held in the soap micelles when the soap is dissolved, so that domestic type carbolic soaps do not show much greater anti-germ properties than soap alone. Special soaps with much higher concentrations of phenols have been made for hospital and similar duties.

More recently various more potent ingredients have been used; and anti-germ properties have been demonstrated on human skin. Freshness claims require the further step that use of the soap can reduce the odour from body areas such as the armpits by reducing, or preventing, the decay of stale perspiration to produce unpleasant odours. Unfortunately, many of these ingredients have been found liable to have undesirable side effects on some users; and great care is necessary to select one which is permitted and fully safe. Apart from straightforward toxicity (mainly where a gross excess has been included in the product in error), and sensitization, or allergic reaction, problems have arisen through:

- the presence of impurities in the germicide as bought.
- breakdown to produce damaging compounds when the soap is stored for long periods at relatively high temperatures, and, particularly, when scrap is re-boiled and returned to the product.
- photosensitization, that is skin, and possibly more general, reaction when a person who has washed with the soap is exposed to bright light.

The names, and formulae, of the main compounds which have been used are shown in Fig. 6.9, but it is emphasized that at least several are no longer acceptable. Mixtures said to be synergistic have also been used.

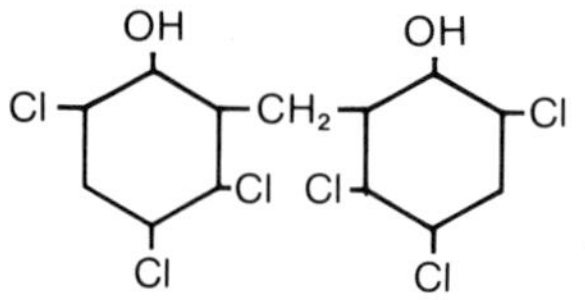

2,2′-methylene bis (3,4,6-trichlorophenol)
Hexachlorophene, GII

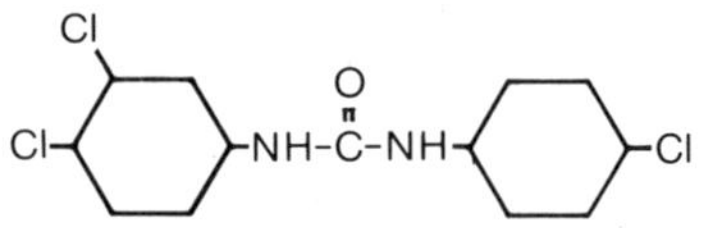

3,4,4′-trichlorocarbanilide TCC

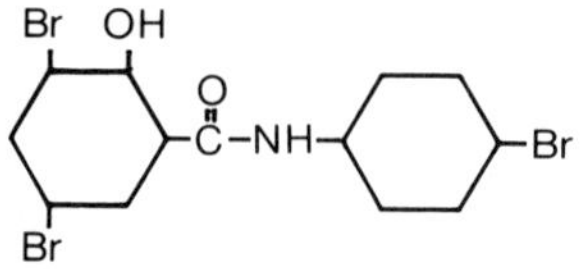

3,4′,5-tribromosalicylanilide
TBS

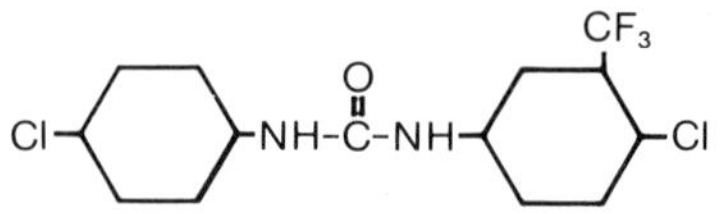

4,4′-dichloro-3′-(trifluoromethyl)carbanilide
Ingasan CF_3

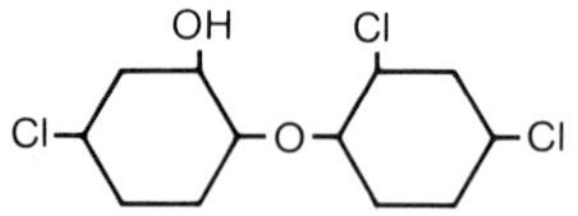

2-hydroxy-2′-4,4′-trichlorodiphenyl ether
Ingasan DP 300

Tetrachlorosalicylanilide T_4CS

Fig. 6.9 – Names and formulae of various germicides.

Various medicated soaps are also made, particularly to deal with certain skin complaints, especially for sale in tropical countries. In the UK, special regulations apply to such products, including requirements for the testing of batches and for documentation. Ingredients used include mercuric iodide and tetraethyl thiuram monosulphide (sold by ICI as Tetmosol). Mercuric iodide may no longer be acceptable.

6.5.6 Superfatting agents

Many superfatting agents, including lanolin, petroleum jelly, coconut oil, and stearin (commercial stearic acid), are used in soaps; they are intended to increase mildness in use. As mentioned earlier, an important type of superfatted soap uses a fat blend with a high proportion of coconut oil to tallow; and is superfatted with 5 – 10% of fatty acids similar to the main fat blend. The maximum effect upon the lather is obtained with about 8% FFA. An unsuperfatted soap made with 30 – 60% of coconut, or palm kernel, oil tends to be more irritant (less mild) than a conventional toilet soap, but the inclusion of the free fatty acid, and consequent elimination of the free alkali, brings mildness back to about that of a 20% CNO 80% tallow soap. The advantage of this type of soap, more expensive than the 20/80 kind, is that it gives a copious, creamy, lather which has found favour with many consumers. A TFM of about 82% is usually needed. This type of soap is particularly sensitive to processing temperature and noodle coolers may be advisable. To aid this cooling, noodles with a special shape to increase their surface/mass ratio are sometimes used.

6.5.7 Scum dispersants, leading to NSD tablets and bars

As mentioned in the market survey, much technical and marketing effort has been expended on the development of toilet tablets in which soap is replaced partially, or completely, by non-soapy surfactants and other ingredients, but, so far, with little commercial success outside the USA. Objectives are:

- to eliminate scum and 'bath-tub ring'.
- to produce an immediate lather on bath, or wash basin, water.
- with some formulations, such as the Monsavon 'PH3', to produce a personal washing tablet which is slightly acidic rather than slightly alkaline.

Problems are:

- cost, with many consumers not prepared to pay the higher price relative to soap.
- to produce tablets which are not too dense and which do not absorb excessive water and become soft.

Many formulations have been patented and/or tried. Some are more, or less, soap-free, but others contain substantial proportions of soap. Stearin is an important constituent of some tablets. Some additional information is given in Bailey (1979) and in Davidsohn and Milwidsky (1978), and also in various patents.

NSD bars for use instead of household soap bars have been successful in some countries, particularly the Philippines. Typical ingredients are sodium alkylbenzene sulphonate, sodium tripolyphosphate, sodium silicate, and sodium sulphate; plus a plasticizer. Waxes and starch may be included to improve texture and feel. The blends are milled and plodded.

6.5.8 Miscellaneous ingredients

Many other ingredients are, or have been, added to toilet soaps to improve texture, feel, lather, and benefits to the skin of the user; the real merits of some are dubious. Jungermann (1982) reviews recent developments which include the use of citronellylsenecioate (also known as the citronellyl ester of 3,3-dimethyl acrylic, or 3-methyl crotonic acid), as a non-germicidal deodorant.

6.6 EFFLORESCENCE

After storage some soaps develop a surface deposit known as efflorescence. This can be of several types, in particular:

- a heavy white deposit, often fibrous in appearance, which comprises inorganic salts, probably mainly sodium sesquicarbonate, and which develops on soaps run with alkaline fillers. Changes in the fillings used may help to reduce this efflorescence. Glued-up soaps are said to be particularly prone to the defect, with fillings based on neutral silicate worse than those using alkaline silicate.
- a white fibrous deposit which can form on extruded genuine household soaps. This deposit comprises filaments of soap which are pushed out through weak spots in the tablet surface; that is the filaments grow from the base like hairs. This type of efflorescence develops when there is a fairly rapid rate of drying from the soap surface; and it can usually be avoided by the use of well sealed wrappers of a material with a low moisture vapour permeability. A useful accelerated test for proneness to develop soap hair efflorescence is to cause a rapid flow of water through a relatively thin piece of the soap. This can be done by closing the end of a large test-tube which contains some water with a small piece of soap and storing the tube in a vertical position in a dry atmosphere. In this way, the upper surface dries out in the low humidity atmosphere, whilst the lower surface picks up water from the nearly saturated air below it. Traces of heavy metals on the soap surface, from stamping lubricant or contact with materials such as carton board, can tend to induce the efflloresence.
- a thin white film which can arise on toilet and other soaps. One cause is the use of excessive quantities of stamping lubricant containing salt; and some versions of magnesium silicate preservative are also sometimes thought to have an adverse tendency.

6.7 MODERN PROCESSES FOR THE PRODUCTION OF HOUSEHOLD SOAPS

6.7.1

The main unit of the now standard plant for the production of household soap is the vacuum chiller/extruder shown in Fig. 6.2(b) and discussed in section 6.3.2. Broadly similar plants are offered by Mazzoni, Meccaniche Moderne (Comis type), and Weber + Seelander. Suitable lengths can be cut from the continuous bar extruded by the vacuum plodder using either a wire cutter swung backwards and forwards at the plodder nozzle, or a simple device shown in Fig. 6.10, sometimes known as a 'hacker-off'. This comprises a cutting wire on a small frame which is connected by a rod of the required bar length to a small plate; the cutter and the plate being at right angles to the rod. The operator holds the plate against the moving end of the bar which is supported on a short piece of narrow wooden roller conveyor, and when sufficient has been extruded, pushes the cutter through the soap. Bars cut in one of these ways can be handled like bars produced from framed soap. Fully mechanized lines, including billet cutter, conditioning tunnel, stamper, wrapping and/or cartonning machine, case packer and case sealer, as shown diagrammatically in Fig. 6.11, are now common; variations to suit individual needs are possible. Some of these operations related to the production of tablets are discussed more fully in section 6.8 concerned with toilet soaps.

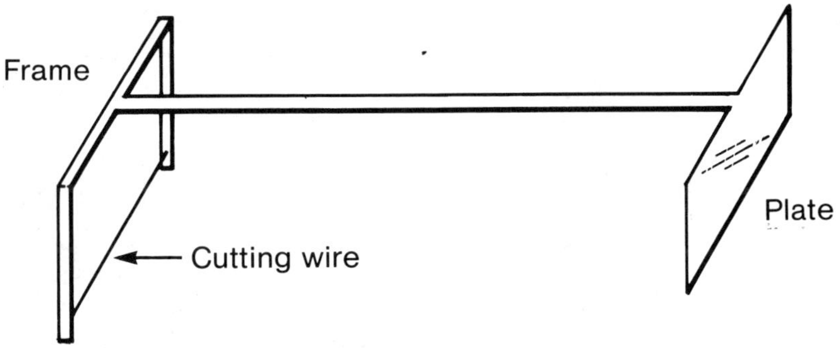

Fig. 6.10 – 'Hacker-off'.

6.7.2

As mentioned in section 6.4, in the 1940s Unilever developed independently a process for the production of household soaps which also used a vacuum plodder, but in which the soap was solidified on chilling drums and worked in mills. This process performed well with high palm oil fat blends, but was less satisfactory when the hard fat was tallow. A rather high salt content was important; and it was found that a tubular heater to give about 1% evaporation prior to the chilling drum was markedly beneficial with high tallow blends. It is not

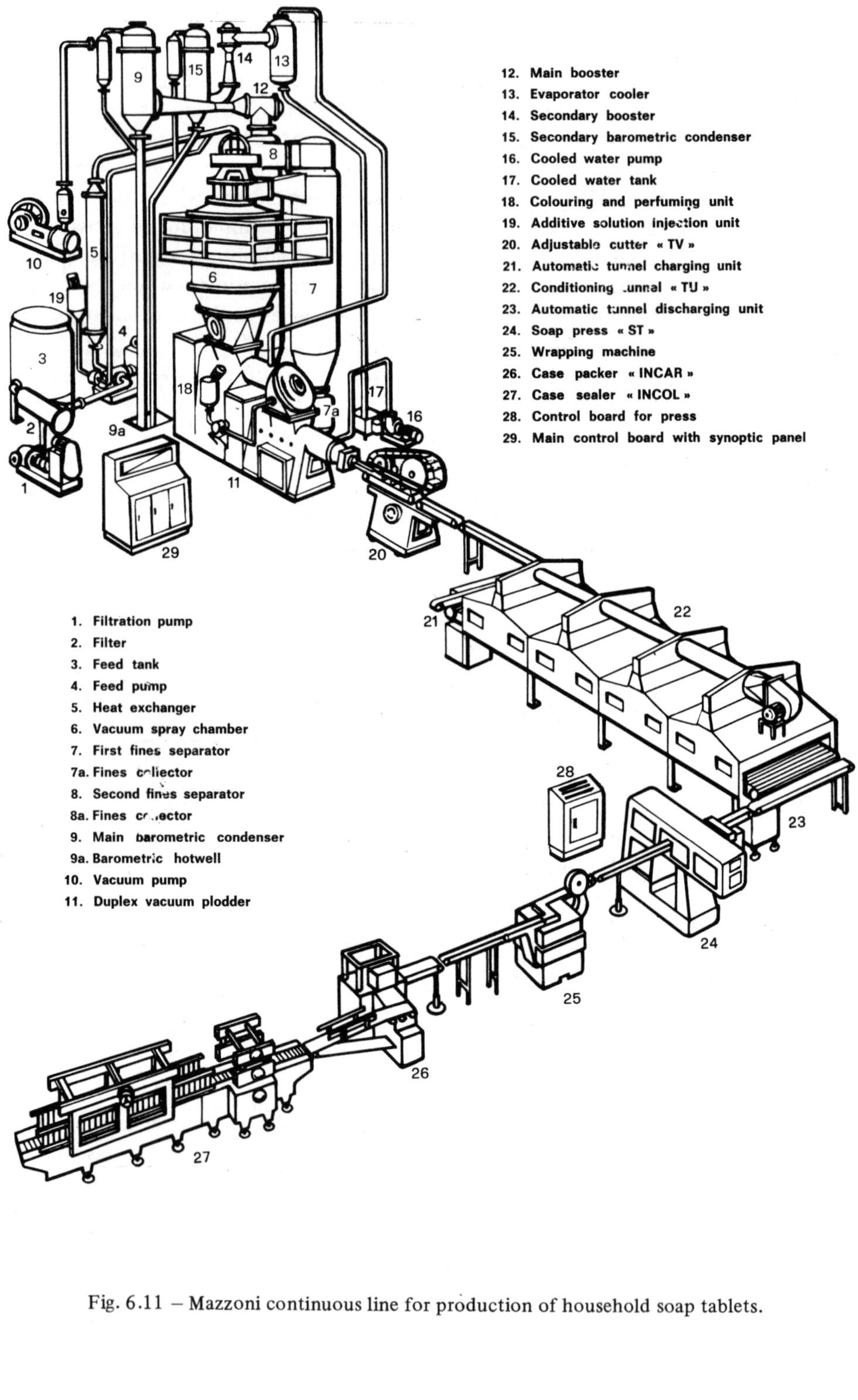

Fig. 6.11 – Mazzoni continuous line for production of household soap tablets.

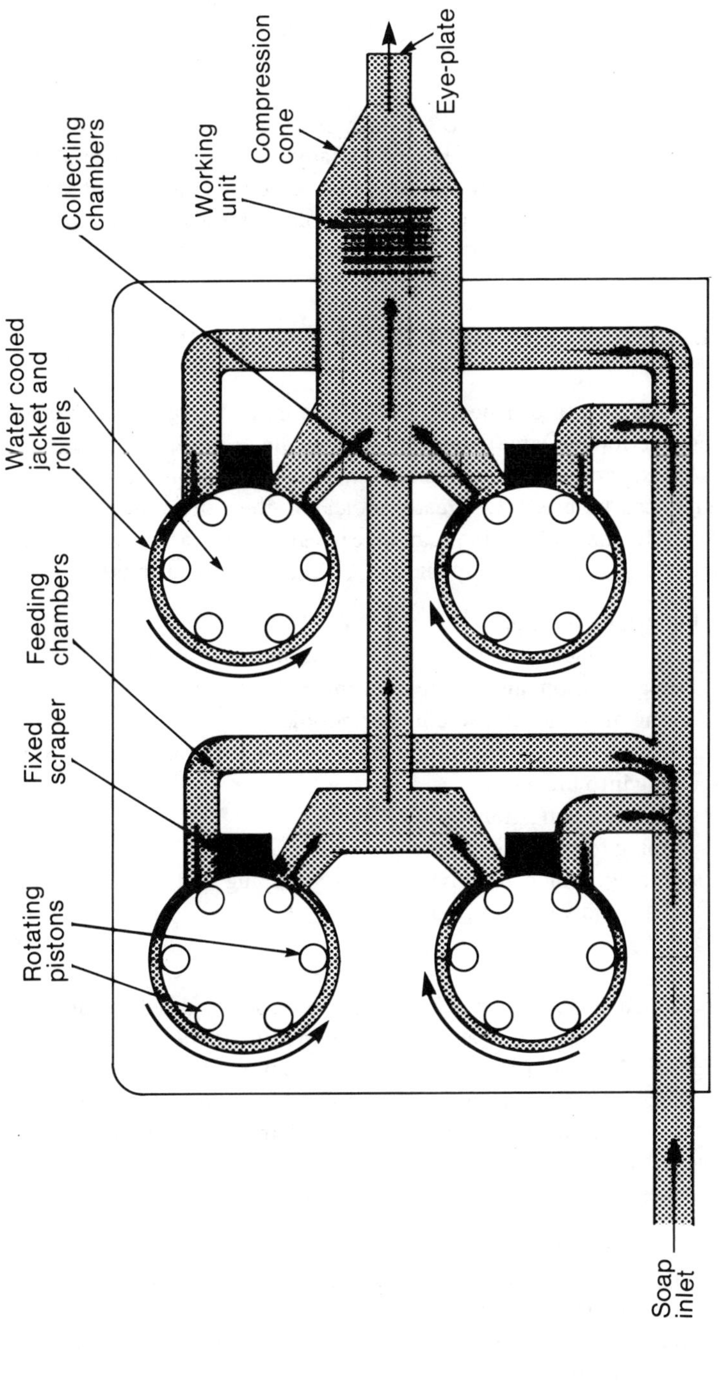

Fig. 6.12 – Meccaniche Moderne extrusion/chiller for household soap.

clear whether this was due to the small reduction in moisture content of the soap, or whether the temperature from which the liquid soap was chilled was the major factor. The more versatile vacuum chilling units are now preferred; and these do involve rapid chilling over a fairly large temperature range.

6.7.3

Prior to their adoption of the Comis system, Meccaniche Moderne developed an extrusion unit shown diagrammatically in Fig. 6.12. In this the liquid soap was cooled and forced out through a barrel, cone and eye plate by rotors which acted as pumps as well as coolers. This plant has also been supersceded by the vacuum chilling system.

6.8 EQUIPMENT FOR TOILET SOAP PRODUCTION

6.8.1

Toilet soap plants can be fully continuous; and largely mechanized units for the production of large tonnages of a single type of tablet are possible; or a series of separate units connected manually as required to make a range of products can be used. As already indicated, there are a number of possible ways in which the main operations can be carried out; but a line will include equipment for most, but not necessarily all, of the following operations:

- chilling and drying.
- rough mixing of dried base with perfume and other ingredients.
- homogenization and plastic working.
- cooling of homogenized chips or noodles.
- extrusion of a continuous bar by a vacuum, or non-vacuum plodder.
- cutting into billets.
- cooling and conditioning of billets.
- stamping to produce tablets.
- packaging, which usually includes wrapping and/or cartonning, as well as final assembly in cases.

6.8.2

As explained in section 6.3.2 soap is now usually dried in some form of flash drier. The commonest type is the vacuum chiller/drier shown diagrammatically in Figs. 6.2(a) and (b), and in the photograph Fig. 6.13. Atmospheric flash driers can be built-up as indicated in Fig. 6.14. Fig. 6.15 shows the Mazzoni two-stage drier, in which the soap is heated and flashed to atmospheric pressure, followed by a second stage in which it is re-heated and flashed to vacuum. Advantages to off-set the additional cost of a two stage drier are:

- a large overall reduction in moisture content is possible, which is particularly useful with some specailized products, for example some which contain NSD ingredients.
- cooling water requirement is about half that for a single stage drier, because of the vapours vented to atmosphere.

Fig. 6.13 – Photograph of Mazzoni plant arranged for the production of dried soap noodles.

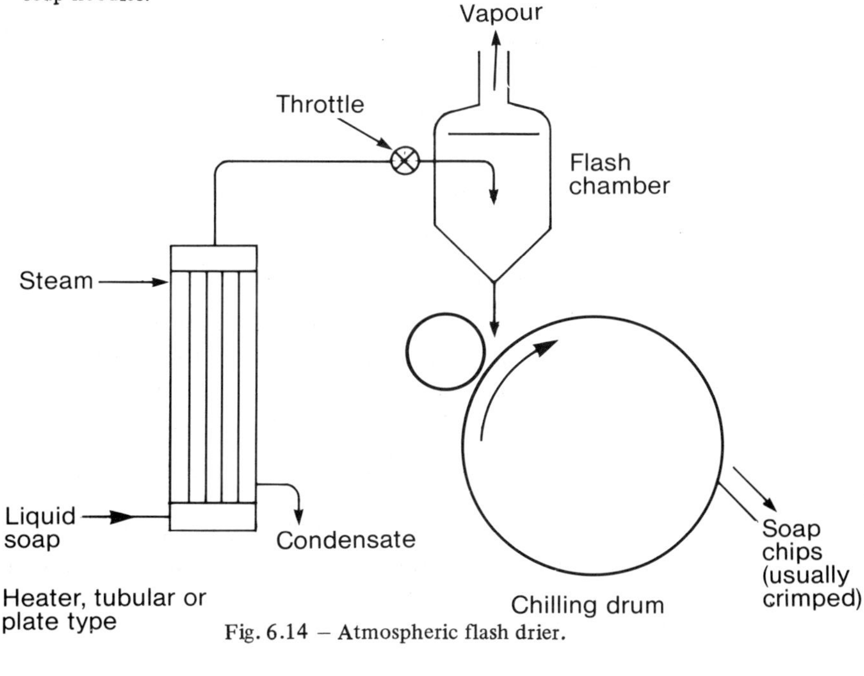

Fig. 6.14 – Atmospheric flash drier.

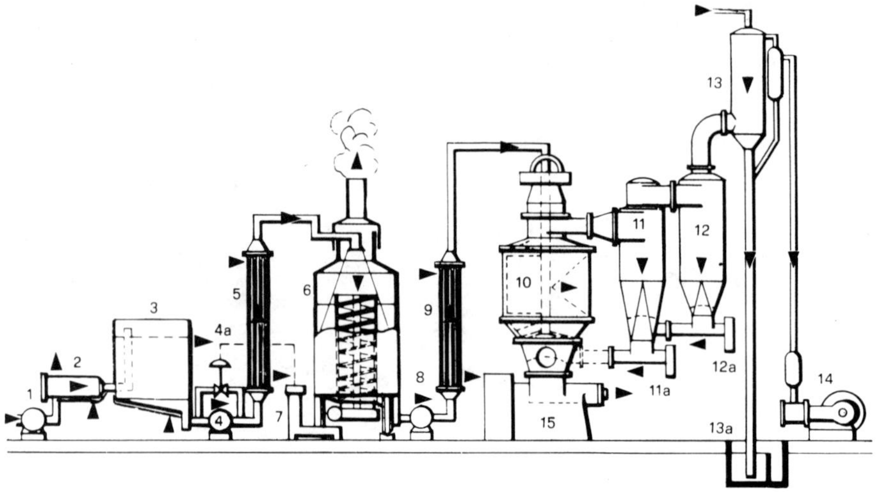

Fig. 6.15 – Mazzoni multistage drier: 1. Filtration pump: 2. Filter: 3. Feed tank: 4. First-stage feed pump: 4a. Recycle pneumatic valve: 5. First stage heat exchanger: 6. Flash chamber: 7. Scale: 8. Second-stage feed pump: 9. Second-stage heat exchanger: 10. Vacuum spray chamber: 11. First fines separator: 11a. Fines extruder plodder: 12. Second fines separator: 12a. Fines worm conveyor: 13. Barometric condenser: 13a. Barometric hotwell: 14. Vacuum pump: 15. Simplex plodder for soap base extrusion.

- Mazzoni claim less powder entrainment and less need to clean the heaters and condenser.

6.8.3

The various ingredients are added to the dried soap and roughly mixed in a conventional arm, or ribbon, mixer. In small scale operations a known mass of soap is charged to a batch mixer and weighed, or measured, quantities of the ingredients are added by hand. An adequate guard, often in the form of a grid with a trip switch, is important to avoid risk of accidents. A similar system, with the addition of an autoweigher and conveyor which provides a charge of soap at the touch of a switch, is commonly used in larger installations. The modern trend is to mechanized continuous toilet soap lines for which two alternative systems of blending are available:

- the automatic batch system in which increments of soap, liquids and solids are measured by weight or volume, mixed, and passed on to a hopper.
- relatively long worm type mixers through which the soap is passed continuously at a predetermined rate and to which streams of the various ingredients are added by dosing pumps for the liquids and weigh band feeders for the solids.

The automatic batch system is more accurate and has advantages when critical ingredients, such as germicidal slurries which are difficult to dose continuously at the required low rate, are included. The continuous system is probably rather simpler.

6.8.4

Homogenization and plastic working, using refiners and/or mills, and extrusion by plodders are discussed in section 6.3.3. To produce good toilet soap, it is particularly important to ensure complete homogeneity; and this includes the incorporation of returned scrap which may be drier and cooler than the rest of the material.

6.8.5

The next operation is to cut the extruded bar into billets of the required length. In small scale operation bar lengths can be cut by hand, and then converted to billets by being pushed through a wire frame. Some early automatic cutters were mechanized versions of this principle, but modern cutters, of the type sometimes known as the Simon cutter, but now made by several plant manufacturers, use the principle outlined in Fig. 6.16. Solid knives on a chain descend gradually and, with a correct type of drive, produce vertical cuts. Some machines are driven electrically with a hydraulic coupling to compensate for variations in the rate of extrusion; others use air motors, with adjustment of the supply of compressed air to serve a similar purpose. As well as single track machines, types are available to cope with two bars extruded in parallel. Fig. 6.17 is a photograph of a Weber + Seelander cutter.

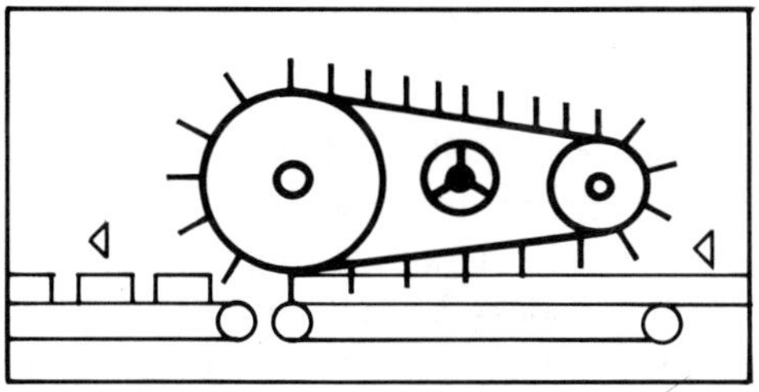

Fig. 6.16 – Billet cutter.

Different billet lengths are required for different soaps and there are cutters in which:

- adjustment within a certain range is by a hand wheel which can be used whilst the machine is running.
- the chain must be removed from the machine and adjusted link by link. The extreme form is a cutter in which the chain must be changed for each billet length.

Fig. 6.17 – Photograph of Weber + Seelander billet cutter.

6.8.6

Conditioning tunnels carry billets through a chamber in which they are subjected to currents of air, either atmospheric or chilled. Opinions differ as to the need for conditioning units for normal toilet soaps, but they are particularly useful for billets which are sticky because of non-soapy detergent ingredients or heavy superfatting, and for laundry soaps. The conditioning process:

- allows changes in internal structure to occur. This is particularly important for household soaps, in which increases in hardness tend to take place in the first few minutes after extrusion.
- cools the billets with consequent increase in hardness, although the low thermal conductivity of soap means that considerable time is needed for much change in the interior of the piece.
- dries, and cools, the surfaces, which reduces their stickiness.

6.8.7

Bars of laundry soap are often given an imprint on one surface which does not significantly change the dimensions of the bar. This can be done manually with a die of the same dimensions as the bar surface which is hit with a mallet.

Similar results can be achieved by machines. A simple device is a wheel with the required design which embosses the lettering, and other imprint, on the soap surface as it rotates with the moving bar in a way similar to the action of a modern cutter. Tablets of household and of toilet soap are produced using dies and stampers which can be divided into two major types:

- those which produce tablets with a side-band by the use of a pair of dies operating inside a collar, or shell. With this method the whole of the soap in the billet actually fed to the stamper remains in the tablet, so that the weight of the tablet is the same as that of the billet, apart from any evaporation between the production of the billet and the stamping operation.
- those which produce tablets without a side band, called capacity, pin, or squeeze dies. These dies use billets larger than the final tablets, come almost together, and separate an off-cut which must be removed from the tablet and the machine. Traditionally, pins locate the dies so that they meet correctly and ejectors operated by the stamper usually help to push out the tablets. The volume of the tablet is controlled by the capacity of the cavity enclosed by the two dies when they come together; and its weight is determined by this volume and the density of the soap. The gap between the dies in the closed position is about 0.002 inches (0.05 mm); and this can be varied a little for weight control purposes. Larger changes in weight normally require re-sinking of the dies, although it may be possible to make a small reduction by grinding the die edges. So far as is known capacity dies are not used for household soaps.

A wide range of stamping machines, or stampers, is available, ranging from hand stampers to high speed automatic machines. For many years machines known to those in the industry as the Jones 'K' type were almost universal for mechanized stamping of both household and toilet tablets. This type, and similar ones now made by various manufacturers, uses a shell with four pockets which rotates in the vertical plane in steps of a quarter turn. The billet is pushed into a pocket and is transferred to the next station where it is shaped by a pair of dies which move in horizontally. Modern machines normally give a double blow, that is the dies come together, move a little apart, and then come together again. This helps the escape of air and improves the quality of stamping. The double blow is quite different from the fault of double stamping in which a second billet enters a pocket from which the stamped tablet has not been removed. After it has been moved to the third station, the stamped tablet is pushed out onto the discharge conveyor. The last station may be used for cleaning. The dies are loaded with powerful springs; and should fit reasonably closely into the pockets in the shell. In this way the soap is squeezed out to fill the cavity formed by the two dies and part of the pocket wall. If the fit between dies and shell pocket is not good an unattractive fringe will appear on the tablets. With expensive soaps any fringe, and, more particularly, that which is liable to be produced with capacity dies, may be trimmed by hand, but this is not possible with popular brands produced by high speed, mechanized lines. On the other hand, too tight a fit between dies and shell does not permit the escape of trapped air, unless other arrangements are made for its release. An escape route can be made by the provision of a small gap between die and shell, on the central part of the long side. Alternatively, the part of the shell through which the dies move, that is not the central part which actually shapes the tablet, can be provided with escape grooves. Some operators drill fine holes (vents) in parts of the dies where air could be trapped, such as lettering which is raised on the tablet and hence recessed in the die. Small threads of soap are squeezed out of the vents, but the amount should be small and the 'pimples' which are left insignificant. Venting in this way increases the sharpness of the lettering. It is important to keep dies clean.

A lubricant, commonly a weak brine, or a solution of glycerine and salt, may be used to reduce any tendency for the soap to stick to the dies, but it should only be applied sparingly to the billets. Excess salt bearing lubricant is liable to cause surface cracking and a type of efflorescence mentioned earlier. A 10% solution of sodium lactate in water is said to be effective and to avoid risk of surface cracking. Modern stampers now usually have facilities for the circulation of refrigerated fluid through the dies; and this cooling, with billet conditioning if required, often makes the use of lubricant unnecessary. Dilute acetic acid was once used as a stamping lubricant for household soaps.

At one time it was thought necessary to select a nearly brick-shaped tablet for production with 'K' type stampers, and this type of tablet is the easiest to handle, but it has been found possible to widen the range of shapes provided they have a side band.

Until fairly recently, capacity dies could only be used with hand, or certain slow speed machine, stampers. Now fully automatic, high speed, machines are available for this duty. They often work vertically and very important features are effective mechanisms for the discharge of the tablets and the removal of the 'flash' or off-cut. The scrap may amount to some 30% of the billet, and it has to be returned to the plodder, usually via a refiner, for re-processing. The development of these machines has enabled cushion shaped tablets, without any side-band, to be handled on high speed mechanized lines and so facilitated the selection of such shapes for mass produced popular brands. Hand, pedal operated, or pneumatic stampers are still used for small production lines and for very fancy, novelty, shapes.

Bronzes, such as gun metal or silicon bronze, are used for dies and shells, often quite satisfactorily, but with high copper alloys there is some risk of slight contamination of the soap surface with copper which is a powerful catalyst for degradation reactions. Stainless steel dies in monel shells are often considered the best materials for the duty. They are expensive initially, but have a long life. Stainless steel dies in stainless steel shells have been avoided by some companies because of the risk of 'galling', or chafing, when one piece of stainless steel is rubbed on another, but it is understood that the combination has been used successfully. Some authorities favour the use of die surfaces which are slightly matt, and like to 'condition' dies by a period of contact with soap. Others prefer to use highly polished stainless steel surfaces.

The best shape of billet to produce a given design of tablet is controversial and a matter for experiment. One school of thought favours the use of a billet as near as possible in shape to the final tablet, others prefer to use billets which have to be completely re-shaped in the stamping operation. The latter may select a cylindrical billet, or, preferably, an oval with flat areas on the long sides. Some factories feed the cut billets into the stamper directly, but others use cutting, or shaping, boxes through which the billets are pushed on their way into the stamper shell. The use of a shaping box permits the use of final billets which fit closely into the shell pocket. This avoids the development of the flow lines which are often seen when straight ended billets directly from a cutter are stamped to tablets with bowed ends, or rounded corners, for which such billets must be shorter than the final tablet. Shaping boxes are also said to improve substantially the uniformity of weight of tablets from 'K' type stampers, but they produce a trim which has to be removed and reprocessed.

Stamping machines, or presses, currently available include:

- from *Mazzoni*
  ST presses – horizontal stampers with revolving shells for tablets with side band. STO presses – vertical stampers for capacity dies. STU presses – universal stampers which operate horizontally and can handle both types of die.
- from *Meccaniche Moderne*
  SAIN machines – for tablets with side bands. SAEV machines – for capacity die shapes. Sarotary machines – universal stampers with facilities for the super-cooling of the dies, said to be capable of handling any type of soap, and other detergent, tablets, of any shape at speeds up to 360 tablets per minute.
- from *Weber + Seelander*
  a range of machines to meet various requirements from simple low speed models to a universal, high speed, type.

Many machines are available in Simplex, Duplex and Triplex forms, which produce respectively one, two, or three tablets at a time. Nominal outputs are often up to about 120, 240 and 360 tablets per minute respectively, although Jones 'K' type machines (Simplex) have been worked at somewhat higher speeds.

Stampers, and other units, never work at 100% efficiency for long periods; and the overall efficiency of a machine, or complete line, must be considered as well as the nominal speed.

6.8.8

Household soap tablets are often sold in cartons or in single wrappers. Wrappers coated with heat sealable, low moisture vapour permeability, lacquers are particularly suitable.

Toilet soaps are sometimes sold naked, or with a number of tablets in a plastic bag, but they are usually wrapped and/or cartonned before being packed into outer cases. A very common form of packaging was the triple wrap, with an inner glassine (or similar) paper, a card stiffener, and a printed outer wrapper. The glassine is now often omitted to give a double wrap. The wrapper can be aluminium foil, or paper, often with a film, or other special finish. Round and oval tablets may be given pleated cellophane wrappers; and many special boxes and other packs are sometimes used, particularly for the more expensive soaps. Discussion of packaging machinery is outside the scope of this book.

6.8.9

As already mentioned, toilet soap plants range from small units with the minimum amount of equipment to large departments including mechanized lines devised to suit particular requirements. A small unit, for which dried soap is purchased, may comprise a mixer, mill, plodder, stamper, and wrapping machine; or a small number of some, or all, these items. Chips, or noodles, are stored between operations in bins. Trays are used for the transfer of billets and tablets from stage to stage; and to deal with fluctuations in the throughputs of the various units. All the major plant manufacturers supply complete mechanized toilet soap units, tailored to suit individual needs; and many variations are possible, including the duplication of some items in parallel when necessary to balance throughputs. An important requirement is a facility for the diversion of billets, or tablets, which are surplus to the immediate needs of the next machine; and to return this material, plus any offcuts, for reprocessing. Figs. 6.18 and 6.19 show diagrammatically two typical lines.

6.9 TRANSPARENT AND TRANSLUCENT SOAPS

Transparent soaps have been known in many countries for a long time; indeed, Pears transparent toilet soap was first made in the UK in 1789. In outline, the traditional Pears process uses the following stages:

- all-in saponification of a tallow, rosin and lauric oil blend in industrial alcohol with caustic soda.
- solidification to bars of the required shape (oval for Pears) in water cooled bar coolers.
- cutting into billets, often after the bars have been grooved to provide chamfers on the billet edges, followed by partial removal of alcohol and water in stoves, prior to stamping to tablets with flat faces.
- prolonged maturing in stoves, with continued evaporation of alcohol and water. During this process the tablets develop the traditional concave shape of Pears soap.

Alcohol is removed from the air extracted from the various drying stages in activated carbon adsorbers; and the weak alcohol liquor stripped from the carbon is rectified in a normal type of still for re-use. Davidsohn, Better and Davidsohn (1953) gives formulae and processes, broadly similar to that just described, for various transparent soaps. Some include castor oil in the fat blend;

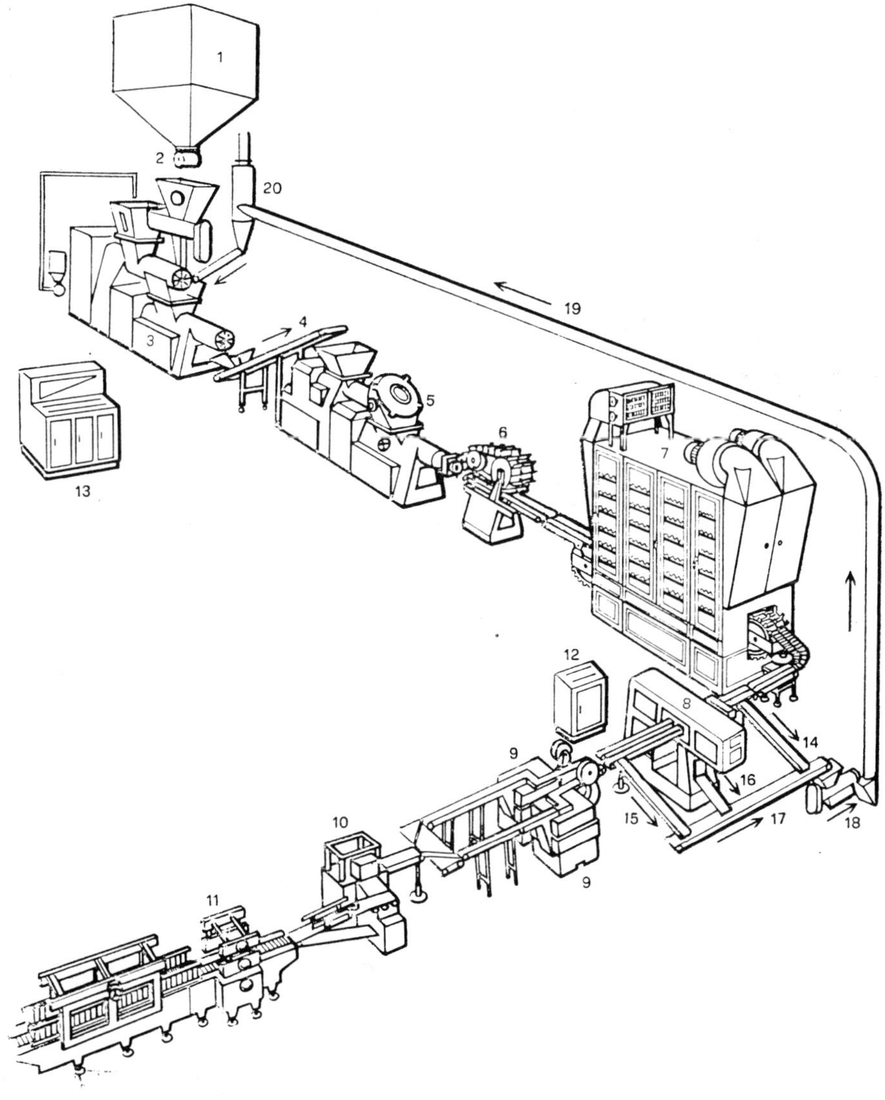

Fig. 6.18 – Mazzoni toilet soap line. 1. Soap bin 'SI': 2. Rotary discharge valve: 3. Mixer-refiner 'MC-Duplo': 4. Belt conveyor 'TSN': 5. Final vacuum plodder 'Duplex' or 'BITRA': 6. Adjustable cutter 'TV-Simplo' or 'TV-Duplo': 7. Conditioning tunnel 'TRV-Duplo': 8. Soap press 'ST' or 'STU': 9. Wrapping machines: 10. Case packer 'INCAR': 11. Case sealer 'INCOI': 12. Control board for press. 13. Main control board with synoptic panel: 14. Conveyor for billets (slugs) expelled before stamping: 15. Conveyor for bars expelled before wrapping: 16. Discharge chute for trimming anf flashing waste: 17. General waste conveyor. 18. Small plodder for pellets: 19. Waste recycling pneumatic conveyor: 20. Fines separator.

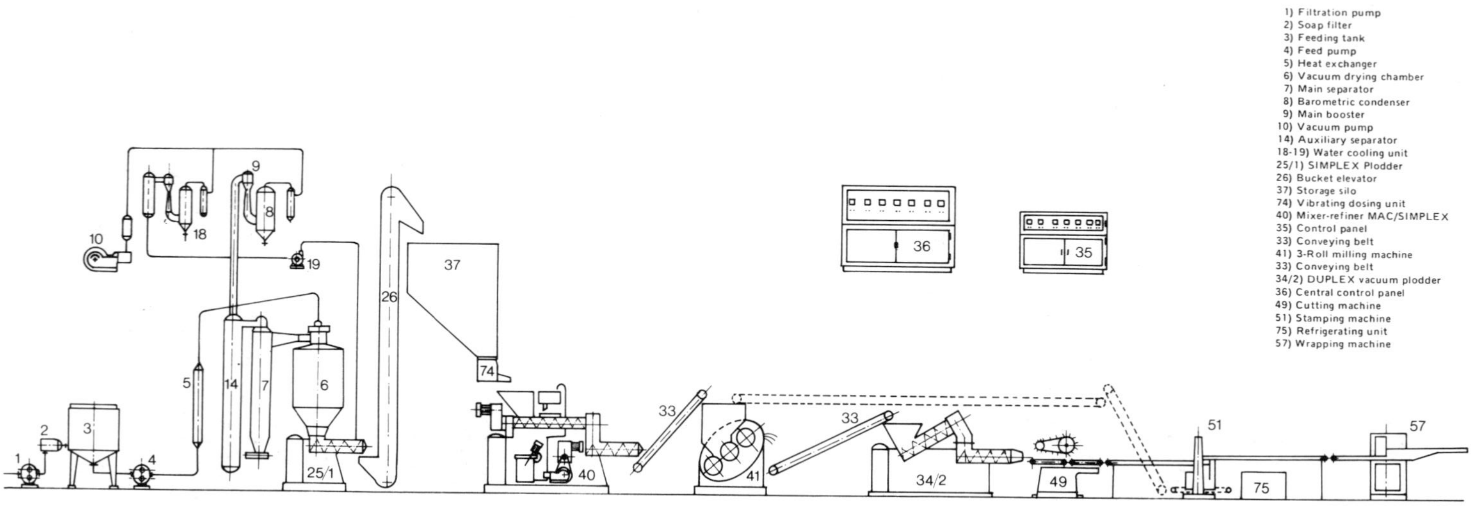

Fig. 6.19 – Meccaniche Moderne toilet soap line. 1. Filtration pump: 2. Soap filter: 3. Feeding tank: 4. Feed pump: 5. Heat exchanger: 6. Vacuum drying chamber: 7. Main separator: 8. Barometric condenser: 9. Main booster: 10. Vacuum pump: 14. Auxilliary separator: 18–19. Water cooling unit: 25/1. Simplex plodder: 26. Bucket elevator: 37. Storage silo: 74. Vibrating dosing unit. 40. Mixer-refiner MAC/Simplex: 35. Control panel: 33. Conveying belt: 41. 3-roll milling machine: 33. Conveying belt: 34/2. Duplex vacuum plodder: 36. Central control panel: 49. Cutting machine: 51. Stamping machine: 75. Refrigerating unit: 57. Wrapping machine.

and sugar solution and other ingredients are included in some final mixes. Cooling is in frames in which the soap is left for long periods, but more alcohol appears to be left in the soaps than in Pears. This is also the case with saddle soaps.

Translucent, and even transparent, soaps can also be made from a toilet soap base by extensive working under the right conditions on mills, or in machines somewhat similar to refiners. A once-popular brand of toilet soap was made by multiple milling at about 75% TFM. Transparent soaps made by a heavy working and extrusion processes may have a tendency to produce an opaque mush in use which can be unslightly, particularly with concave tablets such as Pears on which it has been called the 'fried-egg' effect.

6.10 FLOATING SOAPS

Floating soaps have been made and sold since the beginning of this century, particularly in the USA, where Ivory (Procter and Gamble) is very popular. Ivory is classed as a general-purpose soap used widely for toilet purposes. Swan (Unilever) was also started in the UK at this time, but the brand ceased to be made there many years ago; later a white floating Swan soap became important in the USA.

Several methods have been used to produce floating soaps:

- the level of soap in a crutcher is arranged so that air is whipped in; and in the case of a crutcher with a central worm, the direction of rotation of the worm may be reversed.
- in a process patented by V. Mills (Procter and Gamble), USP 2 295 594 (1942), which is described by McBride (1947), partly dried soap is heated and flashed into a chamber from which it is pumped to a Votator after air has been injected into the pipe line. The Votator is a brine cooled, scraped wall, chilling unit, from which the aerated soap is discharged as a continuous bar.
- in a process patented by F. F. Peace (Lever Brothers Company, or Unilever) USP 2 048 286 (1936) and USP 2 373 593 (1945), soap pellets at about 70°C are worked and aerated in a 'converter' with intermeshed screws from which the mass is discharged continuously to a moulding machine.
- Mazzoni offers, or has offered, an FOM, 'Float-O-Matic' plodder which is fed dried soap base and generates an aerated bar.
- a process in which hollow capsules containing air are inserted into normal soap bars plodded with a central hollow. The billets are then closed by the dies in the stamper. This process is described by J. R. Story, M. G. Norman and E. Jungermann, USP 3 403 106 (1968), assigned to Armour and Company.
- a process in which aluminium powder is mixed with the soap to form hydrogen by reaction with alkali.

6.11 MARBLED TOILET SOAPS

As a marketing attraction plodded soaps are sometimes produced with material of two colours arranged in a pattern, or in random streaks. This can be done by injection of coloured liquids, or of soap of a colour different from that of the base soap. Companies such as Mazzoni offer equipment; and patents have been issued for various systems. Some companies have devised simple procedures in which strips of coloured soap are introduced into white soap at the latest possible point just before the final extrusion of the bar.

6.12 FILLED, OR RUN, SOAPS

As mentioned earlier, low-priced bar soaps can be made by addition to neat soap of water containing a suitable concentration of sodium silicate, or other electrolyte. The required concentration varies with the fat blend and with the nature of the fit used; and for frame or water-cooled soaps it is common practice to make up a few small-scale mixes from each boil, although simple graphs can be produced for each type of soap which at least give a good approximation. Wigner made a detailed study of filled soaps and the following account largely follows his work. Because of the influence of different fat blends, the figures quoted from Wigner (1940) are only indicative and, for application, need to be replaced by experimental data for the base to be used. The use of weak sodium silicate solutions was patented by Sheridan in 1835, but application was restricted until after the invention by Gossage in 1854 of the modern method for the production of sodium silicates of uniform composition at low cost. After that date large quantities of filled soaps were made, much of it for export to undeveloped countries. The importance of these soaps has now greatly diminished, at least in the more developed countries.

Wigner classified filled soaps into the following types:

- *liquored soaps containing more than about 50% TFM.* These products are essentially homogeneous neat soaps; and according to Wigner those made with salt are at least as good as those made with sodium silicate. Referring back to Table 4.2 in Chapter 4 on soapmaking, and omitting the first two lines as slightly anomalous, it will be seen that each soap has a lower TFM (contains more water) and contains more salt than the one above it. After the experimental figures have been smoothed and interpolated graphically, Table 6.7 can be constructed by simple arithmetic. For example, to produce a 61% TFM soap from base at 63% TFM of the composition shown in the first line of Table 6.7 100 parts of 63% base yield $100 \times (63/61) \fallingdotseq 103.28$ parts of 61% TFM soap. Neat soap produced by a fit which gives 61% TFM contains 0.545% NaCl, so that 103.28 parts contain $0.545 \times (103.28/100) = 0.563$ parts NaCl. Hence salt to be added $= 0.563 - 0.434 = 0.129$, and liquor to be added $= 103.28 - 100.00 = 3.28$, so that concentration of salt solution to be added $= 0.129/3.28 \times 100 = 3.9\%$. Such soaps can be produced either by a suitable fit, or by addition of the appropriate salt solution to base at, say, 63% TFM. The direct fitting method can be used down to about 60% TFM, but the mixing method is possible down to around 50% TFM. The figures quoted as an example are for a soluble, washer-type, base and much weaker salt solutions are needed for a less soluble, high tallow or palm oil, base. If alkaline sodium silicate is used instead of salt, 1 part of Na_2O in the silicate is roughly equivalent to 1.3 parts of NaCl.

- *liquored soaps containing less than about 50% TFM, usually 28 – 42%.* Liquored soaps run to less than about 50% TFM are not homogeneous, but on settling for a few hours at crutcher temperature separate into layers equivalent to neat soap and nigre. When the correct concentration of sodium silicate is used there is only a small neat soap layer which can be dispersed satisfactorily in the bulk of the saturated solution (nigre). Adequate hand crutching in the frames is required to minimize separation. Sodium silicate, which has colloidal properties, appears to stablize this dispersion; and this type of run soap cannot be made with salt. If too high a concentration of silicate is used, unacceptable separation will occur in the frames and/or the soap will be rough in texture; whereas with too low a concentration the soap will be soft and pasty. In other words, just sufficient electrolyte is needed to ensure that the nigre phase is saturated with soap. Tallow/rosin, or palm oil/rosin soap bases were found particularly suitable for the manufacture of this class of soap. More concentrated silicate solutions are required than for the higher TFM class, and Wigner quotes the following figures for a certain tallow/rosin soap.

TFM (%)	Alkaline silicate solution ($^{\circ}$Tw)	(% total solids (approx))
28	37	18
30	34.1	17
32	31.2	15
34	28.4	14.5
36	25.6	13
38	22.8	12
40	20	10

Soaps at 42 – 50% TFM are usually difficult to produce satisfactorily.

- *Glued-up soaps*
These soaps are, or were, made from fat blends containing a high proportion of rosin (say 35% rosin, 65% tallow), often at about 56 – 58% TFM. Strong alkaline, or neutral, silicate, together with soda ash (anhydrous sodium carbonate) and sometimes sugar, form the fillings. The mixture becomes very viscous and special facilities, such as portable frames which can be

placed to be filled directly under the crutcher, may be required. The soap is not homogeneous and crutching down to a fairly low temperature is necessary. As mentioned earlier these soaps are liable to effloresce.

- *Mottled soaps*
  Originally mottling was caused by impurities in the caustic soda used to make the soap and was considered a sign that the soap was genuine. Later (but still very many years ago), mottled soaps were made at about 47% TFM using a base containing about 25% lauric oil together with tallow or palm oil, but no rosin. Sodium silicate plus sodium carbonate is used for the filling solution; and the colouring is ultramarine for blue-mottled, or an iron oxide pigment for red-mottled soaps. The frames are only allowed to cool very slowly; and considerable skill is needed to obtain the desired result. When conditions are not right the soap intended to be blue-mottled may be a uniform slatey blue-grey, or the pigment may settle out.

Wigner's work concerned soaps which were cooled in frames, or in slab coolers. Now it may be required to produce filled soaps in spray chilling/extrusion plants. This can be done; but at the lower TFMs the bars are very soft when extruded and it may be difficult to go as low in TFM as is possible when the soap is framed. Because of the softness of the soap very cold water is needed in the extruder jackets; and vacuum chilling of the water is often required. Mazzoni offers special equipment for the extrusion of mottled soaps. A special die removes a thin surface layer from the bar as it is extruded, and the scrapings are collected in a small double worm mixer where the colouring liquid is added by a proportioning pump. The coloured soap is then injected into the non-coloured soap by means of a small plodder and a special distributor to give a mottled bar.

Table 6.7 – Additions of salt solutions required to produce filled soaps (from Wigner, slightly recalculated)

TFM (%)	NaCl (%)	Weight of filled soap from 100 parts 63% TFM base	Total NaCl in filled soap	Parts NaCl to be added	Strength of salt solution (%)
63	0.434	–	–	–	–
62.5	0.451	100.81	0.455	0.021	2.6
62	0.483	101.62	0.491	0.057	3.5
61.5	0.511	102.44	0.523	0.089	3.6
61	0.545	103.28	0.563	0.129	3.9
60.5	0.582	104.15	0.606	0.172	4.1
60	0.625	105.00	0.656	0.222	4.4
59.5	0.670	105.89	0.709	0.275	4.7
59	0.730	106.78	0.780	0.346	5.1

7

Soap powders

7.1 GENERAL

Although pure powdered soaps can be made for various purposes, the term soap powder is normally used for products which contain substantial proportions of builders, particularly sodium carbonate. These products are also known as 'dry soaps'. Apart from its significance as a builder and filler, sodium carbonate plays a key role in the manufacture and physical properties of traditional soap powders. This is because sodium carbonate forms hydrates, $Na_2CO_3 \cdot H_2O$, $Na_2CO_3 \cdot 7H_2O$, and, particularly, $Na_2CO_3 \cdot 10H_2O$, which has a transition temperature at about 32°C. Anhydrous sodium carbonate (soda ash) is added to liquid soap base containing around 30% water and, after being cooled, the mass forms a solid in which much of the water is held by the sodium carbonate in the form of a hydrate, or hydrates. In the earliest process, used in more than one country from at least the middle of the nineteenth century, the soda ash is stirred into the soap until uniform, and the mixture is tipped onto a floor to cool. When ready, the mass is broken with hammers and ground to a powder in a hammer mill. The mill commonly includes a grating through which the powder passes when fine enough, but 'dressing' the powder by passing it through a screen (and re-grinding the oversized 'tailings') is usually desirable to improve the appearance and solubility of the product. Soda ash stored in sacks is prone to cake, and may sometimes need to be ground and screened before use. By the 1920s and 1930s, for large-scale manufacture this process was superseded by a spraying, or blowing, process which gives an improved product, but the traditional floor method is still used by some small operators, particularly in developing countries. Soap powders can be produced by two different spraying processes, spray chilling and spray drying.

7.2 SPRAY-CHILLED POWDERS

The spray-chilling process, which was for many years the most usual method for the manufacture of soap powders, relies on the use of sodium carbonate to take up the water from the soap base as just explained. The ingredients are mixed in an agitated pressure vessel and heated with open steam to a temperature around 150°C. 150°C is equivalent to a steam pressure of 55 psig, but a higher pressure, preferably about 150 psig, is needed to push the steam into the mix and to provide top pressure to force the mix from the mixing vessel to the atomization nozzle. When ready, the mix, or slurry, is sprayed into a chamber through which a current of unheated air is drawn by fans. The chamber can be a vertical tower, but other designs, including simple improvised enclosures, are possible. The air extracted from the top of the chamber is passed through dust collectors, such as vessels containing rings over which brine is circulated. The nozzle, or nozzles, used for the atomization of the slurry can be very simple, because atomization is achieved by the flash evaporation which occurs when the pressure on the superheated slurry is reduced to atmospheric. A short length of small, say 1 inch (2.5 cm) diameter, pipe is adequate. A larger, say 3 inch (7.6 cm), pipe should be used to convey the slurry from the pressure vessel/mixer to the nozzle, with a control valve near the nozzle. The steam usually provides sufficient top pressure to convey the slurry from the pressure vessel to the nozzle, but, if necessary, the pressure can be augmented by a positive rotary pump. The most uniform particle size distribution in the powder seems to be obtained by ensuring that little vaporization occurs in the pipe prior to the nozzle, a condition which gives relatively silent blowing. The higher the temperature of the slurry the finer the powder. Steam is condensed whilst the mix is heated with open steam to the blowing temperature, but this water evaporates when the pressure is released and there is a small net evaporation of around 4%, caused by the sensible heat in the liquid soap base and by evaporation from the particles into the cooling air. In an attempt to reduce the bulk density of the powder, aeration of the slurry by the introduction of compressed air into the slurry in the pressure vessel has been tried and found to have some limited effect. Cooling air above about 25 – 30°C cannot normally be used satisfactorily, and it is desirable for it to be cooler. One plant arrangement, with a pear-shaped horizontal blowing chamber, is shown diagrammatically in Fig. 7.1. The particles cool as they fall through the rising air and should be sufficiently solid to be removed as a powder when they reach the conveyor belt at the base of the chamber. Crystallization of the sodium carbonate hydrate takes a certain time and it is usually necessary to 'weather' the powder after it leaves the chamber and before it is packed. Weathering involves additional cooling and more complete crystallization. If it is not done effectively, the powder is liable to reheat through liberation of latent heat of crystallization and to cake in the packets. Weathering can be carried out by:

- storage in sacks for about 12, or more, hours. Caked lumps, usually only lightly compacted, are broken when the weathered powder is emptied from the sacks, screened and packed.
- transportation slowly on water cooled steel conveyor bands. Thick layers

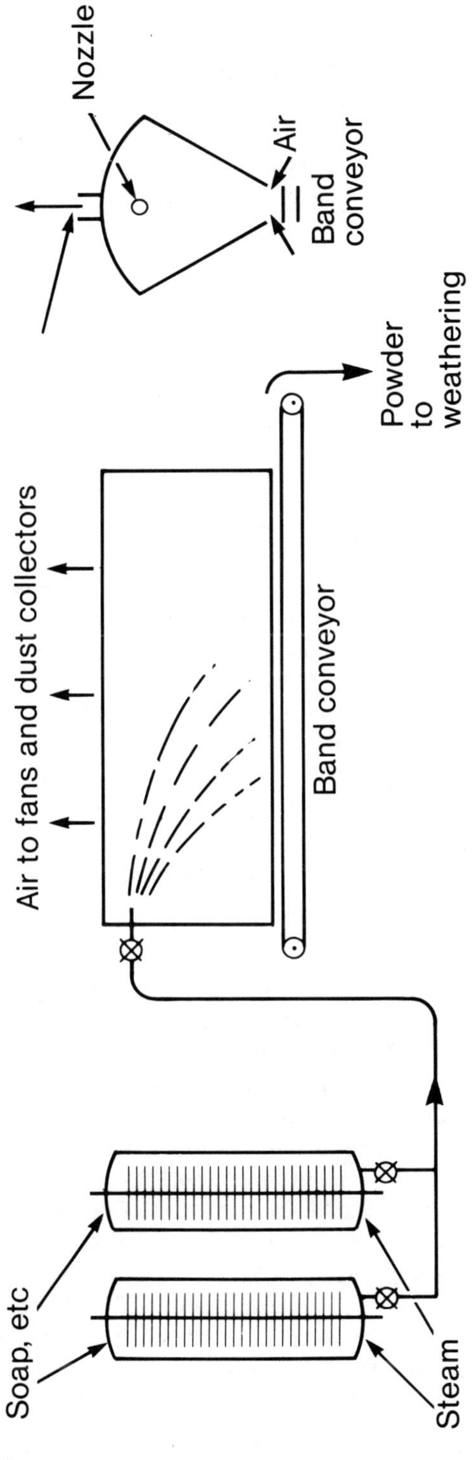

Fig. 7.1 – Plant for production of spray chilled soap powder.

are used to increase the holding time to facilitate crystallization, but the powder is mixed as it falls from one band to the next.

- the use of an air lift which raises the powder to a convenient height for subsequent operations as well as weathers it. (Such air lifts are described in more detail in section 10.3.3.7 concerned with NSD powders).

The air-lift system does not provide much time for the completion of crystallization, but it has been used satisfactorily with many formulations.

Factors which limit formulations which can be sprayed chilled are:

- there must be a sufficiently high ratio of sodium carbonate to water in the powder.
- the mixture of soap base, soda ash and other ingredients must be sufficiently liquid to permit proper mixing, flow to the nozzle, and atomization.

Ingredients used in spray-chilled powders in addition to soap base and soda ash include:

- sodium silicate, commonly added to the slurry as 80°Tw neutral silicate solution. Silicate is useful for its building properties and because it reduces risk of corrosion of metals, particularly aluminium, by the soap powder in use. This is probably less important than it once was because aluminium vessels are now little used in washing operations.
- Optical brighteners. If it is necessary in order to avoid deterioration in the hot liquid mix, the optical brightener can be injected into the pipeline, shortly before the nozzle.
- relatively small percentages of sodium pyro- or ortho-phosphate.
- sodium carboxymethyl cellulose, although this is less necessary with soap than with NSD powders.
- water when the soap base and the silicate do not provide sufficient.

Ingredients which cannot be added to the slurry to be spray chilled, because of risk of thermal damage, can be dry-mixed with the blown base powder, often preferably after it has been weathered. Perfume and sodium perborate are commonly added in this way, and the procedure can also be used for phosphates.

As an empirical rule, it is found desirable not to have more water in the blown powder than would be about sufficient to convert the sodium carbonate to $Na_2CO_3 \cdot 10H_2O$, if all the water was transferred from the soap base and the silicate solution to the sodium carbonate. Some typical mixer charges and blown powder compositions are given in Table 7.1, together with an indication how the simple calculations can be carried out. Even with 100% soap base, 4% net evaporation would only give a product with about 65% TFM, and with allowance for the soda ash necessary to give a dry powdered product, it is not possible to produce a powder with more than about 55% TFM by the spray-chilling process. With the low TFM powders, such as that shown in formula 3, considerable variations are possible. In the production of these low TFM powders, it may be best to combine the soap base and water, plus any necessary electrolyte, in the soapmaking department to give a soapstock which is easy to handle, or to make a similar soapstock in some other way, for example, from by-products. The minimum amount of liquid needed to make mixes for high TFM powders sufficiently fluid to handle depends to some extent upon the details of the plant and upon the fat blend of the soap. A wide range of fat blends can be used for soap powders, but reasonably high proportions of lauric oil and of hard fat (say 20 – 30%, or more, of each) are desirable to give crisp, dry, powders with good lathering and washing properties.

As already mentioned, the powders can be sold as blown, or after admixture with other ingredients, such as perfume, sodium perborate and various phosphates. When NSD powders started to become important in the 1950s, considerable efforts were made to reduce the tendency of the competing soap powders to produce scums when used in hard water. Phosphates were employed for this purpose, sometimes included in the blown powder and sometimes added separately to the base powder. An addition of about 8% of disodium orthophosphate dihydrate, $Na_2HPO_4 \cdot 2H_2O$, to a blown powder such as that shown as formula 1 in Table 7.1 was typical. Sodium carbonate is a precipitative builder but, unless the water is softened with the sodium carbonate before the soap is dissolved, alkaline earth soaps tend to precipitate with soap based systems. An addition of freshly ground calcite (crystalline calcium carbonate), can, under some circumstances, induce more precipitation of the water hardness as carbonate rather than soap, but an attempt to use this commercially was not successful in the market place.

Considering now the weathering of a spray chilled soap powder in a little more detail, the purpose is to obtain a powder which is stable and which will not cake, that is form lumps, or in its extreme form, a solid block, during storage in cartons. Freedom from tendency to cake depends upon the composition of the powder, the temperature and humidity of the atmosphere in which the product is stored, and on the completeness with which the powder has been weathered. The heat to be removed during the weathering process includes both the sensible heat involved in a reduction in temperature and latent heat of crystallization of the sodium carbonate. The particles are cooled from the surface during their fall in the spray chamber and during the weathering process, and if incompletely weathered powder is put into sacks, or packets, the temperature will rise with time. A similar effect is noted if a 'ball' of powder is held round the bulb of a thermometer being used to take the powder temperature. This rise is due to transfer of heat from the interior of the particles and to continued crystallization. It is important to ensure that a temperature permanently below about 32°C is achieved before the powder is packed. Weathering systems which allow appreciable time for continued crystallization, such as sacks or water cooled steel bands, have at least a theoretical advantage, but pneumatic, air-lift, systems are often satisfactory. Inclusion of disodium orthophosphate has been found to increase caking problems, both when it is included in the slurry and when it is added after blowing. Experiments with a slurry containing this phosphate blown in a plant with a pneumatic weathering system showed that the powder left the air lift at an exceptionally low immediate

Table 7.1 – Mixer charges and powder compositions for spray chilled soap powders. Net evaporation is assumed to be 4%

Ingredients added to mixer			H_2O in mixer charge	Notes
	%	parts in 104		
Formula 1. Clothes washing powder for temperate climates				
Soap base (assumed 62% TFM and 30% H_2O)	77	80.1	24.0	TFM of blown powder $= 62 \times 0.801 = 49.7\%$
Soda ash	16	16.6	–	Moles $H_2O = \frac{24.7}{18} = 1.372$
Neutral 80°Tw sodium silicate (36% solids 64% H_2O)	7	7.3	4.7	Moles $Na_2CO_3 = \frac{16.6}{106} = 0.157$
Total	100	104.0	28.7	i.e. $\frac{\text{Moles } H_2O}{\text{Mole } Na_2CO_3} = \frac{1.372}{0.157} = 8.7$
Net water evaporated	–	4.0	4.0	
In blown powder	–	100.0	24.7	or, carbonate is nominally $Na_2CO_3 \cdot 8 \cdot 7\ H_2O$

Formula 2. Clothes washing powder for very hot climates

Soap base	69.5	72.3	21.7	TFM of blown powder $= 62 \times 0.723 = 44.8\%$
Soda ash	23.0	23.9	–	
Water	7.5	7.8	7.8	Moles $H_2O = \frac{25.5}{18} = 1.417$
Total	100.0	104.0	29.5	Moles $Na_2CO_3 = \frac{23.9}{106} = 0.225$
Net water evaporated	–	4.0	4.0	
In blown powder	–	100.0	25.5	i.e. $\frac{\text{Moles } H_2O}{\text{Mole } Na_2CO_3} = \frac{1.417}{0.225} = 6.3$
				or, carbonate is nominally $Na_2CO_3{\cdot}6{\cdot}3\ H_2O$

Formula 3. A cheaper, general-purpose, powder

Soap base	23	23.9	7.2	TFM of blown powder $= 62 \times 0.239 = 14.8\%$
Soda ash	30	31.2	–	
Neutral 80°Tw sodium silicate	7	7.3	4.7	Moles $H_2O = \frac{49.5}{18} = 2.750$
Water	40	41.6	41.6	Moles $Na_2CO_3 = \frac{31.2}{106} = 0.294$
Total	100	104.0	53.5	
Net water evaporated	–	4.0	4.0	i.e. $\frac{\text{Moles } H_2O}{\text{Mole } Na_2CO_3} = \frac{2.750}{0.294} = 9.4$
In blown powder	–	100.0	49.5	or, carbonate is nominally $Na_2CO_3{\cdot}9{\cdot}4\ H_2O$

temperature, but that very little heat was transferred to the air. Large temperature rises occurred subsequently, and the unusual behaviour was presumably due to a delay in the crystallization of the carbonate caused in some way by the phosphate. The tendency for a powder to cake can only be determined by trials under the relevant atmospheric conditions. Accelerated tests at higher than ambient temperatures which are useful, for example, to assess the keeping properties of soaps, can be misleading because of the significance of hydrate transition temperatures. Experiments have shown that:

- powders made to formula 2 in Table 7.1 (without silicate) stored better under severe tropical conditions (typical of those in parts of India) than powder to formula 1.
- powder to formula 1 to which 8% $Na_2HPO_4 \cdot 2H_2O$ was added behaved much worse than the base powder alone under tropical conditions, although the blend was satisfactory under normal UK conditions.

Spray chilled soap powders in ordinary, unprotected, cartonboard cartons tend to lose moisture when stored in atmospheres with relative humidity up to about 75 – 80%.

7.3 SPRAY-DRIED POWDERS

Spray drying differs from spray chilling in that conversion of liquid droplets into solid particles is achieved primarily by the evaporation of a considerable proportion of water from the liquid sprayed. The products normally contain about 50 – 92% anhydrous soap (45 – 83% TFM), and those which are substantially unbuilt are sometimes called powdered soaps. Powdered soap beads, often of very low bulk density, had a vogue some time ago, particularly in the USA, but are seldom made now. Two major companies in the UK (where soap powders continued to be of great importance after this ceased to be so in most other developed countries) devised processes in which built soap powders were made from fatty acids, instead of 62% TFM soap base, using spray drying. This gave greater flexibility as regards composition and improved rate of solution properties, but tonnages of soap powders in the UK are now falling fast because of replacement by low, or controlled, sudsing products.

The spray drying of soap powders is similar to that of NSD powders discussed in detail in Chapter 10. Wurster and Sanger Inc. has offered plants for the spray drying of soap powders which can be modified to produce NSD powders, but the company states that it is no longer active in this field. High soap content powders are more flammable than NSD powders which contain substantial proportions of inorganic ingredients; extra care is, therefore, necessary to minimise risk of fire or explosion.

7.4 OTHER MANUFACTURING PROCESSES

Soap powders, and powdered soaps, can be made in various other ways for a number of uses. These processes commonly involve drum drying, followed by grinding and screening. For example:

- pure soap powders can be produced by drying soap base to a nearly anhydrous state and grinding.
- built soap powders can be made by drying a mixture of soap base and sodium silicate, grinding and mixing with soda ash and, if required, other ingredients.

8

The recovery and refining of glycerine

8.1 GENERAL

As explained in earlier chapters, the glycerol in fats is usually recovered in either soap lyes or sweet waters. Normally, these materials are purified and most of the water they contain evaporated under conditions which do not vaporize any significant quantity of glycerol to give respectively soap lye crude glycerine or saponification, or hydrolyser, crude glycerine. Crude glycerine is normally purified by distillation, sometimes followed by bleaching and other treatments. Section 8.8 describes alternative processes which use ion exclusion and/or ion exchange instead of distillation.

Various processes for the manufacture of synthetic glycerine from sugar by fermentation have been developed, commencing with work in Germany during the 1914 – 18 war; but the modern synthetic glycerine industry started with a Shell process based on propylene. Several companies using a number of processes now have plants, with apparently a total capacity in excess of demand. Considerable problems were experienced initially, particularly as regards purification, but the quality of modern synthetic glycerine is very high. Indeed, refiners of natural glycerine have sometimes found it difficult to match.

The manufacture of synthetic glycerine is outside the scope of this book; but it can be said that the natural material should always be able to compete. This is because glycerol must be liberated when fats are saponified, or split, and the marginal cost of recovering it is likely to be below the cost of making the synthetic equivalent.

8.2 SOME ANALYTICAL TESTS AND SPECIFICATIONS

It is convenient at this point to outline the main analytical tests used for glycerines and to consider some implications of the results which they give.

8.2.1 Glycerol content

Glycerol content can be estimated by several methods. The standard method is now oxidation with periodic acid (*sodium periodate* in acidic solution) as described in BS 5711:1979 and other publications. Glycerol is oxidized to formic acid according to the equation

$$\begin{array}{l} CH_2-CH-CH_2 \; + \; 2H10_4 \rightarrow 2H10_3 + 2H\cdot CHO + H\cdot COOH + H_2O \\ \;|\qquad\;\; |\qquad\; | \\ OH\quad OH\quad OH \end{array}$$

glycerol — periodic acid — iodic acid — formaldehyde — formic acid — water

The quantity of glycerol originally present in the sample is determined by titration with standard caustic soda of the formic acid formed under standard conditions. This test estimates as glycerol any compound which contains a group

$$\begin{array}{l} -CH-CH-CH- \\ \;\;\;|\qquad\; |\qquad\; | \\ \;OH\quad OH\quad OH \end{array}$$

with three adjacent carbons each with a hydroxyl group attached. Such compounds include sugars and sugar alcohols (sorbitol, etc.) which are sometimes found in glycerines which have been contaminated with glycerine substitutes used in some food industries, or added in an attempt to defraud. On the other hand, TMG (trimethylene glycol, 1,3 dihydroxypropane, or propane 1,3 diol) and polyglycerols are not estimated as glycerol by this test. As considered later, TMG appears in some crude glycerines made from low-grade fats which have been stored badly, or, more particularly, from lyes which have been stored under conditions which permit destruction of glycerol by micro-organisms. As well as indicating a destruction of glycerol, TMG in a crude glycerine can cause problems in the refining operations. Polyglycerols are formed by heating glycerol under certain conditions, particularly in the presence of alkaline, and other, catalysts. The reactions are of the type

$$\begin{array}{l} CH_2-CH-CH_2 + CH_2-CH-CH_2 \rightarrow CH_2-CH-CH_2-O-CH_2-CH-CH_2 + H_2O \\ \;|\qquad\; |\qquad\; |\qquad\quad |\qquad\; |\qquad\; |\qquad\quad\; |\qquad\; |\qquad\qquad\qquad\qquad\quad\; |\qquad |\\ OH\;\; OH\;\; OH\quad\; OH\;\; OH\;\; OH\quad\; OH\;\; OH\qquad\qquad\qquad\qquad OH\; OH \end{array}$$

glycerol — glycerol — di-glycerol

Tri-glycerols, and higher members of the polyglycerol family, are formed by similar condensation reactions in which a molecule of water is eliminated between two hydroxyl groups to form an ether linkage. The reaction can occur between two primary alcohol, $-CH_2OH$, groups as shown, or a secondary alcohol group $>CH-OH$ can be involved, so that many isomeric polyglycerols are possible.

Formation of polyglycerols, or, more strictly, avoidance of their formation, is important in glycerine distillation, and estimation of polyglycerols in still residues (foots) is often useful.

For many years before the development of the periodate method, the *acetin method* was standard for the estimation of glycerol. This involves acetylation of the glycerol with acetic anhydride and determination by hydrolysis with standard caustic soda, of the quantity of the ester tri-acetin formed. The acetin method includes as glycerol other hydroxyl compounds likely to be present, such as sugar, TMG and polyglycerols. Because of the elimination of hydroxyl groups when polyglycerols are formed, the acetin method does not return polyglycerols as the full amount of glycerol from which they were produced. Thus:

- two molecules of glycerol contain 6 hydroxyl groups, but the diglycerol formed from them has only 4 hydroxyl groups.
- similarly tri-glycerol produced from three molecules of glycerol with 9 hydroxyl groups, contains only 5 hydroxyl groups.

The third important procedure for the estimation of glycerol is the *dichromate method.* This method, used particularly for soaps and lyes, involves removal of soapy, and other, impurities, followed by estimation of oxidizable matter calculated as glycerol. It includes all oxidizable compounds not removed in the preliminary purification of the sample. Because the condensation reactions in which polyglycerols are formed by elimination of water do not change the oxygen requirement for complete oxidation to carbon dioxide and water, the figure obtained for glycerol percent includes any polyglycerols expressed in terms of the mass of glycerol from which they were formed.

For many purposes, the glycerol content of refined glycerines is estimated by determination of *specific gravity* (15°C/15°C), or *relative density* measured at 20°C/20°C, and reference to tables giving the properties of pure glycerol/water mixtures. The presence of other substances can interfere with the accuracy of results obtained.

8.2.2 Non-volatile inorganic and organic residues, particularly in crude glycerines

The content of inorganic matter is determined as *percentage ash.* In the case of refined grades a little sulphuric acid is added during the determination to give *sulphated ash.*

For many years, important organic impurities in crude glycerines were estimated by determination of *total residue %,* the residue left after vaporization of the glycerol under carefully controlled conditions in an oven at 160°C. The *organic residue (OR)* was then total residue % – ash %.

More recently, organic residue has largely been replaced by *MONG %* (matter organic non-glycerol) which is defined as 100 – (ash + glycerol + water) %. Glycerol is determined by periodate and water by the Karl Fischer method. Determination of total residue is tedious and rather subject to error, so that

MONG is more convenient. The figures found for OR and MONG are very similar for many crude glycerines, but there are differences which are significant under some circumstances:

- TMG is included in MONG, but not in OR, because it is more volatile than glycerol. Hence MONG should equal OR + TMG.
- any volatile alcohols, occasionally present in samples of glycerine including material from the alcoholysis of fats to form methyl, or ethyl, esters of the fatty acids, are included in MONG, but not OR.
- sugars, and other polyhydroxy compounds which increase the apparent glycerol content of a sample, artificially reduce MONG.
- polyglycerols and other relatively non-volatile organic compounds should be included in both MONG and OR, but substantial proportions of diglycerol, which is slowly volatile under total residue conditions, may make difficult the determination of the total residue end point (that is the stage at which the residue is deemed to have reached constant weight).

8.2.3 Nitrogen

Organic nitrogen which comes from proteinaceous materials is sometimes found in quantities up to 0.5% as nitrogen in poor quality saponification crude glycerines made from low grade animal fats, although 0.025 – 0.15% is more usual. Crude glycerines with a high organic nitrogen content tend to give distillates with objectionable odours and colours. Many nitrogenous compounds have been found at low concentrations in such distillates, with 2-piperidone, a toxic substance, the most important. Washing the fats with sulphuric acid before they are split does not help much, but earth bleaching is often useful.

8.2.4 Arsenic

A limit for arsenic content is found in many glycerine specifications. At one time it was troublesome because of use in the factories which produced the crude glycerines of sulphuric acid contaminated with arsenic; but there is now seldom any problem.

8.2.5 Sodium chlorate

Sodium chlorate is an unusual, and unexpected, constituent of crude glycerines, but it has sometimes been found in significant quantities which have caused explosions during distillation in some types of still. It arises from the use of caustic soda made in electrolytic cells which are incorrectly designed or operated. If crude glycerine containing sodium chlorate has to be processed, the chlorate can be reduced under strongly acidic conditions, for example with spent pickling acid derived from steel works which contains ferrous chloride and hydrochloric acid (see later for the possible use of such material in the purification of soap lyes).

8.2.6 Silver tests for reducing substances in refined glycerines

Most specifications for refined glycerines include a test for reducing substances, such as aldehydes, which is based on the darkening which occurs with ammoniacal silver nitrate. In some cases comparison is made with the darkening produced by a standard quantity of dextrose (glucose). Different silver tests vary in sensitivity, so that conditions laid down in a particular specification must be followed precisely. Substances such as soluble phosphates derived from some activated carbons, can interfere with certain tests.

8.2.7 Fatty acids and esters in refined glycerines

The fatty acid and ester content of a refined glycerine is usually important; and most specifications include a test to measure the quantity of caustic soda reacted when the sample is boiled with an excess under standard conditions. For chemically pure grade this was known as soda absorption expressed in ml of N/10 NaOH absorbed in a standard test. Now it is commonly called saponification equivalent for all grades of refined glycerine and expressed as percentage Na_2O absorbed, or as milli-equivalents per 100 g. Fatty acids and esters in a distillate usually affect its odour and taste adversely.

8.2.8 Chlorides in refined glycerines

Most specifications include a test for sodium chloride based on reaction with silver nitrate. This may be expressed as maximum parts per million permitted, or in terms that the turbidity produced in the test is nil, or less than in a control. The sensitivity of this test makes important the almost complete exclusion of entrainment of still liquor when soap lye crude containing a high percentage of salt is being distilled.

Some specifications also include a limit for organic chlorides which is related primarily to synthetic glycerines.

Table 8.1 – British Standard Specifications for crude glycerines

	BS 2621:1979 Soap lye crude glycerol[a]	BS 2622:1979 Hydrolyser crude glycerol[a]
Glycerol % m/m (periodate) (min)	80.0	88.0
Ash % m/m (max)	10.0	1.0
MONG % m/m (max)	2.5	1.5
Water % m/m (Karl Fischer) (max)	10.0	–
Propane 1,3 diol (TMG) % m/m (max)	0.5	0.5
Arsenic as As ppm or mg/kg (max)	2.0	2.0
Sugars (max)	Nil	Nil

[a]The British Standards Institution uses the word 'glycerol' where 'glycerine' would be used according to the usage in this book.

Soap lye crude glycerol is described as a viscous liquid prepared by evaporation of the purified lyes obtained from the manufacture of soap.

Similarly, hydrolyser crude glycerol shall be a viscous liquid prepared by evaporation of the sweet waters obtained from the hydrolysis of fats under pressure or in the presence of catalysts.

Table 8.2 – British Standard Specifications for refined glycerines

	BS 2623:1979 Technical grade glycerol	BS 2624:1979 Dynamite grade glycerol	BS 2625:1979 Chemically pure glycerol
Glycerol % m/m (periodate)	> 99.0	> 99.0	> 99.0
Relative density 20°C/20°C	←——	> 1.261 and < 1.264	——→
Sulphated ash %	< 0.010	< 0.010	< 0.010
Acidity or alkalinity milli-equivalents/100 g.	< 0.32	< 0.32	< 0.064
Saponification equivalent milli-equivalents/100 g.	–	< 0.64	< 0.64
Total heavy metals as lead (mg/kg) or ppm)	–	–	< 5.0
Lead as Pb (mg/kg or ppm)	–	–	< 1.0
Iron as Fe (mg/kg or ppm)	< 2.0	–	–
Arsenic as As (mg/kg or ppm)	–	–	< 2.0
Chlorides as NaCl%	< 0.010	< 0.010	No turbidity in standard test
Organic chlorides	–	–	Turbidity not greater than standard
Reducing substances (silver test)	–	Not darker than standard	Not darker than stanard
Colour, Lovibond* (133 mm cell)	< 5.0 Y and 1.2 R	As technical grade	–
Sugars	–	–	Nil
Odour and taste	Free from abnormal odour when tested as described in BS 5711	Shall not emit an acrid odour when tested as described in BS 5711	Free from odour in test described. Shall taste sweet and be free from any burnt flavour

*Also expressed in chromaticity co-ordinates.

Summary of preambles to these specifications

Technical glycerol

Clear viscous liquid, consisting essentially of glycerol.
Free from visible foreign matter.

Dynamite glycerol

As for Technical glycerol.

Chemically pure glycerol

Glycerol complying with BS 2625:1979 will also meet the requirements of the *European Pharmacopoeia* (1st edition).

Chemically pure glycerol shall be a clear colourless viscous liquid consisting essentially of glycerol and free from visible foreign matter.

Table 8.3 – Summary of specifications, courtesy Unichema International

Product	glycerine content by specific gravity % min	Specific gravity in air		at °C	Colour max. APHA	odour	acidity/ alkalinity eq. as Na_2O % max	fatty acids and esters eq. as Na_2O % max	sulphated ash % max	chloride as NaCl ppm max	organic chloride ppm max	reducing substances ***	sugars ***	arsenic as As ** ppm max	heavy metals as Pb** ppm max	water content % max
					ASTM-D1209	DAB-8/ BS 5711 Part 19	BS 5711 Part 5	BS 5711 Part 21	BS 5711 Part 7	BS 5711 Part 12	BS 5711 Part 14			BS 5711 Part 10		Karl Fischer
PRICERINE 9083 Chemically pure glycerol	99.5	1.262		20/20	10	odourless	neutral	0.02	0.01	2	passes < 5	neg.	nil	1	1***	0.5
PRICERINE 9081 Chemically pure glycerol	86.5	1.228		20/20	10	odourless	neutral	0.02	0.01	2	passes < 5	neg.	nil	1	1***	
PRICERINE 9071 Nitration grade glycerol	99.5	1.262		20/20	25	no acrid odour at 100°C	0.01	0.02	0.01	5		neg. acc. to NTD 21	nil	1	1***	0.5
Comparable requirements*		min	max													
DAB-8 (1978)	98	1.258	1.263	20/20	(a)	(b)	0.0025/0	0.025	0.01	16	35	not directly comparable	not directly comparable		5	
DAB-8 (1978)	84-88	1.221	1.232	20/20	(a)	(b)	0.0025/0	0.025	0.01	16	35				5	
BS 2623 (1979)	99	1.261	1.264	20/20	ca. 100	normal	0.01		0.01	100						
BS 2624 (1979)	99	1.261	1.264	20/20	ca. 180	not acrid	0.01	0.020	0.01	100						
BS 2625 (1979)	99	1.261	1.264	20/20	(a)	(b)	0.002	0.020	0.01	2	28			2	1	
Europ. Pharm. III (1976)	ca. 98	1.258	1.263	20/20	(a)	(b)	0.0025/0	0.025	0.01	16	35				5	
Europ. Pharm. III (1976)	84-88	1.221	1.232	20/20	(a)	(b)	0.0025/0	0.025	0.01	16	35				5	
Dutch Pharm. VI	98	1.258	1.264	20/20	ca. 20	(c)		0.015	0.01	75				5	10	
Dutch Pharm. VI	88	1.228	1.233	20/20	ca. 20	(c)		0.015	0.01	75				5	10	
French Pharm. (1976)	ca. 98	1.258	1.263	20/20	(a)	(b)	0.0025	0.025	0.01	16	35				5	
Scand. Pharm.	84.5-89	1.220	1.232	20/4	50	(b)		0.036	0.05	7.5	5				6	11-15.5
British Pharm. (1973)	98	1.255	1.260	20/20	ca. 20	(b)	(d)	0.020	0.01	2	28			2	1	
USP XIX	95	1.249		25/25	15-20	normal	(d)	0.031	0.01	15	30			1.5	5	
Food Chem. Codex (1973)	95	1.249		25/25	15-20	normal		0.036	0.01		30			3	5	
Am. Chem. Soc. (1968)	95	1.245		25/26	10		(d)	0.018	(0.005)	7.5					2	
Nobel Test D 21	98.5	1.262		15.5/15.5	ca. 50	normal	(d)	0.1	0.05	100						1 5

(a) 'colourless'
(b) odourless
(c) nearly odourless
(d) neutral to litmus

* Not all of these figures are published in the appropriate Standard/Pharmacopoeias. To allow comparison they have been converted by calculation and/or tests.
** Typical figures = 0
*** Unichema International test method: meets requirements of all major Standards/Pharmacopoeias.

8.2.9 Other tests

Apart from tests for colour, odour, sugars, lead, and other heavy metals included in general specifications, some purchasers of glycerine impose tests specially relevant to the purpose for which they use the material.

8.2.10 Specifications

Table 8.1 gives the main points in BS 2621:1979 and BS 2622:1979 concerned with crude glycerines. Table 8.2 similarly tabulates the main points in BS 2623: 1979, BS 2624:1979 and BS 2625:1979 which relate to refined grades of glycerine. BS 5711, parts 0 – 22, describes methods of sampling and the various tests used in specifications plus some additional tests.†

Table 8.3, courtesy of Unichema International, summarizes specifications for three grades of Pricerine glycerine sold by the company and the requirements of various pharmacopoeias and other standards. It will be noted that Unichema has, by calculation and by tests, shown the requirements in as comparable a form as possible. For example, the saponification equivalent of not more than 0.64 mill-equivalents per 100 g. (as shown in BS 2624 and BS 2625) is converted to 0.02% Na_2O by multiplication by the factor 0.031.

8.3 PURIFICATION OF SOAP LYES AND SWEET WATERS

8.3.1

As discussed in section 4.9, soap lyes are now commonly neutralized with fatty matter before they leave the soapmaking department. They still contain some alkali, as well as soapy matter and other organic impurities, and a substantial percentage of salt (around 10%). If the older system of cooling and skimming, but not neutralization, is used, the lyes contain a higher percentage of free alkali than in neutralized lyes. Sweet waters from high pressure splitters operated without catalyst only contain organic impurities of a fatty nature, plus some other compounds from the fats used. If the fats are split with the aid of a zinc oxide, or other alkaline, catalyst, the zinc, or other metal, leaves in the fatty acids, as a soap, and the sweet waters are little affected. Twitchell sweet waters contain some sulphuric acid, or sulphates, as well as organic impurities. The purification, or treatment, of the lyes, or sweet waters, is intended to remove organic compounds other than glycerol as completely as possible; and the success with which this is achieved is reflected in the MONG, or organic residue, of the crude glycerine produced.

8.3.2

The steps in the usual treatment of soap lyes are:

- *acidification* to a pH of about 4.6 – 4.8, although this is often omitted if the lyes have been well neutralized with fatty acids. Acidification is preferably with hydrochloric, rather than sulphuric, acid to avoid intro-

†See section 1.8 for the conditions under which quotations from British Standard Specifications are included.

duction of sulphates into the system. It liberates fatty acids and those which are insoluble in water, although of relatively low molecular mass, separate on standing. They are removed from the settler, often periodically, and are sometimes returned to a low-grade soap, perhaps after filtration. These skimmed fatty acids are poor soapmaking material; they are low in molecular mass and contain hydroxy acids produced by oxidation, and they tend to give sodium salts which have little value and/or again dissolve in the next lyes.

- purification by addition of a soluble iron or aluminium salt, a stage normally called *first treatment*. The pH is kept at, or adjusted to, about 4.5 – 5.0 (often 4.6 – 4.8); and the mix is well agitated by compressed air blown in through perforations in a pipe near the base of the tank. This air also ensures that if iron is used it is oxidized to the ferric state and blowing should, if necessary, be continued until the mix is brown rather than greenish black. Oxidation is particularly necessary if sodium hydrosulphite is used in the soapmaking process. The precipitate may be ferric hydroxide, or aluminium hydroxide, on which fatty acids, proteinaceous matter, and other impurities are adsorbed; or basic salts of the low molecular mass fatty acids may be present, the precise chemistry is obscure. The two main precipitants are solutions of ferric chloride, $FeCl_3 \cdot 6H_2O$, or of aluminium sulphate, sometimes called alum, $Al_2(SO_4)_3 \cdot 18H_2O$. Ferrous chloride with excess acid, which is the residue left after hydrochloric acid has been used to pickle steel, can be used, preferably after the iron has been oxidized to the ferric state by bubbling in chlorine gas carefully through a porous diffuser. Excess acid can be reduced in concentration by allowing the spent pickling acid to stand in contact with scrap iron wire when additional iron goes into solution. It is preferable to use chlorides, rather than sulphates, in all additions to soap lyes, because otherwise sodium sulphate will build up in the recovered salt which is re-used in soapmaking, and because sulphates are liable to increase problems with scale in the evaporators. Aluminium chloride is, however, not a convenient material to handle; and when aluminium treatment is used the aluminium is usually added as a sulphate. Sodium aluminate can also be used as a source of aluminium, sometimes in conjunction with ferric chloride. Opinions differ as to the relative merits of iron and aluminium for the purification of soap lyes, but iron has the advantage that it can usually be added as chloride. The quantity of precipitant needed varies with the composition of the lyes, but about 0.1 – 0.2% of the hydrated ferric salt (added as a concentrated solution) is common, although more, up to about 0.4% may be required. The completion of the first treatment process is often checked by the filtration of a small sample and addition of a few drops of diluted precipitant to the clear filtrate. If the treatment process is complete, no precipitate should be produced, but a good end-point is not always obtained; the pH at which the test is carried out is important. The adequacy of the treatment is proved ultimately by the MONG of the crude glycerine produced. Lyes from soaps with a high proportion of good tallow are more easily treated than those from high lauric oil blends, or those from low-grade fats. A very high glycerol concentration in the liquor can also make effective treatment more difficult.

- the first treatment mud is removed by filtration, normally with filter presses or leaf filters. It is often advantageous to allow the tank contents to settle, to filter first the lower layer, and to put the nearly clear upper layers through a clean press, or presses.
- the filtrate is made slightly alkaline with caustic soda, in what is known as the *second treatment.* A pH of about 9 is common when the precipitant is iron, but a slightly lower alkalinity may be preferred with aluminium. Mixing and oxidation are again with air. A relatively small quantity of precipitate separates in the second treatment. It contains the excess iron or aluminium, silica, and a little organic matter.
- finally, the treatment process is completed by a second filtration.

Variations in the details of the treatment process are possible, including estimation of the quantity of precipitant required by small scale tests. Single treatment processes have been devised in which, after precipitation of impurities with iron, or aluminium, as in the double treatment process, the mix is made slightly alkaline with sodium carbonate, or sodium bicarbonate, before being filtered once only. In the author's opinion, this process involves a compromise best avoided if possible.

Calcium chloride, followed by a second stage using sodium carbonate, has been used to treat soap lyes, but the purification is less good than with iron or aluminium.

8.3.3

Procedures for the purification of sweet waters are more varied. Sweet waters from modern high pressure splitting plants contain some nitrogenous bodies from proteinaceous matter in the fats, the nature and amount depending upon the quality of the raw material and the pretreatment it is given. They also contain some fatty matter, usually mainly fatty acid plus monoglyceride dispersed in the form of fine droplets which have not separated in the settling zone at the base of the splitting column, plus any which is dissolved. In some cases the only treatment used is additional settling, normally by continuous slow flow through a baffled tank. It has been found that a very small, carefully controlled, quantity of a suitable polyelectrolyte, as used in the treatment of water and sewage effluent, can effect a substantial improvement in the quality of the crude glycerine produced. In one application, an addition of 3 ppm of Zetag 63 (from Allied Colloids Ltd, Bradford, England) to sweet waters containing 10 – 13% glycerol at 60 – 80°C reduced the MONG of the crude glycerine from 3% to 1%. The need for careful investigation and control is indicated by the fact that 10 ppm of Zetag 63 spoiled the separation. Simple continuous settling is particularly convenient when the splitting and glycerine recovery plants are integrated so as to make maximum use of the heat in the streams of fatty acid and sweet water leaving the splitter; but other, or additional, processes are used when necessary to achieve adequate purification. In particular:

- the traditional treatment for Twitchell sweet waters, which contain sulphuric acid as well as organic impurities, is also applied to high pressure sweet waters. This uses an addition of slaked lime, $Ca(OH)_2$ (which removes some soluble sulphate as well as soaps, etc.), followed by removal of the excess calcium, usually with sodium carbonate. Sulphuric acid can be used for the second stage, but calcium sulphate is more soluble than calcium carbonate; and can lead to scaling problems in the evaporator.
- a variation using barium chloride and sulphuric acid. Barium chloride is much more expensive than lime, but barium salts, including the soaps, are more insoluble than the corresponding calcium salts.
- treatment with ferric chloride, or aluminium sulphate, as for soap lyes. Use of aluminium sulphate has been known to result in scaling problems in the evaporator.
- further settling of fatty matter from the crude glycerine after concentration of the sweet waters.

Saponification crude is often considered superior to soap lye crude, because of its much lower content of inorganic salts; and, as discussed later, this has substantial advantages with most, but not all, types of glycerine still. However, because of the low-grade fats from which saponification crude is often obtained, and/or the performance of the purification used for the sweet waters, serious problems can arise with the quality of the refined grades produced from saponification crude. Sweet waters are sometimes combined with soap lyes and the mixture treated as soap lyes. This may help in the removal of troublesome organic impurities, but the mixed crude is saturated with salt, and iron treatment becomes difficult if the glycerol content of the liquor is too high (23% has been handled satisfactorily).

8.4 EVAPORATION OF PURIFIED LIQUORS TO CRUDE GLYCERINE

8.4.1 General

Many variations are possible in the design of plants for evaporation to crude glycerine. Readers unfamiliar with the unit operation of evaporation may find it useful to consult one of the standard chemical engineering textbooks, or the book *Efficient Use of Steam* by Oliver Lyle (originally published by Her Majesty's Stationery Office in 1947; the new edition, edited by P. M. Goodall, published by Butterworth Scientific Ltd, 1980). A short, but comprehensive, review by the present author, (Woollatt, 1982) may also be helpful.

Points to be considered in the design of an evaporation plant to produce crude glycerine include:

- the capacity required in terms of water to be evaporated, crude glycerine produced, and, for soap lyes, salt to be separated.
- thermal efficiency and maximum consumption of steam.
- type of heating unit.
- arrangements for separation of salt, when relevant.

- avoidance of undue losses of glycerol.
- vacuum equipment and other ancillaries.
- method of control.
- materials of construction.

8.4.2 Water to be evaporated, salt to be separated, and crude glycerine produced

A simple procedure for the estimation of the relationships between the relevant quantities is illustrated by an example of a soap lye containing 4% glycerol, 11% salt and 85% water. Minor ingredients and losses are neglected.

	Lye	Crude glycerine	Lye to produce 100 parts of crude glycerine contains		To be separated
Glycerol	4	83		83	–
Salt	11	8	$\frac{11 \times 83}{4} =$	228	228–8 = 220
Water	85	9	$\frac{85 \times 83}{4} =$	1764	1764–9 = 1755
Total	100	100		2075	1975

Hence, to produce 1 part of 100% glycerol in crude:

- use $\frac{2075}{83}$ = 25.0 parts lye
- make $\frac{100}{83}$ = 1.2 parts crude
- evaporate $\frac{1755}{83}$ = 21.1 parts water
- separate $\frac{220}{83}$ = 2.7 parts salt

(make + evaporate + separate: 25.0 parts)

The results of similar calculations for lyes containing 4, 6, 10 and 15% glycerol respectively, plus 11% salt, evaporated to give a crude glycerine containing 83% glycerol, 8% salt and 9% water are shown in Table 8.4. The figures show how dramatically even a small increase in lye strength reduces the quantity of water to be evaporated, with consequent reduction in steam usage and evaporating capacity needed. It also reduces the quantity of salt to be recirculated. Similar calculations can be used to estimate quantities for the evaporation of treated lyes to semi-crude strength (see later), or of sweet waters which do not separate salt.

Table 8.4 – Quantities required and/or separated to produce 1 part of 100% glycerol in crude glycerine at 83% glycerol

% glycerol in lye	Parts lye used	Water evaporated	Salt separated	Crude glycerine produced
4	25.0	21.1	2.7	1.2
6	16.65	13.7	1.75	1.2
10	10.0	7.8	1.0	1.2
15	6.65	4.8	0.65	1.2

8.4.3 Thermal efficiency

In addition to steps to reduce the quantity of water to be evaporated which are discussed in Chapters 4 and 5, methods are available to minimize the usage of steam required to carry out the necessary evaporation. One kg of steam will evaporate approximately 1 kg of water fed at its boiling point, but less than 1 kg when the feed is at a lower temperature and sensible heat, which does not immediately produce vapour, has to be supplied before boiling commences. The boiling point of water, or any other liquid, falls as the pressure under which it is held is reduced. These simple facts provide the key to several ways in which the thermal efficiency of an evaporation system can be improved. Thermal efficiency is often and conveniently expressed in terms of kg of water evaporated per kg of steam supplied (called 'kg/kg', or in England more usually 'lb/lb').

The main ways in which thermal efficiency can be improved are:

- *multiple effect operation* in which the vapour from one evaporating vessel, or 'effect', becomes the heating steam in another effect operating under a lower pressure. In this way, the latent heat is used more than once.
- *vapour recompression* in which the vapour from a vessel is compressed, by a steam jet or by mechanical compressor, and returned as heating steam to the same evaporating unit. It will be appreciated that for the latent heat to be re-usable the pressure must be increased; and that a simple addition of superheat to the vapour will not do.
- selection of the optimum *method of feeding* the liquor to the evaporator; and/or the use of the heat exchangers. As discussed later, heat exchangers can be used internally between effects, or to recover heat from outgoing streams of vapour, condensate, or product.
- integration with other plants in the factory.

Before these procedures are considered, it is necessary to discuss boiling point elevation and temperature differences. Boiling point elevation is the difference between the boiling point of the solution in an effect and the boiling point of water under the pressure in the vapour space of the effect. It is due to two causes:

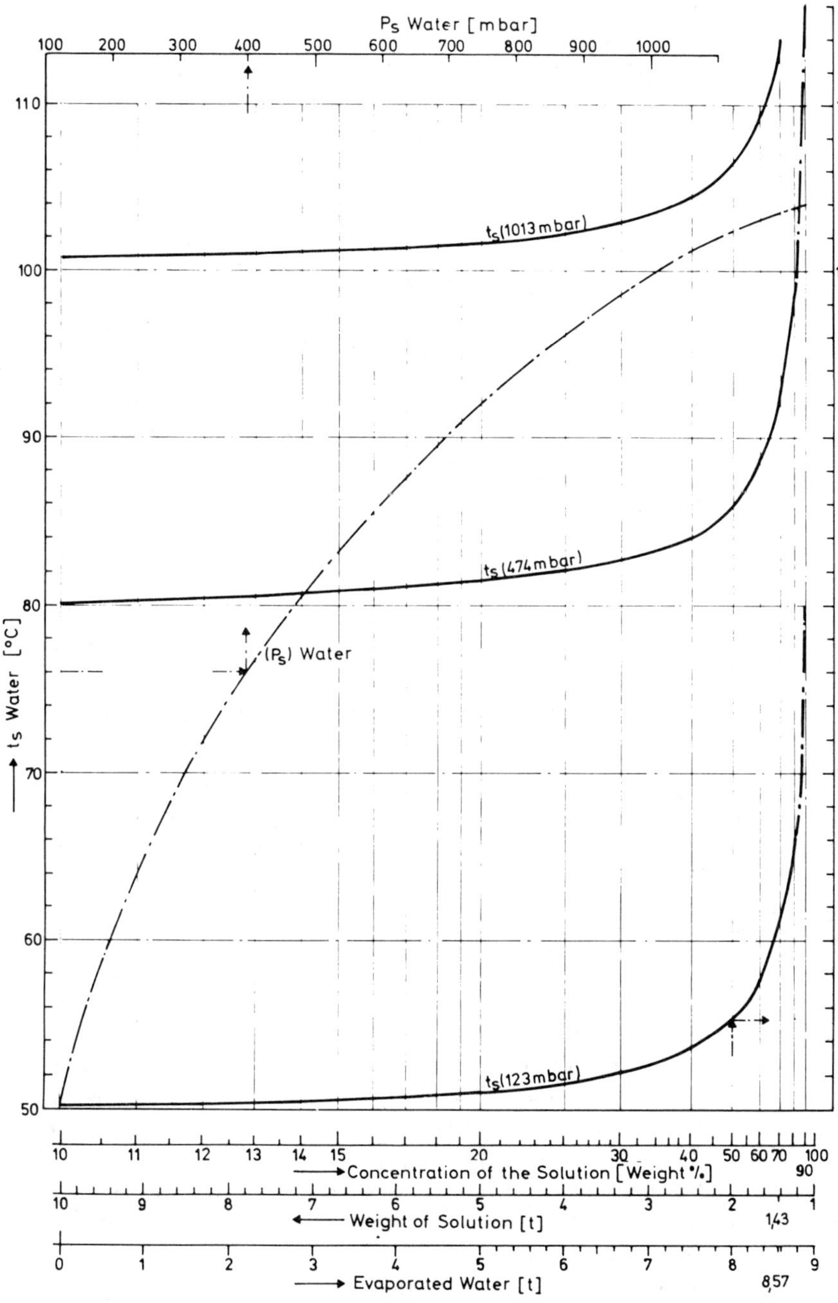

Fig. 8.1 – Boiling diagram for sweet water solutions.

- *boiling point elevation due to dissolved substances,* as discussed in section 5.4.2 in relation to fatty acid distillation, which increases as the concentration of dissolved substances increases. For semi-crude soap lye glycerine at 40 – 45% glycerol it is about 15°C; and for soap lye crude at 83% glycerol about 37°C. Fig. 8.1, courtesy Lurgi, gives similar information in a different form for concentrated sweet waters.
- *boiling point elevation due to hydrostatic head.* When a liquid is heated at a distance below its surface, the pressure is that in the vapour space plus that equivalent to the head of liquid above the point at which the heat is supplied. 3 ft of glycerine of specific gravity 1.2 is equivalent to a hydrostatic pressure of $3 \times 12 \times (1.2/13.4) = 3.2$ inches of mercury, so that with a pressure of 2 inches Hg in the vapour space, the pressure at a depth of 3 feet is 5.2 inches Hg. The relative effect is clearly smaller when the pressure in the vapour space is higher. Boiling only occurs when the temperature reaches the boiling point under the prevailing pressure, but it is possible for the liquid to be heated to above its boiling point at the pressure in the vapour space and for it to flash when the liquid is circulated to a zone of lower pressure.

The apparent temperature difference, Δt_{ap}, is defined as the difference between the condensing temperature of the steam in the heater and the temperature at which water boils under the pressure in the vapour space. Consider, for example, a single effect evaporator in which the heating steam condenses at 5 psig, 109°C, and the vapour space has a pressure of 2 inches mercury at which water boils at 38°C. The apparent temperature difference is $109 - 38 = 71$°C. If the liquid being boiled is crude glycerine with a boiling point elevation, bp_{elev} including any hydrostatic head effect, of 37°C, it will boil at $38 + 37 = 75$°C; and the actual temperature difference, Δt_{act}, is $109 - 75 = 34$°C (or, alternatively, $71 - 37 = 34$°C, the apparent temperature difference minus the boiling point elevation). This situation is depicted as case 1 in Fig. 8.2. If a double effect evaporator is used in which vapour from the first effect, which contains partially concentrated liquor with a boiling point elevation of 10°C boiling at 91°C, is used to boil crude glycerine under a pressure of 2 inches Hg in the second effect, the conditions are as shown as case 2. The various temperatures and pressures adjust themselves automatically to fulfil both heat transfer and heat balance requirements, so that the figures shown in Fig. 8.2 are approximate only and vary with conditions.

A positive Δt_{act} is essential in each effect in order to achieve heat flow and evaporation. With a natural circulation heater, particularly when the liquid is viscous, a Δt_{act} is required larger than the minimum which can be used with a forced (pump) circulation type of heater.

Case 3 in Fig. 8.2 shows approximate conditions for evaporation from lye to semi-crude glycerine in double effect using low pressure steam, a system which was once widely used.

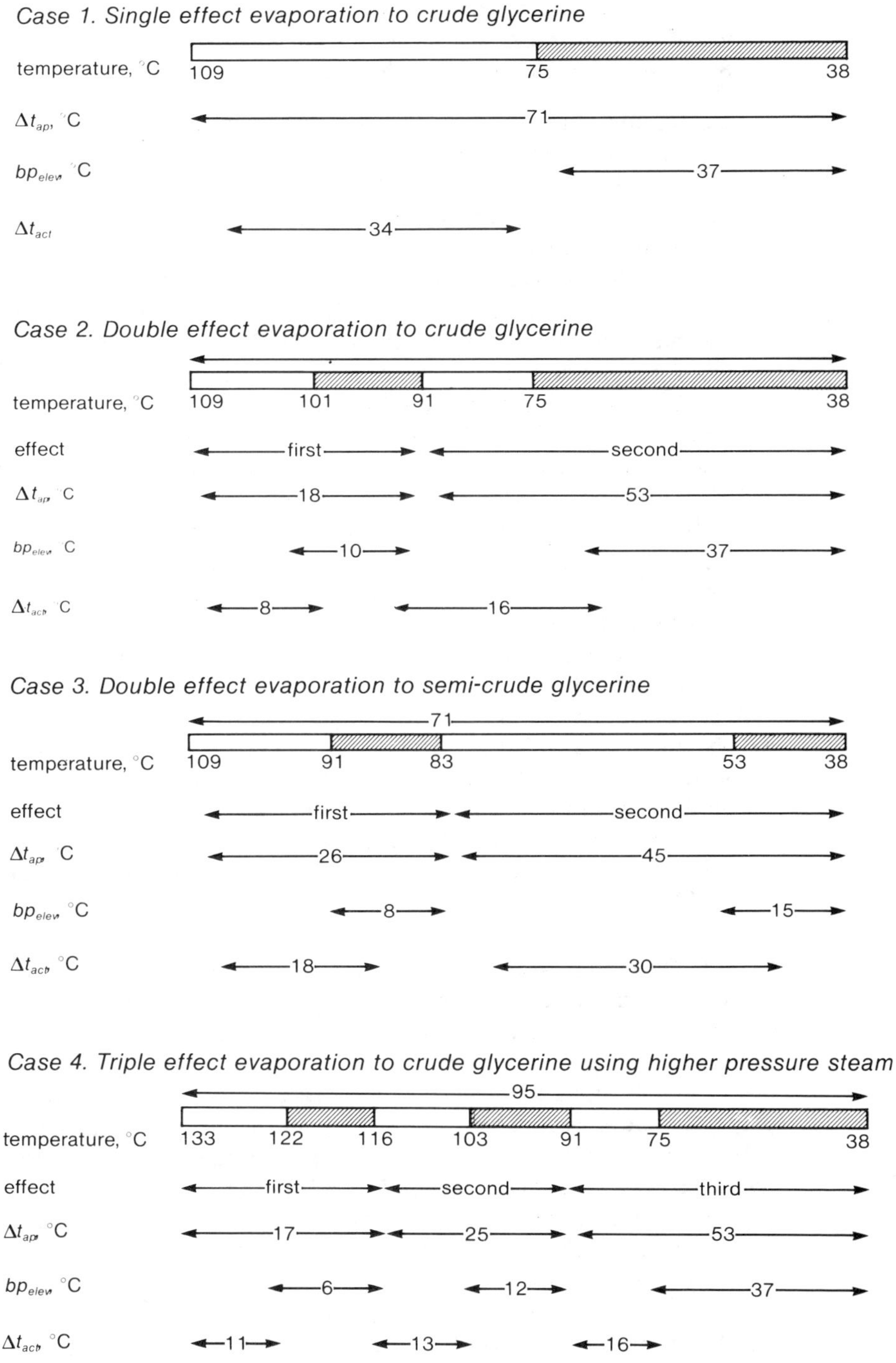

Fig. 8.2 – Temperature conditions for single, double and triple effect evaporation to soap lye crude or semi-crude, glycerine.

Case 4 shows possible conditions for evaporation from lye to crude glycerine in triple effect using steam which condenses at 30 psig, 133°C.

Some general points related to these topics are:

- boiling point elevation due to hydrostatic head varies with the design of the heaters and is assumed to be included in the figures for bp_{elev} used in Fig. 8.2. It does, however, reduce the effective temperature difference, even if the liquor is heated without boiling and only flashes when it arrives near to the liquor surface.
- boiling point elevation due to dissolved substances depends only on the composition of the liquor. It is high for crude glycerine; and this is a disadvantage of continuous operation in which the liquor in the effect from which the product leaves must be maintained at the product concentration. If an evaporator is operated semi-batch, that is the unit is fed with fresh liquor but the bulk of the product is not discharged until its concentration reaches the desired level, the average boiling point elevation is lower and the average Δt_{act} higher, other things equal, than with continuous operation. At one time semi-batch operation to concentrate lye to semi-crude and, separately, semi-crude to crude glycerine was normal, but the usual advantages of continuous operation are now commonly considered to outweigh this disadvantage.
- forced circulation heaters, now often used in all effects, are particularly valuable for effects in which crude glycerine concentration is reached, as they permit operation with a low Δt_{act}, even when the liquid is quite viscous.
- the need for an adequate Δt_{act} in each effect limits the number of effects which are physically possible with given terminal conditions. In general, an increase in the number of effects leads to a reduction in steam consumption, but to an increase in capital cost. There is an economic optimum which may well use less than the maximum number of effects physically possible.

Double effect evaporators to concentrate lye to semi-crude were traditional for many years; and further concentration to crude glycerine was carried out in a separate single effect finisher, or in the second effect of the double effect unit operated either as a single effect using a connection directly to the steam supply, or as a double effect using water in the first effect. These plants used steam at only just above atmospheric pressure; and the supply commonly included exhaust from steam pumps and small steam engines. Increased emphasis on fuel economy around the 1940s and 1950s led to the construction of at least one quadruple effect natural circulation evaporator to produce semi-crude glycerine from soap lyes. This was fed with steam at 90 psig and the pressure in the first effect heater rose to 30 – 50 psig depending upon conditions. Initially, this unit was operated semi-batch like its double effect predecessors, but it was found preferable to work it continuously. Triple effect evaporators have been, and are, also used for soap lyes and sweet waters. They now commonly use forced circulation and take the concentration to crude glycerine strength.

The concentration of glycerol in the lye, or sweet water, to be evaporated is an important factor in the design of the plant.

Vapour recompression evaporation with an electrically driven mechanical compressor is only likely to be economical when the compression ratio (that is, the ratio of the pressure of the heating steam to that of the vapour it produces by evaporation) is low, which means a low Δt_{app}. A low Δt_{act} can be achieved by the use of a forced circulation heater with a generous area of heating surface, but bp_{elev} is inevitably high if the unit is to produce crude glycerine; and mechanical recompression is not suitable for this duty. Jet recompression, combined with multiple effect operation, is valuable; and is now widely used when steam at boiler pressure, rather than steam including low pressure exhaust, is to be the heating medium. Fig. 8.3 shows the main features of a triple effect system with jet recompression of part of the vapour leaving the first effect. Again, recompression is most effective when the pressure ratio is small, as it can be when the liquor concentration in the first effect is fairly low. When conditions permit satisfactory recompression, a useful improvement in thermal efficiency is achieved without increase in other costs and for only a very small additional capital cost for the jet.

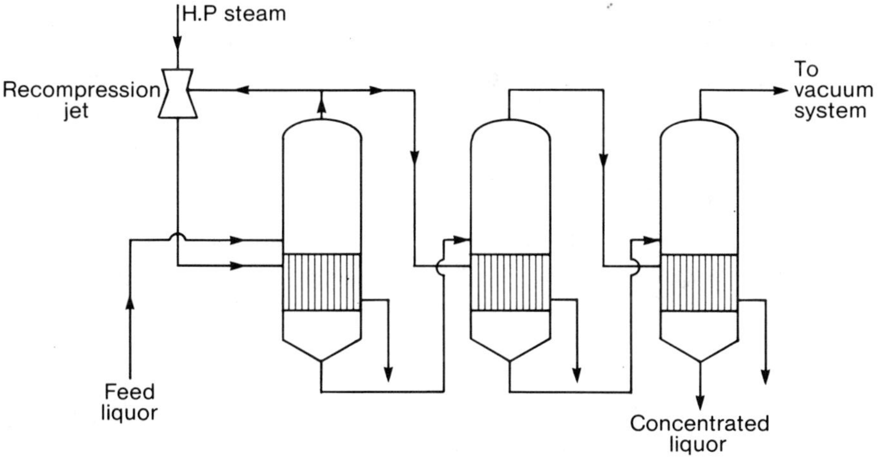

Fig. 8.3 – Principle of triple effect evaporation with vapour recompression.

Parallel feed, that is separate feed to each effect, is suitable for the evaporation of brine to produce salt, but not for glycerine liquors in which the concentration of glycerol increases very substantially as the evaporation proceeds. Feed can be forward, that is 1 – 2 – 3 etc., the effects being numbered in the direction of heat flow; or backward which is the reverse direction, for example 3 – 2 – 1. The direction of feed often has an important influence upon steam consumption, depending upon the temperature of the feed liquor. If the feed is cold, it has to be heated to its boiling point under the pressure in the effect to which it goes. With forward feed, the temperature of the whole of the feed must be raised to the relatively high boiling point in the first effect by new steam and this sensible heat does not produce any vapour in that effect, although the heat in the partly concentrated liquor does produce some vapour by flashing when it is transferred to the second and subsequent effects which operate under lower pressures than the first effect. With backward feed, a cold liquor is heated in stages as the unevaporated portion is transferred from effect to effect. Only the final portion which reaches the first effect is heated with new steam, and the sensible heating in the second and subsequent effects is done by vapours from an earlier effect; that is by latent heat which has already been used one, or more, times in the evaporator. If the feed liquor is very hot, for example above the boiling point in the first effect, no sensible heat is required even with forward feed; and flash vapour is produced in each effect and forms part of the steam used in the next effect. With backward feed, the hot liquor flashes to the pressure in the last effect and the latent heat is not used again, at least in the evaporator. The residual liquor then has to be re-heated to the temperature in its next effect, as for cold feed. These considerations show that it is beneficial thermally to use forward feed with hot liquor and backward feed with cold liquor; at some intermediate temperature the factors balance and there is no difference between backward and forward feed. The thermal advantage of backward feeding with relatively cold feed liquor can nearly be achieved by the use of forward feed with interstage preheaters as indicated in Fig. 8.4. The feed is preheated stage by stage with vapour at as low a temperature as possible to make the maximum re-use of latent heat. A preheater cannot operate without a positive temperature difference in the required direction; and with modern vacuum equipment it is often not possible to use a preheater in the vapours from the last effect (number 1 in the diagram), but any heat which can be recovered from the final vapours is clear gain. Heat transferred to the feed in interstage preheaters is not available for use in the next effect, but the overall result is beneficial with cool feed for

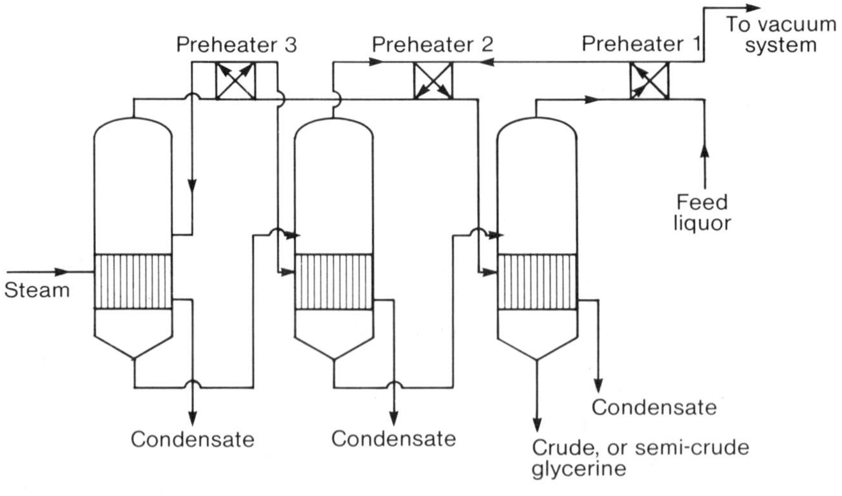

Fig. 8.4 – Evaporation with forward feed and interstage preheaters.

the reasons discussed in relation to forward and backward feed. Heat can also often be recovered from other streams leaving the effects, particularly the condensates. These condensates can be handled in several ways, bearing in mind the fact that condensate leaves the first effect heater at a higher temperature than that from the second effect, and so on:

- condensates can be allowed to flash on reduction in pressure on transfer from one effect heater to the next, or in separate vessels. The flashed vapour is used as steam at a suitable point in the system, and the residual hot, or warm, water remains available. This condensate can be re-used as process water in the soapmaking or fat-splitting process, as well as a source of heat.
- unflashed, or flashed, condensate can be used in a heat exchanger.

Other factors to be considered in the selection of a system of feeding are:

- with forward feed the most concentrated liquor is evaporated at the lowest temperature, so that its viscosity is higher than that of the fully concentrated liquor using backward feed. This adversely affects heat transfer, particularly with natural circulation heaters.
- more sensible heat leaves in the crude, or semi-crude, with backward than with forward feed; but some of this may be recoverable in heat exchangers.
- liquor can normally be transferred from effect to effect by pressure difference with forward feed, although a pump may be needed to introduce the feed liquor into the first effect. With backward feed it is normally possible to suck the feed liquor into the last effect, but pumps are needed for transfer from effect to effect, unless unusual levels are adopted which provide sufficient gravity head to more than offset the adverse pressure differences between the vapour spaces.
- the salt crystals which separate during the evaporation of soap lyes are rather fine; and they tend to become finer with rise in the temperature at which they are separated and with increase in the glycerol content of the liquor. Fine crystals are more difficult to wash effectively than coarser ones. The same factors also tend to induce 'salting' of the heater tubes, a phenomenon discussed later. Salting can be a serious problem with natural circulation soap lye evaporators under some conditions and forward feed may then be advisable.

When the feed liquor is relatively cool it is usual to use forward feed with interstage preheaters. For special reasons, it is also possible to use other arrangements. For example, the quadruple effect evaporator mentioned earlier was for many years fed 1 – 2 – 4 – 3.

As mentioned earlier, and for the reasons just discussed in relation to methods of feeding, it is commonly a serious error to assume 1 kg of live steam will evaporate n kg of water in an n-effect evaporator, unless the feed is exceptionally hot or a recompression jet is used. A good estimate of the performance to be expected from an evaporator, including its evaporative capacity, can be made with the aid of detailed heat balances over each stage of the process. Because of the dynamic equilibria maintained within a unit, the temperatures and pressures adjust themselves automatically to meet both heat balance and heat transfer criteria in each section of the plant. For a plant which does not exist, it may be necessary to make provisional assumptions as to intermediate temperatures, or likely heat transfer coefficients; and to make adjustments, if necessary, to reconcile the two sets of criteria. Such calculations are not difficult, but manual calculation can be tedious, particularly when several effects are involved. Plant manufacturers will provide estimates in response to serious enquiries, doubtless by the use of computer programmes, but it is sometimes useful for an operator to estimate what is likely to result from changes such as the inclusion of an additional heat exchanger, or a different feed arrangement. In making such estimates, it must be remembered that any change is likely to induce other changes to maintain the dynamic equilibria.

As a first approximation it is sometimes assumed that an n-effect evaporator costs n times as much as a single effect plant for the same duty, because the overall temperature difference is split between the effects with a consequent need for additional heating surface. This is also subject to considerable error. Estimates obtained some years ago from a well-known plant manufacturer for single and double effect units for the same, rather small, duty showed that the cost of the double effect was substantially less than twice that for the single effect. At least in part this was because of the larger final entrainment separator, condenser, and vacuum equipment needed for the single effect unit in which the whole of the water evaporated passes to the condenser.

The discussion of thermal efficiency has so far considered the evaporator alone, but it is sometimes worthwhile to involve other plant. Steam from another unit which is not at high enough pressure to be used as first effect steam may be able to be introduced between effects. Similarly, vapour can sometimes be bled between effects for use elsewhere. Different plant units commonly do not synchronize at all times, so that it is usually necessary to provide an alternative steam supply, probably through a reducing valve. Surplus heat from a high pressure fat splitting plant can often be used in the evaporation of the sweet water.

8.4.4 Types of heating unit

Long tube climbing film heaters, of the Kestner type, in which much vaporization takes place in the tubes, are used for sweet waters, but are not suitable for soap lyes, because the tubes would block with salt. Traditionally, short tube natural circulation evaporators were usual for soap lyes. These have a bundle of vertical tubes (say 1 – 2 m long) arranged in a steam chest to give what is called a calandria. Commonly calandrias built into the main evaporator bodies, and provided with downcomers of various designs, such as a sector, a large central hole, or a series of tubes larger than the boiling tubes, are used. The liquor circulates up the small tubes and returns through the downcomers. It is also possible to have natural circulation evaporators in which the liquor rises through the vertical tubes in an external heater and returns via the liquor reservoir in the main body. In order to obtain the maximum rate of heat transfer, and hence of evaporation, it is necessary to work with a static level of liquid below the top of the tubes, so that some vaporization takes place in the tubes giving an increased rate of circulation due to the vapour lift effect. With soap lyes care must be taken to avoid too low a static level, as otherwise salt will deposit in the tubes in effects in which the liquor is saturated with salt. Salting of the tubes can sometimes occur when an evaporator is being operated in a way which would normally be satisfactory; this is presumably due to some small changes in the composition of the purified lye. It is also sometimes possible to operate natural circulation evaporators with a static level well above the upper tube plate of the calandria, particularly when finishing batchwise to crude strength. In this case, the liquid is heated in the tubes to above its boiling point at the pressure in the vapour space and flashes as it rises to a region of lower hydrostatic pressure. It is usually necessary to avoid operation at some intermediate levels at which 'priming', or boiling over, is liable to occur.

Evaporator effects in which salt separates usually need to be washed out at regular intervals with water, or with other liquor which is not saturated with salt. When a plant is operated semi-batch, rather than continuously, the charge of concentrated crude, or semi-crude, containing some solid salt, is discharged from the last effect; and it is common practice to start the next batch by refilling with unsaturated lye, or even water, to dissolve deposited salt. If the tubes become blocked with salt, it is necessary to soak with water and to boil as long as is needed to clear them. Mechanical methods are sometimes used, but care is necessary to avoid damage and this procedure is best avoided if possible. Various insoluble scales can build up slowly in the tubes of a glycerine evaporator and these normally have to be removed mechanically. Scales can contain silica, but they are particularly common when lye, sweet water, or other liquor containing sulphate and calcium is introduced, for example, if the plant is washed out with hard water. Calcium carbonate scales are also encountered; these can sometimes be removed by careful use of acid. The modern tendency is to use external calandrias with forced circulation by pump. This system improves heat transfer, and so permits the use of less heating surface and/or a lower Δt_{act}. External heaters with forced circulation minimize the risk of salting and scaling; and an external heater can be arranged to permit mechanical descaling more readily than an internal calandria. As mentioned earlier, forced circulation is particularly useful for the final effect of an evaporator in which crude glycerine is produced.

Whatever type of heater is used, it is necessary for satisfactory heat transfer to remove effectively both condensate and non-condensible gases, mainly air. When the pressure in the steam chest is above atmospheric, ordinary steam traps can be used for the condensate; and air can be discharged, with a little steam, through a slightly opened vent valve, or cock. For steam chests at below atmospheric pressure, the air must be vented through a small connection to a vapour space at a sufficiently low pressure; and a pumping trap, or pump, used for the condensate. A convenient system for the condensate is shown in Fig. 8.5. The centrifugal pump runs continuously, but discharge of condensate is intermittent. The condensate is virtually at its boiling point at the pressure in the steam chest and can only be pumped when there is a sufficient head of liquid in the receiver to prevent vaporization (cavitation) in the pump. The level in the receiver rises until the pump starts to discharge and then falls until flow virtually ceases; the cycle being repeated indefinitely.

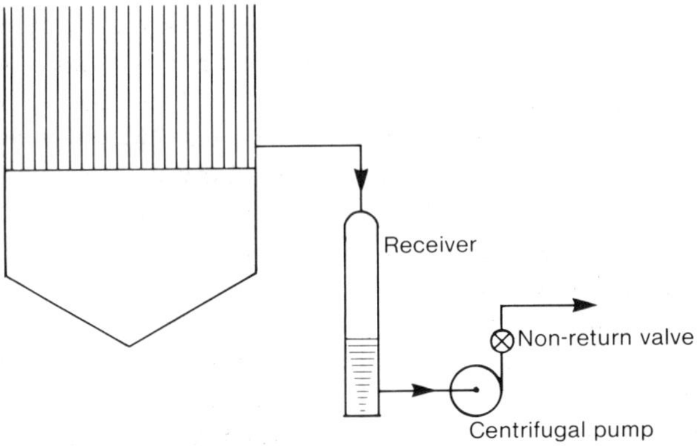

Fig. 8.5 – Arrangement for discharge of condensate from a vessel at sub-atmospheric pressure.

8.4.5 Removal and recovery of salt in the evaporation of soap lyes

All effects in which the lyes have become saturated, and solid salt separates, need a facility for the removal of the solid as soon as possible after it is formed and the crystals have grown to a reasonable size, otherwise the liquor may become unduly thick. Effects which do not normally produce salt may be fitted with simple vessels to collect sludge and any abnormal salt. Three systems are, or have been, used:

- salt filters, or dry salt boxes.
- wet salt boxes.
- devices through which the effect contents are circulated, a thickened slurry is discharged, and substantially clear liquor is returned to the effect.

Dry salt boxes were standard, but they are messy, labour intensive, and do not permit adequate removal of glycerol from the salt to meet the requirements of efficient modern soapmaking. In the larger, and the more efficient, factories they were replaced many years ago, but they are still in use in some places. The other two systems remove the solid salt from the evaporator in the form of a slurry in a more, or less, concentrated glycerine liquor; and the separation process is completed in some form of centrifuge.

The operation of dry and wet salt boxes can be explained by reference to Fig. 8.6. In both types salt settles into the quiet zone of the effect body below the calandria and down into the salt box. A sufficiently deep cone and settling zone is important; and, if the plant is to be operated semi-batch, sufficient liquor volume is necessary in the last effect to hold an adequate batch of crude, or semi-crude, glycerine. Consider now a dry salt box as shown in Fig. 8.6(a). The salt settles onto the grid and filter gauze. When sufficient salt has collected, valve A is closed, valve B is opened, and the liquor is drawn back into the effect, helped by the pressure of the steam introduced through the valve at the top of the box. The bed of salt is washed with lye, and/or water, and the liquor is again displaced by steam. The steam valve and valve B are closed, the vent to atmosphere opened, and the door on the box opened. The salt is shovelled out for return to the soapmaking department as solid salt, or as brine. The door of the box is closed, the box is evacuated through the balance valve C, and valve A opened to commence another cycle. It is particularly important to ensure that the box is fully evacuated before valve A is opened; otherwise expansion of the air from the box will cause the liquor in the effect to prime. It is also important to ensure that air cannot leak in through the valves; lubricated cocks have been found useful for these duties.

Wet salt boxes can be cylindrical drums, or vessels of any other convenient shape. They are operated in a manner similar to, but somewhat more simply than, the dry boxes. When sufficient salt has accumulated, as judged by inspection through sight glasses, or in accordance with a time schedule, valve A is closed, valve B and the vent are opened and the slurry is run out and/or blown out with steam after the vent has been reclosed. The box is evacuated via valve C before it is reconnected to the effect.

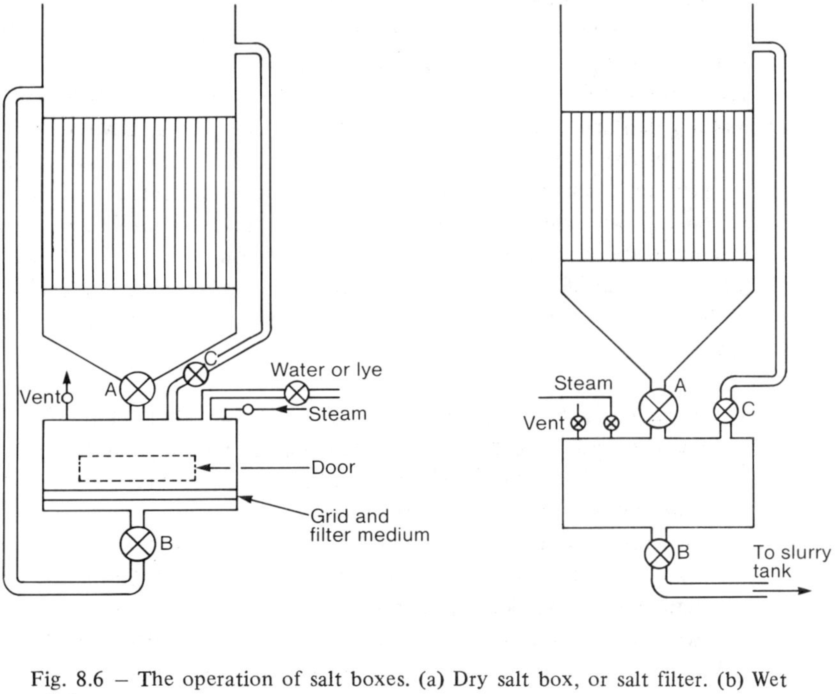

Fig. 8.6 – The operation of salt boxes. (a) Dry salt box, or salt filter. (b) Wet salt box.

Pairs of boxes, dry or wet, and used alternately, are common on effects in which large quantities of salt are produced. If only one box is fitted, salt accumulates in the cone of the effect whilst the box is disconnected. Discharge by continuous pumping from the base of the effect is arranged as shown in Fig. 8.10 later for a Wurster and Sanger evaporator which uses a baffled settling vessel. Thickened slurry is discharged from the base of the vessel and the rest of the liquor is returned to the effect and/or transferred to the next effect. Sufficient pressure is maintained in the circulation loop by the pump and the valves in the return and transfer lines to enable the discharge of thickened slurry to be controlled by a valve without another pump. Hydroclones (liquid cyclones) instead of the baffled vessels have been found very satisfactory separating devices for systems of this kind. Care must be taken not to interfere with the circulation through the heater tubes if a continuous pumping system is introduced into a natural circulation evaporator. An attempt to do this many years ago produced a well thickened slurry, but led to unacceptable salting of the calandria tubes. Return of the circulated liquor to the base of the effect, as in the Wurster and Sanger example, with sufficient volume below the calandria, may prevent this difficulty. The system works well with forced circulation effects and is probably best modern practice.

It is necessary to separate the solid salt in the slurry removed from the evaporator from the liquor, and to wash the salt as free from glycerol as is economically feasible. Some washing with lye may be possible, but mainly fresh water is used; and this water has to be evaporated, which sets a limit to the amount of washing which is worthwhile. Traditionally, basket centrifuges operated batchwise are used, and the salt is discharged as a crystalline powder. It is returned to the soapmaking department as such, or after solution in water, preferably condensate, to form a near saturated brine. Modern soapmaking plants use only brine; and systems have been developed in which the salt is dissolved in the centrifuge itself. Alternatively, the salt is dissolved in a separate unit, for example like that described in section 1.5.1. When sulphate is introduced into the system, by the use of aluminium sulphate or sulphuric acid, it is necessary to discard some recovered salt from time to time to prevent undue build up of sodium sulphate in the salt. The batch centrifuges used are normally of the suspended basket type; and some points regarding their use are discussed in relation to Fig. 8.7:

- perforate basket machines are used; and the filter medium is a fine weave metal gauze backed by a coarse gauze to facilitate drainage. A high speed to give a high centrifugal force is desirable to remove the liquors from the solid salt as completely as possible.
- the slurry should be fed with confidence in intermittent bursts having a high rate of flow. A 4″ pipe with a control cock operated with a long handle is convenient. This rate of feed makes the slurry run up the basket wall to the top and gives a salt cake of fairly uniform thickness. Too high a feed rate, or unduly long bursts, cause slurry to overflow the rim of the basket and to carry solid salt over into the centrifuged liquor. Actually

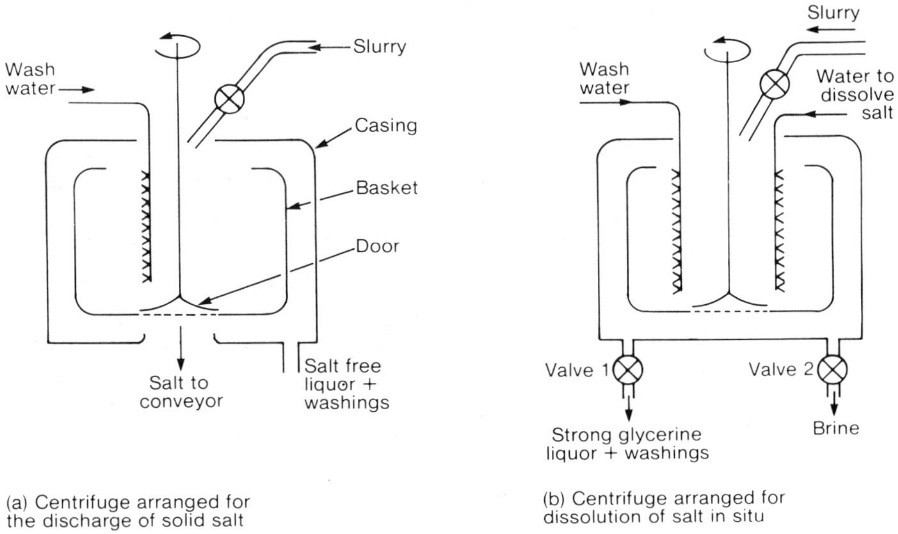

Fig. 8.7 – Batch salt centrifuge, (a) arranged for the discharge of solid salt, (b) arranged for dissolution of salt *in situ*.

some salt always gets into the clear liquor and arrangements must be made to deal with it; as also from any vessels in which soap lye crude, or semi-crude, is held. Too low a feed rate gives a very uneven salt cake, with a large thickness at the base and a small one at the top; this is because most of the liquor has time to penetrate the cake before the slurry can rise far up the basket wall.

- when most of the liquor has been removed, the salt cake is washed with a controlled quantity of water and spinning is continued to remove the wash water as completely as possible. Sometimes two short washes, separated by a period of spinning, are given. If a fixed perforated pipe is used for wash water, care must be taken to ensure that it does not interfere with the discharge of the salt and a spray on a flexible hose is an alternative. Timers to help the operatives to adhere to a cycle laid down by the management are useful. In order to reduce the glycerol content of the salt to the required level with a minimum use of wash water, careful control is essential, particularly of the time during which the water is sprayed.
- with the system shown in Fig. 8.7(a), the machine is stopped, the door over the bottom outlet raised and held up on a hook, and the salt dug out to fall into a conveyor, or receptable, below. A narrow spade with the cutting edge cut at an angle is useful for this work. When lower speed, and less efficient, centrifuges were used, a rope placed round the bottom of the basket before the slurry was fed provided a simple means to dislodge the salt cake. The end of the rope had a hook which held it to the upper rim of

the basket; and this end was pulled to bring down the salt at the end of the cycle. However, well washed and dried salt in modern machines is normally too hard for this technique to be used.

- with the system shown in Fig. 8.7(b), valve 1 is open and valve 2 closed during most of the cycle for the glycerine liquor and the washings to flow away. When the salt has been washed and dried, valve 1 is closed, valve 2 opened, and a controlled flow of water introduced to dissolve the salt. With proper adjustment of the conditions the salt can substantially all be dissolved to produce nearly saturated brine. Care is necessary to ensure that the casing is arranged to allow the two streams of liquid to drain with minimum contamination one with the other. Also that the valves close tightly, and that the operators control the valves and sprays with care. With proper precautions, this system can give very satisfactory results without the heavy manual labour required with system (a).

In large factories, the modern trend is to automatic centrifuges. Several types have been used, but some have not been very successful because either they could not handle fine glycerine salt satisfactorily, or because they need too much wash water to reduce the glycerol in salt to the required level. The best type is probably one in which the operation is very short cycle, automatic batch. The salt is discharged with a knife; and it is a useful refinement for the knife mechanism to be arranged so that the cut leaves a slightly different thickness of salt in consecutive cycles in order to minimize the formation of a 'glaze' of very fine broken crystals which may impair drainage. Machine manufacturers' literature should be consulted for details of the various types of machine available.

The glycerol left in salt varies considerably with the system employed, the quantity of wash water used, and the care with which the operation is carried out. With centrifuges 0.4 – 0.6% glycerol on salt should be possible, but figures from 0.2 – 1.0% are quoted. With dry salt boxes, 1% is sometimes claimed, but 2– 4% is probably more usual (corresponding figures calculated on saturated brine are about one quarter of those for salt).

8.4.6 Minimization of losses of glycerol

During evaporation, liquor containing glycerol and salt can be lost by being carried out in liquid/solid form with the vapours and discharged in a condensate. The carry-over can be as droplets, foam, or, if a serious prime occurs, as bulk liquid. To minimize this loss:

- adequate vapour space should be provided above the level of the boiling liquid.
- suitable separators, and external 'catchalls', usually cyclonic, should be included in the system. It is useful to provide sight-glasses in external separators, and/or the return lines, so that the plant operator can see immediately when any serious carry-over occurs and take remedial action.
- the liquor levels in the effects should be controlled with care, automatically or manually.

- the treatment of the lyes, or sweet water, should ensure that the liquors have no undue tendency to foam.

External catchalls are sometimes fitted only to the effect in which the concentration of glycerol is highest. Devices which measure roughly the electrical conductivity of the condensate from the evaporated vapours are very useful indicators of carry-over. They actually indicate the presence of salt, or other electrolyte, but this is relevant as salts and glycerol are carried over together. The main problem is to obtain a continuous flow of condensed vapour from the last effect when, as is common, a direct contact condenser is used in the vacuum system; and the condensate is mixed with cooling water. It is not essential to obtain a representative sample of the vapours plus any entrained liquor; all that is necessary is to get a sample which will indicate when any abnormal amount of salt is present in the vapours, and this can often be achieved by the use of a small partial condenser in the vapour line. For example, a ring carrying cold water placed above a collecting tray, in a vertical vapour main, as shown in Fig. 8.8, can be used. There is a continuous flow of condensate through the box containing the electrodes which are connected to a suitable low AC voltage and a measuring instrument or, better, recorder. An audible alarm can be included to draw immediate attention to any substantial increase in conductivity.

Losses of glycerol during evaporation can occur in other ways, for example, if a leak develops in a calandria. The effects are normally left containing water during shut-downs (to dissolve deposited salt) and, if a leak is present, some salt solution can be drained from the steam chest of the calandria after the plant has stood for a period of time.

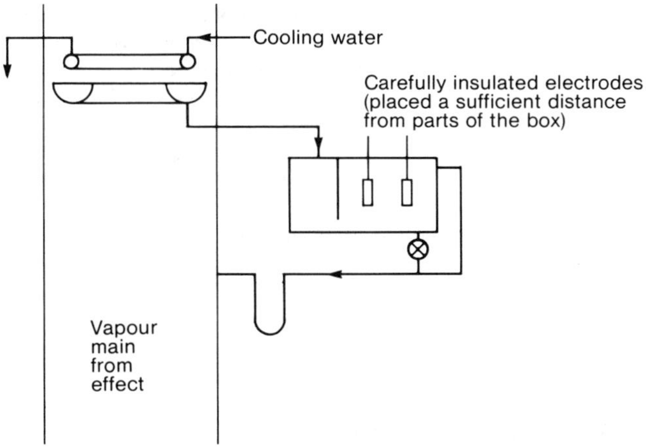

Fig. 8.8 – Partial condenser for entrainment indicator.

Fig. 8.9 shows the flow of glycerol through a soap factory with glycerine recovery; and indicates the sources of loss. The diagram is drawn for a double-effect evaporator with direct contact condenser; and refers to closed coils in soap pans. Such coils are now obsolete, but they can be a serious source of loss if present. The diagram can readily be modified to suit other plant arrangements and similar ones prepared for the splitting route. It is useful to determine the recovery in each section of the factory, particularly if the overall loss is high, but, this is usually not done easily; and, for routine purposes, those with limited resources may find it best to concentrate on checks on the controllable losses, and the important glycerol in re-cycled salt. Care is necessary to ensure that truly average samples of materials such as muds are tested. Misleading samples of treatment mud can be taken from near the point at which water is introduced into the filter presses.

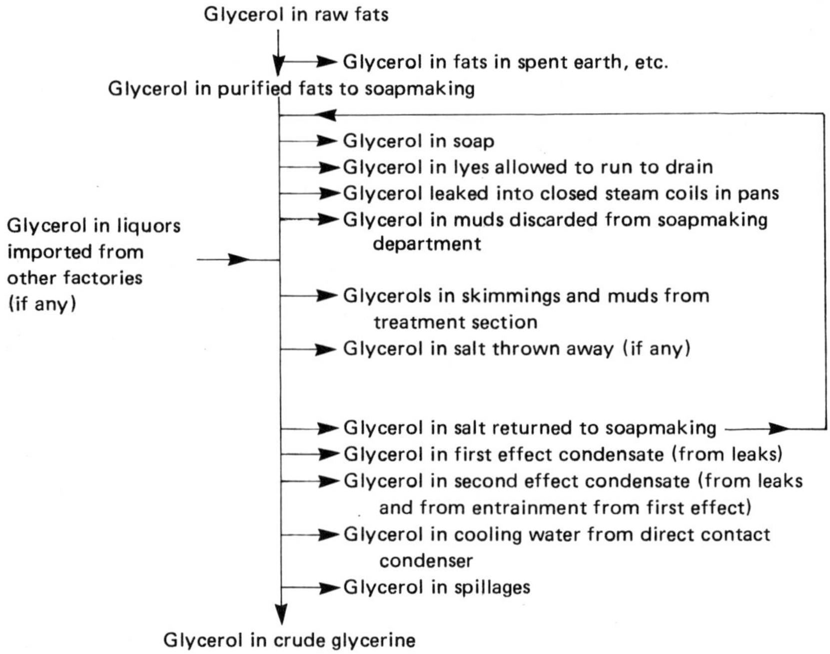

Fig. 8.9 – Flow of glycerol through a soap and glycerine factory.

8.4.7 Vacuum maintenance

Principles of vacuum maintenance are discussed in Appendix 2. Glycerine evaporators now normally operate at 2 – 4 inches Hg absolute pressure (28 – 26 inches Hg vacuum with 30 inches Hg barometric pressure); and a high-level direct contact condenser with barometric leg is commonly used. The non-condensible gas can be removed with a two-stage ejector system, with an intermediate condenser, or by a mechanical vacuum pump. Because of the level of pressure needed in a glycerine evaporator, no booster jets are normally installed.

The direct contact main condenser can be replaced by a surface condenser if the condensate from the last effect is required, for example, for use with other condensates in the splitting section of a fatty acid plant.

8.4.8 Evaporator control

Traditionally glycerine evaporators were controlled manually, but standard instrumental control techniques are now used to make operation increasingly automatic. Possible difficulties due to dissolved and solid salt must be remembered with soap lye evaporators. Also, the influence of vapour on the apparent density of boiling liquid in, for example, a natural circulation evaporator.

A reliable indication of the concentration of glycerol in the material being discharged from an effect producing crude glycerine is required. This is frequently based upon the relationship between boiling point elevation and concentration; in other words, between boiling point under a known pressure and concentration. A very simple way for a plant operator to control a manually operated finisher, or other unit producing crude glycerine, is to provide:

- a barometer to measure atmospheric pressure (b inches Hg).
- a mercury U-tube with an open end to the atmosphere and the other end connected to the vapour space in the effect. This gives the vacuum under which the liquor is boiling, x inches Hg; that is the difference between the atmospheric pressure and the pressure in the vapour space. The absolute pressure in that vapour space is thus $(b - x)$ inches Hg. Alternatively, gauges which measure this absolute pressure directly can be used.
- a thermometer set to measure the temperature of the boiling liquid. Because of the effect of hydrostatic head, this temperature will vary somewhat with the quantity of liquor in the effect and the point at which the thermometer is placed.

Because of the influence of hydrostatic head, and of impurities in the crude glycerine being produced, it is necessary to calibrate the system by the analysis of samples taken in relation to various sets of readings. Roughly, a boiling temperature of about 77°C with an absolute pressure of 2 inches Hg indicates 83% glycerol in a soap lye crude glycerine. It has also been found that an extra 0.1 inch Hg pressure requires an extra 1°C in boiling liquor temperature. Hence, if 77°C is found to be correct for 2.0 inches Hg, 79°C will be about right for 2.2 inches Hg pressure, and 74°C for 1.7 inches Hg pressure.

8.4.9 Materials of construction for glycerine treatment and evaporation plants

Traditionally, mild steel, wrought iron, or cast iron tanks and pipes have been used for soap lye treatment plants, with cast iron filter presses clothed with cotton cloths. Particularly for the early stages of treatment carried out under acidic conditions, the presence of chlorides and blowing with air give highly corrosive conditions; and the life of tanks is often short. The lining of such tanks with rubber, cement, or some acid resistant plastic coating, is highly desirable. Cast iron filter presses have a reasonably long life, although it is often useful to scrape the grooves in the plates from time to time. Expensive polypropylene filter plates are now sometimes used; and it may be desirable to assess the economics of using polyester cloths.

For many years, cast-iron was used for the bodies of soap lye evaporators, with arsenical copper tubes in the calandrias and heat exchangers. Such evaporators lasted a very long time; examples are known where the cast-iron bodies were very difficult to break-up after a life of about forty years, although they had been re-tubed 3 – 4 times during this period. These evaporators used exhaust steam, plus make-up, at about 4 psig in the main, and less in the actual calandrias. The adverse effect of higher temperatures, resulting from the use of higher pressure steam, is shown dramatically by experience with the quadruple-effect evaporator mentioned earlier. The arsenical copper tubes in the first effect calandria lasted only 1 – 2 years; and the rate of corrosion fell with reduction in temperature from first to last effect. Experiments showed that monel metal was a particularly good material for tubes, but that the less expensive cupronickels (including the cheapest 90:10 type) were adequate for the duty. The cast-iron bodies also behaved less well than expected, but this may have been due, at least in part, to the quality of war time cast iron. Special cast irons, such as Ni-resist, have also been used for the bodies of soap lye evaporators. Copper has been used successfully for the bodies of sweet water evaporators.

These traditional materials are still recommended by some plant manufacturers, but others suggest alternatives, including mild steel. Sacrificial anodes of aluminium have sometimes been found beneficial in reducing the corrosion of mild steel, or cast-iron, evaporators. The present author would be chary of recommending mild steel for soap lye evaporators, but a number of such plants appear to be giving acceptable service. Plant manufacturers who make soap lye evaporators in mild steel normally also offer plants in more resistant materials, expected to give a longer life, at substantially higher prices.

The baskets of centrifuges used for salt separation have to withstand high stresses as well as corrosive conditions. Mild steel was normally used, but, with modern, more efficient, machines, failures, particularly at welds, are known. High nickel alloys (or, perhaps certain stainless steels), with the necessary mechanical properties as well as corrosion resistance, should normally be selected. Design to avoid high stresses is very important, whatever metal is selected; and rivetted, rather than welded, side seams may be advantageous.

8.4.10 Complete evaporation plants

It will be appreciated that many combinations of plant elements are possible. Figs. 8.10 to 8.14 outline some examples from recent plant manufacturers brochures but, as in other fields, most manufacturers offer alternative designs to suit various requirements.

Fig. 8.10 shows a *Wurster and Sanger* backward feed, double-effect, evaporator, for saop lyes, directly connected to a single-effect finisher. Heating is in natural circulation calandrias; and extraction of salt slurry is by continuous pumping through separating vessels. Salt is recovered from the slurry in a batch centrifuge, or centrifuges.

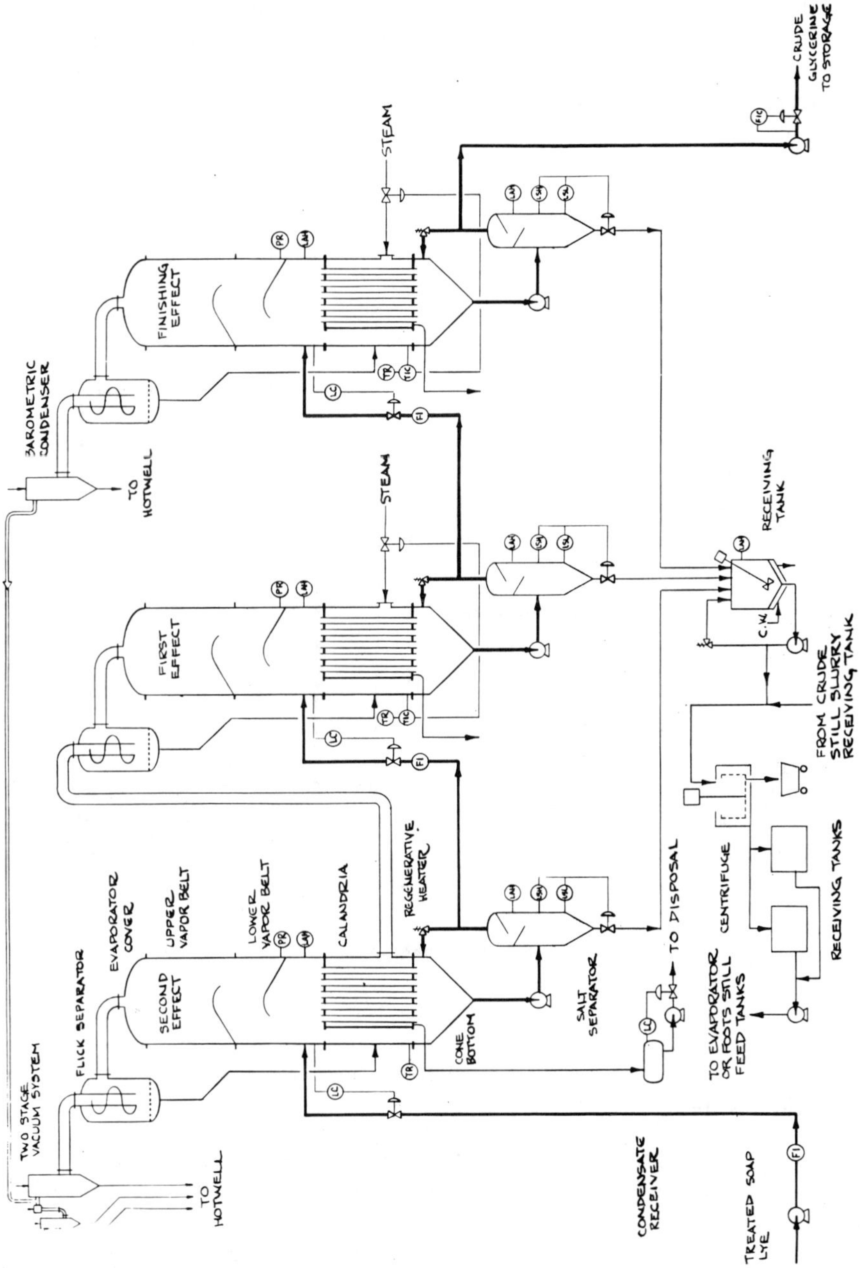

Fig. 8.10 – Wurster and Sanger evaporator for soap lyes.

Fig. 8.11 represents a *Gianazza* evaporator for soap lyes. This unit is double-effect with forward feed; and uses jet recompression of part of the vapour from the first effect. External calandrias are used, with natural circulation for the first effect and forced for the second which produces crude glycerine. Salt slurry is extracted by pneumatically operated salt boxes, and from the crude glycerine in a static separator, for separation in a centrifuge. BO 1 is a direct-contact condenser with barometric leg and PV 01 a mechanical vacuum pump.

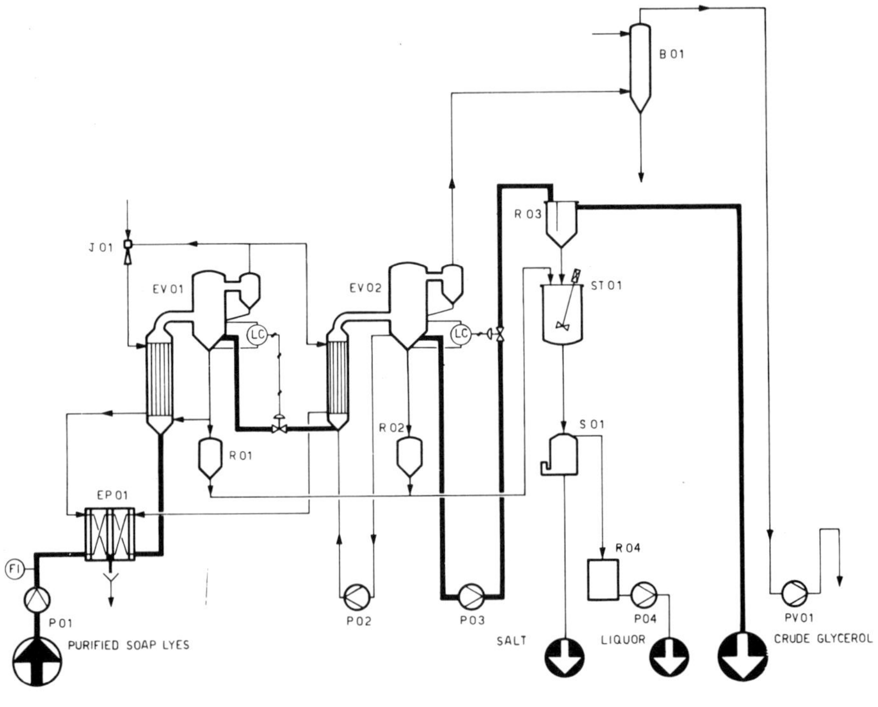

Fig. 8.11 – Gianazza evaporator for soap lyes.

Fig. 8.12 shows a *Lurgi* double effect sweet water evaporator. Special features are:

- jet recompression of part of the vapour from the first effect.
- multi-compartment, long tube, climbing film calandrias, arranged so that the liquid flow is once through each compartment in series, with the vapour discharge from each into a single space. The last part of the second effect is separated from the rest and is fed with steam directly from the recompression jet; that is the final concentration is in single effect. As an example, Lurgi quotes concentration from sweet water at about 10% glycerol to 25% in the first effect, to 70% in the second effect, and to 88% in the final stage.

- a surface condenser is used for the final vapours to increase the quantity of condensate available for re-use.
- the feed is preheated by the first effect condensates (including that from the final stage).

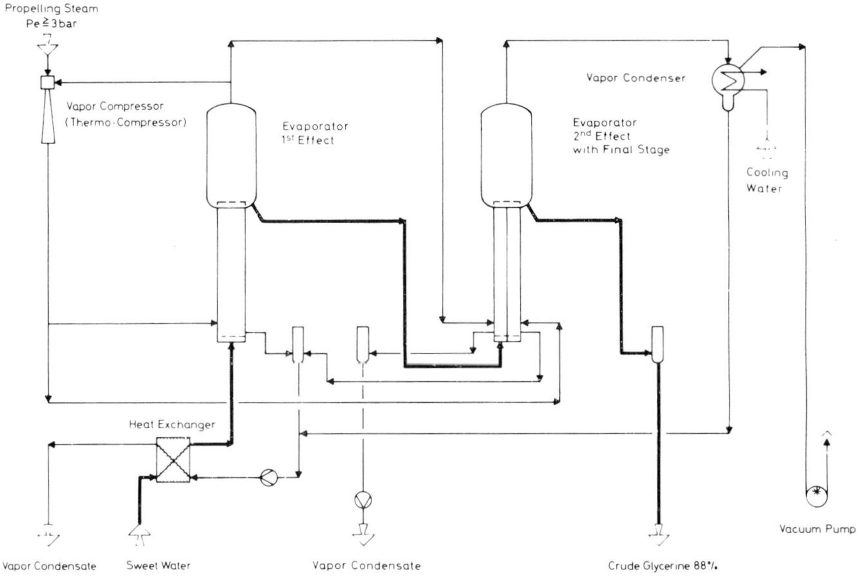

Fig. 8.12 – Lurgi double-effect sweet water evaporator.

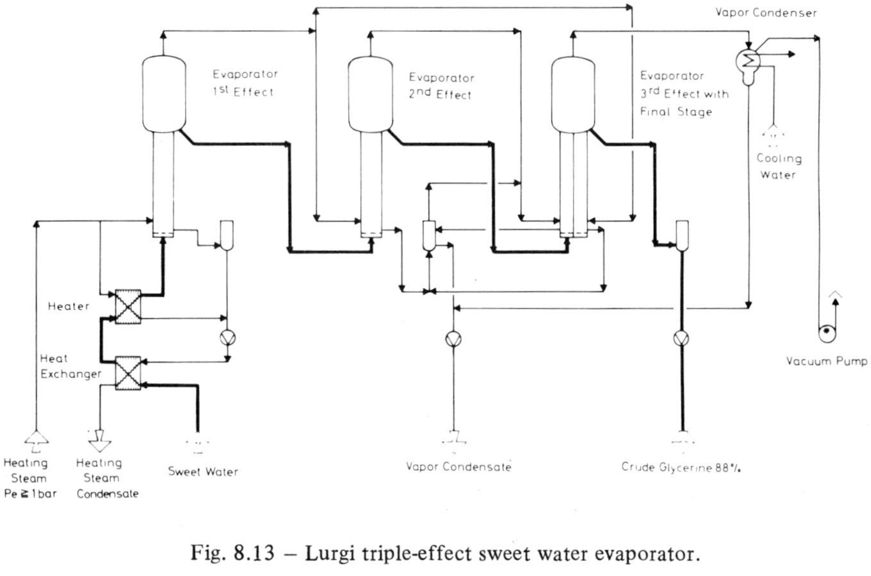

Fig. 8.13 – Lurgi triple-effect sweet water evaporator.

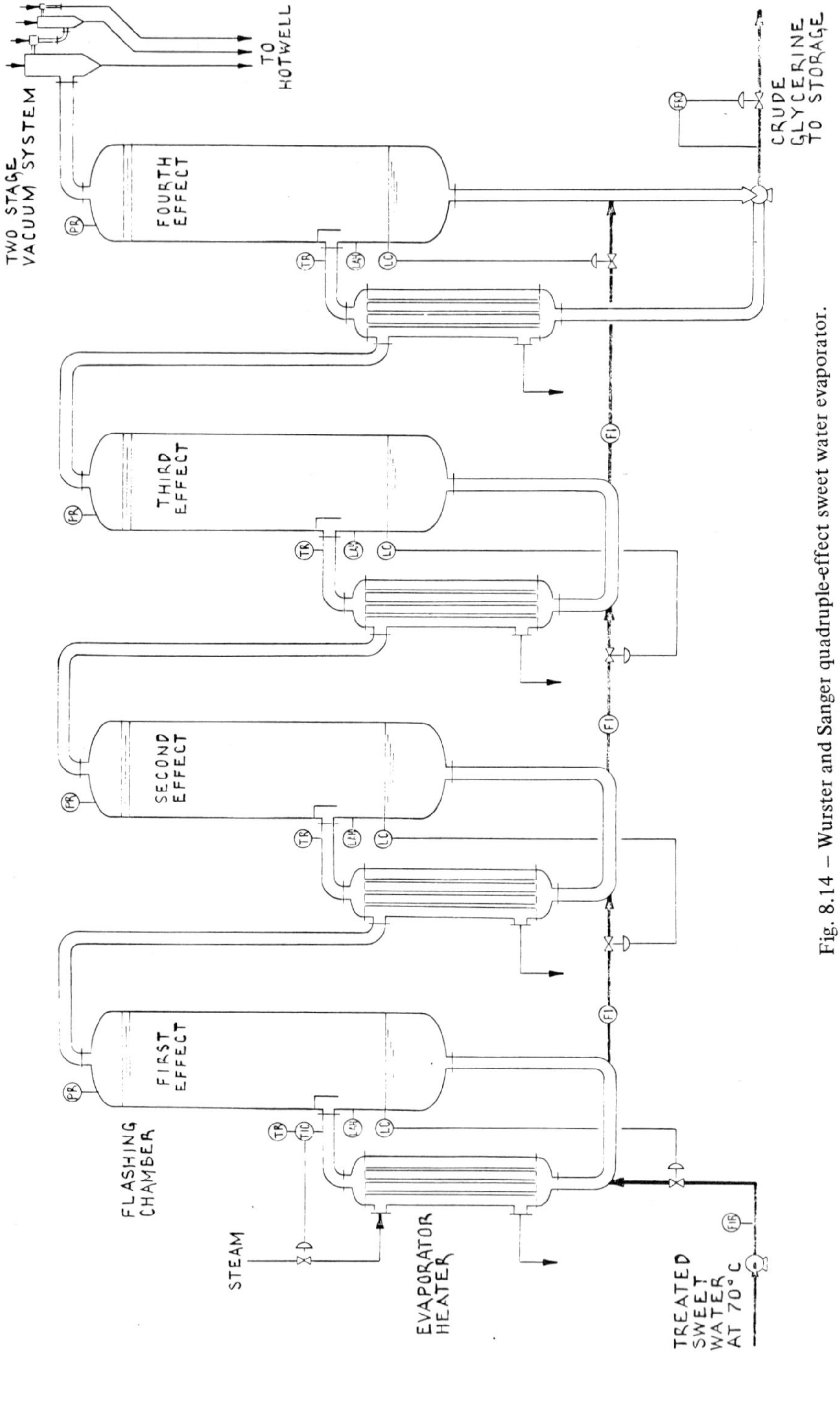

Fig. 8.14 – Wurster and Sanger quadruple-effect sweet water evaporator.

Fig. 8.13 shows a Lurgi triple effect plant for sweet water which is generally similar to the previous example. No recompression jet is shown, but one could be used if the available heating steam is at a suitable pressure. The feed is pre-heated by condensate, followed by live steam. A separate finishing section, operating in double effect, is included in the third effect shell. Fig. 8.14 depicts a Wurster and Sanger quadruple effect evaporator for sweet waters. Feed is forward. External calandrias are used with natural circulation for the first three effects and forced circulation for the fourth effect.

8.5 TREATMENT OF CRUDE GLYCERINE TO REMOVE LOW MOLECULAR MASS FATTY ACIDS

The quality of soap lye crude glycerine can be improved by steaming it in an acid condition, so as to remove volatile fatty acids. This can be done by acidification before the later stages of evaporation, but, with normal equipment, corrosion problems are likely. The crude glycerine is acidified to about pH 3.5 with sulphuric, or hydrochloric acid. Sulphuric acid is cheaper than hydrochloric, but has the usual disadvantage that it adds sulphates to the system, although this is less important than when salt to be re-cycled can be affected. Both mineral acids have similar effects on corrosion, because substantial concentrations of chloride ions are already present. Type 316 stainless steel (16 – 18% Cr, 10 – 14% Ni plus 2 – 3% Mo) has been found to give a reasonable plant life, although some pitting occurs. The main design and operating problem is to adjust conditions so that a substantial part of the low molecular mass fatty acid is removed without the loss of too much glycerol. A continuous unit which has been used satisfactorily is shown diagrammatically in Fig. 8.15. The process can also be carried out batchwise.

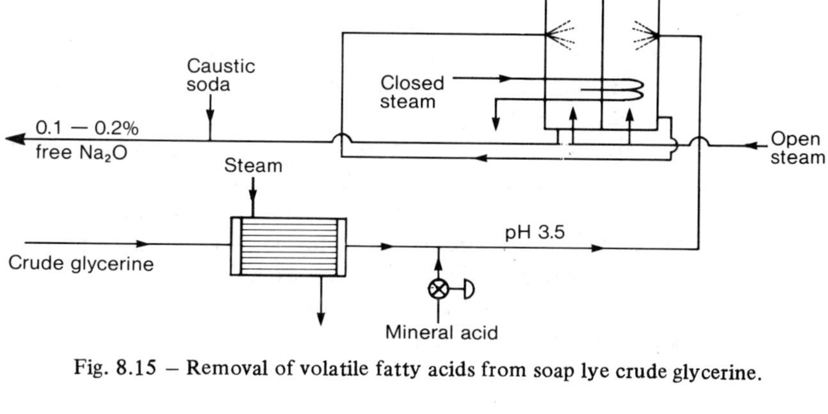

Fig. 8.15 – Removal of volatile fatty acids from soap lye crude glycerine.

8.6 DISTILLATION OF GLYCERINE

8.6.1 General

Under atmospheric pressure pure glycerol boils with decomposition at about 290°C and, as for fatty acids, it is necessary to use reduced pressure, and/or open, or carrier, steam to reduce the boiling temperature in glycerine stills. Principles of distillaton discussed for fatty acids in section 5.4.2 apply to glycerine distillation, but there are also important differences between the processes for fatty acids and for glycerine. In particular:

- the nature of the non-volatile residue, which from soap lye crude glycerine includes large proportions of salt and soapy matter.
- the fatty acid vapours contain a considerable range of compounds of varying volatility, whereas in glycerine distillation the vapours are predominantly glycerol and water, although the relatively small concentrations of organic impurities are important.
- glycerol and water are completely miscible in the liquid phase, whereas fatty acids and water are not. Also glycerines do not solidify at temperatures likely to be encountered in still condensates.

8.6.2 Chemical aspects of glycerine distillation

Several important, but undesirable, reactions tend to occur in glycerine liquors held at still temperatures:

- formation of volatile glyceryl esters by reaction between glycerol and the soaps of low molecular mass fatty acids.

$$\underset{\text{glycerol}}{C_3H_5(OH)_3} + \underset{\text{soap}}{R\cdot COONa} \rightarrow \underset{\text{mono-glyceride}}{C_3H_5(OH)_2{-}O\cdot CO\cdot R} + \underset{\text{caustic soda}}{NaOH}$$

- formation of polyglycerols, as discussed in section 8.2.1.
- formation of objectionable organic nitrogen compounds from proteinaceous residues in the crude glycerine.
- decomposition of glycerol to highly volatile compounds normally lost in the vacuum system.

Esters, fatty acids (if any), and the volatile nitrogenous bodies effect adversely the quality of distillates in which they are condensed. Polyglycerols are substantially non-volatile under distillation conditions; but they are undesirable because they represent a loss of glycerol, and because they modify the residue in ways which can be adverse. The condensation reactions by which polyglycerols are formed are not reversible and the ether links they contain are not broken under still conditions. Careful materials balances determined for glycerine distillations normally show a small unaccounted for loss, which may be associated with radical decomposition of glycerol to gaseous products.

Ester formation is a balanced reaction which is suppressed by the presence of caustic soda in the still vaporizer; and polyglycerol formation is catalysed by alkalis, with caustic soda more potent than the milder alkalis sodium carbonate and soaps. It is, therefore, important to carry out the distillation with as little caustic soda in the still as will permit the production of distillates satisfactory as regards content of esters and fatty acids (measured by the saponification equivalent, but also likely to affect taste and odour). Physical conditions of distillation are very important. Increases in the time for which, and the temperature at which, glycerine is held in the still vaporizer both increase the formation of polyglycerols. There is an interaction between the two conditions, so that a very short time at a relatively high temperature may be less damaging than a longer time at a lower temperature. More severe physical conditions also increase the influence of alkaline catalysts, so that mild alkalis may not be very significant if the physical conditions are mild, but become important when they are more severe. It has also been shown that with a given crude glycerine the distillate is much lower in saponification equivalent when distillation is very rapid. Similar factors probably affect the formation of volatile nitrogeneous bodies. The introduction of a small quantity of water into the still in, or with, the crude glycerine may also help to suppress polyglycerol formation, even though the water is rapidly vaporized. Under some conditions, the use of open steam may also help to reduce ester formation. When distillates need to be re-distilled, it is important to saponify any esters in them before the second distillation. This can be done by agitating the hot distillate with air after addition of caustic soda; and a suitable quantity of caustic soda is needed in the still during re-distillation to suppress re-formation of esters.

Formation of polyglycerols is commonly a major source of loss of yield in glycerine distillation; and it may be useful to estimate the polyglycerol in the still residues, which are often called foots. Elaborate separation procedures have been developed for special investigations, but the polyglycerol content is usually deduced from a periodate and an acetin, or dichromate, test on the foots, or, better, from all three tests. As explained in section 8.2.1, the periodate test does not respond to polyglycerols and, unless sugar or other unusual substances are present, records only glycerol. The acetin test includes polyglycerols, so that the difference between glycerol estimated by acetin and glycerol estimated by periodate is a measure of the polyglycerol content of the sample. To convert this difference to the glycerol from which the polyglycerols were formed, it must be multiplied by a factor which depends on the composition of the polyglycerols. If di-glycerol is assumed, the factor is $6/4 = 1.5$, and if tri-glycerol, it is $9/5 = 1.8$. To obtain the general relationship, when n moles of glycerol condense, $(n - 1)$ ether links are formed with a loss of $(n - 1) \times 2$ hydroxyl groups. Hence, $3\,n$ hydroxyl groups form a polyglycerol containing $3\,n - 2\,(n - 1)$ hydroxyls. In practice, a factor around 1.8 is probably reasonable. Similarly, the difference between glycerol percent by dichromate and glycerol percent by periodate is a measure of the polyglycerols present, and in this case no factor is required. One still known to the author was giving a very poor distillation yield when found to be operating at 60 – 70 mm Hg absolute pressure in the still instead of the 10 – 14 mm Hg for which it was designed. This raised the boiling liquor temperature to 170–180°C, substantial quantities of caustic soda had to be used, and the yield was only about 75%. A sample of liquid residue run out of the still, not a true average sample of the foots, tested 7% glycerol by periodate and 37% by dichromate. It, therefore, contained 7% glycerol and 30% polyglycerol calculated as glycerol; and showed qualitatively that major conversion of glycerol to polyglycerol had occurred and was a major cause of the very bad yield. In order to relate the polyglycerol found in foots to distillation loss due to this cause, it is necessary to test a true average sample of the residue and to find the quantity of this residue formed from a known quantity of crude glycerine by measurement, or by calculation from the salt contents of the two materials.

8.6.3 Vaporization principles

The vapour pressure of pure glycerol is as shown in the table

temperature °C	120	130	140	145	150	155	160	165	170	175
vapour pressure mm Hg	0.75	1.35	2.45	3.2	4.3	5.5	7.4	9.2	11.5	14.7
temperature °C	180	190	200							
vapour pressure mm Hg	18.5	29	45							

The more relevant part of this information is shown graphically in Fig. 8.16.

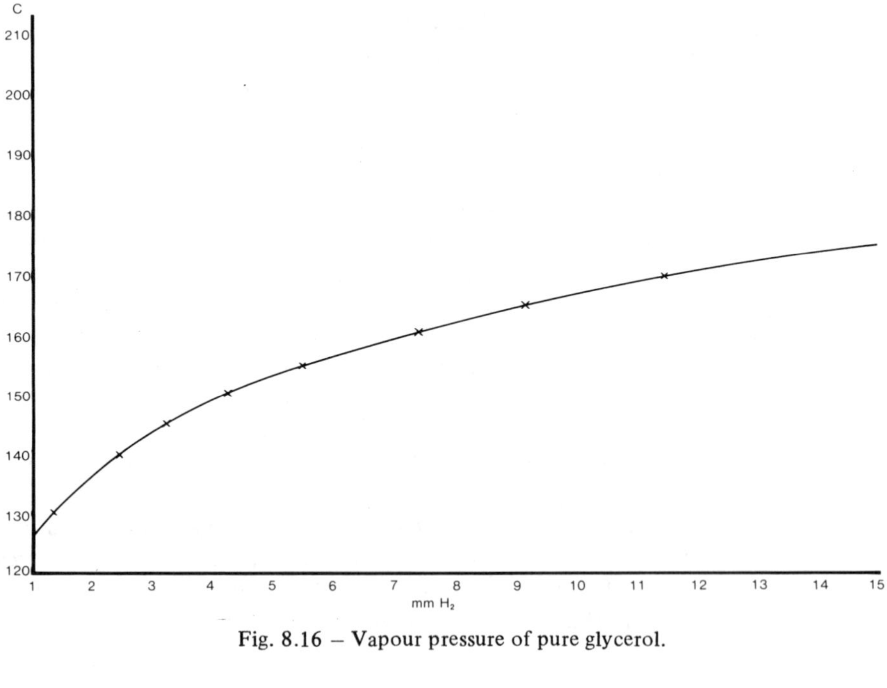

Fig. 8.16 – Vapour pressure of pure glycerol.

The detailed discussion of distillation principles given in section 5.4.2 is not repeated here. Glycerol and water are miscible in the liquid phase, so that the left-hand side of Fig. 5.13 does not apply; but little, or no, water remains in the liquid in a glycerine still and the right-hand side of the steam distillation diagram is relevant. That is $p_G + p_W = P_T$, where p_G is the vapour pressure of glycerol, p_W the vapour pressure of water, and P_T the total pressure in the still. The vapour pressure of glycerol in equilibrium with a still liquor is lower than that of pure glycerol at the same temperature, because of boiling point elevation due to dissolved substances. This elevation varies with the composition of the still liquor, which, in turn, varies with the composition of the crude glycerine fed and the degree of concentration which has been achieved. In the case of soap lye crude, from which salt leaves solution as the distillation proceeds, a figure based upon experience of roughly 10 – 15°C is assumed for the main part of the distillation in the rest of this discussion. That is to say that the vapour pressure of glycerol in equilibrium with a still liquor at 155°C is taken to be about that for pure glycerol at 140 – 145°C, or about 2.5 – 3.2 mm Hg.

Substituting 92, the molecular mass of glycerol, and including the vaporization efficiency E, the equation derived in section 5.4.2 for steam distillation can be re-written, lb of open steam/lb glycerol distilled = $((P_T - Ep_G) \times 18)/(Ep_G \times 92)$. Table 8.5 gives estimated usages of open steam for a range of conditions with the provisional assumption that E = 1. These estimates, which show a wide variation with conditions, refer only to the major part of the distillation run; considerably more open steam is used in the final stages when the residue is being worked down. Crude glycerines contain a small proportion of water; in some stills this is removed in a preliminary dehydrator, but in others it enters the main vaporizer, in which case it contributes to the open steam.

In a vaporizer which uses open steam as well as reduced pressure, the distillation conditions are established by automatic adjustment to meet three sets of requirements simultaneously:

- heat transfer from the closed steam, or other heating medium.
- heat balance, which determines the rate of vaporization of the glycerol plus that of any liquid water fed to the main still.
- the partial pressure relationship just discussed.

If, for example, the rate of heat transfer is increased, the additional heat will vaporize more glycerol. With a fixed rate of open steam supply this reduces the open steam/glycerol vaporized ratio, which causes the boiling liquor temperature to increase to the level required by the new ratio. Similarly, if the flow of open steam is increased, and the vapour passages and steam condensing equipment are such that the total pressure P_T is not raised significantly, the increase in the open steam/glycerol vaporized ratio results in a fall in the boiling liquor temperature.

So far it has been assumed that E = 1, that is that the open steam becomes saturated with glycerol as it passes through the glycerine liquor in the vaporizer. Miner and Dalton (1953), says that because of failure of the open steam to

Table 8.5 – Open steam usage for glycerine distillation under various conditions

Approximate temperature of still liquor °C	150–155	160–165	170–175	180–185
Assumed equivalent temperature of pure glycerol °C	140	150	160	170
Vapour pressure glycerol p_G, mm Hg	2.45	4.3	7.4	11.5
Partial pressure water, $p_W = P_T - p_G$				
If $P_T = 15$, $p_W = 15 - p_G$	12.55	10.7	7.6	3.5
$= 30$, $p_W = 30 - p_G$	27.55	25.7	22.6	18.5
$= 60$, $p_W = 60 - p_G$	57.55	55.7	52.6	48.5
$\dfrac{\text{Mass of water vapour}}{\text{Mass of glycerol vaporized}} = \dfrac{(P_T - p_G)}{p_G \times 92} \times 18$				
If $P_T = 15$	1.00	0.49	0.20	0.06
$= 30$	2.20	1.17	0.60	0.31
$= 60$	4.60	2.53	1.39	0.83

become saturated with glycerol, the actual usage of open steam is much higher than that calculated with $E = 1$. The author's experience is that with allowance for boiling point elevation and excluding the work-down period, the calculations using $E = 1$ are in fair agreement with experience, for at least some stills. Analysis of results for stills quoted by Miner and Dalton also seems to provide some approximate support for this view as follows.

Assuming the soap lye crude used contained 83% glycerol and 7% water, and that the distillation yield was 95%, 1 lb of distillate came from crude glycerine containing $0.07/(0.83 \times 0.95) = 0.09$ lb of water.

For a Garrigue still, Miner and Dalton quotes:

Heating steam in closed coils 175 – 200 psig (192 – 198°C)
Distillation temperature about 177°C.
Vapour leaving still 166 – 171°C.
Still pressure 40 – 50 mm Hg.
Open steam used during run 0.5 – 0.6 lb/lb glycerol.
Total open steam (including boil down and finishing of foots) 1 – 2 lb/lb glycerol.

From these figures it is assumed that the equivalent boiling temperature for pure glycerol was about 168°C, with a p_G of 11 mm Hg. With the mean operating pressure of 45 mm Hg, this gives (lb open steam)/(lb glycerol distilled) = $((45 - 11) \times 18)/(11 \times 92) = 0.60$. With allowance for the water in the crude glycerine fed, the ratio becomes $0.60 - 0.09 = 0.51$.

For an Ittner still, Miner and Dalton quotes:

Distillation temperature 166°C
Still pressure 10 – 15 mm Hg
Open steam used 0.25 lb/lb glycerol

It is assumed that the equivalent boiling temperature for pure glycerol was about 155°C, with $p_G = 5.5$ mm Hg. With the mean operating pressure of about 13 mm Hg, the ratio $= ((13 - 5.5) \times 18)/(5.5 \times 92) = 0.27$, which, with allowance for water in the crude, becomes 0.18.

Water in crude glycerine obviously becomes of increasing importance as the total usage of open steam falls.

8.6.4 Vaporization equipment

Soap lye crude glycerine contains some 10% of salt and soapy matter; and the disposal of this non-volatile material influences the choice of vaporization equipment which can be used, and makes it usual with many types to operate semi-batch, rather than continuously. This problem is much reduced with saponification crudes, particularly those made from high pressure split sweet waters which contain little inorganic matter. For this reason, saponification crudes are often considered superior to soap lye crude, but, especially when produced from low grade fats, they are liable to contain nitrogenous and other impurities which can lead to problems with the quality of the distillates made from them.

As a batch of soap lye crude glycerine is distilled semi-batchwise, with additional crude fed to replace the glycerol vaporized, the concentrations in the residue of sodium chloride, other inorganic materials, and MONG increase; and the viscosity of the residue rises. The recovery of as much glycerol from this residue as is economically worthwhile is a major problem; and several procedures are possible:

- traditionally the main distillation was stopped whilst the residue was still liquid enough to flow out of the still. The residue was then diluted and purified as much as possible using treatments similar to those adopted for lyes or sweet waters. The purified liquor was re-evaporated to give 'foots crude', which was distilled separately to produce a low grade distillate. A variant was to return the still foots to the soap lyes before treatment. This was simple, but is not recommended because of the build-up of impurities in the system, despite some purification in the re-treatment.
- some plants remove residues from the main still in liquid form for the rest of the glycerol to be stripped in a separate foots still. This enables the main still to be operated more, or less, continuously, although it needs to

be washed out from time to time. Wurster and Sanger offer stills of this type in which solid salt is separated in a centrifuge through which the liquor is passed on its way from the main still to the foots still. Apart from the cost of the additional equipment (offset by the additional capacity of the main still), the use of a foots still tends to increase the time during which the material is exposed to distillation temperatures. Separation of salt and other solids in a centrifuge helps the final stages of vaporization, but exposure of hot liquor to air may entail some risk of oxidation of glycerol.

- for many years it has often been found best to continue the main distillation until little more glycerol can be vaporized. With many still designs this involves a loss of glycerol in foots discarded to drain of at least about 2% of that fed. If a reasonably good quality soap lye crude glycerine has been distilled, and conditions are such that only a little polyglycerol is formed, the residue is so stiff that it cannot be run out of the still; and it has to be dissolved by being boiled with water. If the quality of the crude glycerine is poor, polyglycerol formation is high, or the final steaming has not been complete, some of the residue can be run out as liquid foots. When liquid foots are formed, investigation is desirable to ascertain the cause and, if possible, to arrange remedial action for future distillations.

In most stills the residue after as much glycerol as possible has been vaporized from a soap lye crude is a toffee-like material, but, as described later, it has been found recently that under some circumstances a dry, powdery, residue can be obtained from a once-through vaporizer.

When re-distillation of once distilled glycerine is used to produce chemically pure grade, a procedure which is no longer common, a liquid residue containing glycerol, polyglycerols, caustic soda, and soapy matter is left. From time to time, this needs to be transferred to a first distillation of crude glycerine.

Three major types of vaporizer are used in the distillation of crude glycerine:

- pot stills, and variants, in which open and closed steam are used to provide natural circulation vaporization.
- stills with external heaters and forced circulation.
- scraped wall, once through, vaporizers.

Types 1 and 2 can be used for saponification as well as soap lye crudes, but type 3 cannot because the necessary dry, powdery, residue is not produced, unless a liquid residue can be discharged satisfactorily.

Natural circulation pot vaporizers are used in Van Ruymbeke stills, a type which was very widely employed for many years; and in various other designs. In essence, they comprise pots fitted with closed heating coils (usually steam) and a perforated ring, or cross, near the bottom for the introduction of open steam. The Van Ruymbeke still operated at around 60 mm Hg; and hence required about 1 – 1.5 lb open steam/lb glycerol vaporized, with a boiling temperature around 170 – 175°C. Steam in the closed coils at 175 – 200 psig (condensing at 192 – 198°C) gives a reasonable temperature difference to induce heat flow and a good rate of distillation. One major installation with several Van Ruymbeke stills for the distillation of soap lye crude glycerine, plus two smaller stills for re-distillation to produce chemically pure grade, operated satisfactorily for many years with steam from Lancashire boilers at 150 – 160 psig (185 – 188°C) which gave a rather small temperature difference. Special arrangements to make the best use of the available steam are shown diagrammatically in Fig. 8.17. Features are:

- a good flow of steam through the coils, with efficient removal of condensate and air, was induced by use of surplus steam which was recovered from the 90 psig collector.
- superheated steam was used in the crude still coils. This is contrary to the usual view that superheat in the steam used in heating coils reduces rather than increases heat transfer, but this use of superheat was found to increase distillation rate, presumably because it increased the temperature of the upper parts of the coils which were in contact with froth, as well as more compact liquor. Saturated steam was found advisable for re-distillation, when the boiling point elevation was lower.

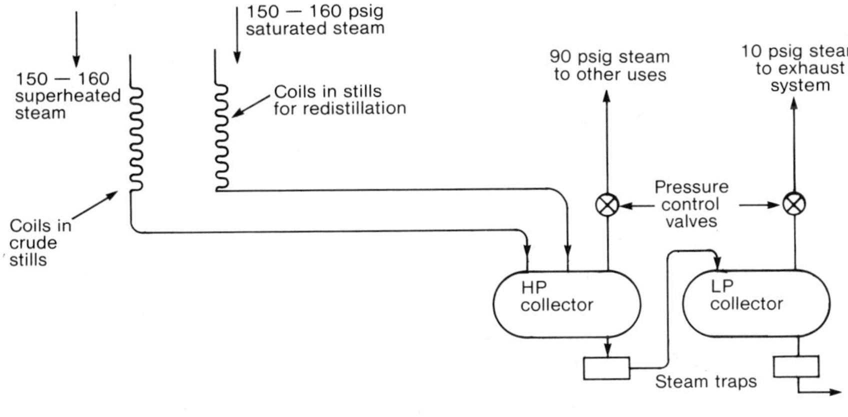

Fig. 8.17 – Steam supply arrangements for a group of Van Ruymbeke stills.

Classical Van Ruymbeke stills included a 'superheater' for the open steam. This comprised a tube and shell heat exchanger in which the flow of open steam towards the perforated distributor in the still was exposed to heat from full pressure steam. The superheater was presumably intended to avoid the introduction of liquid water into the hot glycerine liquor where it might cause primes. It can, however, be shown by theoretical analysis, using the Fanno line technique, that expansion of steam through a control valve and pipe is very different from that through a nozzle, because of the 're-heating' effect of pipe wall friction as the linear velocity becomes very high; and that liquid water is unlikely to be formed. At least one operator found over 50 years ago that these superheaters were unnecessary, and they were removed without any ill effect.

In addition to its role as a carrier fluid, an important function of the open steam in pot-still vaporizers is to provide agitation, with proper circulation of the liquor over the closed coils without local overheating. Some plant manufacturers have introduced the open steam through jet-type vapour-lift devices to improve the circulation; for example up the centre of the pot and down over the annular heating coils. This is particularly useful when the still is designed to work at a relatively low absolute pressure and with a low open steam ratio.

It is important to ensure that flow passages through condensers and ducts are adequate to take the maximum requirement of open steam without undue pressure drop. Stills have been constructed in which this was not done, so that an adequate flow of open steam could not be used without a substantial increase in still pressure.

Several important modern stills use *external heaters and forced circulation* which permit operation at low pressures of about 2 – 4 mm Hg. Often no open steam is used during the main part of the distillation, but it is also possible to inject some open steam into the vessel to and from which the liquor is circulated; in which case a rather higher still pressure is usual. One early type was a Scott still for which the soap lye crude was dried before being fed into the still in which a pressure of about 2 mm Hg was maintained. According to Miner and Dalton (1953) this still operated with a liquor temperature of 177°C, which is surprisingly high for a pressure of 2 mm Hg, even with no water in the crude glycerine as fed to the vaporizer. Information from another source showed that a still of this type working in the 1930s required a large addition of caustic soda to produce distillate with an acceptable saponification equivalent and that the yield was not good. The large amount of caustic soda used, and the polyglycerol formed, would tend to account for the high boiling temperature; and the high boiling temperature, plus the excessive caustic soda, for the poor yield.

Since that time very successful stills with vaporizers of this type have been developed. One 'family' of such stills stemmed from a pilot plant constructed in about 1940 by W. J. Frazer and Co., then of Dagenham, England. This led to a full scale plant by Frazer designed in conjunction with Unilever. As reported by L. V. Cocks, Chapter 6 of Newman (1968), additional stills of this type built by Scott to functional designs by Unilever were installed in the early 1960s. Similar vaporizers have been used by others. These Scott-Unilever stills operate at 3 – 4 mm Hg without open steam, but with a little water left in the crude glycerine. The boiling liquor temperature is around 155°C and the yield is superior to that with Van Ruymbeke stills (in the author's opinion some of the yields for various stills quoted in the literature must be treated with reservation). To maintain an absolute pressure in the vaporizer of 3 – 4 mm Hg, the vapour system, spray arresters, ducts and condensers, must be designed to have very low vapour velocities, and hence low resistance to flow. As explained in Appendix 2, booster jets are used to compress the water vapour and non-condensible gas leaving the last glycerine condenser; and the capacity of the booster jet, or jets, is often the factor which limits the capacity of the still. This means that the water content of the crude glycerine should normally be kept low. The significance of the presence of a little water in the crude glycerine as fed to the vaporizer on polyglycerol and ester formation is not clear, but it may be important. To enable the distillation to be completed in the main vaporizer, without a foots still, closed and open steam coils are fitted in the vessel, together with a damper in the circulation pipe, as shown in Fig. 8.18 for the Frazer-Scott-Unilever vaporizer. During the main part of the distillation the butterfly damper is open and the liquor is circulated at a high rate by the centrifugal, or axial flow, pump through the external heater. It is returned to the vessel through a tangential inlet onto a ledge round which it swirls; vapour flashes off and the residual liquor runs back into the reservoir to be recirculated with the new feed. The closed and open steam coils in the still vessel are not used during the main part of the distillation. When sufficient crude has been fed, and the liquor worked down until its viscosity becomes too high for continued operation in this way, the damper is closed and the liquor in the circulating system is transferred by the small pump to the main reservoir. The residue is then worked to dryness using the closed and open steam coils, as for a pot still. During this final stage the pressure in the still and the liquor temperature are higher than for the main run.

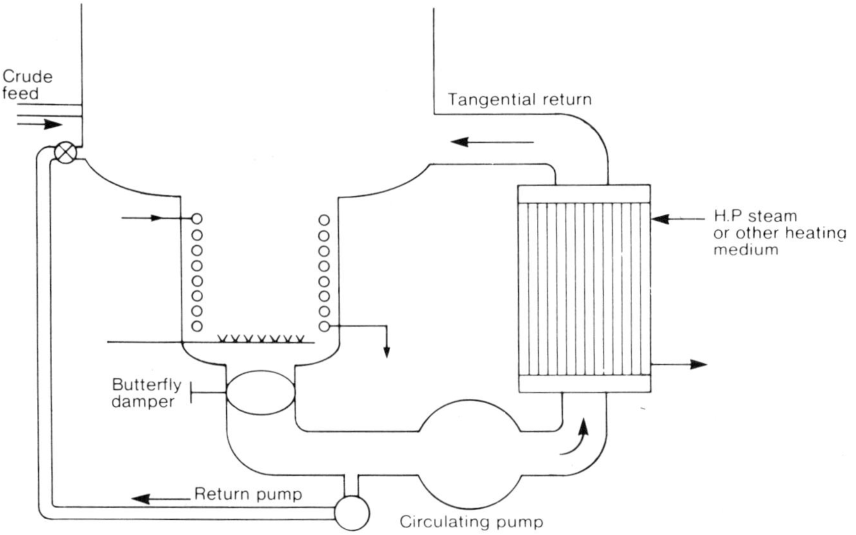

Fig. 8.18 – Forced circulation vaporizer. Frazer–Scott–Unilever still.

A more recent development is the use of a *'once-through' scraped wall vaporizer* for the distillation of good quality soap lye crude glycerines. When conditions are right the glycerol is vaporized very rapidly and almost completely, without formation of polyglycerols, and a dry, powdery, residue is discharged from the base of the unit. The Luwa scraped wall vaporizer has been used for this duty. Gianazza now offer a type GDTF 'Turbotherm' still (see Fig. 8.25 later) in which a scraped wall vaporizer, as shown in Fig. 8.19, is used. In the Gianazza still the residue is collected in a large vessel from which

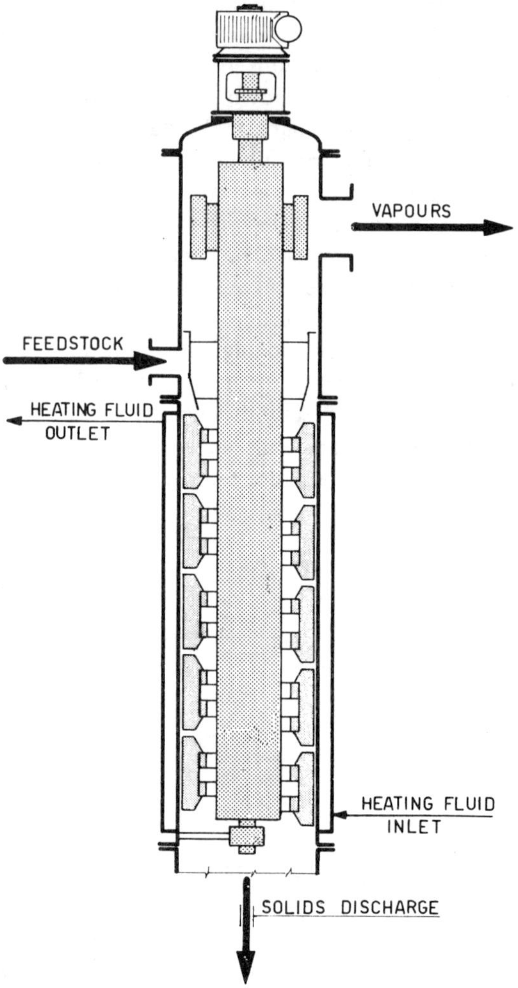

Fig. 8.19 – Scraped wall vaporizer for Gianazza GDTF still.

it is said to be discharged once, or twice, a week. It is also possible to remove the residue continuously through an automatic air-lock system and screw conveyor. The vaporizer is heated with steam at, say, 250 psig (208°C); or with hot oil, or other high temperature heat transfer medium. A low pressure of a few mm Hg is maintained as for other stills not using open steam. Because of the very short residence time it seems probable that the working pressure may be less critical than with other types of vaporizer. Because of its almost instantaneous vaporization, this type of unit has great advantages when it works well. The yield can be exceptionally high, because virtually no polyglycerol is formed, and vaporization of glycerol is nearly complete. The formation of esters and other undesirable compounds is also substantially lower than with other vaporizers using the same crude glycerine. On the other hand, it can only work if a dry, powdery residue is formed; patches of toffee-like material can cause major problems. This requires the unit to be kept in good mechanical condition and capable of uniform heat transfer. It also requires a good soap lye crude glycerine with a sufficiently high ratio of salt to MONG; at least 3.5 is said to be a useful criterion. It has also been found that particularly good crude glycerines, which produce a very dusty residue at the base of the vaporizer, can cause some difficulties due to carry over of dust, because there is little organic matter to bind the particles of residue. The dry residue produced by this type of vaporizer may be more readily disposed of than the rather foul smelling and polluting liquors obtained when other types of vaporizer are washed out.

8.6.5 Spray arresters

Traces of liquor, or solid residue, carried over in the vapours from the still can spoil a distillate, particularly the first (and best) when more than one fraction is collected. The strict limits for chlorides mentioned in the specifications discussed earlier make this particularly critical with soap lye crudes.

The main problem is to remove spray without excessive flow resistance, especially in stills which operate at very low pressures. Thin beds of rings having a large cross-section, so as to minimize linear velocity and hence resistance, are now usual. Such beds are often situated in the upper part of the still vessel; and they should be provided with water sprays to enable them to be washed when the still is cleaned. Older stills, such as the Van Ruymbeke, which operated without booster jets, commonly used catch-all vessels.

Particularly when foots are worked down to dryness and considerable boiling out with water is needed to remove the residue, carry over is liable to occur during the wash out. Some stills include a large valve in the vapour line which can be closed to isolate the vaporizer from the condensing system during the washout. Such valves can be valuable; but care must be taken to avoid constriction of the vapour passages and the possibility of inleakage of air.

8.6.6 Condensation, condensers, stripping columns, concentrators, and deodorizers

Apart from any carry-over of droplets, the vapours leaving a glycerine still contain glycerol and water, plus small concentrations of esters, TMG, and other organic impurities. The proportion of water to glycerol depends upon the quantity of open steam used, including water in the crude glycerine as fed to the vaporizer. Because of the large difference in volatility of water and glycerol, separation of these two components by fractional condensation is relatively easy; and is the main way in which refined glycerines are produced. Fractionating columns with reflux and reboil are not usual; and are generally undesirable because of the pressure drop they introduce and the extra time for which the alkaline crude glycerine is kept at distillation temperature.

There is a pressure drop from the vaporizer through the system to the vacuum maintenance equipment, so that P_T falls on passage from section to section, but at any point $P_T = p_G + p_W + p_I + p_N$ where

p_G = partial pressure glycerol
p_W = partial pressure water
p_I = partial pressure organic impurities
and p_N = partial pressure non-condensible gas

As a condensate rich in glycerol separates, p_G falls and the other partial pressures rise to maintain the total pressure.

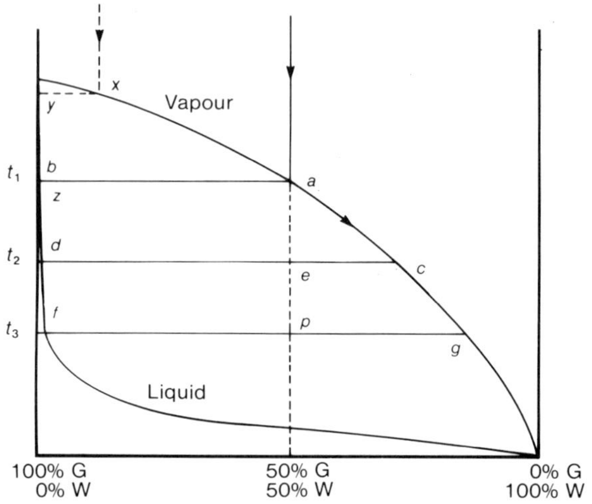

Fig. 8.20 – Vapour/liquid equilibrium for glycerol and water.

Consider first only glycerol and water, and disregard changes in total pressure. The vapour-liquid equilibrium diagram is roughly as shown in Fig. 8.20. The shape of the curve mirrors the large difference in volatility of the components. If a superheated vapour containing 50% glycerol is cooled carefully, condensation will start at point a where the 50% vertical cuts the vapour line; and the drop of liquid which separates will, under equilibrium conditions, have composition *b* (that is it is nearly pure glycerol). As more heat is removed, further liquid will separate with compositions along the line *bd*; and the residual vapour composition, and temperature, will fall along the line *ac*. If the cooling in the first condenser stops at temperature t_2, the liquid condensed will have an average composition between *b* and *d*; and the remaining vapour composition *c*. Similarly, on further cooling to t_3, another fraction with average composition between *d* and *f* separates, leaving vapour of composition *g*. To remove practically all the glycerol from the vapour *g*, the temperature has to be lowered substantially more in a cold water condenser and the condensate contains a considerable percentage of water. The fall in pressure through a real still does not affect this qualitative discussion. The organic impurities are mainly more volatile than glycerol, but considerably less volatile than water. They, therefore, condense with the glycerol, but some fractionation occurs, so that a first fraction, *bd*, contains less impurities than a second *df*, and the highest concentration of impurities is in the final, relatively weak, fraction. In modern, high vacuum, stills practically all the glycerol remaining after the product condenses is condensed as an impure fraction containing, say, 60% glycerol before the water vapour and non-condensible gas pass on to the vacuum maintenance system. In some older stills, such as the traditional Van Ruymbeke, the condensers for vacuum maintenance were also the final condensers for weak glycerine which produced what was sometimes known as 'weaks' containing about 2 – 6% glycerol. Many years ago it found in some factories that much of the steam needed to evaporate the water from weaks could be saved by the introduction of a scrubbing tower as shown in Fig. 8.22 discussed later. This tower recovers nearly all the residual glycerol as an impure fraction containing about 60% glycerol, so that only a negligible quantity passes into the vacuum maintenance system.

Stills have been made which use several vaporizers, and in which the vapour, mainly water, left after the product has been condensed from the vapours from the first vaporizer is used as open steam in the next vaporizer; and so on through the system. This system is used in the Wood still.

So far equilibrium between vapour and condensate has been assumed. When surface condensers are used they should ideally have a large heat transfer area; and so work with a small temperature difference which gives condensation under near equilibrium conditions. If a small surface area and a large temperature difference are used to achieve the same heat flow, local overcooling can occur with less fractionation. This point can be illustrated by the extreme, fortunately not real, case in which a pocket of vapour is condensed completely by itself; the condensate would then have the same composition as the vapour. It is possible to arrange contact between liquid and vapour after the initial condensation. Interchange of components then occurs in a move towards equilibrium. Hence counter-current condensers may be beneficial; and stripping, or washing, columns can be used to improve distillate quality. In such columns, distillate flows downwards in counter-current to rising vapours, usually directly from the still, which vapours are superior in quality to others left after some condensate has been separated. Stripping columns are sometimes used to correct for quality deficiencies due to condensation with a large temperature difference.

Some stills, particularly small ones, have only one product condenser, but there are often advantages in having two. The best distillate, from the first condenser, is then available for the top quality product and the somewhat less good distillate from the second product condenser can be diverted to the less exacting grades. This may enable distillation to be carried out with less caustic soda in the vaporizer and so tend to a better yield. Fig. 8.20 shows that a little water condenses with the glycerol; and a comparison of routes *abdf* and *xyz* indicates that the amount is larger in stills using open steam than in modern high vacuum stills in which the vapours from the vaporizer contain little water. Concentration of the distillates is sometimes, but by no means always, needed to meet specification. This can be done in a separate evaporator, but stills in which concentration of distillate is needed may include vessels in which the distillate is subjected to heat from closed coils whilst under vacuum. Such vessels sometimes also use open steam, when they are known as deodorizers and are intended also to remove relatively volatile organic impurities which have condensed in the distillate. Older type stills, such as the Garrigue which produce large quantities of weaks, sometimes include built-in concentrators which also produce open steam for use in the vaporizer.

TMG is more volatile than glycerol and concentrates in the later fractions. The final, impure fraction from the cold water condenser is normally returned for redistillation, after treatment with caustic soda and possibly concentration; and if the crude glycerine being used contains a relatively high concentration of TMG this compound may build-up in the system to the extent that the distillates cannot meet the high glycerol clauses in some specifications, such as the 99% in the British Standards, even if all water is eliminated. Unless customers can be found who do not mind the TMG, the best procedure is usually to collect the impure last fractions until sufficient is available for re-distillation alone, from which the cold condenser fraction is discarded with some loss of glycerol. If necessary, it may also be possible to push more TMG into the cold condenser fraction, plus additional glycerol, by an increase in the temperature of the vapour as it leaves the last product condenser. It is much more satisfactory for the producer of the crude glycerine to avoid formation of TMG, which, as stated earlier, also indicates a loss of glycerol by decomposition.

The condensers used in glycerine stills fall into three main types, although variants of each are possible:

- air cooled cylinders.
- surface condensers, using a hot cooling medium for the main product condenser, or condensers.
- direct contact condensers.

Air cooled cylinders are now obsolete, but they were used in Van Ruymbeke and the even earlier direct-fired stills. The vapours from the vaporizer pass through a lagged catchall and then through a series of large unlagged cylinders. Some 8 – 12 cylinders are usual and they can be arranged in line, or in tiers. Heat is lost by natural convection and radiation from the surface of the cylinders. Distillation has to be continued for several hours after the commencement of a cycle to bring the system up to the temperatures necessary to produce satisfactory condensate. During this period the liquid condensed is returned to the still; and this helps to remove impurities carried over into the cylinders during the washing out of the foots. When the return to still is closed, condensate collects in the cylinders, but, with some designs, 'drop-cylinders' are provided for removal of condensate during the run. Miner and Dalton (1953), following Webb (1927), quotes the following temperatures for the sequence of cylinders in the condensing system of a Van Ruymbeke still.

Cylinder	1	2	3	4	5	6	7	8	9	10	11	12
Temperature °C	197	195	191	185	178	170	160	149	122	106	90	72

Distillation was said to commence at 199 – 203°C, using closed coils fed with steam at 225 psig (203°C). The distillation, and early condensing cylinder, temperatures seem rather high, but the figures illustrate:

- the favourable temperatures for near equilibrium condensation.
- the way the condensing temperature falls slowly at first and then at an accelerating rate as the vapours are denuded of glycerol.

Objections to air condensers are the space required and the lack of possible control. However, with some designs, particularly the linear arrangement of cylinders, the condensates from the individual cylinders can be combined to suit the operator's requirements. For example, the best fraction might normally be taken from cylinders 1 – 5; but if a larger quantity of slightly lower quality is needed, a change might be made to cylinders 1 – 7.

Surface condensers using cooling media at relatively high temperatures were first used by Garrigue, US Patent 774 171 (1904); and condensers of this type have been, and are, used in many more recent stills. The main features of one version of the Garrigue still are shown in Fig. 8.23, discussed later. Miner and Dalton (1953) quotes the following temperatures for the vapours at various points in a Garrigue condensing system:

Vaporizer	→	Superheater	→	First condenser	→	Second condenser	→	Cold water condenser	→	Vacuum system
166 – 171°C				149 – 154°C		121 – 124°C		60°C		29°C

Figures for the condensates from a modern still working at 3 mm Hg without open steam, other than from water in the crude glycerine, are:

Vaporizer	→	First condenser	→	Second condenser	→	Cold water condenser	→	Vacuum system
		↓		↓		↓		
		130°C		75°C		15°C		

The actual temperatures needed at the various stages depend upon the distillation conditions and the quality requirements, but it will be evident that when two product condensers are used a cooling medium temperature above 100°C is normally required for the first condenser in order to achieve the appropriate cooling of the vapours with a modest temperature difference. In different designs this is obtained by:

- condensers in which heat is removed by water boiling at a pressure, and temperature, controlled by conditions in an auxiliary steam condenser.
- circulation of hot water under pressure.
- circulation of a cooling liquid, such as glycerine at a suitable concentration, with a boiling point higher than that of water.

For the second product condenser, hot water is normally adequate. A cold water condenser separates most of the remaining glycerol as an impure fraction.

Direct contact towers containing a mesh, or other low pressure drop packing can be used as condensers by the circulation of suitably cooled glycerine condensate as shown diagrammatically in Fig. 8.21. Advantages of this system are:

- the condenser also functions as a stripping column in which the quality of the distillate is improved by counter-current contact with the best available vapours. A separate stripping section for the glycerine leaving the unit can, if required, be installed below the condensing section round which condensate is circulated.
- it is cheaper than a large surface condenser.
- control is simple; and considerable flexibility can be achieved.

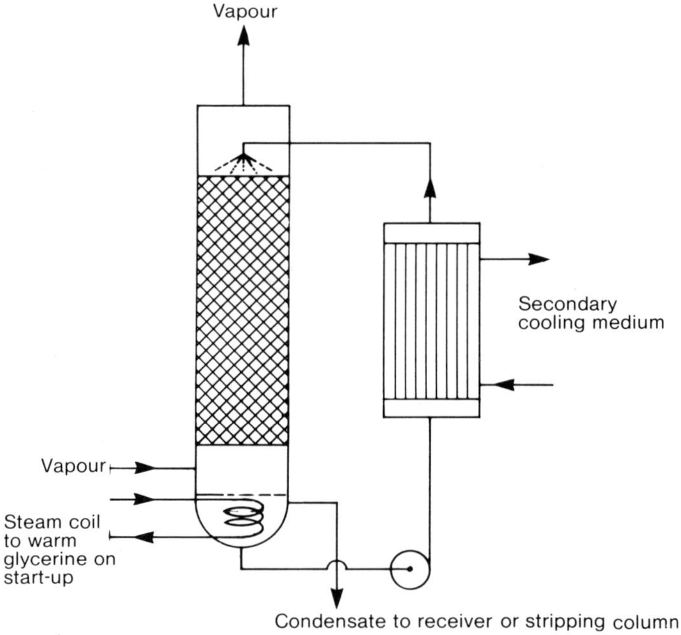

Fig. 8.21 – Direct contact condenser.

The problem is to achieve adequate contact between liquid and vapour without excessive pressure drop. This may make the system more suitable for use with vaporizers which do not require very low absolute pressures of 2 – 4 mm Hg, such as those operating at 10 – 15 mm Hg with a little open steam, and those with once through, scraped wall vaporizers.

Points in the design of a direct contact condenser include:

- the quantity of heat to be removed from the circulated glycerine is controlled by the kg/hr of vapour to be condensed. The higher the flow of circulated glycerine, the smaller the decrease in temperature required to achieve this heat removal. The extent to which the vapours need to be cooled is controlled by the vapour–liquid equilibrium discussed earlier.

The temperature of the circulated liquid, which rises as it passes down the condenser, must be sufficiently below that of the vapour at all points in order to induce the required heat flow, but should be kept as near as possible to that required for equilibrium.

- the diameter of the column must be selected so that the vapour velocity is sufficiently low to give an acceptable pressure drop with an adequate depth of packing; but also so that the liquid flow is enough to wet the packing properly without any risk of flooding.

8.6.7 Materials of construction for glycerine stills

The older stills, such as the Van Ruymbeke, were normally made of mild steel, with only a few parts, such as the perforated open steam distributor, and the valves, of more resistant metal, such as phosphor bronze. Some discolouration of distillate occurred. For the production of chemically pure grade by redistillation in mild steel equipment it is advisable whenever possible to use a separate still for the second distillation. Without the need to boil the foots from a distillation of crude glycerine with water, the condensing cylinders in the second still retain a film of concentrated glycerine which minimizes contamination.

Most modern stills use stainless steel, or aluminium, for all parts in contact with the vapours and distillates. Some designers consider it worthwhile also to make the vaporizer of stainless steel.

8.6.8 Complete stills

The various possible elements considered in previous sections can be selected and combined in many ways to form complete glycerine stills; and many designs have been used and described. Some plant manufacturers have used widely differing designs at various times, so that mention of a plant manufacturer does not necessarily indicate the nature of a still. In this section, only two stills of particular historical importance, plus some designs from recent manufacturers brochures are outlined. Some other early designs are described in Miner and Dalton (1953), and in its predecessor, Lawrie (1928).

Fig. 8.22 shows one form of *Van Ruymbeke still.* The vaporizer is a pot with closed and open steam coils as described in 8.6.4. Traditional stills did not include a scrubbing tower; and the weaks from the cold water condensers had to be evaporated to recover glycerol. The inclusion of a packed scrubbing tower, fed with a small quantity of cold water sufficient to reduce the temperature of the vapours leaving the tower to an appropriate level, recovers practically all the glycerol leaving the air cylinders, as explained in section 8.6.6; the weaks are then discarded to sewer.

Fig. 8.23 shows one version of the *Garrigue still* which was widely used in the USA and other countries from about 1904 onwards. The vaporizer is similar to that of the Van Ruymbeke still. Surface condensers are used to produce product by fractional condensation. Weaks, sometimes called sweet waters, are concentrated in a built-in single effect evaporator which provides the open steam for the main vaporizer. This open steam is superheated in a heat exchanger

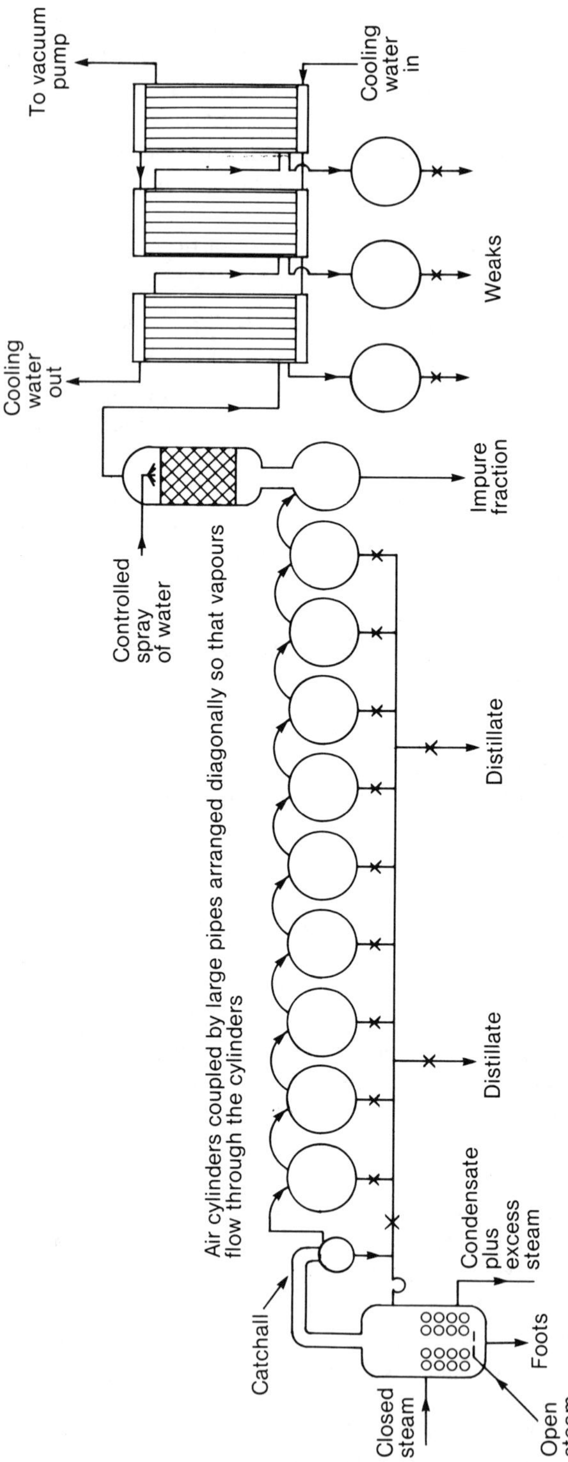

Fig. 8.22 – Van Ruymbeke still.

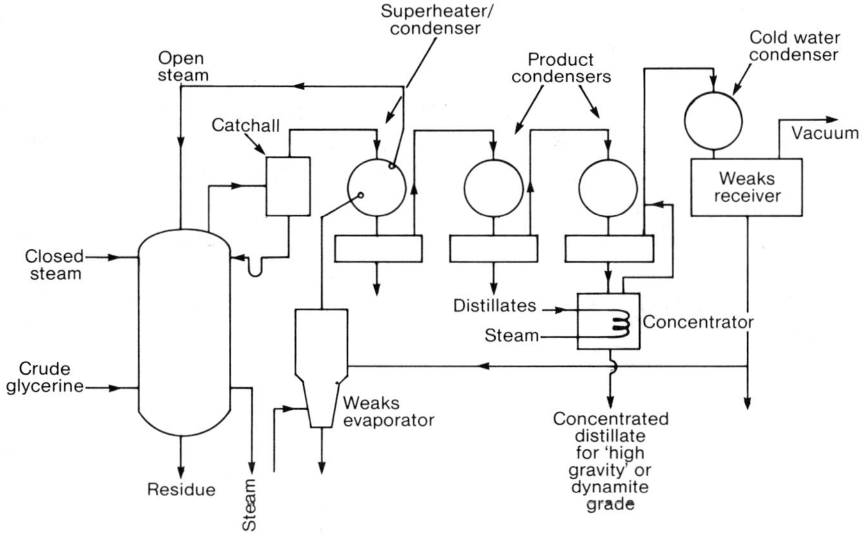

Fig. 8.23 – Garrigue still.

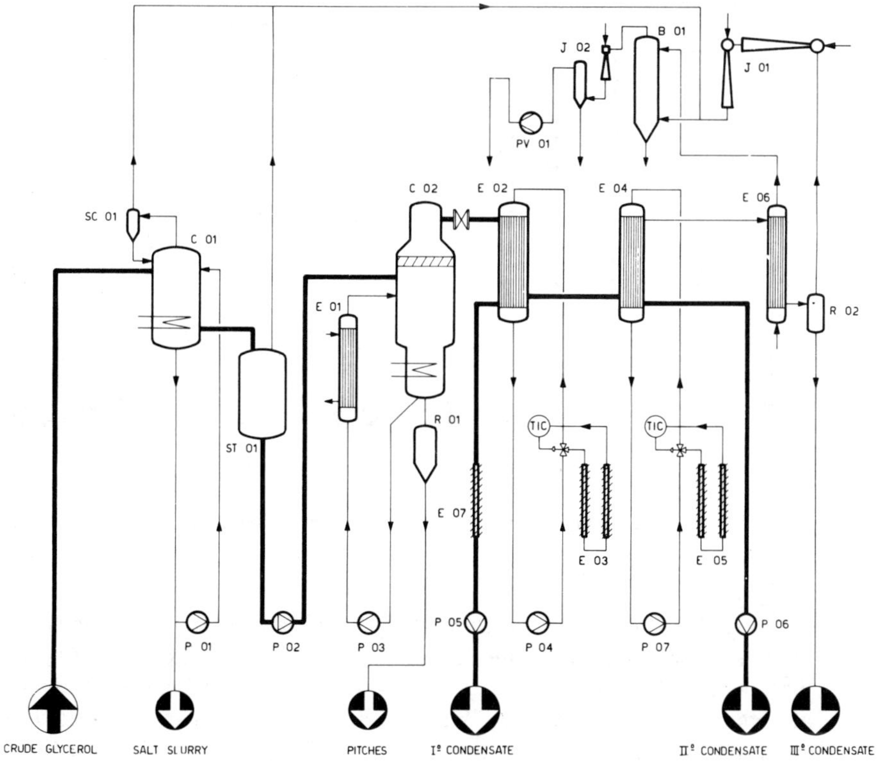

Fig. 8.24 – Gianazza type HPS glycerine still.

which also serves as the first product condenser. A steam heated concentrator is included to bring the third product fraction up to the strength required for dynamite ('high gravity') grade.

Fig. 8.24 shows the *Gianazza HPS still* which is very similar to the *Frazer Scott, Unilever* family of stills, except that a vessel is included in which the crude glycerine is dehydrated before it is fed to the forced circulation vaporizer. Salt slurry, which separates as the crude glycerine loses water in the dehydrator, is returned to the crude glycerine department. Condensation is in surface condensers which use a circulated fluid cooled in auxiliary coolers. The *Gianazza GDTF 'Turbotherm' still* is shown in Fig. 8.25. This includes a scraped-wall, once-through, vaporizer, as discussed in section 8.6.4; otherwise it is similar to the Gianzza type HPS.

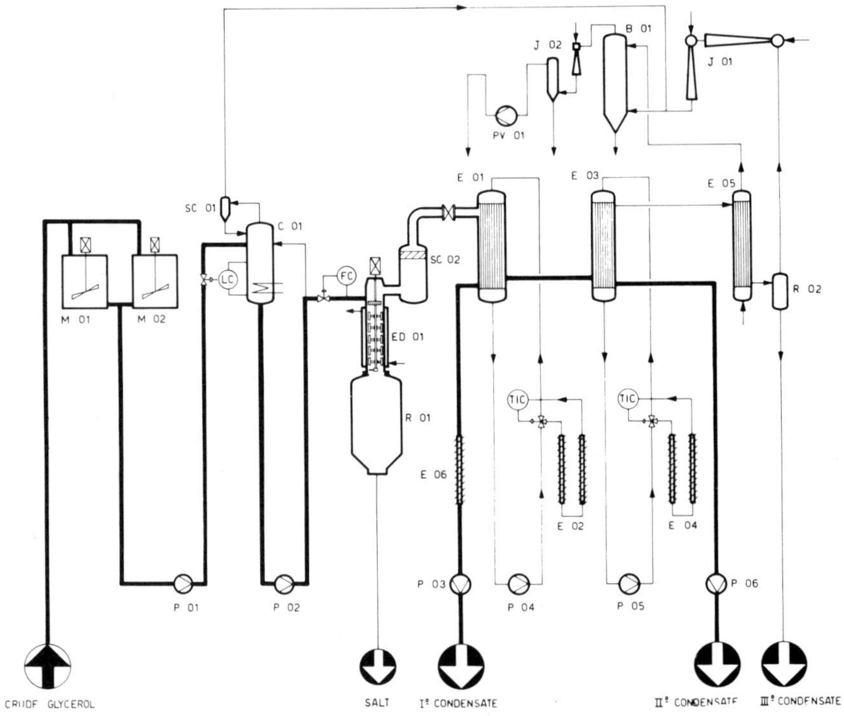

Fig. 8.25 – Gianazza type GDTF 'Turbotherm' glycerine still.

Fig. 8.26 shows a recent *Lurgi* design. Preheated crude glycerine is vaporized in a vessel with one, or more, legs, each equipped with closed heating elements over which the liquor rises by the vapour lift principle induced by open steam injected at the base of the leg. With this steam also acting as carrier fluid, vaporization is at about 10 mm Hg total pressure. This vaporizer is similar in principle to that used in Lurgi fatty acid stills. In some large units the multi-compartment vaporizer distils soap lye crude continuously, a liquid residue being withdrawn to be worked down in a separate foots still. Distillate from the foots still is redistilled. Other Lurgi stills of this type are operated on the semi-batch principle. The main condenser is of the direct contact type using glycerine cooled slightly in a heat exchanger in which it boils water. In addition to the stripping which occurs in the condenser, the product leaves through a separate stripping section of the tower in which it contacts the vapours from the vaporizer. Practically all the glycerol left in the vapours leaving the main condenser is separated as a 90 – 94% glycerol fraction in a cold glycerine direct contact condenser. This fraction, which needs to be reprocessed, amounts to 6 – 10% of the total glycerol fed. A section in which the glycerine is bleached with active carbon is also shown in the diagram.

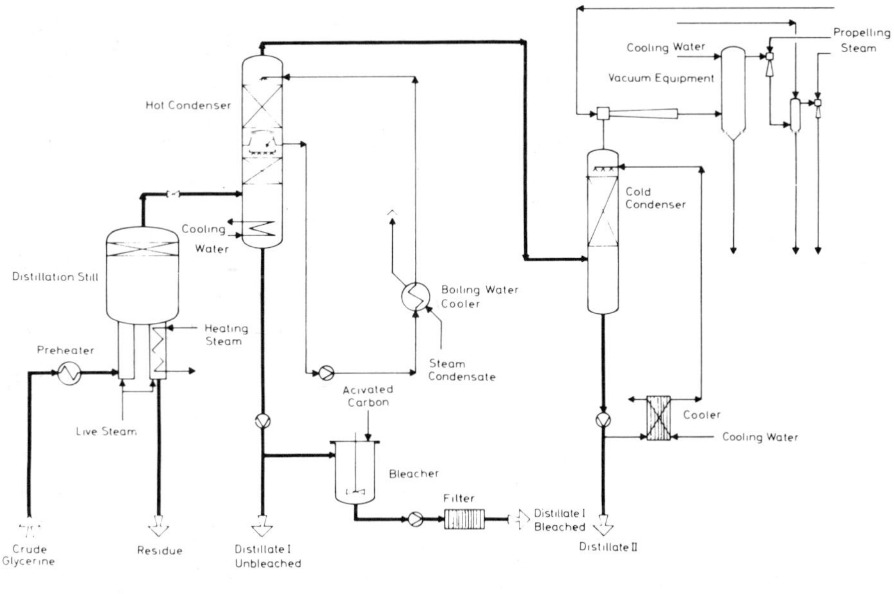

Fig. 8.26 – Lurgi glycerine still.

The latest *Wurster and Sanger still* about which information is available, is shown in Fig. 8.27. This rather complicated design comprises as its main components:

- a forced circulation heater with open steam supplied to the flash vessel.
- another circulating system in which thickened slurry is separated, adjusted in alkalinity by addition of hydrochloric acid, and, when soap lye crude is being processed, centrifuged to remove salt before the liquor is passed to a separate foots still. The foots still also uses an external forced circulation heater, as well as open steam, and is fitted with a salt-box type of residue collector.
- a direct contact main condenser, followed by a scavenging surface condenser.

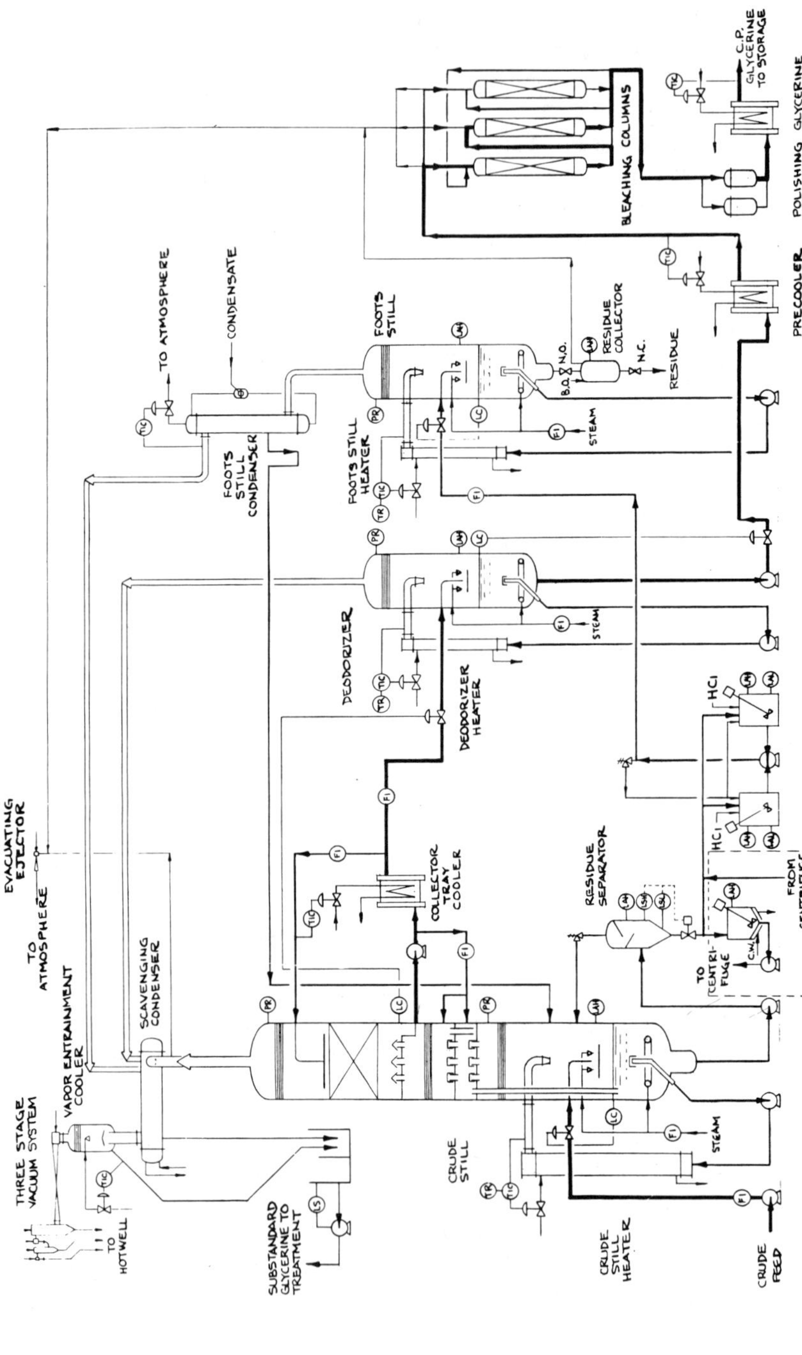

Fig. 8.27 – Wurster and Sanger glycerine still.

- a deodorizer/drier in which the glycerine condensate is heated and treated with open steam to remove water and other volatile impurities.
- a continuous bleaching unit.

8.7 BLEACHING AND THE TREATMENT OF DISTILLATES WITH ION-EXCHANGE RESINS TO UPGRADE QUALITY

Traditionally, distilled glycerines are bleached batchwise by an addition of 0.2 – 2.0% of activated carbon. The mixture at about 80°C is stirred, held for 1 – 2 hours, and filtered through a press. The carbon retains a high proportion of glycerol even after it has been steamed; and it may be worthwhile to slurry the used carbon with condensed water, and to re-filter, wash, steam and/or air blow in the press to recover weak glycerol. It is also possible to re-activate the washed carbon by stoving it in lidded crucibles in an oven (although to a lower level of activity than that of new activated carbon). This recovery process is not necessarily economic. Some bleaching can also be achieved by pumping distillates through the carbon cake in a filter press. As shown in Fig. 8.27, it is possible to bleach glycerines in continuous columns which contain beds of granular carbon.

A relatively recent process is to improve the quality of distillates by the use of ion-exchange resins instead of, or after, activated carbon. At one time it was thought necessary to dilute the distillate to about 80% glycerol, and to reconcentrate by evaporation of water after treatment with the resins, but it has since been found that useful purification can be achieved by the use of appropriate resins on undiluted distillates. Macrorecticular resins (which have large discrete pores capable of removing relatively large molecular mass compounds from the liquid) are normally best for this duty which demands the removal of organic, rather than ionic, impurities. Examples from the Rohm and Haas range are:

- Amberlite 200, or 200C – strong cation resin.
- Amberlite IRA-93 – weak base anion resin.
- Amberlite IRA-900 – strong base anion resin.

Deodorization, using heat and open steam, under vacuum, is often necessary after a distillate has been passed through beds of resins to remove odours imparted by the resins.

Selection of the best combination of resin beds, possibly following an activated carbon treatment, requires experiments, and/or consultation with a resin manufacturer.

8.8 PRODUCTION OF REFINED GRADES OF GLYCERINE WITHOUT DISTILLATION

8.8.1 Ion exclusion

The ion exclusion process for the partial purification of glycerine liquors was developed by Dow Chemical Co., and is described in a review paper by D'Souza

(1979). It is based on the fact that in a bed of granular resin (50 – 100 mesh) of a suitable type, such as Dowex 50 WX8 (an 8% cross-linked, medium porosity resin consisting of sulphonated styrene-di-vinyl benzene co-polymer), fed with an impure glycerine solution, there are three phases; the resin, the liquor within the resin granules (called the resin liquor), and the surrounding liquor (called the interstitial liquor). Most low molecular mass substances of nonionic, or slightly ionic, nature (such as glycerol) diffuse freely in and out of the resin liquor and the interstitial liquor. Ionic substances, such as sodium chloride and sodium sulphate, are repelled by the ionic charges of the resin, and exist at a much lower concentration in the resin liquor than in the interstitial liquor. Thus when a mixture of ionic and non-ionic solutes is passed through a resin column and is followed by a water rinse, ionized solutes are excluded from the resin granules and leave the column first. The nonionic material is held back and appears in the rinse water. By the use of properly balanced flows of impure glycerin and of rinse water a partial purification can be achieved, as shown in Fig. 8.28. Purification can be completed by ion-exchange, as discussed in the next section. It is said that US Pharmacopoeia grade can be produced from saponification crude glycerine without any distillation.

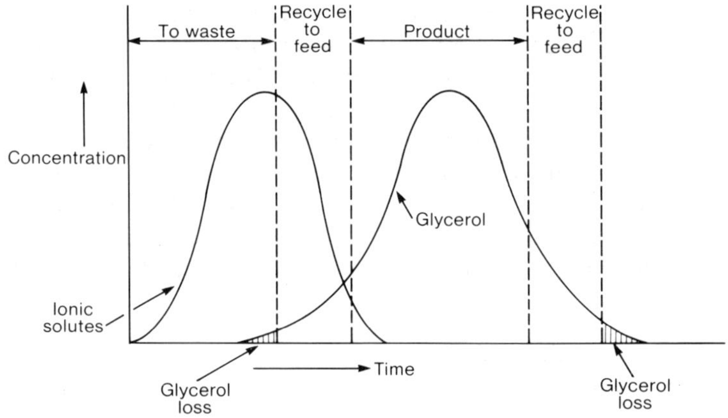

Fig. 8.28 – Partial purification of glycerine liquors by ion-exclusion.

8.8.2 Ion exchange

Glycerine liquors can be purified by the use of ion exchange resins in a process similar to that used for the de-mineralization of water. The cations, Na^+, Ca^{2+} and Mg^{2+} are exchanged for H^+ in a bed of cation resin (regenerated with hydrochloric or sulphuric acid); and the anions Cl^- and SO_4^{2-} (or their acids) are removed by an anion resin (regenerated with caustic soda). Organic impurities are also removed. Because of the quantities of resins and of regenerating chemicals needed, the process is impracticable for soap lyes, but it was at one time considered at least marginally economical for soap lye crude glycerine diluted to 25 – 30% glycerol (much of the salt having been removed in the evaporation of purified lyes to crude glycerine). The weak base anion resins are most economical for the removal of the strong acids, hydrochloric and sulphuric, but they need to be 'backed' by a strong base anion resin to remove weak organic acids.

In 1951, Lever Bros. Co. installed a plant in Los Angeles, USA, designed by Illinois Water Treatment Co., and described by Busby and Grosvenor (1952). The plant was designed to handle diluted soap lye crude glycerine. It comprised three pairs of cation–anion resin beds in sequence, followed by a mixed cation–anion resin bed to complete the purification. Evaporation of the purified liquor produced glycerine of chemically pure quality.

So far as is known, neither ion exchange or ion exclusion is widely used as an alternative to the distillation of glycerine.

8.8.3 Reverse osmosis using semi-permeable membranes

Experiments have shown that weak glycerol liquors can be concentrated some ten-fold by forcing water through a suitable cellulose acetate membrane under a pressure of 1500 psig (103 bar), with a recovery of 96% in the concentrate (Sourirajan and Kimura, 1967). This is an application of reverse osmosis in which preferential sorbtion of water by the membrane, as well as flow through capillary pores, is involved. Many papers on reverse osmosis, which was developed for the de-salination of sea water, include a review for chemical engineers by Sourirajan and Matsuura (1982). This use of reverse osmosis might, perhaps, provide a basis for the recovery of glycerol from liquors which would otherwise be too weak to be processed.

9

Sulphonation and sulphation

9.1 GENERAL

The main types of anionic surfactants are discussed in section 2.3.1. Most can be made by the sulphonation of suitable hydrocarbons, or the sulphation of various hydroxy compounds, using relatively simple processes which are commonly carried out in detergent factories. A few types, particularly the alkyl sulphonates, require more complex procedures, are less often produced in detergent factories, and their manufacture is not described here. As explained earlier, the alkylaryl sulphonates have the sulphur atom linked directly to a carbon atom in the benzene nucleus; that is they are of the general formula $R{-}C_6H_4{-}S(=O)_2{-}ONa$. The sulphates are the salts of half esters of sulphuric acid, $(O{=})_2S(OR)(ONa)$ An important practical difference is that the ester linkage, S–OR, is readily hydrolysed under acidic conditions, so that the acid ester, $(O{=})_2S(OR)(OH)$, formed in the sulphation process must be neutralized immediately. Otherwise, the sulphonation and sulphation processes are usually similar; and in many cases, the same plant can be used for either. To avoid tedious repetition, the term sulphonation is taken to include sulphation unless otherwise specified.

Several sulphonating reagents are, or have been, used, most of which are powerful chemicals with dehydrating and oxidizing properties, as well as sulphonating and acidic ones. This means that various side reactions, including deterioration in colour due to charring, are likely to occur, unless proper precautions are taken. These precautions include no attempt to carry reaction of the feed stock too near to completion, avoidance of local excess of reagent, too high a temperature (heat must be removed because of the exothermic nature of the reactions), and dilution of the reagent, particularly gaseous sulphur trioxide. Some raw materials deteriorate much more readily during sulphonation than others.

The main reactions for the more important cases can be summarized as follows:

reagent chlorsulphonic acid

Chlorsulphonic acid has been used to sulphate long chain alcohols. It is not suitable when the reactant contains an aromatic nucleus, because chlorination of the nucleus is likely to occur.

$R\cdot OH + Cl{-}SO_2{-}OH \rightarrow R{-}O{-}SO_2{-}OH + HCl.$

reagent sulphuric acid, or oleum (fuming sulphuric acid)

Liquid sulphuric acid, or oleum, is used for the sulphonation of alkyl aryl hydrocarbons $R{-}C_6H_4{-}H + H_2SO_4 = R\cdot C_6H_4{-}SO_3H + H_2O$.

reagent gaseous sulphur trioxide.

Sulphur trioxide, moderated by dilution with air, or otherwise, is used for all types of reactant, but some designs of reactor are not suitable for the more sensitive materials. The main reactions are:

- with *alkylaryl hydrocarbons* the main reaction is $R{-}C_6H_4{-}H + SO_3 \rightarrow R\cdot C_6H_4\cdot SO_3H$, but, particularly if an attempt is made to carry the sulphonation too far, other reactions can also occur:

$$\underset{}{2R\cdot C_6H_4\cdot H + 2SO_3 \rightarrow} \underset{\text{sulphonic anhydride}}{(R\cdot C_6H_4\cdot SO_2)_2O} + H_2O$$

$$2R\cdot C_6H_4\cdot H + SO_3 \rightarrow \underset{\text{sulphone}}{R\cdot C_6H_4\cdot SO_2{-}C_6H_4\cdot R} + H_2O$$

$$R\cdot C_6H_3\begin{matrix}/H\\ \backslash H\end{matrix} + 2SO_3 \rightarrow \underset{\text{disulphonic acid}}{R\cdot C_6H_3\begin{matrix}/SO_3H\\ \backslash SO_3H\end{matrix}}$$

- with *alcohols and other hydroxy compounds* the reaction is $R\cdot OH + SO_3 \rightarrow R\cdot O\cdot SO_3\cdot H$.

- with *long chain alpha-olefins* there are two main reactions which occur in roughly equal proportions:

$$R \cdot CH_2 \cdot CH{=}CH_2 + SO_3 \rightarrow R \cdot CH_2{-}CH{=}CH{-}SO_3H$$

alpha-olefin sulphonate

$$R \cdot CH_2 \cdot CH{=}CH_2 + SO_3 \rightarrow R{-}\underset{\displaystyle O\text{———}}{\overset{|}{CH}}{-}CH_2{-}\underset{\displaystyle SO_2}{\overset{|}{CH_2}}$$

alkyl sultone

Minor quantities of a disultone may also be formed. The sultones are hydrolysed with alkali under pressure in a special section of the plant.

$$R{-}\underset{\displaystyle O\text{———}}{\overset{|}{CH}}{-}CH_2{-}\underset{\displaystyle SO_2}{\overset{|}{CH_2}} + NaOH \rightarrow R{-}\underset{\displaystyle OH}{\overset{|}{CH}}{-}CH_2{-}CH_2{-}SO_3Na$$

alkyl hydroxy sulphonate

- with reagent *sulphamic acid* (amido sulphonic acid) the reaction with certain hydroxy compounds is

$$R{-}[O{-}CH_2{-}CH_2]_n{-}OH + HO{-}SO_2{-}NH_2 \rightarrow R \cdot [O \cdot CH_2{-}CH_2]_n{-}O \cdot SO_3NH_4$$

alkyl phenol ethylene oxide condensate — sulphamic acid — ammonium salt

Sulphamic acid has been used particularly with alkylphenol ethylene oxide condensates because it avoids substitution in the benzene nucleus which is liable to occur with sulphur trioxide or chlorsulphonic acid. The reaction gives the ammonium salt which has been an important ingredient in certain near-neutral liquid formulations. It is possible to add caustic soda and to drive off ammonia to obtain the sodium salt, but this seems inconvenient and generally unattractive.

Methods using gaseous sulphur trioxide are now dominant for most raw materials. The use of concentrated sulphuric acid has disadvantages discussed later, but it may still be useful for certain purposes because the process can be carried out in simple equipment using easily handled materials. The chlorsulphonic acid process was quite important for long chain alcohols, but the hydrochloric acid liberated causes corrosion and effluent recovery problems; and glass, or glass-lined, equipment is required.

It is sometimes convenient to sulphonate two types of material, such as an alkylaryl hydrocarbon and an alcohol, together; but some plant manufacturers consider it better to use two reactors in parallel. Chemithon suggests that a superior product can be made in this way; the alcohol is under-reacted and then mixed with fully reacted sulphonic acid. The latter contains dissolved sulphur trioxide and sulphonic anhydride which complete the sulphation. When a hydrotrope is required for use with an anionic surfactant, it is sometimes possible to sulphonate a suitable, relatively small, proportion of toluene, or xylene, together with a long side chain detergent alkylate. Care must be taken to avoid problems due to the volatility of the toluene, or xylene; and Chemithon suggests the addition of a little sulphone inhibitor, such as acetic anhydride.

9.2 SOME PHYSICAL PROPERTIES OF THE SULPHUR TRIOXIDE/ WATER SYSTEM

Sulphur trioxide is the anhydride of sulphuric acid; but it also forms a number of other hydrates which exist only at low temperatures. These hydrates and the eutectics between them result in the complex freezing point/composition curve shown approximately in the lower part of Fig. 9.1. Oleum, or fuming sulphuric acid, comprises H_2SO_4 plus additional, or free, SO_3. A 40% oleum contains 60% H_2SO_4 plus 40% additional SO_3. Commercial oleums commonly contain about 20% additional SO_3, or about 60 – 65% additional SO_3, because the low freezing points at these concentrations facilitate transportation in liquid form.

The upper part of Fig. 9.1 shows approximately how the boiling point at atmospheric pressure varies with composition. More precise information concerning boiling points at about 755 mmHg over part of the range is:

$\%H_2SO_4$	61.7	81.5	96.2	98.5	99.9	–	–	–
$H_2SO_4 + \%SO_3$	–	–	–	–	–	3.6	26.2	63.2
Boiling point °C	140	202	292	317	273	212	125	60

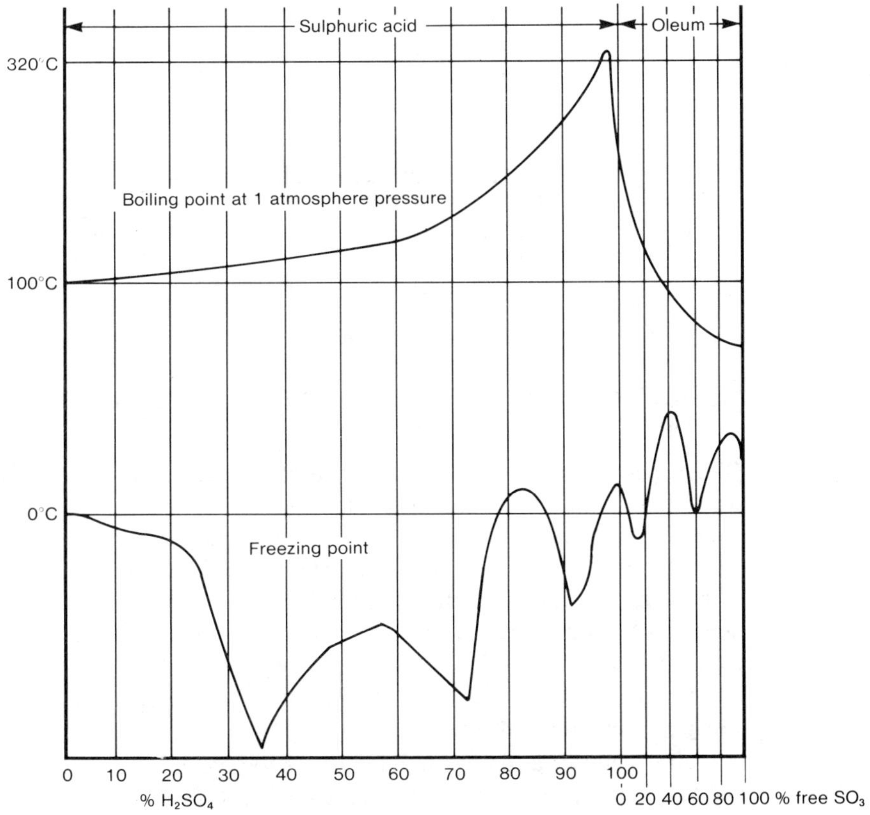

Fig. 9.1 – Freezing and boiling points of sulphuric acids and oleums.

The sharp maximum represents a constant boiling mixture, or azeotrope. The composition and boiling point of the azeotrope vary somewhat with pressure, but at atmospheric pressure the composition is about 98.3% H_2SO_4. Fig. 9.2 shows diagrammatically the vapour-liquid composition curves for the sulphur trioxide/water system. The azeotrope A behaves as a separate component. The curves divide into two parts; the right-hand side represents the distillation behaviour of azeotrope and sulphur trioxide, and the left-hand side that of azeotrope and water. The azeotrope itself boils without change in composition; hence the name constant boiling mixture. As discussed in section 9.4.3, these matters are important when oleum is used as a source of gaseous sulphur trioxide.

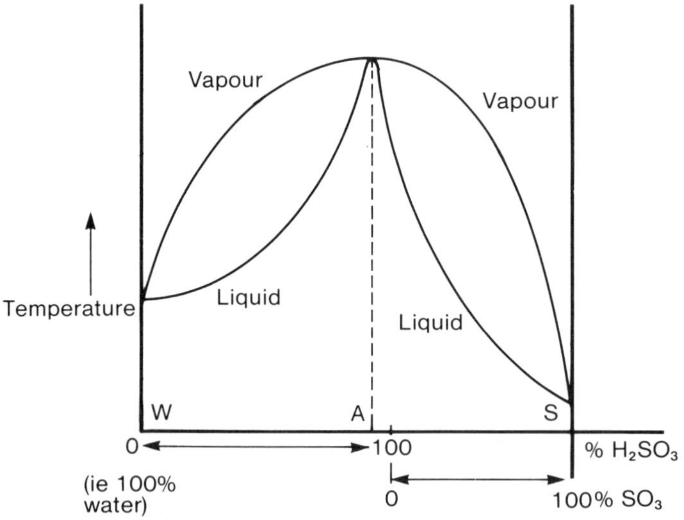

Fig. 9.2 – Vapour/liquid compositions for the sulphur trioxide/water system.

9.3 SULPHONATION WITH CONCENTRATED SULPHURIC ACID OR OLEUM

Oleum and concentrated sulphuric acid can be handled in mild steel, but more dilute sulphuric acid cannot. For sulphonators using liquid acid, more resistant materials, such as a suitable stainless steel or glass-lined steel, are usually needed. This process is often carried out batchwise, using a mixer to bring the hydrocarbon and the acid into intimate contact. Efficient cooling is important; and for larger units an external cooler with pump circulation is best. Continuous plants for sulphonation with liquid acid have been devised, but they are not now very important in view of the dominance under most circumstances of processes using gaseous sulphur trioxide.

60% oleum is a fuming liquid which is too harsh to be used in sulphonation; and the liquid phase process is carried out with oleum containing 20%, or less, free SO_3, or with 98% sulphuric acid. As explained in section 9.1, water is produced in the reaction; and it is necessary to use a large excess of acid to maintain a sufficiently high acid concentration to carry the reaction near enough to completion. Factors which affect the degree of conversion achieved are excess of acid (substantially more being required with 98% sulphuric acid than with oleum), time and temperature; and to some extent they are alternatives. It is often better to accept a conversion which leaves about 1.2 – 1.5% unsulphonated material calculated on 100% surfactant, rather than to use more drastic conditions to obtain another 0.5% conversion with risk of poorer colour and odour. Cooling is necessary to keep the temperature of the reaction mix below about 50°C, although, when sulphonation is with 98% H_2SO_4, it may be advisable to warm the alkylate to about 40°C to ensure smooth initiation of the reaction. Because of the excess of acid which has to be used, direct neutralization of the reaction mix with caustic soda gives a paste with a large ratio of sodium sulphate to sodium sulphonate. Particularly for use in liquid detergent formulations this is not acceptable; and it is usual to add some 10% of water to the acid reaction mix and to allow the blend to settle for some hours. Care must be taken to avoid the development of too high a temperature as the water is added. A lower layer of black sulphuric acid at about 75 – 80% H_2SO_4 and an upper layer of sulphonic acid which contains only 5 – 6% H_2SO_4 separate. If a specially low content of sodium sulphate in the paste is needed for liquid detergent formulations, Davidsohn and Milwidsky (1978) describe and recommend a two stage separation process. Disposal of the spent acid may present problems. Sometimes it is neutralized with caustic soda, or sodium carbonate, to produce a sodium sulphate which is dirty, but may be usable in some NSD powder slurries.

Some points in relation to the sulphuric acid and oleum processes can be illustrated by some approximate calculations for two examples; in the first concentrated sulphuric acid is used, in the second 20% oleum. The reaction is assumed 100% complete, and possible side reactions are neglected. It is also assumed that no sulphonic acid passes into the spent acid and that the sulphuric acid left in the sulphonic acid has the same strength as the spent acid.

Example 1

Sulphonation of alkyl aryl hydrocarbon of mean molecular mass 246 (theoretical for $C_{12}H_{25}\cdot C_6H_5$) with 98% H_2SO_4.

$$\underset{246}{C_{12}H_{25}\text{—}C_6H_5} + \underset{98}{H_2SO_4} \rightarrow \underset{326}{R\cdot SO_3H} + \underset{18}{H_2O}$$

1 part of alkylate requires 98/246 = 0.398 parts 100% H_2SO_4 = 0.406 parts 98% H_2SO_4, and water produced = 18/246 = 0.073 parts.

Actually about 1.6 parts of 98% H_2SO_4 (which contain 1.57 parts H_2SO_4 + 0.03 parts water) are used. Sulphonic acid produced = 326/246 = 1.325 parts. The composition of the reaction mix is thus:

Sulphonic acid	1.325 parts	
Unused H_2SO_4 (100%) = 1.57–0.398	= 1.172 parts	91.9%
Water = 0.03 + 0.073	= 0.103 parts	H_2SO_4

If this mix is neutralized with caustic soda

$$\underset{326}{R\cdot SO_3H} + \underset{40}{NaOH} \rightarrow \underset{348}{R\cdot SO_3Na} + \underset{18}{H_2O}$$

and

$$\underset{98}{H_2SO_4} + \underset{80}{2NaOH} \rightarrow \underset{142}{Na_2SO_4} + \underset{36}{2H_2O}$$

so that the neutralized mix contains 1.325 x 348/326 = 1.413 parts sodium sulphonate plus 1.172 x 142/98 = 1.697 parts sodium sulphate. The water content of the paste equals water in the sulphonic acid mix, plus water produced in the neutralizations, plus water in the caustic soda solution, minus water evaporated during neutralization. Considerable heat is evolved so that careful neutralization is necessary.

An addition of 10% water to the acid reaction mix increases the total from 2.600 to 2.860 parts. It contains 0.363 parts water and 1.172 parts H_2SO_4, equivalent to a sulphuric acid concentration of 76.4%. The settled sulphonic acid layer contains 5 – 6, say 5.5, % H_2SO_4, equivalent to 5.5/0.764 = 7.2% of 76.4% sulphuric acid. Hence the sulphonic acid layer from 1 part of alkylate contains

sulphonic acid	1.325 parts
76.4% sulphuric acid 1.325 x 7.2/92.8 =	0.103 parts
(that is 0.079 parts H_2SO_4 plus 0.024 parts water)	
	1.428

The spent acid layer = 2.860 – 1.428 = 1.432 parts, containing 1.094 parts H_2SO_4 plus 0.338 parts water. After neutralization with caustic soda solution, the sulphonate paste contains 1.413 parts sodium sulphonate, as before, plus 0.079 x 142/98 = 0.144 parts of sodium sulphate. If neutralized the spent acid layer produces 1.094 x 142/98 = 1.585 parts sodium sulphate. The results are consolidated in Table 9.1; some minor discrepancies in the balances, due to rounding of figures, may be noted in this Table and in Table 9.2.

Example 2. Sulphonation of the same alkylate with 20% oleum

1 part of alkylate produces 1.325 parts of sulphonic acid, as before, and requires 80/246 = 0.325 parts SO_3. 20% oleum contains 80% H_2SO_4 + 20% free SO_3, that is

(0.8 x 80/98) + 0.2	= 0.853 parts SO_3
0.8 x 18/98	= 0.147 parts H_2O
	1.000

Theoretical requirement of oleum = 0.325/0.853 = 0.381 parts. In practice,

Table 9.1
Sulphonation alkylate (MM 246) with 98% sulphuric acid (all figures are parts mass, except where percentage is indicated)

	Materials used		Reaction mix	Dilution and spent acid separation			After neutralization with caustic soda solution		
				Water added	Sulphonic acid layer	Spent acid layer	Whole reaction mix	Sulphonic acid layer	Spent acid layer
Alkylate	1.00	—	—	—	—	—	—	—	—
98% H_2SO_4	1.60	—	—	—	—	—	—	—	—
100% H_2SO_4	—	1.57	1.17	—	0.08	1.09	—	—	—
H_2O	—	0.03	0.10	0.26	0.02	0.34	*	*	—
Sulphonic acid	—	—	1.33	—	1.33	—	—	—	—
Sodium sulphonate	—	—	—	—	—	—	1.41	1.41	—
Sodium sulphate	—	—	—	—	—	—	1.70	0.11	1.59
								1.70	
Total	2.60	—	2.60	0.26	1.43	1.43	—	—	—
			2.86		2.86				
H_2SO_4 as % of $H_2SO_4 + H_2O$		98.0	92.1	—	76.4	76.2	—	—	—
Ratio $\frac{Na_2SO_4}{\text{Na sulphonate}}$	—	—	—	—	—	—	1.21	0.08	—
% active detergent (AD) in paste	—	—	—	—	—	—	*	*	—

*These figures depend upon the concentration of caustic soda solution used in the neutralizations.

Table 9.2
Sulphonation of alkylate (MM 246) with 20% oleum

	Materials used	Reaction mix		Dilution and spent acid separation			After neutralization with caustic soda solution		
				Water added	Sulphonic acid layer	Spent acid layer	Whole reaction mix	Sulphonic acid layer	Spent acid layer
Alkylate	1.00	—	—	—					
20% oleum	1.10	—	—	—					
100% H_2SO_4	—	—	0.75	—	0.08	0.67			
SO_3	—	0.61	—	—	—	—			
H_2O	—	0.16	0.02	0.21	0.03	0.20	*	*	—
Sulphonic acid	—	1.33	1.33	—	1.33	—	—	—	—
Sodium sulphonate	—	—	—	—	—	—	1.41	1.41	—
Sodium sulphate	—	—	—	—	—	—	1.09	0.11	0.97
								1.08	
Total	2.10	2.10	2.10	0.21	1.44	0.87	—	—	—
			2.31		2.31				
H_2SO_4 as % of $H_2SO_4 + H_2O$	—	—	96.9	—	—	—	77.0	—	-
Ratio $\frac{Na_2SO_4}{\text{Na sulphonate}}$	—	—	—	—	—	—	0.77	0.08	—
% AD in paste	—	—	—	—	—	—	*	*	—

*These figures depend upon the concentration of caustic soda solution used in the neutralizations.

about 1.1 parts oleum are used, which contain 0.938 parts SO_3 plus 0.162 parts H_2O. The water of reaction (which comes from H_2SO_4 but not from SO_3) equals 0.073 x 0.653/0.853 = 0.056 parts (see formation of water in Example 1). Hence, the composition of the acid reaction mix is

sulphonic acid	= 1.325 parts
SO_3 = 0.938 − 0.325	= 0.613 parts
H_2O	= 0.162† parts.
	2.100

†no allowance is needed for water of reaction as the composition of the oleum is taken in terms of SO_3 and H_2O.

Alternatively, the composition of the reaction mix is

sulphonic acid	= 1.325 parts
H_2SO_4 = 0.613 x 98/80	= 0.751 parts
additional water = 0.162 − (0.613 x 18/80)	= 0.024 parts
	2.100

The concentration of the sulphuric acid in the mix is 0.751/0.775 x 100 = 96.9%, rather higher than the 91.9% found in Example 1. It would seem that, other things equal, it should be possible to achieve the same final concentration of H_2SO_4 in the mix. The difference found may be due to the fact that the ratios of concentrated sulphuric acid, and of oleum, to alkylate assumed in these calculations are approximate only and subject to some variation. The rest of the calculation is similar to that in example 1; and the results are summarized in Table 9.2.

9.4 SULPHONATION WITH SULPHUR TRIOXIDE

9.4.1 General

Sulphur trioxide sulphonates organic materials very rapidly, but, because of its powerful dehydrating and oxidizing properties, it is usually used in a diluted form. Processes have been developed in which sulphur trioxide is used at low temperatures dissolved in liquid sulphur dioxide, and in which its partial pressure is reduced by the use of vacuum, but most use a mixture of sulphur trioxide and air. As described in section 9.4.4, Albright and Wilson have recently developed and patented a process using undiluted sulphur trioxide gas.

Processes using mixtures of sulphur trioxide and air require the following elements:

- equipment to move and dry the air.

- equipment for the generation of the sulphur trioxide gas mixed as required with air.
- a sulphonation reactor.
- equipment for the after treatment of the reaction mix, which may include:
    - ageing.
    - hydrolysis of anhydrides and sulphones.
    - neutralization.
    - hydrolysis of sultones when alpha-olefins are being processed.
    - bleaching.
- equipment in which the effluent air is purified to a standard which is acceptable with regard to possible nuisance.

A low concentration of 4 – 8% SO_3, by volume, in air is often used, the level depending upon the reactor design and the material to be sulphonated; some particularly sensitive materials may require even lower concentrations down to 2%. A large volume of air is, therefore, needed, which has an important influence upon both capital and operating costs. About twice as much air is required with a concentration of 4% as when it is 8%; in some processes air can be recycled which reduces the cost of drying but not of moving the air.

9.4.2 Air driers

It is important to dry the air thoroughly to avoid formation of acid mist when the sulphur trioxide is contacted. This is commonly carried out by compression, cooling, and removal of condensed water; followed by passage through beds of desiccant such as silica gel and/or molecular sieves. Two vessels with beds of desiccant are normally provided in parallel, so that one is in use whilst the other is regenerated with hot air. When the sulphur trioxide is produced from sulphur, surplus heat from the converter is commonly used to dry the desiccant. Fig. 9.3 shows diagrammatically a typical air drying system.

When the sulphur trioxide is generated from oleum, it may sometimes be acceptable and convenient to dry the air by passage up a packed tower down which flows some of the stripped oleum (see later).

9.4.3 Generation of sulphur trioxide gas

Three methods are commonly used to generate sulphur trioxide for sulphonation:

- vaporization of stabilized liquid sulphur trioxide.
- distillation of 60 – 65% oleum.
- burning of sulphur to give a mixture of sulphur dioxide and air (strictly air which has been denuded of some of its oxygen), followed by catalytic conversion of the sulphur dioxide to sulphur trioxide as in a normal sulphuric acid plant.

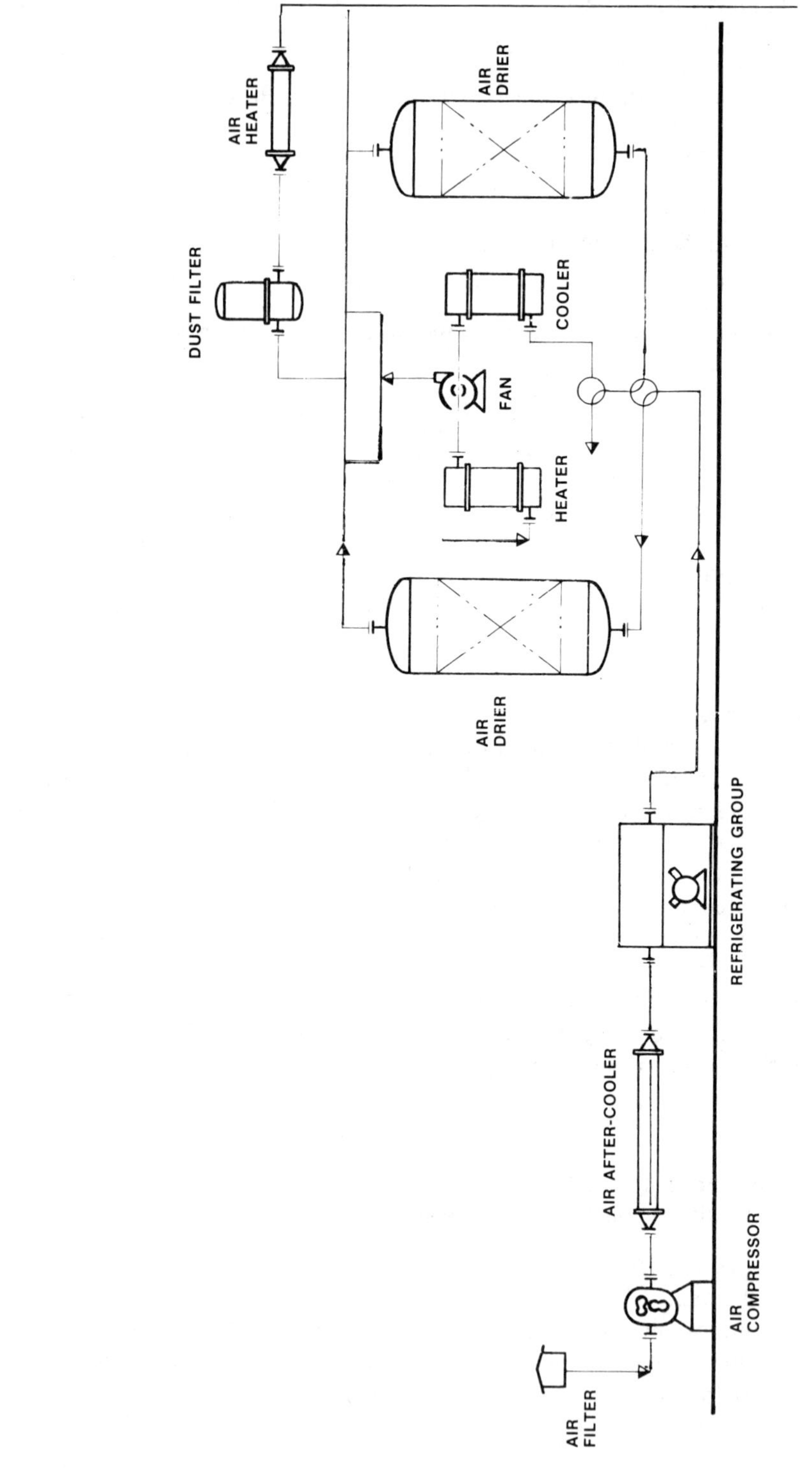

Fig. 9.3 – System to dry air for sulphonation (courtesy of Meccaniche Moderne – Allied Chemical Corporation).

The system adopted needs to provide a means for the accurate control of the rate at which sulphur trioxide is fed to the reactor, as well to produce a gas of the required composition. Sulphur burning is now much the most usual method. Points of comparision between the methods of generation are:

- capital cost is highest for sulphur burning. Meccaniche Moderne/Allied Chemical Corporation say that oleum stripping comes next, with liquid sulphur trioxide the cheapest; but, as mentioned later, very simple oleum stripping units can be used under some circumstances.
- sulphur is relatively cheap and is easily handled. Stablized liquid sulphur trioxide is normally the most expensive raw material. When oleum is used it is necessary to find a use for the stripped acid. In some cases, it is possible to return it to the oleum manufacturer, so that the sulphuric acid/weak oleum serves essentially as a means for the transport of the sulphur trioxide. It may sometimes be possible to buy 60–65% oleum, strip sulphur trioxide for use as a sulphonating gas, and to utilize the residual oleum at about 20% as a sulphonating liquid.
- in general, overall costs for the use of sulphur amply justify the higher initial cost of the plant.

Plant manufacturers can normally supply whichever system a customer requires. Chemithon mentions the possibilities of using either liquid sulphur dioxide with a converter to oxidize it to sulphur trioxide, or of taking converter gas from a standard sulphuric acid plant, but it is not thought that these methods are often likely to be attractive. Proper control is the problem if gas is taken from a large sulphuric acid plant.

Unstabilized liquid sulphur trioxide can be used when the sulphonation plant is adjacent to a sulphuric acid plant from which liquid sulphur trioxide can be obtained, so that minimal storage is necessary. Because sulphur trioxide can convert to its alpha and beta forms which are solids at normal temperatures, it is usually necessary to buy the liquid in the gamma form stabilized with boron trifluoride or carbon tetrachloride. Such liquid is sold under various trade names. Fig. 9.4, courtesy Chemithon Corporation, shows a unit for the production of sulphonating gas from liquid sulphur trioxide. This diagram shows the use of air recycled from the sulphonator after it has been cleaned by an electrostatic precipitator, but this, and similar systems, can also be used with fresh dried air. The metering pump controls the rate of flow of sulphur trioxide to the sulphonator.

From the vapour/liquid equilibrium diagram shown in Fig. 9.2 it will be seen that concentrated oleum, usually 60 – 65%, can, in theory, be fractionally distilled to give sulphur trioxide and the azeotrope 98% sulphuric acid; and the large difference in the boiling points of the constituents indicates that the separation is relatively easy. This would require a high temperature in the reboiler, the boiling point of the azeotrope under atmospheric pressure being about 320°C, and an adequate fractionating column to produce sulphur trioxide sufficiently free from sulphuric acid. The significance of the sulphuric acid content of the gas varies. It should be low for some sulphonations, for example when the surfactant is to be used in liquid products, but, as described later, a

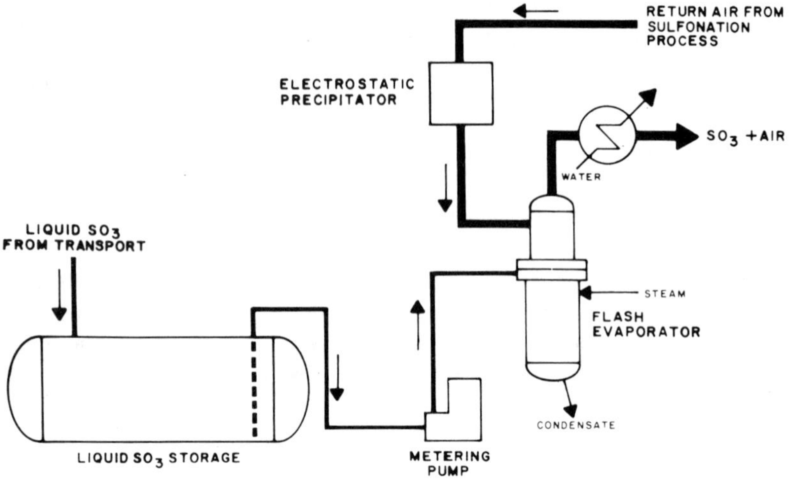

SO_3 Gas From Liquid SO_3

Fig. 9.4 – Vaporizer for liquid sulphur trioxide.

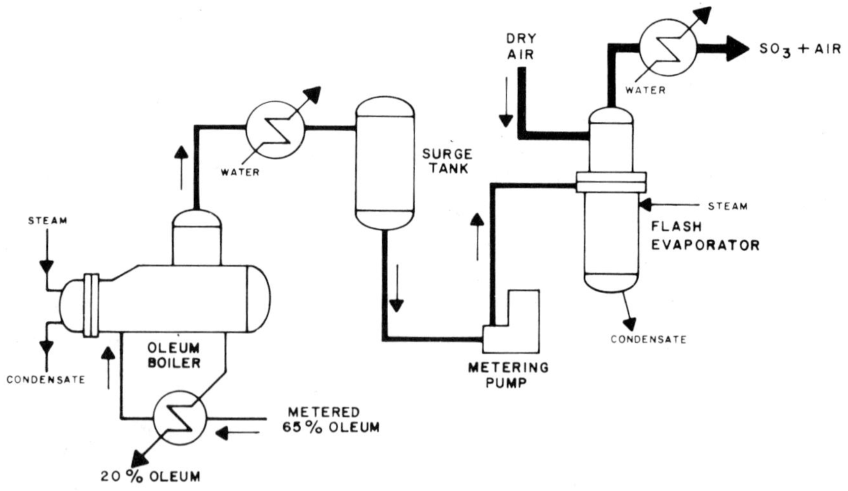

Fig. 9.5 – Equipment for production of sulphur trioxide gas from oleum.

proportion of sulphuric acid is sometimes added deliberately when alkyl aryl sulphonate is being produced for powder manufacture. It is common practice to use steam to heat the oleum and to discharge the stripped acid at about 20% oleum strength. If the sulphur trioxide fed to the reactor is controlled directly by the rate at which the 60 – 65% oleum is fed to the still, it is important to ensure that the boiling temperature of the discharged acid, and hence its composition, is maintained at a predetermined constant value. It is to be noted that at its boiling point 20% oleum is not a specific entity, but just one composition in a series of solutions of sulphur trioxide in sulphuric acid. Some plant manufacturers prefer to condense the sulphur trioxide and to meter it to a vaporizer, as shown in Fig. 9.5, courtesy Chemithon Corporation; that is to use a procedure similar to that employed when liquid sulphur trioxide is purchased. With this system the quantity of gas stripped from the oleum has only to be controlled to maintain an acceptable level in the liquid sulphur trioxide surge tank.

For many years, a simple type of oleum still and air drier, as shown in Fig. 9.6, was used with batch, and then with cascade continuous sulphonators, sulphonating alkyl aryl hydrocarbons. 60% oleum was fed to a horizontal cast-iron cylinder, heated by gas flames, through a rotary valve. The still was kept half-full of hot acid; and the heat input was arranged to give a constant exit temperature of about 250°C. The sulphur trioxide gas left the top of the still through a ribbed atmospheric condenser/gas cooler from which any liquid reflux returned to the still. Part of the stripped acid was used to dry the air in a tower which contained ceramic packing. The combined spent acid was returned to the oleum manufacturer. This system met well the needs of the plants in which it was used, but it might not provide the accuracy of metering and dryness of gas required by some more modern plants and materials to be sulphonated.

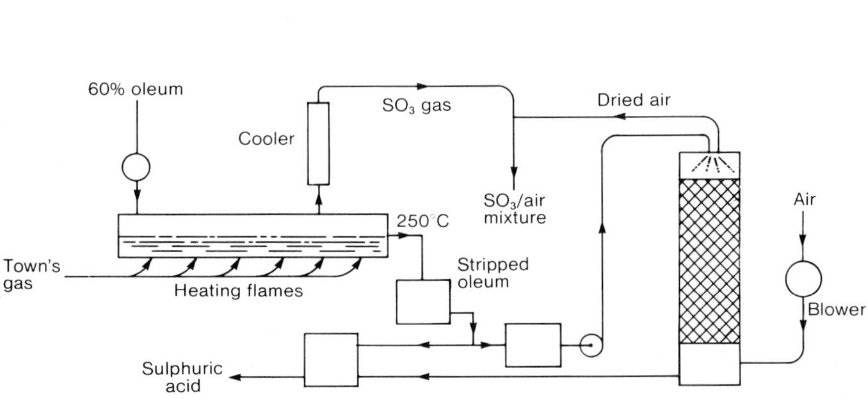

Fig. 9.6 – Simple system to strip oleum and to dry air with sulphuric acid.

Plants which use sulphur as their raw material include what is a small sulphuric acid making plant. A unit in which sulphur trioxide is absorbed in concentrated sulphuric acid, and subsequently diluted with water to produce additional acid, is normally included, so that the sulphur burner and converter can operate steadily, whether or not gas is required by the sulphonator. Some sulphuric acid is inevitably made in this way, and more can be produced, if required, by operating the sulphuric acid section of the plant for additional hours. It is not normally advisable to increase the capacity of the sulphuric acid part of a plant, so as to produce sulphuric acid and sulphonic acid at the same time, because the delivery of sulphur trioxide to the sulphonator is normally controlled by the feed of liquid sulphur to the burner. A typical plant is shown diagrammatically in Fig. 9.7, but this diagram does not include the absorption tower for surplus SO_3. The converter shown includes three beds of vanadium pentoxide catalyst; some other plants use four. The SO_2–SO_3– air gas is cooled between stages and the cooling air is used to regenerate the desiccant in the air drying section of the plant. Surplus hot air may also be used in other plants such as a spray drier. Steam may be generated by the hot gas from the sulphur furnace.

The converter achieves about 98% conversion and the final gas contains about 7% SO_3, which is convenient for most sulphonation processes, after dilution with more air if necessary.

9.4.4 Sulphonation reactors

Many widely differing types of reactor have been developed. Desirable features are:

- exposure of the liquid to be sulphonated to the sulphur trioxide for as short a time as possible.
- removal of the large heat of reaction without the development in the mix of either too high or too low a temperature; liable respectively to cause undesirable side reactions, or an undue increase in viscosity.
- accurate proportionation of sulphur trioxide and organic liquid to be sulphonated, without risk of local excess of sulphur trioxide at any point. A local excess damages the material; and in an extreme case can produce a coke which has to be removed from the plant.

Early reactors comprised vessels which contained cooling coils and held a quantity of liquid to be sulphonated. They included a mixing device to disperse the gas in the liquid, as well as to circulate the liquid over the cooling coils and to keep it uniform in composition. Turbo mixers were found particularly suitable. Around 1950 *Unilever*, used single vessels of this type with an oleum stripper and air drier as shown in Fig. 9.6 for the batch sulphonation of alkylaryl hydrocarbons. *A. Davidsohn* reports an early batch plant in which the SO_3/air mixture was produced from sulphur (Davidsohn and Milwidsky, 1978); and other detergent manufacturers doubtless developed plants. In the early 1950s an important further development by *Unilever* was the design and construction of continuous units comprising a cascade of five vessels through which the liquid

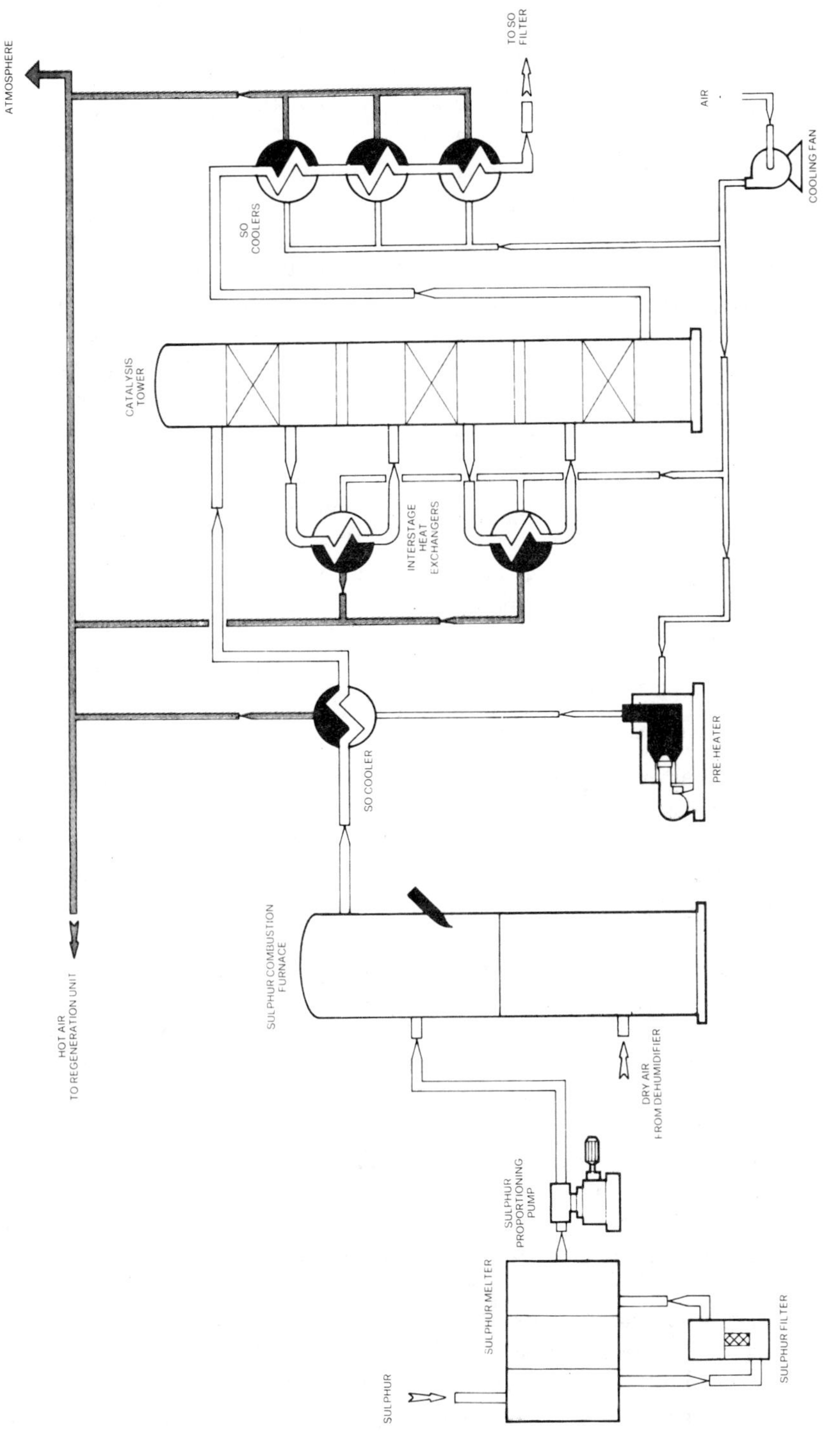

Fig. 9.7 – Production of sulphur trioxide from sulphur (courtesy of Ballestra SpA).

flowed in series. Each sulphonation vessel was fed with gas from a separate oleum still, apart from the fourth and fifth stages which were fed from a single still. This enabled the bulk of the reaction to be carried out with relatively concentrated gas (up to 15 – 20% SO_3) in the early stages, because there is little risk of damage whilst a substantial excess of unreacted hydrocarbon is present. The quantity and concentration of sulphur trioxide was reduced from stage to stage, as the reaction proceeded and the viscosity of the mix increased with adverse affect upon heat transfer. Only a small amount of sulphonation was carried out in the final stage to facilitate control of the end point, and to minimize damage with adverse affect upon the colour of the product. Initially, it was thought that a stepped stator close over the turbo-impeller would be beneficial in causing intense turbulence as the gas was introduced through a perforated ring of pipe fitted to the inner part of the annular stator. Although this embellishment was used for a number of years, later experience showed that it was a mistake because:

- the heat generated by the power input added very considerably to the total quantity of heat to be removed. It was found better to remove the stator and to induce a greater flow of air with a lower power consumption. The sulphonation reaction is rapid, so that additional turbulence and shearing is unnecessary.
- the holes through which the gas was introduced were liable to become blocked with coke. A simple gas inlet pipe near to the eye of the impeller is adequate and better.

When making alkylbenzene sulphonate for use in powders, in which some sodium sulphate is required, it was found useful to feed a 'heel' of stripped oleum (say 10% of the alkylate) into the first vessel with the alkylate. This is thought to moderate the chemical activity of the sulphur trioxide and to reduce the viscosity of the sulphonic acid mix.

Around 1960, *Ballestra* patented the *'Sulphurex' process* which used a cascade of 2–5 vessels, generally similar to those just described, in conjunction with equipment for the generation of sulphur trioxide from sulphur. Sulphurex plants have been widely used throughout the world; and a modern version of the reactors and ageing vessels is shown diagrammatically in Fig. 9.8. This shows a plant with four sulphonation stages followed by an ageing vessel and stabilizer, the functions of which are discussed in section 9.4.5. Fig. 9.9 (courtesy of Soap, Cosmetics and Chemical Specialities) shows the construction of a reactor in rather more detail.

Whilst reactors of the stirred tank type just described are widely used for the sulphonation of alkylaryl hydrocarbons, more delicate materials, and the production of alkylaryl sulphonates of good colour with little, or no, bleaching, usually require reactors which approach more nearly the criteria mentioned at the beginning of this section. One important group comprises the film reactors made to different designs by several plant manufacturers. These reactors operate once through; and the liquid to be sulphonated flows downwards in a thin film on the inside of a water-cooled tube, or tubes, or on both sides of a cooled

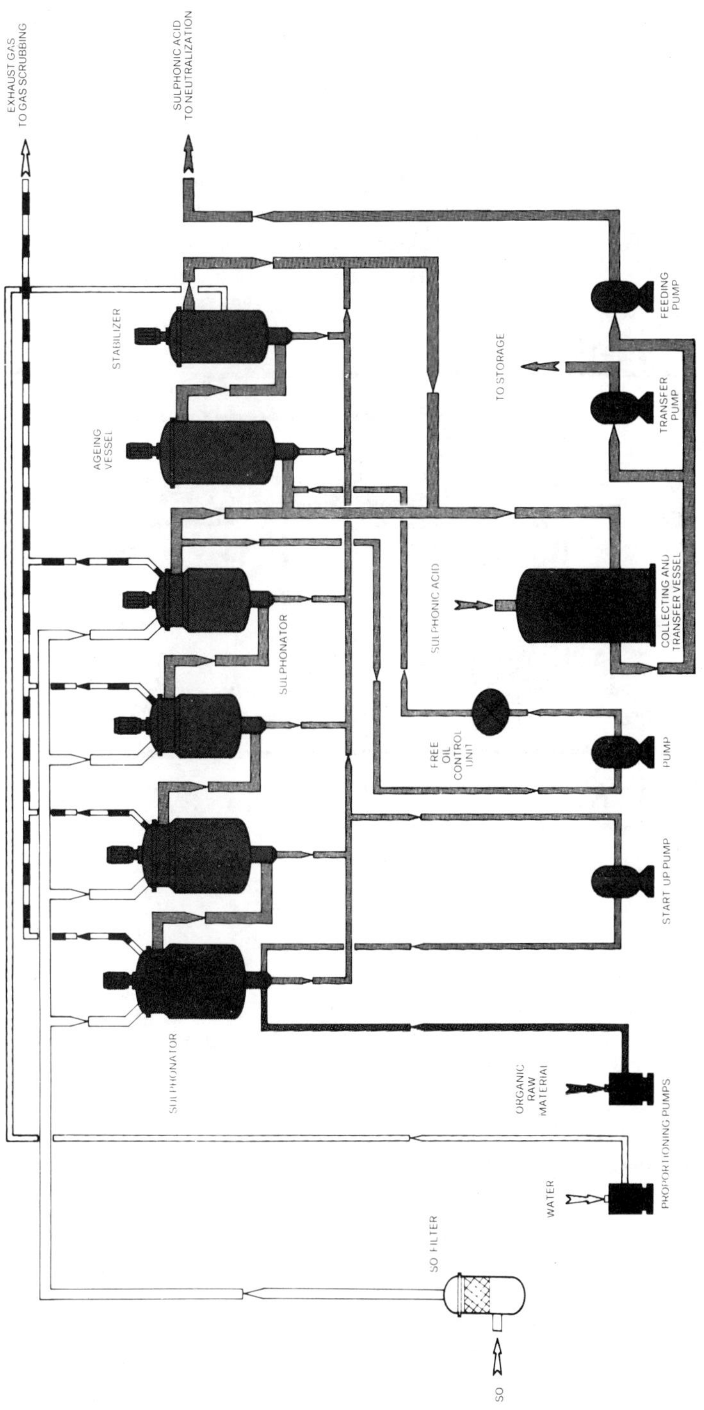

Fig. 9.8 – Ballestra 'Sulphurex' reactors.

annulus. The sulphur trioxide/air gas passes downward in turbulent flow concurrent withe the liquid. The sulphonic acid and the residual gas are separated in some form of inertial device at the base of the reactor and/or in a cyclone. The liquid is only exposed to the sulphonating gas for a short time; and heat is removed directly from the reaction zone. Great care is necessary to ensure that gas and liquid are properly proportioned at all points; if this is not achieved over-sulphonation may occur locally, with deterioration of colour and, possibly, a build up of coke which may necessitate a plant shutdown for cleaning. Lack of uniform distribution may also lead to zones in which sulphonation is incomplete. Nevertheless, film reactors have been very successful and are widely used for high quality sulphonation.

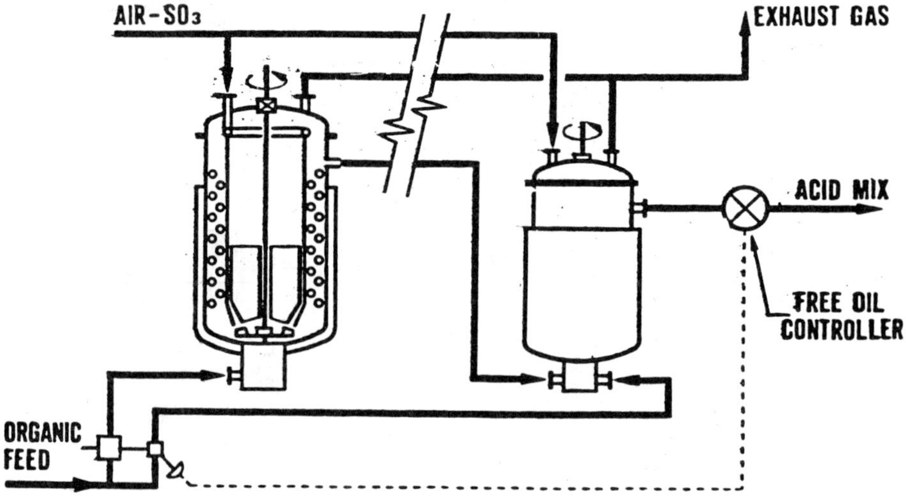

Fig. 9.9 – Detail of Ballestra 'Sulphurex' reactor.

Figure 9.10 shows diagrammatically the design of the *Mazzoni SOCS reactor* which uses liquid films on the inner surfaces of a group of tubes. A special feature of this reactor is the use of equalizing air which enters through passages with a lower pressure drop than that for the flow of the sulphur trioxide mixture through the nozzles. This helps to balance the flow of sulphur trioxide/air through the various nozzles and also shields the liquid from the highest concentration of sulphur trioxide and spreads the reaction down the tubes.

The most important differences between the various film reactors probably lie in the design of the reaction tubes, and of the heads which distribute the liquid and gas. Figure 9.11 shows the head of the *Stephan multitubular reactor.* The *Allied Chemical Corporation/Meccaniche Moderne unit* uses both inner and outer walls of annuli. As shown in Fig. 9.12 the head of the *Chemithon reactor* (also supplied by Lurgi) uses both walls of an annulus, and includes a high speed rotor in the upper part of the annulus to distribute and mix the reactants. Figure 9.13 shows diagrammatically the *Ballestra Sulphurex F multi-tubular film reactor.*

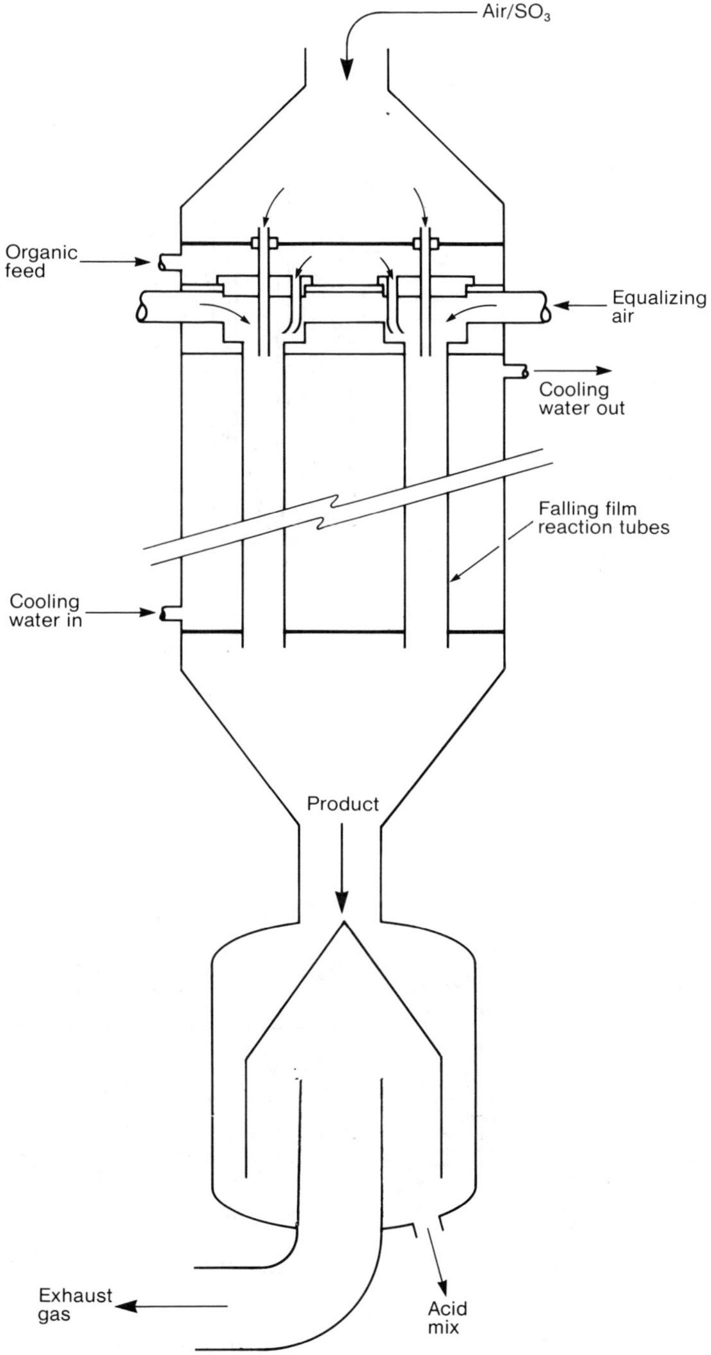

Fig. 9.10 – Mazzoni sulphonation reactor.

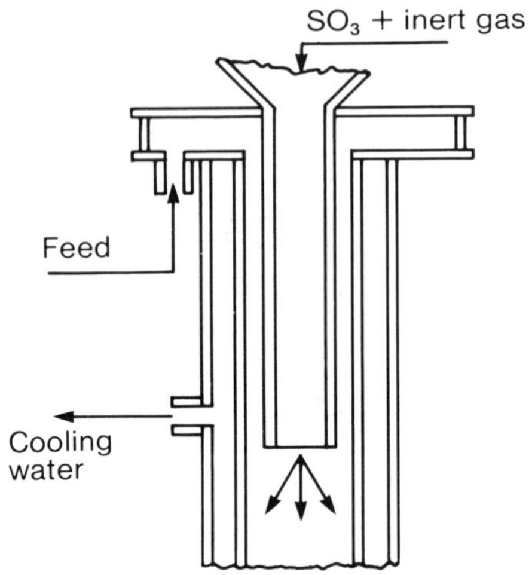

Fig. 9.11 – Head of Stepan reactor.

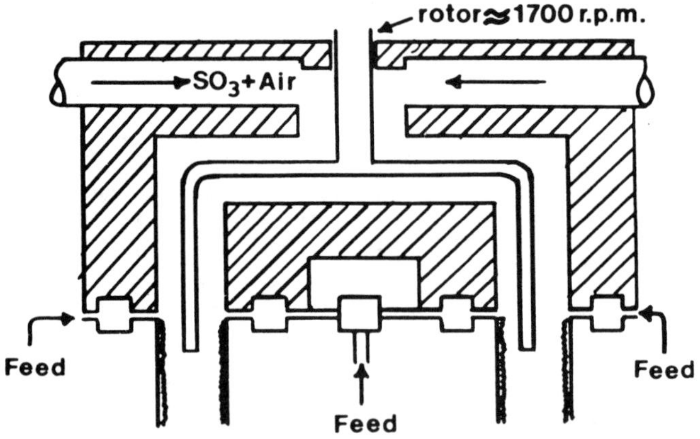

Fig. 9.12 – Head of Chemithon reactor.

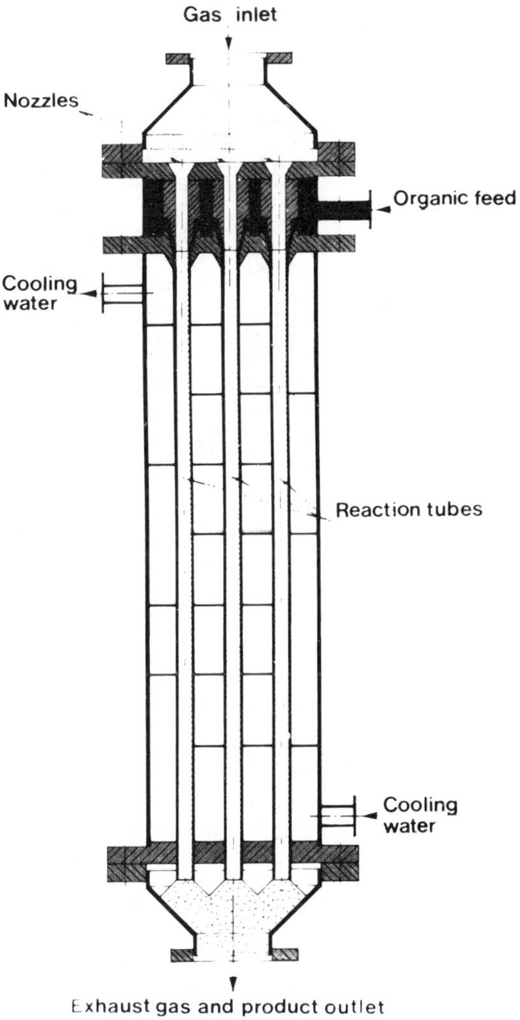

Fig. 9.13 – Ballestra multitube (type F) reactor.

The *Berol reactor* (Fig. 9.14) consists of a cooled vertical cylindrical shell with a coaxial inner revolving water cooled drum. The drum carries a spiral arrangement of pins which serve to cause turbulence and to minimize dead zones in the annulus. The liquid to be sulphonated enters at the bottom of the annulus and the gas is introduced through two rows of nozzles arranged up the lower half of the shell. The air left after the sulphur trioxide introduced through the lower nozzles has reacted dilutes the gas higher up the reactor (a similar dilution occurs in the falling film types). The flow of gas, and the action of the pins on the rotating drum, cause the liquid to flow upward with little back-mixing.

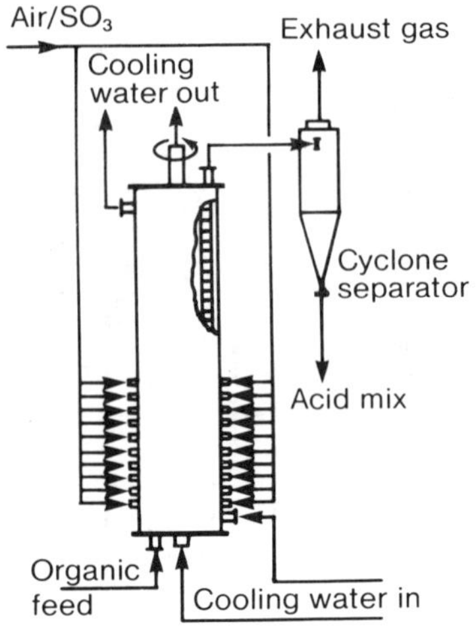

Fig. 9.14 – Berol reactor.

An agitated zone reactor, called the *Flowmaster,* similar in some respects to the Berol reactor, was developed by *Marchon Products Limited* (now Albright and Wilson Limited), BP 969 517 (1964), and produced for them by Baker Perkins Chemical Machinery Limited. The main features are shown diagrammatically in Fig. 9.15. The reactor comprises a horizontal cylindrical vessel divided into a series of compartments by hollow static discs. The reaction mix passes horizontally through small tubes which pass through the discs, where it is cooled by water which enters the bottom of the discs and leaves from the top. The cooling in the various sections can be adjusted as required. The contents of the compartments are mixed by the rotating impellers; and SO_3/air gas is introduced into the earlier compartments. No gas is introduced into the later, maturing, compartments, where reaction continues with sulphur trioxide remaining from earlier stages. Immediately after the reaction is completed, the mixture is:

- quenched with water introduced as shown,
- neutralized, or
- degassed, for example in a vacuum cyclone.

Special features of this reactor are:

- the ratio of cooling surface to contents is very high, so that very effective cooling is possible.
- the residence time is very short, only one or two seconds.
- there is no excess of sulphur trioxide until the later stages are reached; and here it is diluted by the air introduced in the earlier stages. This large volume of gas reduces further the residence time in the later stages where the material being sulphonated is most vulnerable.

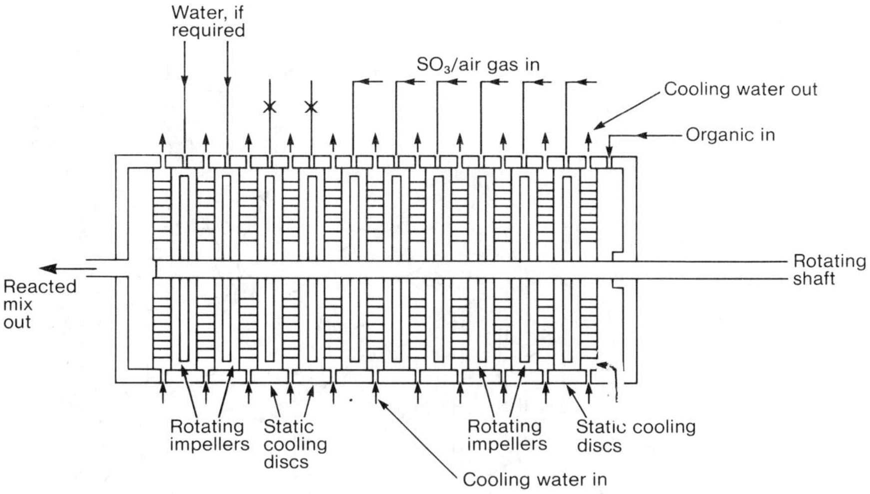

Fig. 9.15 – Albright and Wilson 'Flowmaster' reactor.

These features make the reactor suitable for delicate materials, such as alcohols. An advantage relative to film reactors is that it is not subject to broken films which, in film reactors, can lead to local over sulphonation and a build up of coke.

Another approach is to spray the liquid in atomized form into a converging stream of sulphur trioxide/air. This is used in the *Chemithon Impact Jet reactor* shown in Fig. 9.16. Again the residence time is very short and the reactor is said to give good quality product with higher concentrations of SO_3 than those used in the film reactors.

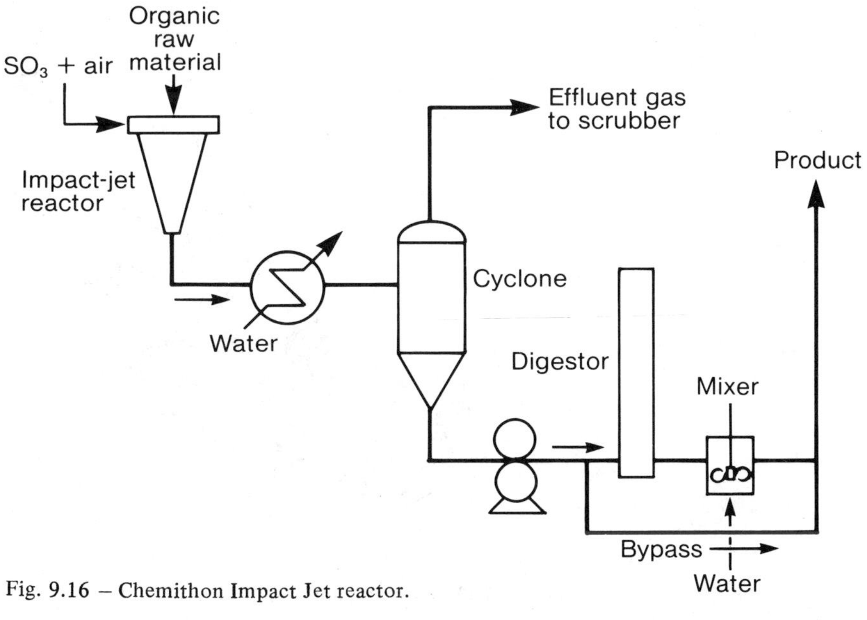

Fig. 9.16 – Chemithon Impact Jet reactor.

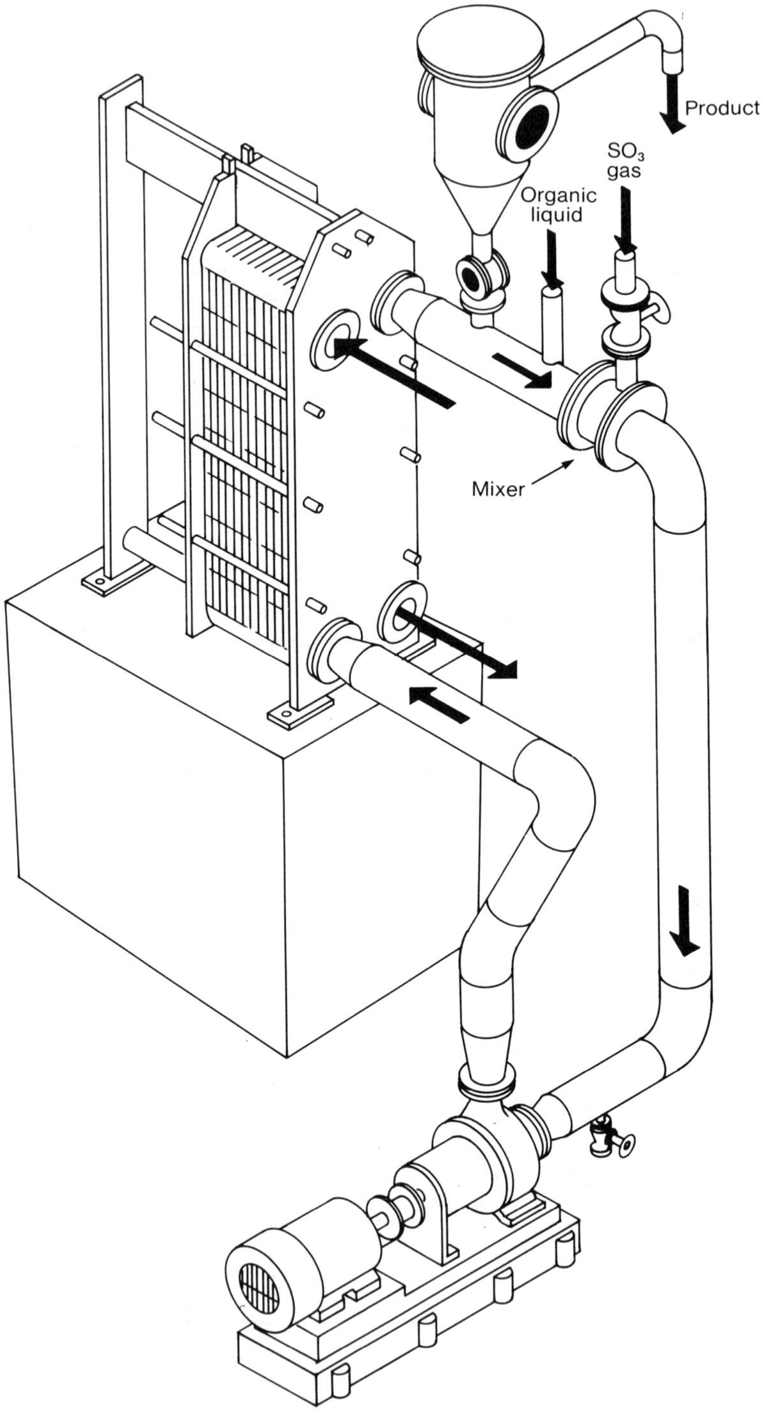

Fig. 9.17(a) – Albright and Wilson loop reactor.

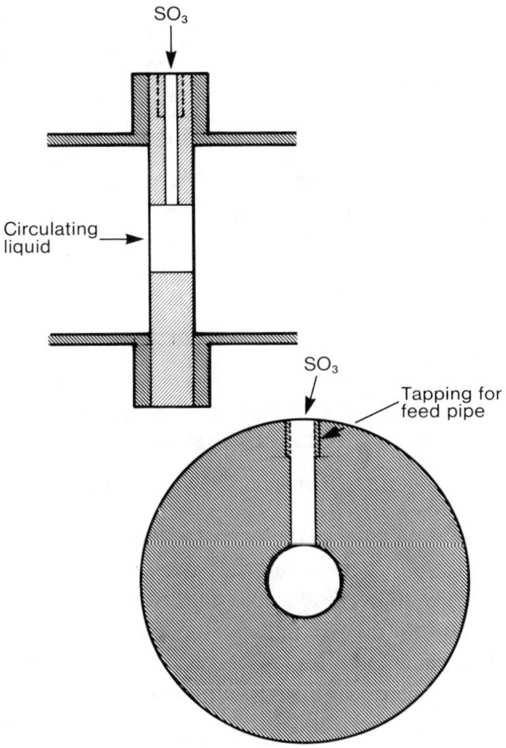

Fig. 9.17(b) – Orifice mixer for Albright and Wilson loop reactor.

An important recent development is the *Albright and Wilson Loop Reactor,* BP 1 563 994 (April 1980). As shown in Fig. 9.17(a) this uses a loop round which reaction mix is pumped through a cooler. The liquid to be sulphonated and undiluted sulphur trioxide gas are introduced; and a very small portion of the circulating material is bled-off as product. The recycle is preferably between 120 and 350 times the quantity of product; and this large recycle presumably replaces the diluent air used in more conventional processes. This is another example of the use of the dominant bath principle. The flow round the circulating system must be turbulent with a Reynolds number ($D\rho\ v/\mu$, where D = pipe diameter, ρ = fluid density, v = fluid linear velocity, and μ = fluid viscosity, in any consistent units) over 3000 and preferably 10 000 – 15 000. An adequate pump is required; and, if necessary, the viscosity of the reaction mix can be reduced by an increase in operating temperature, or by the use of an inert diluent, such as carbon tetrachloride. As it has to be separated later, the use of a diluent is best avoided if possible. Various types of mixer can be used, but the patent recommends as simple and effective an orifice plate through which the sulphur trioxide is introduced into the throat as shown in Fig. 9.17(b). Inclusion of a sulphone inhibitor, such as acetic acid, or acetic anhydride, may be advisable.

This system, which can easily be scaled up, is said to work well and has the great advantage that no air has to be dried, pushed through the plant, and scrubbed free of noxious vapours and droplets.

In addition to brochures issued by the various plant manufacturers and the patent literature, Lanteri (1978) and Kapur *et al.*, (1978) provide useful reviews of modern sulphonation processes. The author is indebted to Soap, Cosmetics and Chemical Specialities for Figs. 9.9, 9.13, 9.14 and to the Journal of the American Oil Chemists Society for Figs. 9.11 and 9.12.

9.4.5 After treatment of the sulphonated mix

As shown in Fig. 9.8, most plants include an ageing vessel in which dissolved sulphur trioxide reacts with residual hydrocarbon; and a stabilizer to which a little water is added to hydrolyse sulphonic anhydrides.

$$(R - SO_2)_2O + H_2O = 2R{\cdot}SO_2{\cdot}OH$$

The sulphonic acid is then neutralized by the addition of the stoichiometric quantity of alkali, usually caustic soda, but sometimes another base, such as triethanolamine. With continuous sulphonators, neutralization is commonly carried out continuously using a loop which may include a cooler. Figure 9.18 shows the Ballestra unit. Like others, this uses a pH device to control the addition of neutralizing agent; and water can normally be added to produce paste at the required surfactant concentration. Some neutralization units can be arranged to use concentrated caustic soda so as to give high active pastes; in this case a portion of the material may be taken from the main circulation loop through a by-pass loop in which water is added before the sample is passed through the pH device. Batch neutralization can be carried out in groups of pots used in sequence to take sulphonic acid from a continuous sulphonator. Batch systems have been designed to use sodium carbonate instead of caustic soda, but carbonate is not normally suitable for continuous neutralizers. Even with batch neutralization it is usually necessary to complete the reaction with caustic soda after the carbon dioxide has been removed. It might be possible to use sodium carbonate as in the continuous neutralization of fatty acids.

The slurries used to produce blown NSD powders usually include sodium silicate and, at least with batch neutralizers, it may be convenient to add silicate to the neutralized paste to be used in powder manufacture.

Some earlier, stirred tank, reactors gave a dark coloured sulphonic acid which neutralized to a buff-coloured paste which needed to be bleached to produce white, or bright coloured, detergent products. The bleach was normally done with sodium hypochlorite and it is necessary to ensure that the paste is alkaline before the hypochlorite is added. It is also necessary to eliminate any excess available chlorine from the paste before it is brought into contact with optical brightener. This can if necessary be done with anti-chlors such as sodium sulphite or sodium perborate. Sodium hypochlorite bleaching of sulphonate pastes is now rather out of favour because of the possibility that chlorsultones may be formed, which may be bad for the skin, or even carcinogenetic. The trend is to use hydrogen peroxide, which bleaches more slowly than hypochorite; but, if possible, it is best to sulphonate so as to eliminate the need to bleach at all.

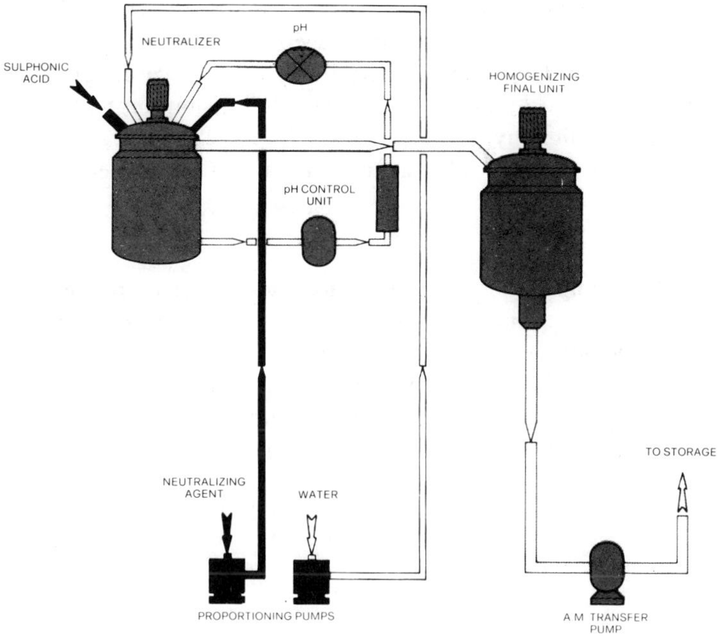

Fig. 9.18 – Ballestra neutralization unit.

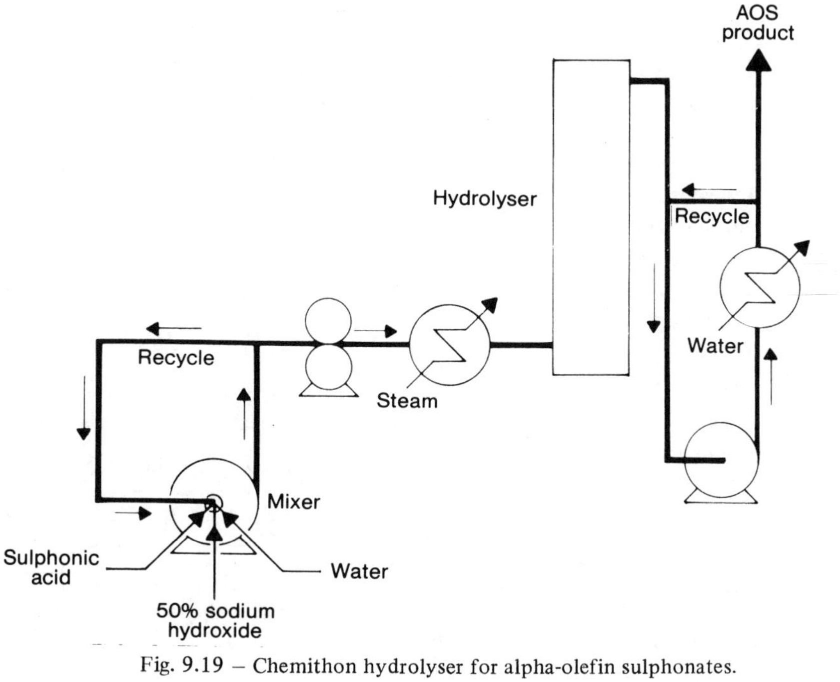

Fig. 9.19 – Chemithon hydrolyser for alpha-olefin sulphonates.

As mentioned earlier, the sulphuric esters produced when hydroxy compounds are sulphated hydrolyse readily under acidic conditions, and any ageing, or stabilizing, vessels are by-passed when these materials are being processed to enable neutralization to be carried out as rapidly as possible.

With alpha-olefins a special section is needed in which the mix including alkali is raised to elevated temperatures, under the necessary pressure, to hydrolyse the sultone to hydroxy sulphonate. The Chemithon unit is shown in Fig. 9.19. A little sodium hypochlorite, or hydrogen peroxide, can be injected into the line carrying the cooled sulphonate to give a continuous bleach.

9.4.6 Compositions of sulphonated and sulphated products

Manufacturers indicate the compositions of the pastes which can be obtained from various raw materials using their plants. Such results can vary with the precise quality of the raw material, and with the operating procedure adopted; for example, colour can usually be improved at the expense of unreacted oil. Table 9.3 gives some indication of the figures quoted for a range of modern plants.

Table 9.3

Typical compositions of sulphonate and sulphate pastes from some important raw materials

Raw material	Surfactant in paste %	Unreacted oil. % on 100% active	Inorganic salts. % on 100% active	Colour klett[a]
Linear alkylbenzene	45–50	1.5	1.5–3.0	40 B[b] 40 – 80
Branched alkylbenzene	50–70	1.0–2.0	0.6–2.0	25–60
Nominal C_{12} alcohol	30–60	1.5–3.0	0.8–2.5	30–60
Nominal C_{16}–C_{18} alcohol	30–40	3.0–4.5	1.5–3.0	40 B[b] 60–120
C_{12} alcohol ethoxylate	30–70	1.5–3.0	0.6–1.5	10 B[b] 5–30
Alpha-olefin	40	1.5–3.0	1.2–2.0	40–80 B[b] 80

[a] Klett colours are for 5% surfactant solutions, 40 mm light path and No 42 blue filter.
[b] Figures marked B are for bleached samples. Special equipment and/or procedure is needed for the very high surfactant pastes.

9.4.7 Treatment of effluent air

The air leaving a sulphonator contains:

- sulphonic acid mist.
- unreacted sulphur trioxide.
- sulphur dioxide.

The cleaning system adopted depends upon a number of factors, including legal and other requirements for the quality of gaseous effluents discharged to atmosphere. Sulphonic acid mist can be removed with cyclones and/or fibre glass demisters; but some plants use electrostatic precipitators. Some sulphur dioxide is produced by reduction of sulphur trioxide by the organic matter being sulphonated, but the concentration is normally considerably higher from plants which burn sulphur than from others, because of incomplete reaction to sulphur trioxide in the catalytic converter. Caustic soda scrubbers are commonly used to absorb oxides of sulphur. Spent scrubbing liquor containing sodium sulphite from the sulphur dioxide may be usable as an anti-chlor, or as a source of sulphate for powder manufacture. As mentioned earlier, the air, after electrostatic demisting, is sometimes re-cycled. It is also sometimes possible to put the effluent gas through an oil-fired furnace. This converts the sulphur trioxide to sulphur dioxide, and does not reduce the total emission of sulphur oxides, but it does eliminate visible mist.

9.4.8 Complete sulphonation plants using sulphur trioxide

Selected elements can be linked to produce a wide variety of plants to suit particular requirements. The elements discussed in previous sections are largely self-contained and connection is straightforward, apart from the need for properly integrated control.

10

The manufacture of NSD (non-soapy detergent) powders

10.1 GENERAL POWDER CHARACTERISTICS

Although the performance of a powder depends upon its composition, and the concentration and other conditions under which it is used, some physical properties of the powder are also important. In particular:

- bulk density.
- particle size distribution, rate of solution, and flow properties.
- storage properties, including freedom from undue tendency to cake.
- colour.
- odour.

Some of these properties are interrelated.

Bulk density is the mass of a given volume of the powder. It is commonly expressed in terms of gram/ml (or kg/litre), or lb/ft^3. The figure in lb/ft^3 equals that in gram/ml $\times$ 62.3; the figure in gram/ml is also equal to the density relative to water, or specific gravity. Because of the way in which powders can compact, on shaking or otherwise, the bulk density of a powder depends very considerably upon the way in which it is measured.

It is particularly important to produce powder at a consistent bulk density in order to ensure that the packets are adequately filled by the correct mass, or weight, of product. The choice of bulk density for a product is influenced by a number of factors:

- technical feasibility, as discussed later.
- a balance between the advantage that a low BD may seem to provide the

customer in terms of good value for money, and the disadvantage that it may induce her to use too little product in each wash, as housewives tend to dose by volume. As well as possibly reducing sales, the use of an unduly low concentration of product in the wash liquor is likely to impair performance and cause dissatisfaction.

- the cost of packaging, transport, and storage is higher per ton when the bulk density is low than when it is higher.

At one time very light powders (around 0.13 g/ml, or 8lb/ft^3) were produced. More recently heavy duty powders have commonly been sold with a medium density of 0.3–0.4 g/ml, or 18–25 lb/ft^3. For industrial purposes, and for transport of intermediate materials to be re-processed overseas, reduced packaging, storage, and transport costs outweigh other factors and favour the production of high density powders, or flakes.

Particle size distribution is important in that it is desirable to have a product which is granular and free from fine particles, which would make it dusty and liable to cause sneezing. Rate of solution, which is affected by particle size, is important, particularly with some washing machines which have a very short cycle of operations. Hollow particles favour good solubility, whereas large solid ones detract. Ingredients added to a blown powder, particularly sodium perborate, may be more slowly soluble than the blown base powder. Soap powders sometimes have a tendency for fine particles to clot on the surface of the water, but this is not normally a problem with NSD powders.

A 'crisp' powder with good flow properties is welcomed by housewives, whereas one which is 'creepy' (like slightly damp brown sugar) is not liked. Good flow properties also facilitate proper dosing of the powder to the washing machine or the hand wash.

What matters to the housewife is the condition of the powder when she is about to use it. This is influenced by changes during storage, as well as the condition of the powder when packed. Contrary to soap powders which, as discussed in Chapter 7, tend to lose water under most conditions of storage, NSD powders tend to be hygroscopic and to pick up moisture from the atmosphere unless the air is very dry. Moisture pick-up can cause chemical damage, for example if the product contains enzymes, or it can lead to impairmant of flow properties and to caking. Powder composition, and processing procedures, can be significant; but it is also possible to use packaging which resists pick-up of moisture from the atmosphere. Examples are the use of wax paper wrappers, or liners, in conjunction with normal cartons, the use of cartons made from carton board which has paper laminated to the base-board with a moisture vapour-proof wax, and the use of well sealed bags made of suitable plastic film. In addition to the selection of a plastic which gives a film of low moisture vapour permeability, it is often desirable to use a laminate. Thin films commonly have pin holes; and the use of composite films minimises the chance of through holes. Lacquered films can also be used. Although special moisture-vapour proof cartons have been widely used, much NSD powder has been packed successfully in ordinary carton-board cartons.

Colour affects consumer reaction rather than product performance. A good white is usually desirable when the product is to be sold uncoloured, but basic colour is also important when the finished powder is blue, or some other bright colour. Poor powder colour can be caused by:

- the use of a dark surfactant paste produced by poor sulphonation, or lack of an adequate bleach.
- inclusion of dirty inorganic ingredients, such as sodium sulphate contaminated with iron compounds.
- soot in the drying gas used in a blowing tower.
- contamination with material discoloured by heat whilst being held on parts of a blowing tower wall.

The optical brightener added primarily to improve the washing performance of the product often also brightens the powder.

In the case of blown powders, the spray-drying process tends to remove undesirable odours introduced in the ingredients. The powder is normally perfumed to meet market requirements.

10.2 METHODS OF POWDER PRODUCTION

Methods used in the manufacture of NSD powders are:

- spray drying, or 'blowing'.
- dry mixing.
- adsorption of liquid surfactant onto inorganic salts.
- adsorption with neutralization of sulphonic acid, and possibly fatty acid, by sodium carbonate, or caustic soda solution.
- drum drying and grinding.
- the Monsanto 'Fluff' process.

or, combinations of some of these processes.

Most of the powder sold to housewives, and much for use in industry, is produced by a spray-drying process, although certain ingredients which cannot be blown without decomposition, or loss, such as sodium perborate, enzymes, and perfume, are commonly mixed with a spray-dried base. Spray drying produces powders with desirable consumer properties (granularity, bulk density and solubility); and on the large scale the process is economical. The capital cost of a spray-drying plant is high; and the smallest unit offered by major plant manufacturers is commonly about 0.5–1.0 t/h. The process is not normally viable for productions of less than about 2,000 t/a, and higher quantities may be necessary. Large spray-drying plants commonly use units of 10–20 t/h capacity.

Despite technical limitations discussed later, particularly of bulk density, small manufacturers may have to select one of the other processes. Such potential manufacturers are well-advised to assess carefully supplies of ingredients, and the market and commercial potential of the proposed product before they commit money to an NSD powder project. Surveys carried out by the present author in several developing countries at different times during the period

1974– 1980 indicated that it was difficult, or impossible, to match formulations on sale by the major international companies, even without any substantial capital, or marketing, charges.

Similar costings should naturally be produced before a spray-drying plant is purchased; and it may be found that a sulphonation plant is also needed to give a viable project.

Ingredients, or relatively high density powders made by one of the other processes, may be mixed with a blown powder to:

- avoid loss of, or damage to, ingredients, as already mentioned. An important example is the inclusion of non ionic surfactants which might inhibit the spray drying process because of their volatility, or instability to heat.
- increase the bulk density of the final product to a value higher than can conveniently be obtained with the blown powder alone.
- increase the productive capacity of a given spray drying plant; and/or improve overall thermal efficiency.
- modify the effect of an ethanolamide on the lathering characteristics of a high lathering powder.

10.3 THE SPRAY DRYING PROCESS

10.3.1 General

Spray drying is used in a wide range of industries and many variations are possible. Droplets of liquid are brought into contact with hot gas in some form of chamber, where they dry rapidly to produce a powder. The liquid can be a solution or a slurry. Shampoos and some very early detergent powders were made from solution, but slurries are now used for household and similar powders. Atomization, or the formation of droplets, can be by rotating disc, by two-fluid nozzle in which the liquid is broken up by steam, or compressed air, or by single fluid nozzle in which the liquid to be dried is atomized by pressure alone. Contact between droplets and drying gas can be counter-current, con-current (also called co-current, or parallel flow), or mixed flow. The shape of the drying chamber depends upon the type of flow and the method of atomization; rotating disc atomizers usually require a wide chamber and, often, con-current flow; whereas pressure nozzles are suited by tall counter-current towers. Rotating disc atomizers tend to give fine powders; and two-fluid nozzles are usually used for solutions of relatively low viscosity. Granular detergent powders are normally made in tall towers using single-fluid pressure atomizing nozzles, or jets. Contact is now usually counter-current, but plants are sometimes built so that con-current flow can also be used if a low bulk-density powder is required. Mixed flow, with cold air introduced at the base of the tower, has been suggested for powders which contain a substantial proportion of sodium carbonate and which require some spray chilling, as used for soap powders. Normal counter-current detergent powder towers include an element of mixed flow in that a small proportion of cold air is drawn into the base of the tower where the powder leaves. This would be difficult to avoid because the tower is operated at a very slight negative pressure (a few mm water gauge), so as to avoid the blowing out of hot gas, spray or powder. The cold air provides useful cooling of the powder, so there is no need to attempt any form of sealed powder discharge which would be difficult with warm sticky powder.

Points in the choice between counter-current and con-current contact are:

- con-current contact gives light powders of bulk density about 0.08–0.15 g/ml and a moisture content of 3–8%. The powder contains many hollow beads, some of which tend to break to dust. Con-current contact may be useful for thermally delicate products such as powders with a very high surfactant content including soap, because the hot gas meets the wet slurry and the dried particles are in contact with the coolest gas.
- counter-current contact gives powders of bulk density 0.15–0.45 (more usually 0.3–0.4) g/ml and a moisture content of 6–15%, commonly about 10%. Beads are less fragile and less dust is usually produced. Thermal efficiency is normally higher than with con-current contact, because the gas leaves the tower after contact with wet slurry and hence can have a lower temperature and/or a higher content of water vapour. The hottest gas contacts the driest particles with consequent greater risk of thermal damage.

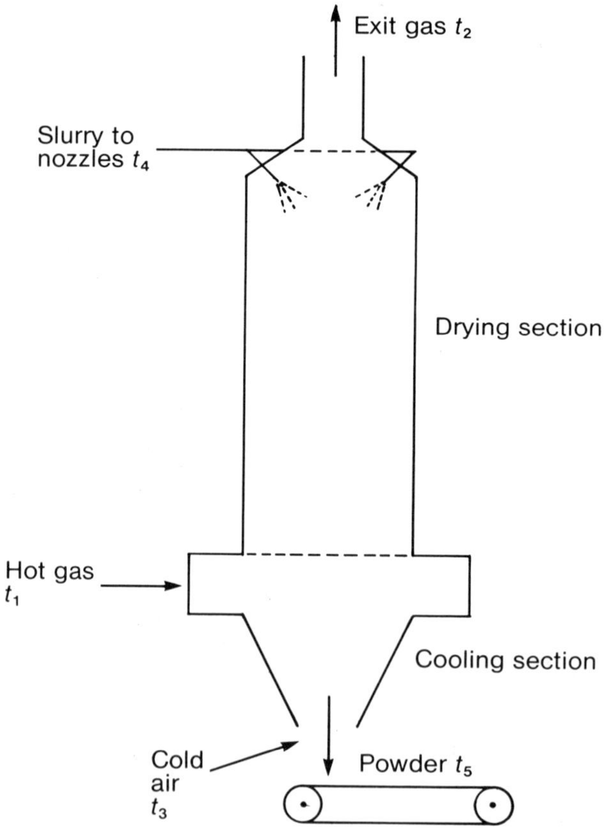

Fig. 10.1 – Counter-current spray drying tower.

As already mentioned, detergent powders are normally made with counter-current contact, using a plant as shown diagrammatically in Fig. 10.1; and the rest of this section is concerned almost entirely with plants of this type. Readers who require a full discussion of all aspects of spray drying are referred to Masters (1972 or 1979).

10.3.2 Principles of spray drying and thermal efficiency

The velocity of the gas up a counter-current tower must not be so high as to carry over much of the product, but, up to that point, an increase in velocity increases the time during which the particles are held in the drying zone, as well as the capacity of the plant. The emissions from the jets induce local downwards currents; and observation shows that particles move erratically in the tower. Masters plots temperatures at various points in certain counter-current and con-current towers, but temperature distributions can vary with a number of factors.

In general, the drying of a droplet/particle of slurry (or solution) in a spray drying chamber follows a similar pattern to that of a slab, or other piece, of damp material in any type of drier using air, as shown diagrammatically in Fig. 10.2. After the heating up period AB, the droplet experiences a constant rate period BC, in which the whole of the surface remains saturated with water. During this stage the water which is evaporated is replaced by movement of other water from the interior of the droplet; and the surface temperature of the droplet is approximately the wet bulb temperature of the drying gas with which it is in contact. When point C is reached, movement of water from the interior is insufficiently rapid to keep all the surface wet and a crust starts to form. After point D the whole of the surface of the droplet, now a particle, is unsaturated, and, if contact is sufficiently long, the particle will eventually reach E, where its moisture content is in equilibrium with the surrounding gas. This point is not normally reached in a spray drier. Stages CD and DE are called falling rate periods. When, as with NSD powders, the liquid is a solution rather than pure water, there is an elevation of wet bulb temperature due to dissolved substances.

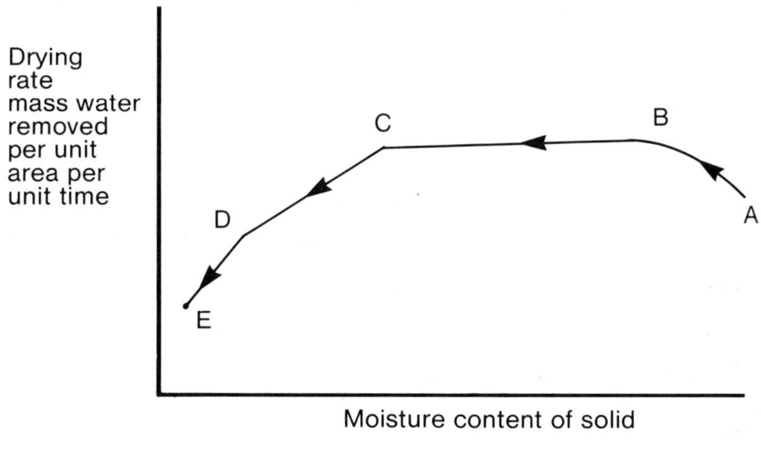

Fig. 10.2 – Drying rate curve.

Again in general, the behaviour of a particle after a crust forms is influenced by the nature of the dried solid and the temperature difference between gas and particle. The temperature of the particle rises above the wet bulb temperature of the gas, and, if the temperature difference is sufficiently high, steam can be generated (and/or air in the slurry expanded). The crust can be plastic or rigid, porous or non-porous. With a relatively plastic and non-porous crust, as is common with detergents, the vapour/air puffs the particle into a hollow shell and then blows a hole through which to escape. Such shells are readily visible in most blown NSD powders. The degree of puffing increases as the temperature difference between particle and drying gas increases. In some cases crust formation starts very early in the process so that there is very little constant rate drying.

This formation of hollow particles is of major importance in the production of a powder with a relatively low bulk density. Masters describes four mechanisms by which hollow particles are formed:

- a surface layer, semi-impervious to vapour flow, forms at the droplet surface. The droplet puffs as vapour is formed and expands with the increase in droplet temperature.
- moisture evaporates at a rate faster then the diffusion of solids in solution back into the droplet interior. On completion of evaporation air voids are present, especially in the case of crystalline products.
- liquid flows with accompanying solids to the droplet surface by capillary action. This leaves the centre of the droplet as a void. This mechanism is applicable to clay slips.
- air entrained in the feed contributes to air spaces within the droplet.

Mechanisms one and four are those of particular significance for NSD powders. As well as puffed hollow spheres, honeycomb-like particles are sometimes found; these may be formed by aggregation of small particles in the tower. Con-current flow with a high gas inlet temperature favours a high degree of puffing, but, if the process is carried too far, the shells become so thin that they tend to break up with formation of dust. With counter-current operation, the gas outlet temperature (t_2 in Fig. 10.1) is particularly important, as it is the temperature of the gas first contacted by the wet droplets.

Important information concerning many drying processes, including spray drying, can be obtained by plotting on a psychometric, or temperature/humidity, chart the changes in the condition of the drying gas which take place in the drying chamber; this is discussed in detail by Masters. In the spray drying of NSD powders the gas temperature is usually above about 100°C throughout the process. Its condition can then be defined in terms of temperature and absolute humidity (kg water vapour/kg air), but relative humidity has no meaning as water vapour and air are completely miscible at above 100°C at atmospheric pressure. Under these circumstances the principle of the adiabatic cooling line, which relates temperature to absolute humidity as the gas evaporates moisture under isolated, or adiabatic, conditions, can be considered more simply in terms of temperature alone. Most of the heat lost by the gas as it passes through the tower goes to supply the latent heat, plus any sensible heat, needed to convert the droplets of slurry to particles of dried powder. This temperature change can be called the adiabatic temperature fall, that is the change which would occur if there were no heat losses. In practice, the temperature drop is a little higher than this because some heat is lost by radiation and convection from the tower walls. The gas usually leaves an NSD tower at about 90–110°C. In the following approximate calculations it is assumed that the gas leaves at 100°C, but that 10°C goes to losses, so that the final temperature under adiabatic conditions is 110°C. The true correction depends upon factors such as the size of the tower, the efficiency of its lagging, and the temperature of the hot gas. As mentioned earlier, some cold air (assumed to be at 20°C) is drawn into the base of the tower; and this is assumed to be 10% of the total gas. It is convenient to consider the total mass of gas drawn through the tower and to allow for the cold portion in assessing the effective, or corrected, inlet temperature, t_c. For example if the inlet gas is at 400°C, t_c equals $(0.9 \times 400) + (0.1 \times 20) =$ 362°C. The quantity of heat supplied in the furnace to heat 90% of the gas from 20°C to 400°C is the same as that which would be required to heat 100% of the gas from 20°C to 362°C. The heat supplied for various values of t_c equals mass of gas × specific heat of gas × $(t_c - 20)$. The heat used to achieve drying equals mass of gas × specific heat of gas × $(t_c - 110)$, so that the thermal efficiency equals $(t_c - 110)/(t_c - 20) \times 100\%$. Table 10.1 shows approximate thermal efficiencies calculated in this way for a range of conditions of interest in the spray drying of NSD powders. It will be seen that thermal efficiency increases as the gas inlet temperature rises. This is because a smaller mass of gas is required to achieve a given amount of drying; and this gas, which leaves at a temperature substantially independent of the inlet temperature, carries with it less heat than the larger mass of gas used when the inlet temperature is lower. This point is illustrated by the heat balances considered later.

Table 10.1 – Estimated thermal efficiencies for various hot gas temperatures.

Inlet gas temperature t, °C	Gas temperature after admixture with 10% cold air °C (t_c)	Thermal efficiency with the corrections discussed in the text $= \frac{t_c - 110}{t_c - 20} \times 100\%$
500	452	79
450	407	77
400	362	74
350	317	70
300	272	64
250	227	57
200	182	44
150	137	23

The maximum gas inlet temperature which can be used is limited by a number of factors:

- the danger of discoloration by heat, or of charring, of the product, particularly of material deposited on the chamber walls near to the gas inlet.
- possible vaporization, or other damage, to certain ingredients, particularly some nonionic surfactants.

  This may lead to a choice between:

  - a lower gas inlet temperature and/or the selection of a more rubust nonionic surfactant, or
  - addition of the nonionic surfactant to the already blown powder, for example in a powder made by the adsorption process.

- problems with bulk density, which may also restrict the choice of slurry water content.
- difficulty in the production of very hot gas if indirectly heated air is used; this does not normally arise with detergent powders for which direct products of combustion are now usually employed.

Inlet temperatures of 250°C–400°C have been common for detergent powders, but improved hot gas inlet design may permit up to about 450°C.

Conventional heavy duty powders are usually dried to about 10% moisture content. At least in part this is because removal of water from sodium tripolyphosphate hexahydrate, which these powders normally contain in substantial proportions, results in undesirable decomposition of the tripolyphosphate. Light duty powders with little, or no phosphate, and often made to a rather low bulk density, may be dried more completely

Simple arithmetic shows the importance of the water content of the slurry in relation to the quantity of water to be evaporated; and hence its effect upon both fuel consumption and plant capacity. Consider a slurry which contains 40% water and 60% other ingredients, described as solids, dried to a powder which contains 10% water. 100 parts slurry contain 60 parts solids plus 40 parts water. Disregarding losses, they produce $60 \times 100/90 = 66.7$ parts powder, containing $66.7 - 60.0 = 6.7$ parts water. Hence water evaporated $= 40.0 - 6.7 = 33.3$ parts; equivalent to $33.3/66.7 = 0.50$ parts water evaporated per part powder produced. Table 10.2 summarizes the results of similar calculations for slurries with various water contents dried to produce powders containing 10% and 5% water.

With fixed inlet and outlet temperatures, the fuel consumption per tonne of powder with a 35% water slurry is about $0.39/0.64 = 61\%$ of that with a 45% water slurry. Also, unless limited by ancillaries, the productive capacity of a plant is $0.64/0.39 = 1.6$ times as great. Even a small change in slurry concentration has a significant effect.

Table 10.2

Water content of slurry (%)	Parts water to be evaporated to produce 1 part of powder	
	Powder 10% H_2O	Powder 5% H_2O
50	0.80	0.90
45	0.64	0.73
40	0.50	0.58
35	0.39	0.46
30	0.29	0.36

Fan performance is normally stated in terms of standard air at 20°C and atmospheric pressure. As explained in Appendix 3, speed and other conditions constant, the volume moved by a fan is independent of temperature, so that the mass flow varies directly with air density, or inversely with absolute temperature (t + 273°C). From the limited information available, the nominal capacity of the exhaust fan commonly installed varies with the inlet gas temperature roughly as shown in Table 10.3

Table 10.3

Gas inlet temperature (°C)	Standard m^3 of air per tonne of powder made from slurry containing 40% water
400	7500
350	10000
300	12500
250	15000

When a range of products which require different gas inlet temperatures are made, it is usual to maintain the full rate of gas flow and to accept different rates of powder production.

The rough heat balances now to be calculated for inlet gas temperatures of 400°C and 300°C respectively, both with an outlet gas temperature of 100°C, seem to show that the gas quantities actually used are lower than those indicated in Table 10.3. This may be because the flow is reduced by a pressure to be overcome higher than that used when the fan is rated, or because of allowance for contingencies. As before it is assumed that the ambient air is at 20°C; and that 10% of cold air is drawn into the base of the tower, so that the corrected gas inlet temperatures are 362°C and 272°C, as shown in Table 10.1.

The density of the gas (assumed air) at atmospheric pressure and 0°C is about 0.08 lb/ft^3, or 1.28 kg/m^3. At 100°C it is 1.28 × 273/373 = 0.94 kg/m^3. The specific heat of air is about 0.24 BTU/lb °F, or 0.24 kg cal/kg °C.

Example 1. Inlet gas 400°C
The 7500 m^3 gas/tonne powder is equivalent to 7500 × 0.94 = 7050 kg gas/tonne powder

	kg cal/tonne powder	
Heat in (above 20°C)		
7050 × (362 − 20) × 0.24 =	579 000	
Heat out (above 20°C)		
Water evaporated, 500 kg from Table 10.2.		
Latent heat required = 500 × 550 =	275 000	(47.5%)
Radiation and convection losses, say 5% of input =	29 000	(5.0%)
In gas exhausted from the tower = 7050 × (100 − 20) × 0.24 =	135 000	(23.3%)
Errors by difference)	140 000	(24.2%)
	579 000	

Example 2. Inlet gas 300°C
From Table 10.3 12500 m^3 gas/tonne powder, equivalent to 12500 × 0.94 = 11750 kg gas/tonne powder

	kg cal/tonne powder	
Heat in		
11750 × (272 − 20) × 0.24 =	711 000	
Heat out		
Water evaporated, 500 kg, as before.		
Latent heat required = 500 × 550 =	275 000	(38.7%)
Radiation and convection =	35 000	(4.9%)
In gas exhausted from the tower = 11 750 × (100 − 20) × 0.24 =	226 000	(31.8%)
Errors (by difference) =	175 000	(24.6%)
	711 000	

In reality, conditions must adjust so that the heat inputs and outputs balance, which supports the view that the mass of air actually used is lower than that indicated by Table 10.3.

The major loss of heat from a spray drying tower is in the exhaust gas, which is far from saturated with water vapour. When this situation arises in other types of drier, such as the band type once widely used for toilet soap, it is usual to recycle part of the air; and so to ensure that the air finally discharged is relatively well saturated, with consequent improvement in thermal efficiency. The same principal applies to spray driers, but there are problems, particularly when the gas to be re-cycled contains some organic particles and organic vapour which are likely to be darkened by excessive heat. Masters (1979) discusses several cases related to spray drying in general:

- completely closed systems with gas tight drying chambers. These employ indirectly heated inert gas, such as nitrogen, and are used with hazardous materials, which are highly inflammable, explosive in the presence of oxygen, or toxic. The water evaporated from the product must be condensed and removed from the circulated gas.
- partial recycle systems, intended to utilize waste heat. Sufficient air must be exhausted to carry away the evaporated water; and it is usually necessary to increase the exit gas temperature to some extent in order to maintain the moisture content of the dried powder at the required level.
- vented closed systems, in which the major part of the gas from the tower is recycled. Some plants with indirect heaters include scubber/condensers in which the water evaporated from the product is removed from the system as liquid. Other which use direct products of combustion must introduce the air needed for combustion and vent at least an equivalent quantity of gas.

Formerly it was not normally considered advisable, or worthwhile, to recycle gas from NSD powder driers, but the increasing need for better thermal efficiency since the relative increases in fuel costs have stimulated interest and practical applications. Ballestra (*Ballestra Technical News* **4**, May 1983) describes two methods for recovery of heat from spray drier exhaust gases; partial recycle, and indirect preheating of air supplied to the furnace, or of water to be used in slurry preparation, by the portion of the gas which is finally exhausted. Fig. 10.3 shows diagrammatically a Ballestra plant which incorporates both features. For clarity, the additional ducting often included a permit con- as well as counter-current operation has been deleted; and other aspects of the plant are considered later. Figs. 10.4 and 10.5, also courtesy of Ballestra, provide data concerning the performance of these items. Fig. 10.4 relates T_f, the furnace outlet temperature, to the fraction of recycled gas, x, for various values of T_i, the tower inlet temperature. Fig. 10.5 relates the thermal efficiency of the system to ΔT_r, the value by which the temperature of the air to the furnace is increased in the scrubber/preheater which also purifies the exhausted air. The thermal efficiencies shown in Fig. 10.5 for $\Delta T_r = 0$, no use of a preheater, are similar to, but not identical with, those estimated in Table 10.1. The use of these diagrams is illustrated by the following example:

> $T_i = 300°C$, with no recycle and no preheat. $T_i = T_f$, and thermal efficiency is about 67%. Recycle to make $T_f = 600°C$ requires x = 0.6, and, with no preheat, thermal efficiency is about 83%. With, in addition, air preheat from 20°C to 50°C, $\Delta T_r = 30°C$, and the efficiency rises to 88%.

Fig. 10.6 shows diagrammatically an actual installation recently completed by Peabody Holmes Limited. It has been found that about 35% of the tower exhaust gases can be recycled with a fuel saving of 10–15%.

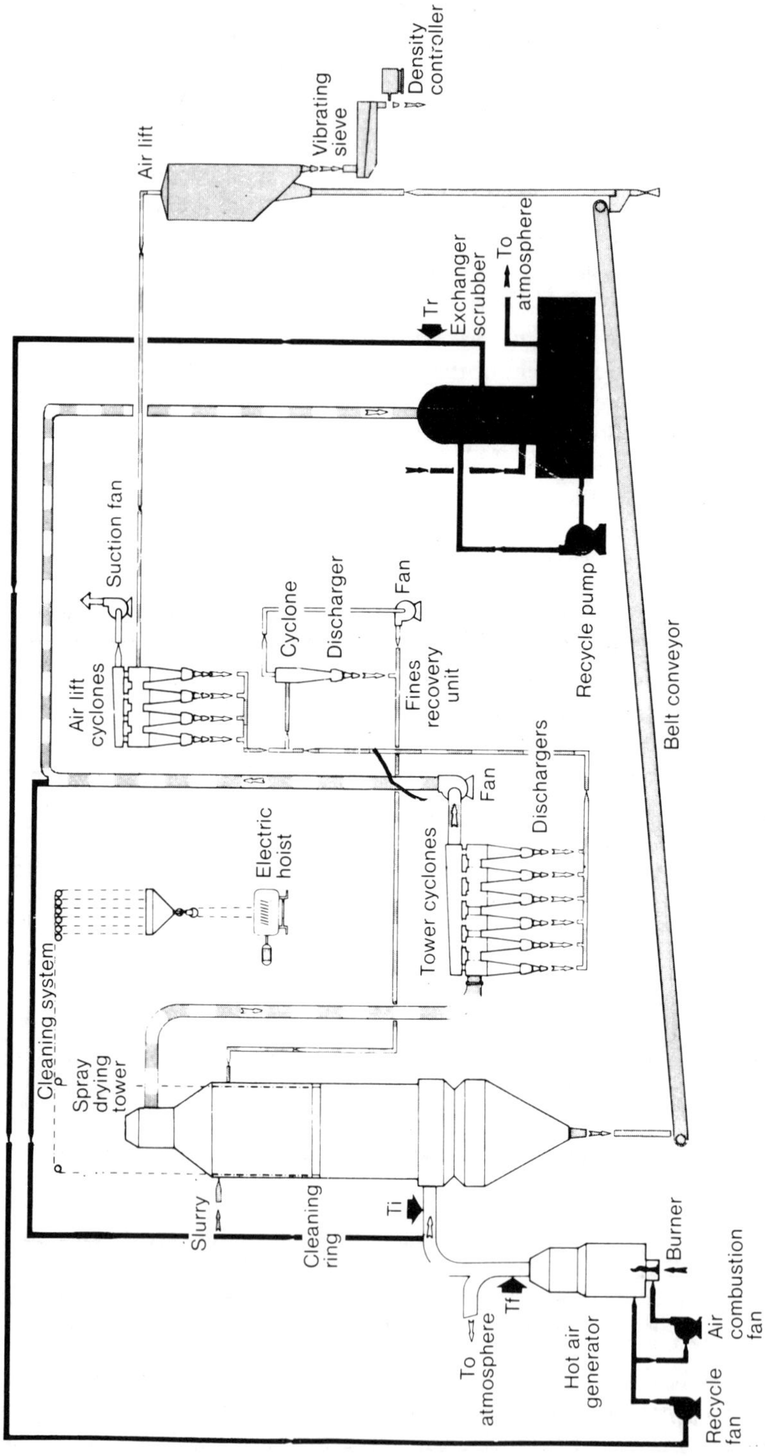

Fig. 10.3 – Ballestra spray drier with heat recovery from the exhaust gases.

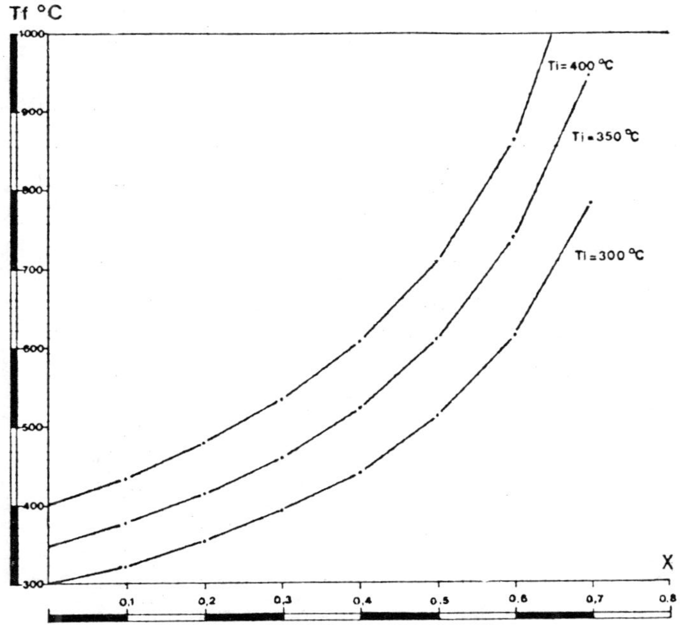

··: *Hot air generator outlet temperature (Tf) as a function of the recycle fraction (x) for different values of tower inlet temperature (Ti).*

Fig. 10.4 – Ballestra diagram for heat recovery from exhaust gases. I.

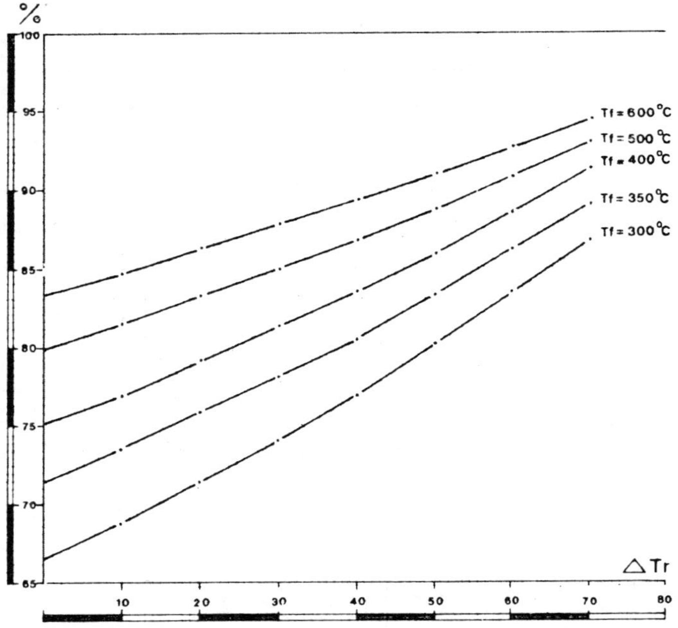

·–: *Efficiency of the system as a function of the hot air generator outlet temperature (Tf) in relation to the temperature increase obtained (△Tr) on pre-heating.*

Fig. 10.5 – Ballestra diagram for heat recovery from exhaust gases. II.

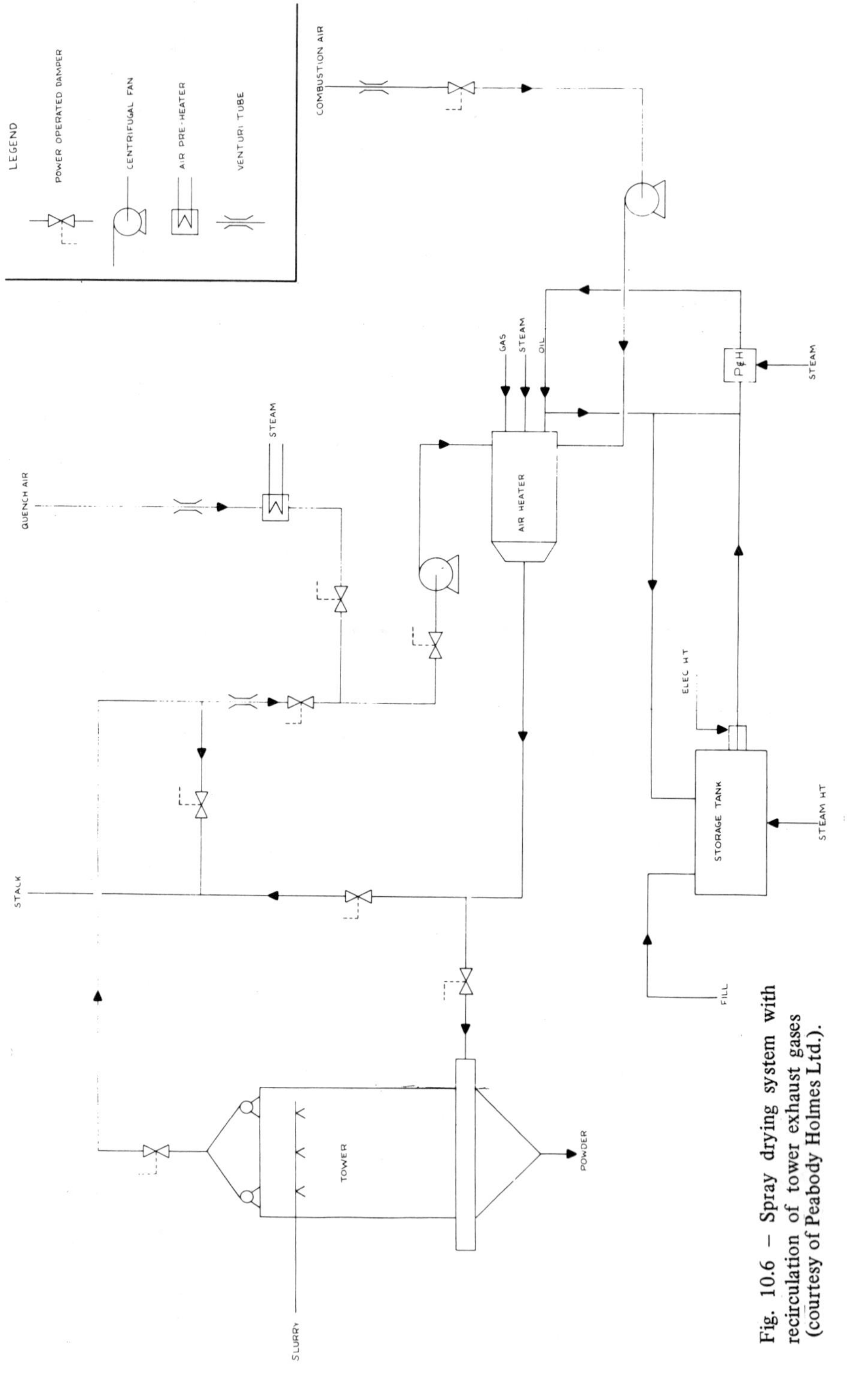

Fig. 10.6 – Spray drying system with recirculation of tower exhaust gases (courtesy of Peabody Holmes Ltd.).

In addition to recovery of heat from the tower exhaust gases, improvements in thermal efficiency have been achieved by improvements in furnace design (discussed in section 10.3.3.6), and by utilization of waste heat from sources outside the spray drier, for example:

- the use of hot air, say at 80°C, from sulphur burning sulphonation plants.
- the use of spare low pressure steam to preheat the air to the furnace.
- the use of boiler flue gases in the hot gas furnace. An installation of this type was supplied by Peabody Engineering Corporation to the Baltimore, USA, factory of Lever Brothers Company and described in *Power,* June, 1983. Oil or gas can be used as the fuel; and boiler flue gas or ambient air can be the quench, or diluting, gas. The decision to use the boiler flue gas in this way was taken as an alternative to the installation of boiler economizers.
- Ballestra has mentioned the recent installation in Italy of a power station using gas turbines, the exhaust gas from which, at 450–480°C, is used in the spray drier. This is an interesting modern analogue of the traditional back-pressure, or pass-out, steam turbine system for the production of power and process steam.

10.3.3 Spray drying practice

10.3.3.1 *General*

Spray drying plants for the production of NSD powders include most, or all, of the following elements:

- equipment to handle the raw materials and to prepare the slurry.
- equipment for the after-treatment of the slurry:
  - ageing.
  - removal of lumps and grit.
  - de-aeration.
  - aeration.
- pumps and nozzles for atomization.
- a tower.
- a hot gas generator; and fans to move the gas and to maintain appropriate pressures throughout the system.
- equipment to collect, cool, and weather; and to convey the powder.
- equipment to dose the blown powder with ingredients which are not spray dried. Powder storage, handling and packing facilities.
- equipment for the separation of dust and volatiles from the exhaust gas. Disposal of recovered fines.

10.3.3.2 *Raw materials handling and slurry preparation*

It is important to produce slurry which is free from lumps, or grit, and of a suitable consistent viscosity. For reasons discussed earlier, it is usually desirable to make the slurry as concentrated as possible, but additional water may sometimes be used to increase puffing and so reduce bulk density. The hydrotope STS, or SXS, is sometimes included to thin the slurry, and so permit the use of a lower slurry water content, as well as to improve powder quality. The properties and method of addition of the sodium tripolyphosphate are important. The basic chemistry of sodium tripolyphosphate is described in section 1.5.10, where its rate of hydration, measured by its temperature rise test, and controlled by its form 1 content and particle size distribution, is discussed. Formation of sodium tripolyphosphate hexahydrate in the slurry is desirable in order to produce a crisp powder, but a tripolyphosphate which hydrates rapidly, as indicated by a high temperature rise test, can cause problems in slurry making and handling, due to the formation of lumps and grit, and thickening of the slurry on standing. These difficulties can be overcome by good slurry making procedures, so that high temperature rise polyphosphate is normally selected by detergent manufacturers who can handle it satisfactorily, whereas lower temperature rise material may be needed by those with poorer equipment. Some suppliers of sodium polyphosphate can supply it in a prehydrated form, a procedure developed by Hoechst-Knapsack.

Slurries can be made batchwise or continuously, but continuous production is now usual, except for small plants. Various designs of mixer can be used for batch slurry making, including types with fixed and moving arms, and some with central worms and outer arms. Local excesses of powder, liable to form lumps, should be avoided. Turbo mixers can be particularly good, but they are specially suitable in continuous systems, because they need a nearly fixed liquid level, such that the ingredients can be dropped into the vortex and so immediately swept through the mixer in a stream of circulating liquid. If the liquid level is too high, the ingredients may not be drawn properly into the circulating liquid, whereas if it is too low, the eye of the impeller may become dry and have powders stick to it. Mixers should usually be designed and operated to minimize aeration of the slurry. In a batch system, one mixer can be used, together with an agitated holding/ageing vessel large enough to hold more than a mixer charge. More usually, two, or more, mixers are installed. Fig. 10.7 shows the Ballestra Dosex S system for continuous slurry preparation which is an improved version of earlier Dosex systems to which it is similar in principle. In the Dosex S system all ingredients, solid and liquid, are weighed in small increments using load cell units with automatic tare, so that correction is made for any material which remains in the weighers when they should have emptied. Mechanical weighers can also be used for solids, and dosing pumps, or other volumetric measuring devices, for liquids. Modern Dosex units include devices by which materials are first fed at a high rate, but, to improve accuracy of weighing, the increment is completed at a low rate of feed. The solid and the liquid ingredients are premixed as shown, and then, together with recovered slurry (see later), fed to the main turbo-mixer, from which the slurry passes to the ageing vessel. Other plant designers achieve similar results in somewhat different ways.

For large plants, the powdered raw materials are normally received in bulk; and are then conveyed to the day silos as shown on the left of Fig. 10.7. Minor solids are often premixed before being dosed; and the fourth unit is for this purpose. Batch slurry making units can also be fed from bulk handling installations; but small plants, and those remote from the chemical plants in which the ingre-

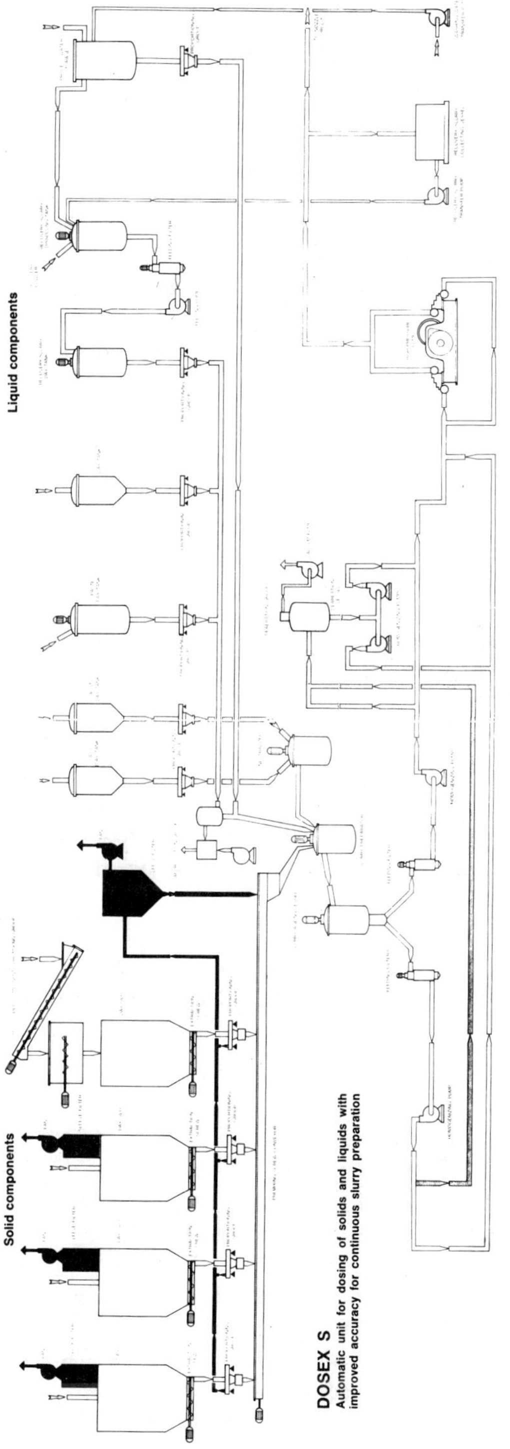

Fig. 10.7 – Ballestra 'Dosex S' continuous slurry preparation system.

dients are made, often use powders received in bags. As mentioned in relation to soap powders, some facility to deal with caked powder may be needed. The anionic surfactants are commonly handled as liquid pastes prepared in an adjoining sulphonation plant, from which they are pumped to the powder plant; but other systems are possible:

- sulphonic acid, and/or fatty acid, can be fed directly to the slurry making system; Fig. 10.7 shows feeders for acid and caustic soda, together with a mixer in which neutralization takes place. Factories which have no sulphonation plant may find it convenient to buy anionic surfactants in the form of sulphonic and, if required, fatty acids. If the paste needs to be bleached particular care is necessary to eliminate any residual available chlorine with this system.
- the surfactant can be bought in the form of high active flakes, or powder, containing about 80–85% surfactant, together with sodium sulphate and a little water.

Most nonionic surfactants can be handled as liquids, as can sodium silicate, although it may be convenient to mix the silicate into the anionic surfactant paste in the sulphonation plant. Lather boosters, such as coconut ethanolamide, can be handled as flake or as liquid.

When slurry is made batchwise, the main liquid ingredients are charged and the powders added to them, as is necessary to maintain a mixable liquid through the process. Some authorities say that the order of addition is so important that it is considered a commercial secret. Order of addition does not arise with continuous mixing systems in which all ingredients are added simultaneously. The mix must be kept alkaline to avoid decomposition of silicates, soaps and some other types of active; a point which is likely to arise mainly if sulphonic, or other, acid is used as an ingredient. Ballestra recommends neutralization to pH 8.5–9.0. Another authority says that it is important to maintain a pH above about 9.6 (1% solution) to avoid grit formation due to a phosphate-silicate complex.

Under most circumstances advantages of continuous mixing are:

- labour is saved, particularly when the scale of operation is large.
- close control of slurry composition by automatic means is possible.
- optimum mixing conditions can be achieved, and, within reason, the time and temperature of the ageing period can be selected and kept uniform. This often permits the use of lower moisture content slurries than from batch mixers, with the advantages discusssed earlier. Thickening, which is usually minimized by continuous slurry making, is often due to continued hydration of the sodium tripolyphosphate, but it can also be caused by hydration of sodium carboxymethyl cellulose, or to the effects of electrolyte on soap.

10.3.3.3 *After-treatment of slurry*

Some ageing of the slurry is desirable for the formation of sodium tripolyphosphate hexahydrate, as well as necessary to provide a small buffer from which the drier can be fed; about 20–30 minutes is often suitable.

The formation of grit and lumps is minimized by good slurry preparation as just discussed, but holes in the atomization nozzles are small, and filtration is needed to avoid undue risk of blockage. Coarse filters are used to protect the system against metal fragments and to remove any large lumps. The main, finer, filters are often self-cleaning; and magnetic filters are sometimes included to remove ferrous particles. In addition, so-called colloid mills, or homogenizing pumps, may be used to break down grit and so to improve the smoothness of the slurry.

Air is introduced into the slurry with the powders, as well as possibly by the mixer; and under some circumstances, such as the neutralization of acids with caustic soda containing a little carbonate, gas may be produced by chemical reaction. Air, or other gas, in the slurry increases the tendency for the powder particles to be puffed, and so to reduce bulk density. A deaerator, in which the slurry is sprayed into an evacuated chamber, as shown diagramatically in Fig. 10.7, is often included in the system to ensure slurry uniformity, and to increase bulk density. Less usually, air is injected into the high pressure line to the nozzles by an air compressor in order to reduce bulk density.

10.3.3.4 *Atomization*

As explained earlier, single-fluid, high-pressure nozzles are almost invariably used in spray drying plants for the production of detergent powders. The pressure is developed by a reciprocating pump which has a number of cylinders (often three, four or six), followed by a surge vessel further to smooth pressure fluctuations. The high-pressure pump has a variable speed drive and its speed determines the rate at which slurry is fed to the tower. The pressure actually developed depends in the usual way upon the rate of flow from the pump and the resistance of the nozzle and pipe-line system.

The pump is commonly designed to develop a maximum of 1,000 psig (69 bar), or even 1,500 psig (103 bar). Protection against overpressure is usually by bursting disc, as relief valves would be likely to become blocked. It is necessary to feed the high-pressure pump with slurry at a substantial positive pressure provided by a low-pressure pump. This pump must automatically supply the volume of slurry being taken at any time by the high-pressure pump. Direct acting steam pumps, such as the traditional Worthington–Simpson type, can be used, but centrifugal pumps are now usual. The low-pressure pump may also act as an homogenizer. Duplicate pumping systems may be installed to provide standby facilities.

A number of nozzles fed by a ring main are normally inserted near the top of the tower as shown diagrammatically in Fig. 10.8. More nozzle connections than are required are provided so that:

- the number of nozzles in use can be selected to suit the requirements of the product being made.

- the sprays can be arranged to cover the whole tower, with particular care to avoid the spraying of slurry onto the tower walls.
- a clean nozzle can rapidly be brought into use to replace one which has become blocked. At one time nozzles on their lances were taken out to be cleaned; now cleaning is often carried out with steam controlled remotely.

The use of several nozzles also minimizes the disturbance to tower conditions when a nozzle becomes blocked. Some outline flow sheets show a return line from the nozzles to slurry preparation; it is not known if this is used during normal operation, or only for cleaning; probably the latter.

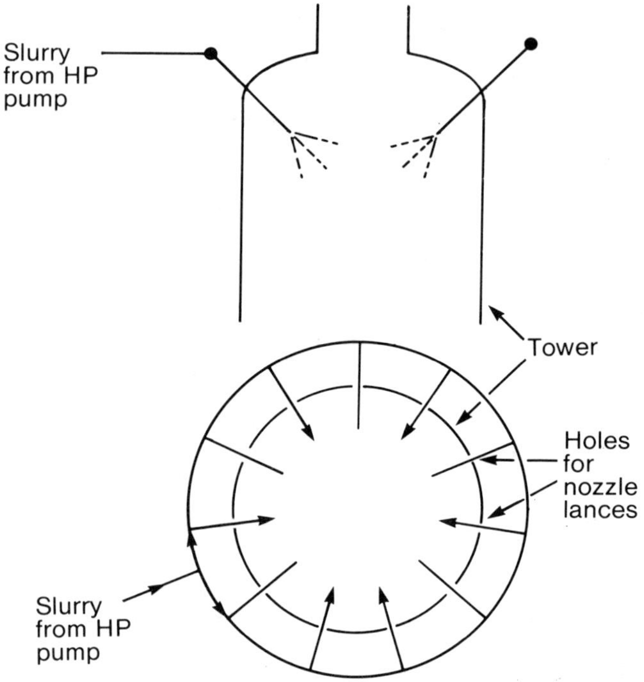

Fig. 10.8 – Arrangement of nozzles at tower top.

One design of nozzle which has been found satisfactory for detergent powders is shown in Fig. 10.9, courtesy of Spraying Systems Company. The slurry enters the whirlchamber through a tangential hole and leaves as a hollow cone through the orifice insert. The conical sheet of liquid breaks into droplets. The nozzle bodies are in stainless steel; and the replaceable and interchangeable whirl chambers and orifice inserts in hard wearing tungsten carbide to cope with the abrasive slurry. These replaceable parts can be chosen with hole sizes to suit particular requirements. Other nozzle designs are possible; and Fig. 10.10, also courtesy of Spraying Systems Company, shows another type. Nozzles can be designed to give various spray angles. Some use hardened steel wear resistant parts rather than tungsten carbide. The capacity of a type TC Whirljet nozzle is controlled by:

- the size of the holes in the orifice and in the whirl chamber.
- the liquid pressure. Capacity increases approximately in proportion to the square root of the pressure.

The size of the liquid droplets formed, which is not the same as the size of the dried particles because of puffing, is controlled by:

- the size of the holes in the orifice and the whirl chamber.
- the liquid pressure.

Fig. 10.9 – Type TC Whirljet spray nozzle *ex* Spraying Systems Company.

- the spray angle. With a given nozzle, this tends to increase to a maximum with increasing pressure, and then to decrease, but sometimes only the decreasing part of the curve is involved.
- the liquid viscosity and surface tension, although these generally have only a small effect.

In general, to produce larger droplets:

- use larger size orifices and whirl chamber holes.
- use lower pressures.
- use narrower spray angles.

Increases in viscosity and in surface tension also favour larger particles.

It is necessary to control both the atomizing conditions and the rate at which the slurry is sprayed. Slurry pressure affects both; and it is sometimes desirable to increase, or decrease, the number of nozzles in use in order to achieve the best conditions. Other conditions fixed, an increase in slurry pressure increases the slurry flow, decreases the gas outlet temperature, increases the water content of the powder, and increases the powder bulk density.

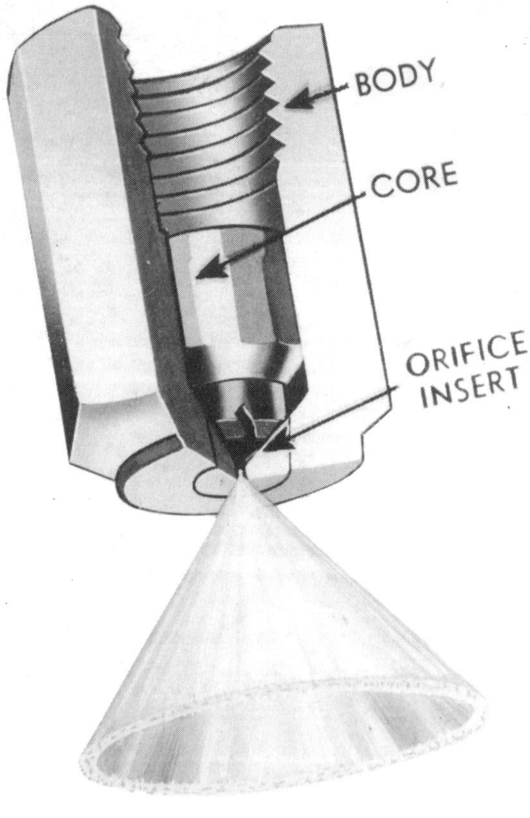

Fig. 10.10 – Type R Spraydry nozzle *ex* Spraying Systems Company.

10.3.3.5 *The tower*

Towers differ in detail, but the designs offered by most, if not all, manufacturers of plant for the production of granular detergent powders are broadly as shown in Fig. 10.1; and they are made of mild steel.

Little detailed design information is generally available. The cross-sectional area of the cylindrical portion for a given output of powder is decided by the volume of drying gas needed and the velocity up the tower. Too high a linear gas velocity would prevent proper descent of the particles; and would lead to an excessive carry over of fine material in the exit gas. Too low a velocity would tend to reduce residence time and impair drying. A velocity of about 0.6–0.8 m/s seems typical. The jets are usually positioned some distance below the top of the cylindrical portion of the tower to minimize carry-over.

Towers are usually quite tall, say 20–35 m overall, but it is difficult to decide the minimum height which would be satisfactory.

As mentioned earlier, some cold air is drawn into the base of the cone below the hot gas inlets. This cools the powder to perhaps 60–70°C.

Despite efforts to avoid the impingement of wet slurry, or damp powder, onto the tower walls, material does normally collect on the walls. This material is liable to darken under prolonged influence of the hot gas particularly near the gas inlets; and if allowed to build-up unduly, heavy falls can occur with disruption of the smooth operation of the plant and contamination of the product with brown particles. Regular cleaning of the tower is necessary. Ballestra usually installs a cleaning ring which can be moved up and down just inside the walls during production as indicated in Fig. 10.3. Other devices used include an air-broom to blow material off the walls; and manual cleaning by men in boatswain's chairs is also possible.

10.3.3.6 *The hot gas generator and the fans*

Indirectly heated air is used in spray driers for delicate foods and pharmaceuticals, but direct products of combustion are employed almost, if not quite, universally for detergent powders, because this system is convenient and cheaper, particularly with the relatively high temperatures required for efficient operation. The author is indebted to Peabody Holmes Limited for much of the information used in this section concerning more recent developments.

Furnaces have been designed capable of handling quite heavy distillate fuel oil, but, with changes in relative fuel prices, natural gas is now usually the most economical as well as the most convenient fuel. Advantages of gas are:

- it is clean and relatively simple to use.
- it contains little, or no, sulphur compounds and hence does not produce any appreciable amounts of sulphur oxides when burnt, as does fuel oil.
- it has a greater turndown capability; that is the furnace can, if required, be worked at a lower fraction of its full capacity.
- it can be used with 'all metal' furnaces.

At one time the furnaces were equipped with stacks. These were used to enable the furnace to be heated up very slowly, so as to minimize damage to the refractories through thermal shock. Dampers were provided, so that hot gas could be fed to the tower when required, but diverted to the stack whenever the drier ceased to produce. The furnace was thus kept in operation unless a prolonged plant stoppage was anticipated. In some cases, safety devices were installed which automatically diverted the furnace gas to the stack if, for example, the tower exit gas temperature exceeded a predetermined value. When ducting is provided to permit con-, as well as counter-current operation, the vertical section to the top of the tower can provide most of the stack. This system of operation involves considerable waste of fuel, including that caused by leakage of hot gas past the dampers. With improvements in refractory quality and greater reliability of spray drying plants, the modern trend is not to install a stack when the fuel is gas. Furnaces are started and stopped very quickly, as the very considerable fuel savings justify annual replacement of refractories, if necessary. If required, water can be sprayed instead of slurry whilst the proper conditions are established in the tower. Particularly when the fuel is oil, it is necessary to carry out the combustion with only a relatively small excess of air, so that the temperature in the combustion zone is high and no smoke is formed. The products of combustion are then diluted with quench air, or other gas as discussed in section 10.3.2, to produce a gas at the temperature required for the spray drier. With oil fuel a stack is still needed to discharge smoky gas produced whilst the furnace is being heated. Soot in the drying gas can damage the colour of the product.

Fig. 10.11, courtesy of Peabody Holmes Limited, shows a fairly conventional horizontal air heater suitable for liquid or gaseous fuel; and Fig. 10.12, also courtesy of Peabody Holmes Limited, a modern 'all-metal' horizontal air-heater. In the 'all-metal' design, which actually includes a refractory ring just around the burner, the very hot gases, diluted as quickly as possible, are carried in a tube of heat-resistant alloy steel. This type of furnace, used only for gaseous fuel, has the following advantages over the traditional refractory lined type:

- rapid start-up and shut-down, with a very fast response to process needs because it does not contain refractory with a large heat capacity.
- lower maintenance costs.

Furnaces can also be of vertical design. The suppliers of complete plants, such as Ballestra, Lurgi, Mazzoni, Meccaniche Moderne, and Wurster and Sanger, include furnaces. Another specialist manufacturer of furnaces, particularly for gas, is John Thurley Limited.

Fans are required to move the drying gas and to maintain appropriate pressures at various key points in the system. The way in which the flow through a fan is established so that the pressure induced by the fan balances the resistance of the system is described in Appendix 3. An exhaust fan is used after the tower and cyclones, together with the fans needed by the furnace. These are often a fan for combustion air and another for quench air, but a hot gas fan to draw the gases through the furnace is sometimes used.

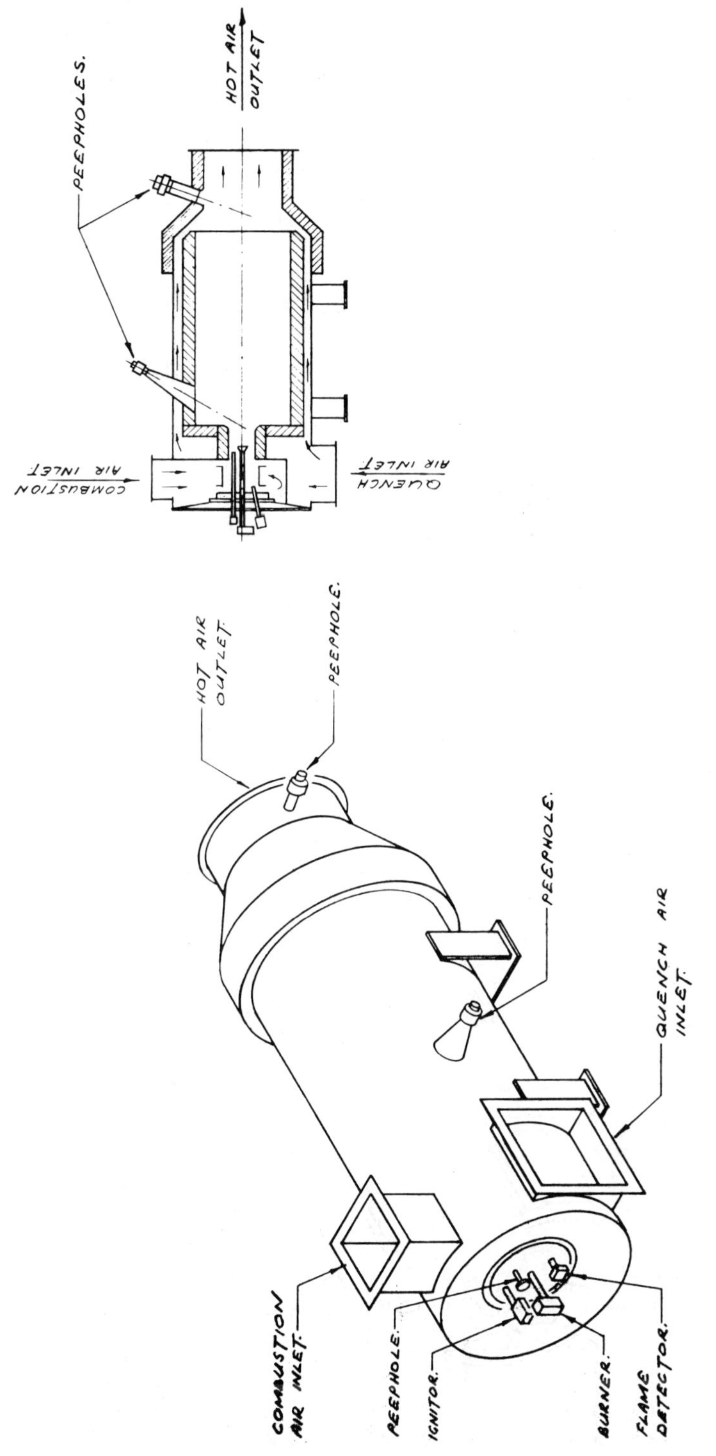

Fig. 10.11 – Peabody Holmes refractory lined furnace.

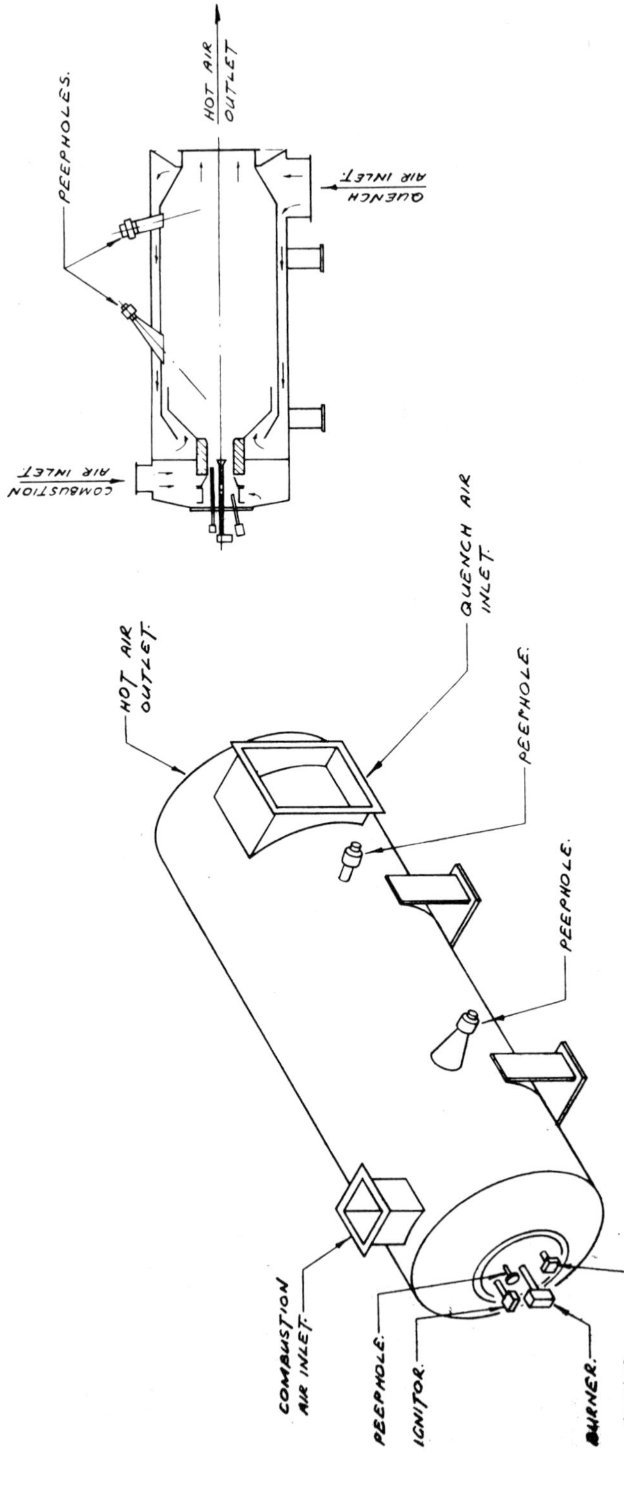

Fig. 10.12 – Peabody Holmes 'all metal' furnace.

10.3.3.7 *Treatment of the powder after the tower*

The powder which leaves the tower at 60–70°C requires to be:

- cooled further and weathered; that is salts allowed to crystallize in hydrate form.
- mixed with ingredients not included in the slurry, such as perfume, sodium perborate, and enzymes.
- screened to remove any coarse material, which needs to be re-slurried (this may be done before, or after, dosing with ingredients, or both).
- conveyed directly to the packaging machines, or to intermediate storage.

A typical arrangement is shown in Fig. 10.13, courtesy of Lurgi. Cooling and weathering is normally done in an air lift which also separates lumps, which are not lifted, and fines which pass the separating chamber to be recovered as discussed later. The powder is also elevated to a level from which it can flow by gravity through much, or all, of the rest of the system. The air lift duct and fan must be selected so that the volume of air per tonne of powder is sufficient to cool the powder to about 35°C; and the linear velocity is high enough to carry it. Masters suggests that a linear velocity of over about 20 m/s is required for most spray dried products. Two fairly old NSD installations known to the author used:

	Linear velocity (m/s)	Volume of air m^3/tonne powder
Plant 1	31	2000
Plant 2	14	2800

More accurate design figures are doubtless available to those involved in the construction of modern plants.

Facilities for subsequent treatment of the cooled powder need to be tailored to the requirements of a particular factory, but points to be considered are:

- reasonably accurate dosing of ingredients to be added is needed. Weigh bands are usual for the base powder and solid ingredients; and volumetric dosers, including positive pumps, for liquids.
- mixing needs to be thorough but gentle, so that delicate hollow particles are not broken unduly. Rotating drums are common, but fluidized bed mixers can also be used.
- care is necessary to minimize segregation of components in the conveying and storage system. Fluidized bed conveying troughs and re-mixers are sometimes useful, but careful design is necessary, as separation can result if the conditions are not right. When belt, or similar, conveyors are used, it is often better to use a cascade system with a re-cycle of surplus powder, rather than one in which powder is dropped, or ploughed, from a single belt into the various packing machine hoppers.
- proper enclosure of, and dust collection from, the powder handling system to minimize dust in the packing room and associated areas is important; when enzymes are used it is vital.

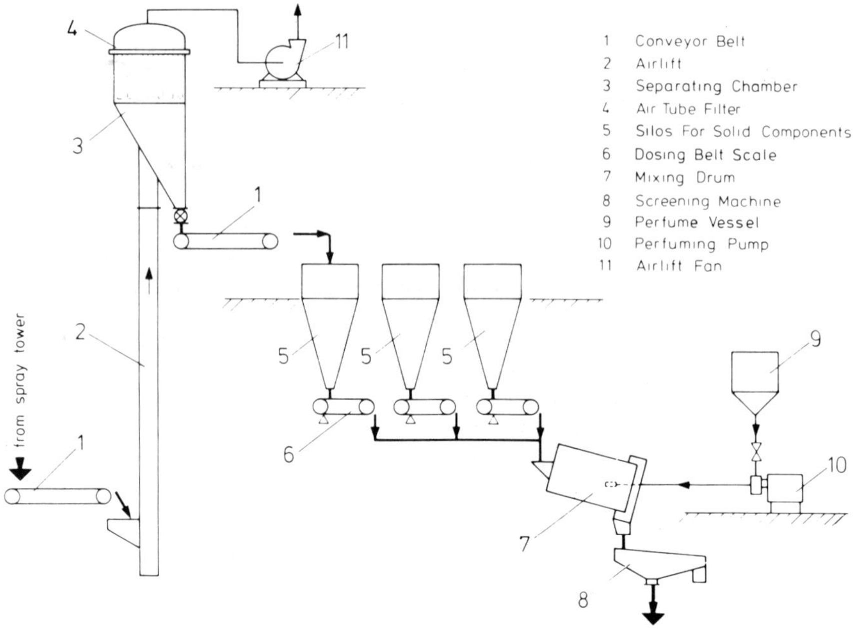

Fig. 10.13 – Post treatment of blown powder.

The design of packing machines is outside the scope of this book. A declared net weight is a legal requirement in many countries, but the machines commonly use volumetric feeders which may be adjusted automatically by some form of weight monitoring device. Careful control of the bulk density of the finished powder is important.

10.3.3.8 *Recovery and disposal of fines; and cleaning of the effluent gases*

The gas from the tower top, at around 100°C, contains:

- fine powder.
- water vapour.
- organic vapour and fume, possibly from unsulphonated matter in the anionic surfactant paste, lather booster, or nonionic surfactant, depending upon the composition of the slurry.
- sulphur dioxide, and a little sulphur trioxide, if the fuel used contains sulphur.

Air containing fine particles, but little of the other consituents, is emitted from the air lift and the powder handling systems. It is normally necessary to treat all these gas streams to recover valuable material, and to prevent nuisance. The minimum standard of purity of the air discharged which must be achieved to meet legal and other requirements depends upon the location of the factory, but standards are tending to rise.

Types of equipment to be considered include:

- cyclones
- bag filters.
- wet scrubbers of various kinds.
- electrostatic precipitators, although these are not usual.

It is normally desirable to recover as much dry powder as possible. This may be:

- mixed with the main product, although it is fine.
- agglomerated with other fine particles in a unit such as the 'Pulmix' discussed in section 10.4.3.
- returned to the tower to be agglomerated by being blown into the wet spray from the nozzles. The Ballestra arrangement to do this is indicated in Fig. 10.14 which also illustrates other points. The problem is to ensure that the fine particles are stuck together sufficiently securely not to be broken apart as the powder is handled.
- used in products such as scouring powders.

Recovery of dry fine powder is commonly achieved in cyclones, which can be of various designs. At one time large, rather low efficiency, but also low-pressure drop, types were often used, sometimes followed by batteries of small diameter high efficiency units. Modern practice tends to batteries of medium size, high efficiency, but rather high-pressure drop, cyclones, as shown in Fig. 10.14. Cyclone design is a complex topic with an extensive literature not considered here; but an important operating point is to ensure that there is no substantial in leakage of air through the powder discharge seal, as this can disturb the performance of a cyclone.

Bag filters also recover powder in dry form and, with the right bag fabric and good maintenance, they tend to be more effective in the removal of very fine particles than most cyclones. Bag filters are often used for the air from air lifts and packaging systems, but they are not advisable for tower exhaust gases, because of the content of water vapour and organic fumes which can lead to tacky deposits on the bags. Various designs of bag filter are available. Types cleaned by pulses of air, or by compressed air from rings which traverse up and down the bags, are now commonly installed, instead of the more traditional type with mechanical shakers and flaps to isolate the section being cleaned.

When tower gases are not sufficiently clean as they leave the cyclones, additional cleaning is usually effected by some kind of wet scrubber. Ballestra uses a tower as shown in Fig. 10.14 and liquor containing about 10–12% solids is recirculated as the main scrubbing agent. Fresh water is added to maintain the required concentration, that is to compensate for evaporation and for the quantity of powder collected. In the Ballestra unit, some of this water is used to provide a final stage of scrubbing. Sodium chloride solution is added when necessary to help control any undue development of foam. When appropriate it is also possible to add caustic soda to absorb sulphur dioxide and trioxide. Some years ago, Type N Rotoclones, as shown in Fig. 10.15, courtesy of AAF

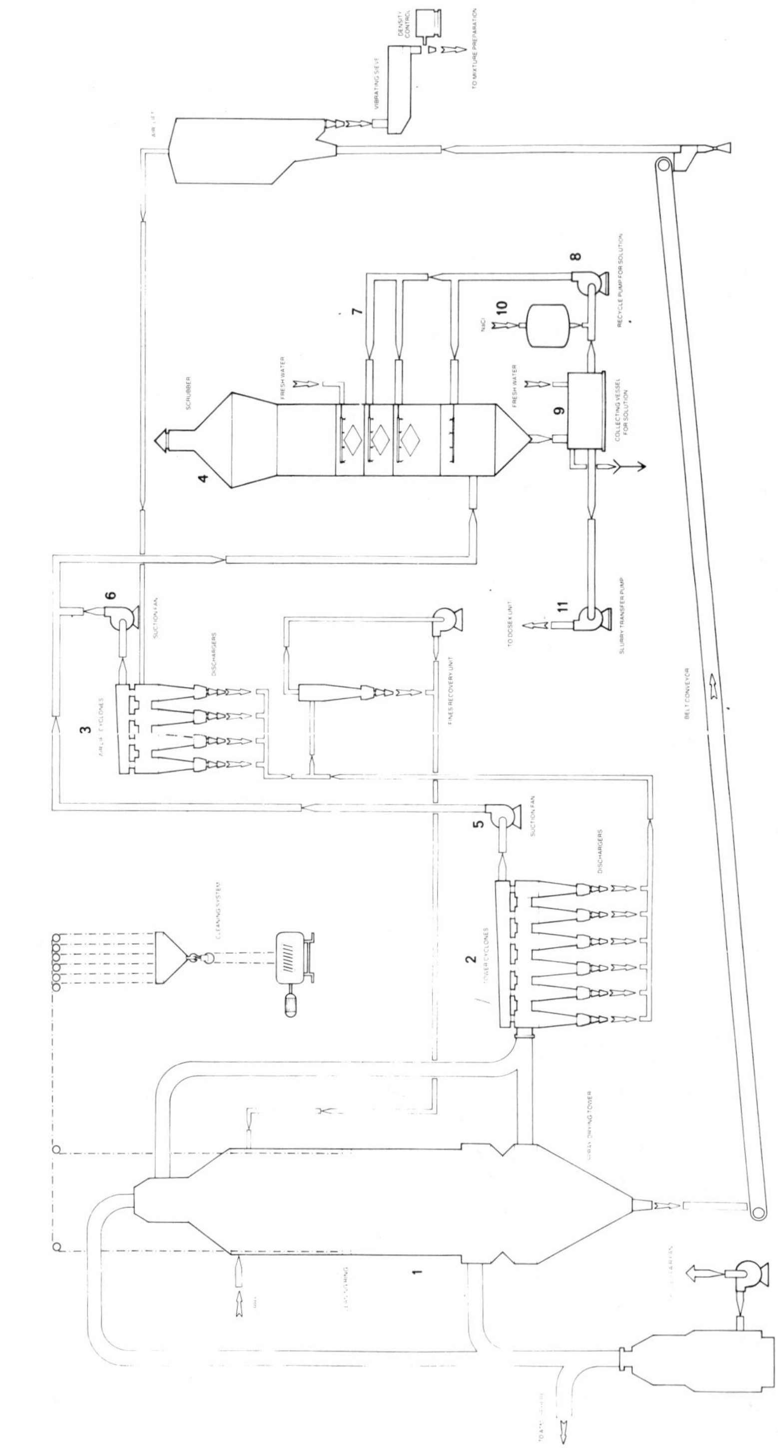

Fig. 10.14 – Ballestra plant showing wet scrubber and other features.

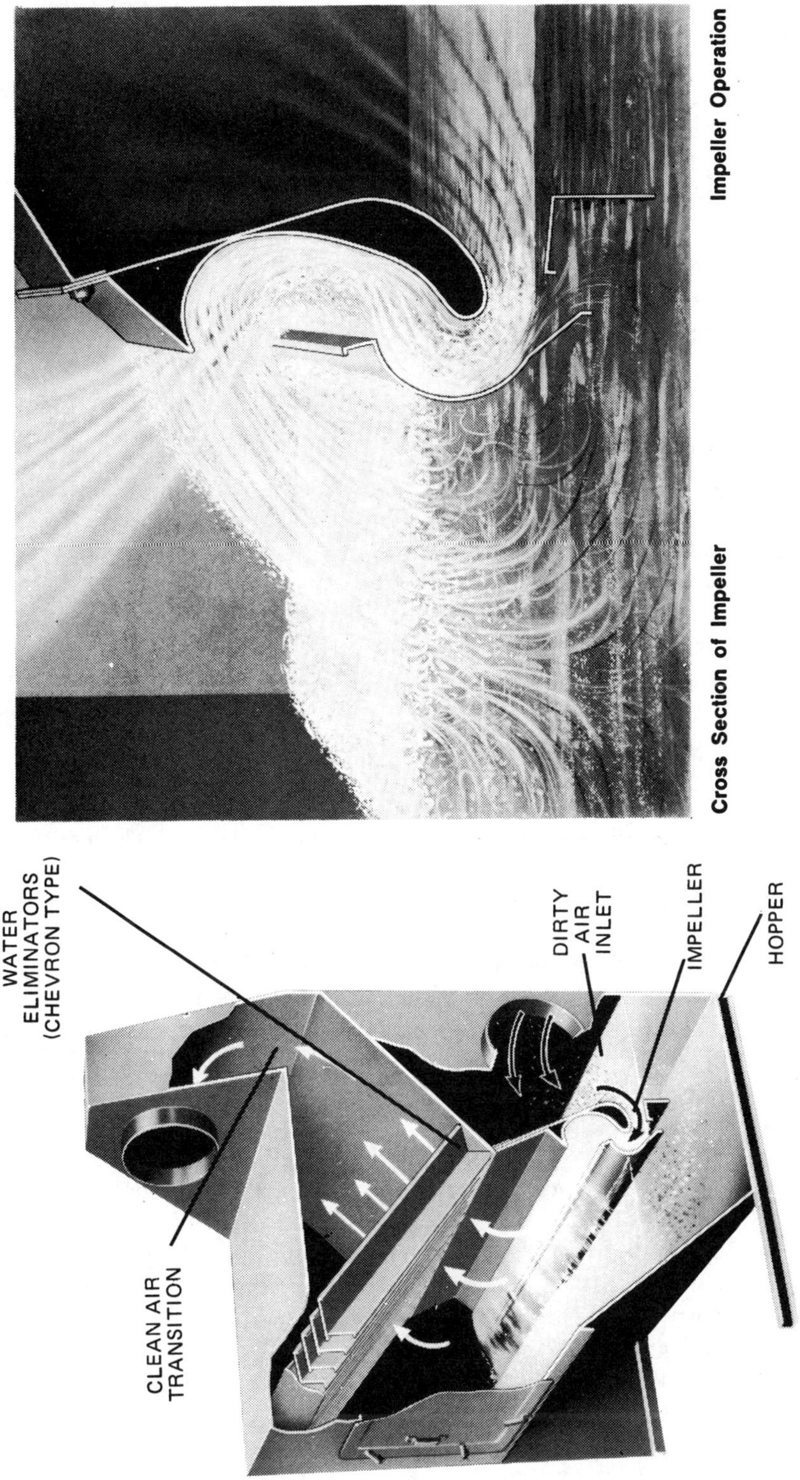

Fig. 10.15 – Rotoclone type 'N' wet scrubber (courtesy of AAF Ltd.).

Limited were used successfully in a number of installations making high lather powders. A high concentration slurry, with about 30% solids, was used to avoid undue foam development. Types of wet cyclone are now sometimes used. The slurry produced in a wet scrubber is returned to the system in which the slurry to be blown is made. It has sometimes been found that powder made when scrubbing slurry is recycled is less bright and crisp than when it is replaced by water. For this, as well as economic reasons, it is advisable to collect as much powder as possible in dry form.

The recovery of heat from the tower exit gases, by direct recycle and indirectly, is discussed in section 10.3.2.

10.3.3.9 *Complete spray drying plants*

Figs. 10.3, 10.7, 10.13 and 10.14 indicate broadly how the sections are combined to form a complete plant, but, as discussed in relation to the different elements, various alternatives exist. In early plants, the elements were controlled manually on the basis of simple instruments; and off-stream measurements of powder moisture and bulk density. The modern trend for large-scale plants is a high degree of automation and computer control.

Various factors which affect bulk density have been considered in earlier sections. They are now collected in the form of a number of generalizations, but there can be interactions and it is not always possible to predict the effect which a change in conditions would have on bulk density:

- counter-current operation gives a higher range of bulk densities than concurrent.
- an increase in high pressure pump speed, with other factors constant, increases the flow of slurry and the atomizing pressure, which, in turn, gives finer droplets. The increased flow of slurry increases the evaporative load on the tower which decreases the outlet gas temperature and increases the moisture content of the powder. The net result is normally to increase bulk density; and Ballestra has long used a bulk density controller based on this principle.
- higher gas temperatures normally increase puffing and reduce bulk density, unless taken to the stage at which the hollow shells break down to dust.
- an increase in the air content of the slurry reduces bulk density; and conversely.
- by itself a reduction in the water content of the slurry tends to reduce puffing and to increase bulk density. Reduced water content increases slurry viscosity, which is also affected by the inclusion of hydrotopes, the degree of hydration of the sodium tripolyphosphate and other aspects of formulation. Average droplet size tends to increase with increase in viscosity. Slurry composition may also affect the toughness and permeability of the particle skin, so that it is difficult to predict the net result of changes in slurry composition.

10.4 PRODUCTION OF NSD POWDERS WITHOUT SPRAY DRYING

10.4.1 General

The alternative processes to spray drying are mentioned in section 10.2, together with some factors related to their adoption.

10.4.2 Dry mixing

A wide range of formulations can be produced by simple mixing of ingredients. Anionic surfactant is commonly added in the form of a high active flake, or powder, containing about 80–85% of sodium dodecylbenzene sulphonate, together with sodium sulphate and a little water. If sodium carboxymethyl cellulose is included, it is usually advisable to mix it with a substantial proportion of an inorganic salt and to grind in a hammer, or similar, mill in order to disperse the rather cotton-wool like material. It may also be useful to premix the optical brightener which is only added at a low concentration. Most types of powder mixer can be used. As with other processes, it is normally necessary to screen the mix, and if required also some ingredients; and to grind the lumps, or tailings.

Dry mixed detergent powders have a high bulk density around 0.7 g/ml, but this depends to some extent upon the constituents used. Light soda ash is normally selected if sodium carbonate is included; and light phosphates may be available. Dry mixed detergent powders tend to be rather dusty; and to have rather poor flow and storage properties. The addition of a small proportion of a desiccant powder, such as silica gel or specially prepared calcium silicate, may improve storage properties, but experiments are needed to find whether or not, the extra cost is justified.

10.4.3 Processes which include the addition of liquids

Limited quantities of certain liquids can be added to powdered ingredients which include salts which form hydrates, particularly anhydrous sodium carbonate. The sodium carbonate takes up water from the liquid rather as in the production of soap powders, although adsorption on other materials with large surfaces may also occur. In the simple adsorption processes 5–10% of anionic surfactant in the form of a low viscosity solution can be added, as can a few percent nonionic surfactant.

It is also possible to use sulphonic acid, and, if required, fatty acid, in neutralization plus adsorption processes. The neutralization is commonly with sodium carbonate present in sufficient quantity to produce sodium bicarbonate and not carbon dioxide.

$$R.SO_3H + Na_2CO_3 = R.SO_3Na + NaHCO_3$$

In other processes caustic soda solution is added to effect the neutralization. Rather special mixers, some of which, such as the Lodige, include disintegrating blades, are desirable. Other types rely more on dispersion of the particles whilst the liquid is being sprayed. Sulphonic acid from an oleum, or concentrated sulphuric acid process, contains a little water and reacts with sodium carbonate immediately; and care is necessary to add it slowly to avoid local excess. Sulphonic acid from a gaseous sulphur trioxide process does not react until a little water is added. One process is then to mix the sulphonic acid with the dry ingredients before the addition of the water. Special mixers can obviate the need to weather the material for 12–24 hours, for example on a floor, and then to grind, as is otherwise necessary. If sodium hypochlorite is used to bleach the mix, particular care must be taken to eliminate any excess available chlorine before optical brightener is added. It may also be advisable to weather, screen and, if necessary, grind before any addition of sodium perborate. The use of neutralization/adsorption processes provides more flexibility in formulation than simple adsorption; and higher levels of surfactant can be achieved. Processes in which liquid is added favour aggregation of fine particles, so that the products should be less dusty then those produced by dry mixing.

More details regarding these processes, including some formulations, are given in Davidsohn and Milwidsky (1978); and in Davidsohn (1965) and Zilske (1965).

Important applications of adsorption, or neutralization/adsorption, processes, either alone or in combination with spray drying, are offered by Lurgi and by Ballestra. The Lurgi 'Pulmix' spray mix drum, which is a drum in the form of a double cone, is shown in Fig. 10.16; and the speed of rotation and the angle of inclination of the drum can be varied. As shown, the main solid ingredients are fed into chamber A through hopper d on the right-hand side of the diagram; and the liquids are sprayed through nozzles e into a haze of powder lifted by the rotation. The powder is peeled from the top of the drum wall by scraper f. The rotation and inclination of the drum cause the powder to flow into the smaller chamber B, where the centrifugal force is not sufficient to lift the particles fully to the top. The lifting and rolling action in chamber B favours aggregation. If required, fine powder, such as silica gel included as a desiccant, can be added to chamber B through the second hopper d on the left-hand side of the diagram. A flow sheet for one application of the Pulmix unit is shown in Fig. 10.17. In this example the Pulmix is used to augment the powder from a spray drier.

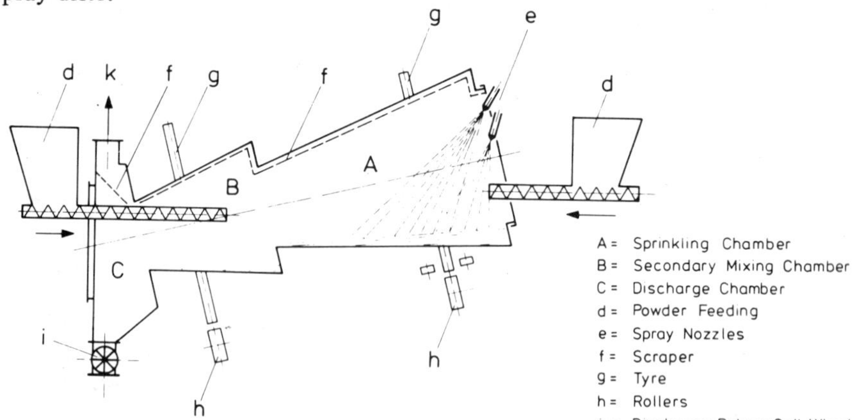

Fig. 10.16 – 'Pulmix' spray mix drum (courtesy of Lurgi).

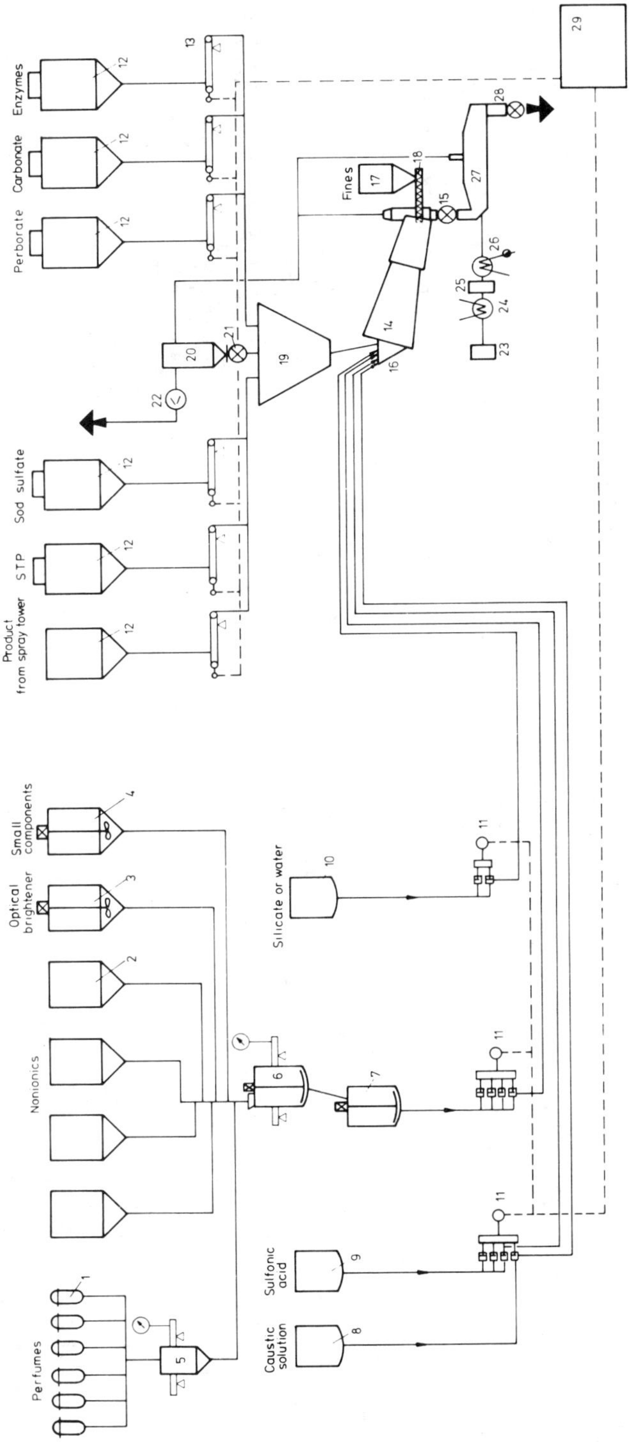

Fig. 10.17 – Powder plant using 'Pulmix' drum (courtesy of Lurgi).

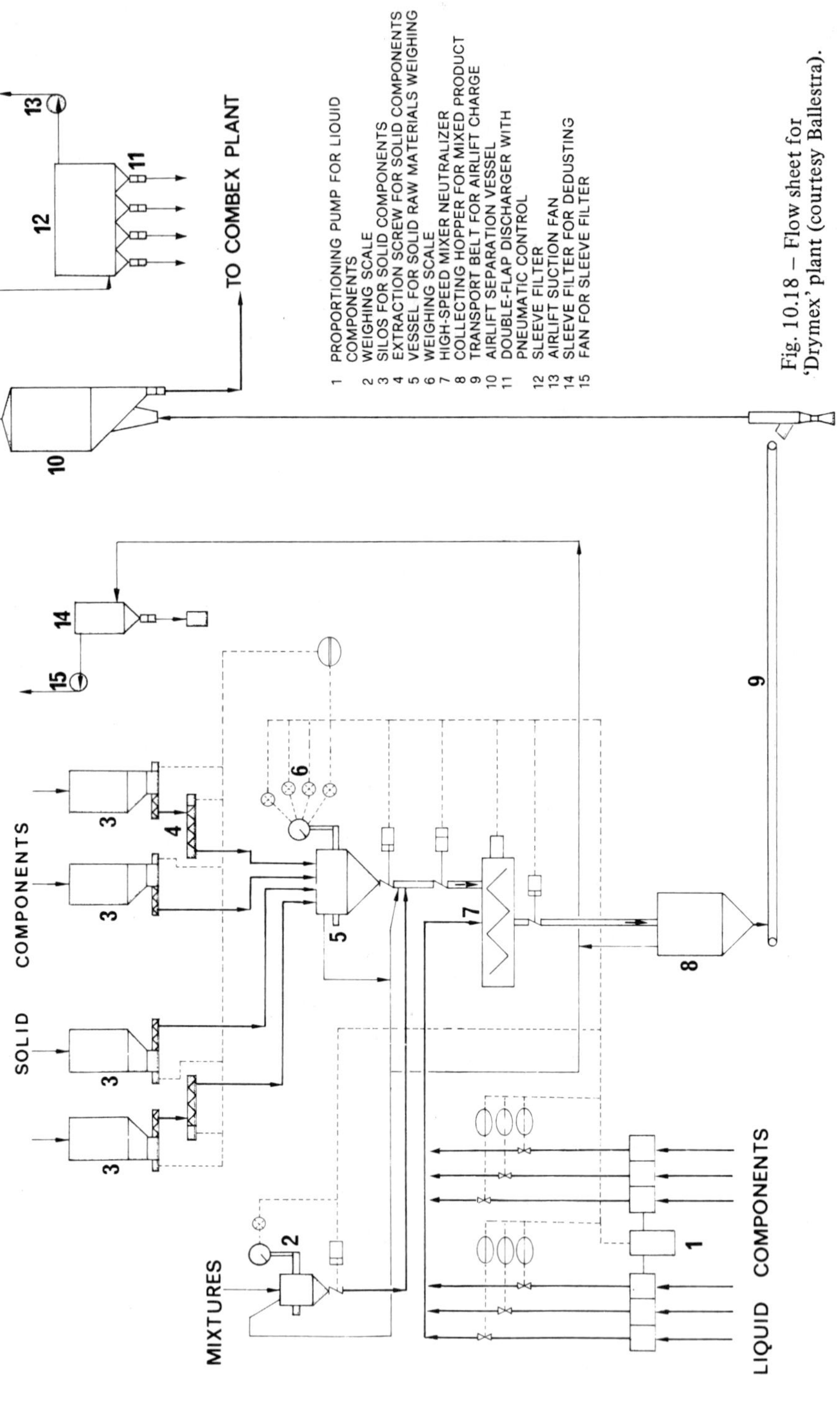

1 PROPORTIONING PUMP FOR LIQUID COMPONENTS
2 WEIGHING SCALE
3 SILOS FOR SOLID COMPONENTS
4 EXTRACTION SCREW FOR SOLID COMPONENTS
5 VESSEL FOR SOLID RAW MATERIALS WEIGHING
6 WEIGHING SCALE
7 HIGH-SPEED MIXER NEUTRALIZER
8 COLLECTING HOPPER FOR MIXED PRODUCT
9 TRANSPORT BELT FOR AIRLIFT CHARGE
10 AIRLIFT SEPARATION VESSEL
11 DOUBLE-FLAP DISCHARGER WITH PNEUMATIC CONTROL
12 SLEEVE FILTER
13 AIRLIFT SUCTION FAN
14 SLEEVE FILTER FOR DEDUSTING
15 FAN FOR SLEEVE FILTER

Fig. 10.18 – Flow sheet for 'Drymex' plant (courtesy Ballestra).

The Ballestra 'Drymex' plant is shown in Fig. 10.18. This uses a distintegrator mixer (item 7 in the diagram). A plant in which powder produced in a Drymex unit is combined with spray dried powder is called by Ballestra the 'Combex' plant.

To provide an approximate estimate of the bulk density of a blend of two powders Ballestra gives the formula, bulk density of the mixture = 100/ $(a/d_1 + b/d_2)$ where

a = percentage (by mass) of powder 1
b = percentage (by mass) of powder 2
d_1 = bulk density of powder 1
d_2 = bulk density of powder 2.

Thus a 50:50 blend of powders having bulk densities of 0.3 and 0.7 respectively has a bulk density of about 100/(50/0.3 + 50/0.7) = 0.42.

10.4.4 The Monsanto 'Fluff' process

This process, described by Shen and Callis (1965) uses sodium trimetaphosphate. The ingredients are slurried together and caustic soda solution is added. This converts the trimetaphosphate to pentasodium tripolyphosphate.

$$Na_3P_3O_9 + 2\ NaOH = Na_5P_3O_{10} + H_2O$$

Under the process conditions, the heat of reaction is sufficient to dry and granulate the mix. So far as is known, the method is not widely used, perhaps because of the limitations imposed by the need to use a substantial proportion of the special phosphate. An attraction is that it gives powder of lower bulk density than other non-spray processes.

10.4.5 Drum drying

Powders can be produced by drum drying followed by grinding, but the process is commonly confined to the manufacture of high active intermediates.

11

The manufacture of liquid detergents and miscellaneous detergent products

11.1 LIQUIDS

11.1.1 General

Important types of liquid detergent and related products are:

- liquid and cream soap products.
- dishwashing and other mobile NSD liquids.
- liquids for the washing of fabrics.
- thick liquids and creams, which usually contain suspended ingredients.
- fabric conditioners.
- bleaches and lavatory cleaning liquids.
- cleaner polishes.
- window cleaning liquids and creams.

The formulation of, and ingredients used in, many of these products are discussed in Chapter 2; and the chemistry of sodium hypochlorite (bleach) is considered in section 1.6.2.

11.1.2 Liquid and cream soap products

These were widely used in commercial, institutional, and public washrooms, and also as shampoos, but at least in the case of shampoos they have now largely been replaced by NSD products. Specifications BS 1545:1969 (liquid toilet soap) and BS 4405:1969 (liquid soap for domestic and industrial purposes) are considered in section 1.8. In order to achieve adequate solubility at low temperatures, the alkali is usually potassium hydroxide and the fats include substantial proportions of lauric and/or unsaturated material. To meet BS 1545 and BS

4405, products must remain clear and bright, and not gel, after 24 hours at 3° ± 1°C. For BS 1545 the TFM is above 15% and for BS 4405 above 10%, for BS 4405 the fatty matter usually contains more than 20% fatty acids of molecular mass 200, or above.

A 37–40% solution of anhydrous coconut oil potassium soap remains clear down to about 5°C, but concentrations of 15–30% TFM are usual. Because of the slightly harsh reaction on the skin of coconut oil soaps, some proportion of unsaturated oil, such as oleine or castor oil, is often included. The soap is normally made by an 'all in' process in a crutcher, so that the glycerol liberated is retained in the product. Additional glycerine and other materials, such as the thickening agents sugar, a Carbowax (polyhydroxyethylene glycol *ex* Carbide and Carbon Chemical Corporation), or sodium carboxymethyl cellulose, may be added. On the other hand, ethanol may be used to improve solubility and reduce viscosity. Many formulations are possible, and Davidsohn, Better and Davidsohn (1953) can be consulted for some, together with manufacturing details.

To produce clear liquid soaps, calcium and magnesium must be avoided, so that the soaps should be made with distilled, or base exchange softened water. To take care of any traces of calcium, magnesium, and heavy metals, a little EDTA is sometimes included as a sequestrant. On the other hand, certain brands deliberately include a small proportion of calcium in a way which gives the product an opalescent appearance.

11.1.3 Dishwashing and other mobile NSD liquids

The ingredients for these products can be blended in simple mixers of most standard types. Contact parts should be of stainless steel, or other corrosion resistant material, to avoid contamination of the product. Tanks can be of reinforced polyester, or of mild steel coated with epoxyresin, as well as of stainless steel.

Companies such as Ballestra offer batch and continuous liquid blending plants, but detergent manufacturers often engineer their own. Fig. 11.1 shows the Ballestra batch plant and Fig. 11.2 the Ballestra continuous plant, which includes a neutralizing section for use when the surfactant is added as sulphonic acid. Plants can be designed to suit particular requirements.

Liquid detergents can be packed in a range of containers including glass bottles and drums, but plastic bottles of various shapes and sizes are now normal for the domestic trade. Types of polythene of varying rigidity are most usual. Some products are packed in rigid bottles; but flexible squeeze bottles, with caps provided with a small hole, are virtually standard for dishwashing liquids.

Packing lines can be very simple, with hand operations and semi-automatic fillers; but with the very large tonnages now being produced by major companies, the trend is to highly mechanized, high-speed, lines. Detailed discussion of such lines, with bottle and cap unscramblers, fillers, cappers, labellers, casepackers and palletizers, is outside the scope of this book, but the following points are of interest:

- fillers normally operate on one of three principles:
  - volumetric fillers in which the liquid is measured in a set of flasks which form part of the machine.
  - level fillers in which the bottles are filled to a set level. Such fillers only fill a standard volume if the bottles are uniform in size.
  - fillers in which the bottles are set under a uniform stream of the liquid for a standard time.

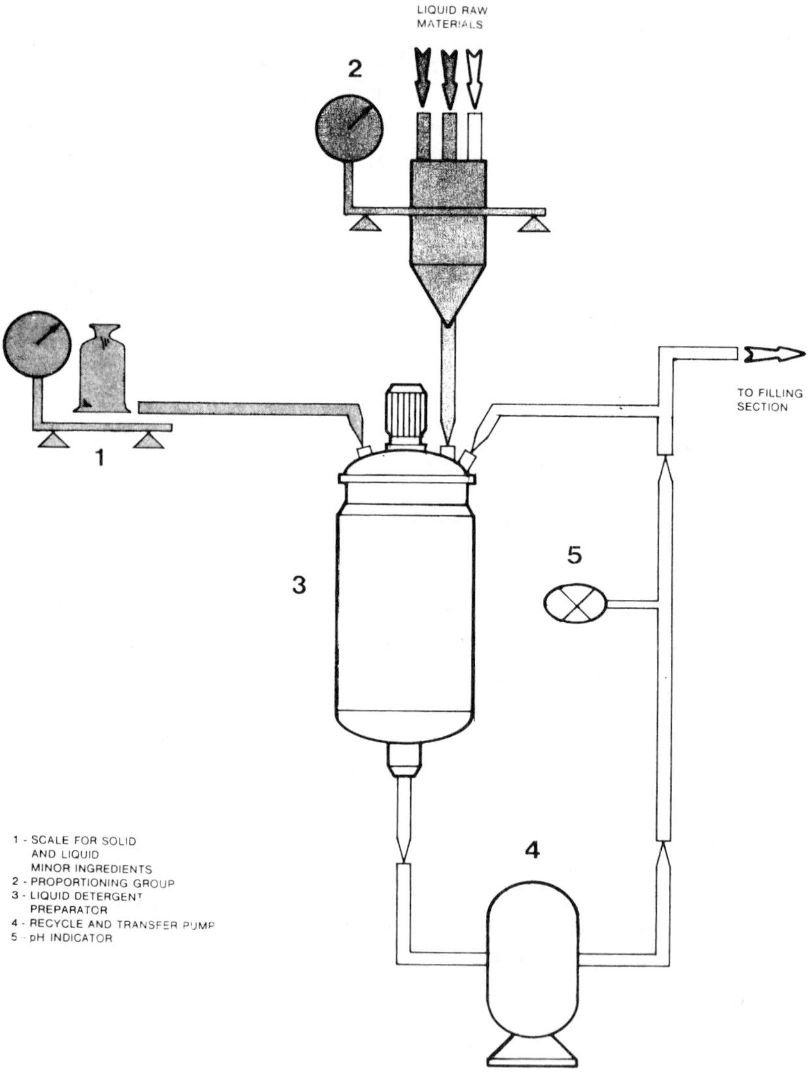

Fig. 11.1 – Ballestra batch plant for liquid production.

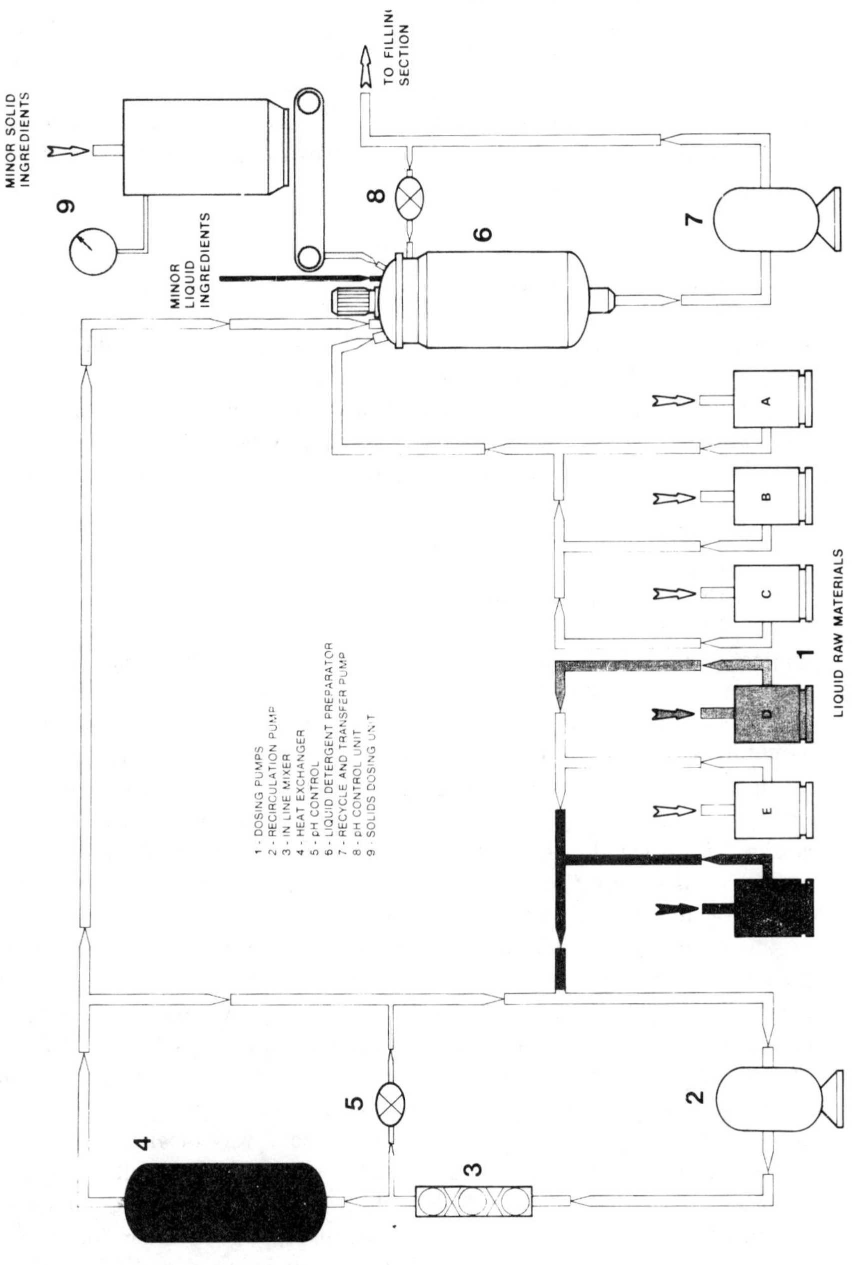

Fig. 11.2 – Ballestra continuous plant for liquid production.

- some plastic bottles which are to be labelled require to be flamed to ensure that the labels stick properly.
- plastic bottles used to contain detergent liquids can suffer from stress cracking, or from panelling. Panelling, or distortion of the bottle due to reduction in the internal pressure during storage, can result from differential transfer of gases through the bottle walls, or from reaction of a component of the head gas, that is oxygen, with the liquid.
- certain liquids can generate gases and so cause distension, or even bursting. As discussed in section 1.6.2, this is particularly a problem with sodium hypochlorite bleach.

11.1.4 Liquids for the washing of fabrics

Of the formulations discussed in section 2.4.7, the unbuilt types, generally only suitable for light duty work, can be made in plant which is the same as, or similar to, that used for dishwashing and other mobile liquids.

Built, heavy duty, liquids often require special mixing facilities tailored to suit particular types of formulation; premixing of parts of the composition may be needed. The suspended sodium tripolyphosphate type falls into the thick liquids class.

11.1.5 Thick liquids and creams

Some thick liquids can be made in simple mixers equipped with suitably heavy mixing gear, which often includes a scraper blade to move accumulations from the base and/or walls. If sulphonic acid is used a cooling system is usually needed to remove heat of neutralization. Particular care is necessary to avoid aeration which can produce foam and, more particularly, undue thickening. As in all liquid mixing where aeration is not specifically needed, care should be taken to avoid a top arm which is not adequately immersed in the liquid and so entraps air. When the ingredients include a powder, such as the fine calcite in some scouring creams, air tends to be introduced with the powder and an enclosed, vacuum, mixer is often desirable.

As mentioned in section 2.4.6, certain active systems, normally blends, are capable of suspending solid particles, which may be inert minerals in scouring creams, or substances such as sodium tripolyphosphate. With tripolyphosphate, coacervates which are complex viscous particles, are formed. For the formation of stable suspensions, from which the solid particles do not separate on standing, the liquid must have non-Newtonian properties with a yield point. Such liquids are sensitive to mechanical action; and the precise way in which the ingredients are added and mixed may be important. Comparatively small changes in the composition of the ingredients, including levels of impurities, are sometimes significant. Various formulae have been published, but the finer points are often trade secrets; and development work is usually required to find a satisfactory composition and method of manufacture. In some cases, it is necessary to make premixes of certain ingredients; and more than one mixer is then almost essential.

11.1.6 Miscellaneous

Points concerning other types of liquid product are discussed in chapters 1 and 2.

11.2 FABRIC WASHING DETERGENTS IN MEASURED DOSES

Heavy duty detergents have been produced in soluble sachets, and as tablets, to simplify dosing to the washing machine, but, so far as is known, the products have not proved sufficiently attractive to achieve much commercial success. It is not easy to produce tablets which are sufficiently robust to be transported to the consumer and yet disperse readily in use. Mallee (1963) discusses the production of tablets, particularly machine aspects.

11.3 SCOURING PRODUCTS

11.3.1 General

Several types of product which contain an abrasive powder, together with organic and inorganic detergents and builders, have been, or are, used:

- solid products, such as mechanics hand soaps intended to wash heavily soiled hands; and products like the old Monkey Brand, once used to scour pans and other hard surfaces. These products can be made by mixing the abrasive powder into the soap, together with any other ingredients, followed by conversion to tablets by the frame process. The soap in Monkey Brand was considerably diluted with water and the billets were stoved. Hand soaps have used powdered pumice, or mild vegetable abrasives, such as sawdust or cornmeal. Older books, such as Davidsohn, Better and Davidsohn (1953), and Levitt (1967) include formulations and specifications current some time ago.
- powders which were major products, but which have substantially been replaced by milder products.
- creams and pastes, which normally use only mild abrasives; and are less liable to dull and otherwise damage modern surfaces.

11.3.2 Abrasive powders

In the selection of an abrasive powder for use in scouring products there needs to be a compromise between scouring performance against difficult soils, such as foodstuffs burnt onto pans, and mildness to surfaces; as already mentioned, the modern trend is towards greater mildness, but an abrasive also needs to be selected to meet the marketing requirements of the product in which it is to be used. Factors which affect performance are:

- particle size. Large particles tend to scratch more than smaller ones, but also to be more effective. Size relative to the deposit of soil is important; and tests show that particles smaller than the thickness of the soil film are relatively ineffective. A narrow range of particle sizes is particularly desirable, because it minimizes both damaging large particles and largely useless very small ones.

- particle hardness. Hard minerals tend to scour more efficiently than soft ones, but also to cause more damage. The level of damage is related to the hardness of the abrasive relative to that of the surface being cleaned; thus the difference between a hard and a soft mineral shows up much more on stainless steel than on the softer aluminium which is usually scratched by both. Pure ground calcite does not scratch stainless steel, but commercial calcites may contain a few harder particles, particularly quartz, which do.
- particle shape. As might be expected, particles in the form of crystals with sharp edges are more effective, other things equal, than those without. For this reason, some types of calcite are particularly good for a relatively soft mineral.

Other factors which need to be considered in the selection of an abrasive are:

- safety in the detergent factory, and in the quarry and grinding plant where the powder is produced. Quartz, a crystalline silica, was very widely used, but it has now been replaced in some countries because of the risk of silicosis. Other forms of crystalline silica, crystobalite and tridymite, which may be present if the silica is treated at high temperatures, are even more dangerous than quartz.
- colour. The colour of a scouring powder has no bearing on its performance, but in many markets a good white colour has a significant effect upon consumer acceptance.
- cost. Fot the product to be competitive, the abrasive used in a scouring powder must be cheap; for scouring creams there may be a little more latitude. This normally requires the use of a site at which the mineral, of good white colour, occurs in massive deposits; and which is situated at a place from which transport costs to the processing factory are low.

The main minerals which are, or have been, used are:

- crystalline silica, normally quartz. For many years quartz was the almost universal choice, but, in the UK at least, it has been replaced for medical reasons. Processes to improve colour were sometimes used; and some of these converted part of the quartz to crystobalite or tridymite.
- feldspar, often called felspar. Slightly softer than quartz, the feldspars are important replacements for quartz in heavy duty scouring products. The potassium aluminium silicate type of feldspar is usually difficult to find in sufficiently massive deposits; and it is commonly associated with quartz. A feldspar of this general type is recovered as a by-product from the Cornish kaolin (china clay) industry and, at one time, this material was used in scouring powders, but, because of its quartz content, it is no longer acceptable in the UK. Deposits of anorthosite, or labradorite, a soda lime silicate, have been found in Norway which are very suitable for use in heavy duty scouring products; and which are massive, of good colour, near the sea, and substantially free from quartz. This material is commonly described as feldspar.

- pumice and volcanic ash. Powdered pumice and volcanic ash have been used in mechanics' hand soaps, but, so far as is known, are not widely used in scouring powders or creams.
- carbonates of calcium and magnesium. 'Calcite (calcium carbonate), dolomite' (calcium magnesium carbonate, $CaCO_3.MgCO_3$), and magnesite (magnesium carbonate) have become increasingly important as minerals for use in scouring products. The sequence of decreasing hardness is magnesite, dolomite, calcite; and magnesite may be chosen for scouring powders. Dolomite and various types of calcium carbonate, chalk, marble, and other limestones, occur in very massive deposits, but it is not easy to obtain qualities which meet the various criteria for a good scouring product material. For example, in the UK the Carboniferous limestones, used widely for other purposes, do not have a sufficiently good colour; and UK chalk is not adequately crystalline in the required particle size range, although whiting produced from chalk has long been used for some cleaning and polishing purposes. Marbles, and similar white, relatively coarsely crystalline, limestones are usually the best source of calcite powder for scouring products. The very finely crystalline waterworks, and other precipitated, chalks are used as abrasives in dentrifices, but are normally too fine for general household products.
- diatomaceous earth (kieselguhr, tripolite etc.). This material, which comprises the silicious skeletons of minute marine organisms, the diatoms, is a very mild abrasive, not normally suitable for scouring products.
- talc is a very soft hydrated magnesium silicate which may be suitable for very mild cleaner-polishes for tiles and paintwork, but which is not normally used in scouring powders or creams.

11.3.3 Production of abrasive powders from the mineral

It is difficult to produce economically the very closely sized powders which are ideal for scouring products; and the powders used commercially commonly include a considerable range of sizes. Typical particle size distributions for a ground quartz, or feldspar, as used in heavy duty scouring powders; and for a grade of calcite commonly used in scouring creams, are given in Table 11.1, but considerable variation is possible. Figures determined for percentages of various particle sizes can be affected by the analytical method used, particularly in the finer part of the range.

Rocks such as quartzite, or feldspar, are crushed in a jaw, or gyratory, crusher and then ground, usually in a tube, or ball, mill, although other types of mill can be used. The product from the mill is separated in a pneumatic size separator. To give a more uniform size range, and to reduce power consumption, closed circuit grinding is usual. In this system, the material leaving the mill contains a high proportion of oversize which is returned from the separator to pass through the mill again. In some plants, two, or more, tube mills, each with a separator, are used in series. The Polysius 'Double-rotator' uses two grinding sections combined in the same mill shell. Tube mills lined with ferrous metal, and using balls of similar material, have a higher capacity and give a lower power consumption per tonne of product than mills with non-ferrous contact parts, but are liable to contaminate the product with ferrous particles and to cause deterioration of colour. It is, therefore, often found necessary to use non-ferrous materials. Non-ferrous balls may be flint pebbles, or made of fused alumina. Linings may be silex or fused alumina; undue damage to such linings, particularly the fused alumina type, is sometimes encountered.

Table 11.1 – Particle size distributions of abrasive powders

	Ground quartz or or feldspar % cumulative	Ground calcite % cumulative
Coarser than 5 μm	84	94
Coarser than 10 μm	72	87
Coarser than 20 μm	53	70
Coarser than 30 μm	40	55
Coarser than 40 μm	30	42
Coarser than 50 μm	23	32
Coarser than 60 μm	18	24
Coarser than 70 μm	14	—
Coarser than 80 μm	11	13
Coarser than 90 μm	—	—
Coarser than 100 μm	—	6
Coarser than 120 μm	—	1

1 μm = 0.001 mm.

Calcite, and related materials, are more readily reduced in size than quartzite and feldspar, but broadly similar procedures are used, although wet methods of size separation may be included. As already mentioned, freedom from particles of hard minerals is important. It is also necessary to ensure that such hard particles are not introduced in other materials used in the product, or during processing.

An improved particle size distribution can often be achieved by grinding rather fine to minimize large particles and then to separate a fine fraction from the product. For this to be commercially acceptable, a market for the fine fraction is usually necessary. Such a procedure is more likely to be possible for a specialist mineral grinder, who has a number of customers for various products, than for a detergent manufacturer.

11.3.4 Other ingredients; and the production of scouring powders

In the UK from 1903 to the early 1950s, the ground abrasive powder was normally blended with soap and sodium carbonate; and the soluble ingredients were commonly added in the form of a relatively low TFM spray-chilled soap powder.

It was found that the higher the ratio of soap to sodium carbonate the more readily the in-use slurry slipped over the surface being cleaned and the less the apparent 'bite'.

Since the 1950s, soap has normally been replaced by sodium alkylbenzene sulphonate; and during the last 20–30 years the major brands in the UK have included a bleach. Many ingredients and formulations have been used, but a typical formulation is:

	%	Common range
Ground abrasive powder	91	90–92
sodium alkylbenzene sulphonate	4	2–4
anhydrous trisodium orthophosphate (or other alkali)	3	2–4
trichlorocyanuric acid (TCCA)	0.8	0.5–1.0
perfume, including stabilizing olefin	0.2	0.2–0.4
sodium sulphate, water, etc.	1	0.5–1.0
	100	

In use, the TCCA reacts with water to give available chlorine, or hypochlorous acid, which provides both bleaching properties, useful particularly against tea and other stains, and also germicidal properties. Care is necessary to avoid premature decomposition of the TCCA during storage. It has been found (BP 848 397 to Colgate-Palmolive) that addition of a perfume containing an olefin in which one of the carbon atoms joined by the double bond is tertiary, such as limonene or pine oil, stabilizes TCCA and related compounds. A minimum of about 0.1% of stabilizing olefin calculated on product is needed. Some immediate loss of available chlorine (the mixing loss) occurs when the ingredients including the TCCA and stabilizing perfume are mixed; and the magnitude of the loss depends upon the quantity and nature of the perfume. Loss of available chlorine is also closely related to moisture pick-up during storage. Scouring powder are normally sold in canisters; and plastic canisters are usually more resistant to transfer of water vapour than those made of carton board covered with aluminium foil, or film; with those covered with printed paper the least effective. The composition of the powder is also important. Alkali is required for product performance; and anhydrous trisodium orthophosphate combines alkalinity with the power to take up water so as to prevent decomposition of the TCCA. Other alkalis, such as sodium carbonate, disodium hydrogen phosphate, sodium pyrophosphate, and sodium tripolyphosphate can be used alone, or in combination, but they are less effective in the protection of TCCA. When a scouring powder of this type is stored, there is an induction period during which little, or no, loss of available chlorine occurs. This is followed by a period during which the available chlorine content falls steadily, eventually to zero. The length of the induction period depends upon the tightness of the container, the humidity of the atmosphere, and the nature and quantity of the alkali present. Other organic compounds which liberate available chlorine on contact with water, such as chloroxal and the chlorhydantoins, can be used, but TCCA is usually preferred. Sodium bromide has been used in conjunction with TCCA as a bleach booster. Sodium phosphate hypochlorite, SPH, roughly $(Na_3PO_3.12H_2O)_5NaOCl$, has been used in scouring powders. Contrary to TCCA, this compound requires its water of crystallization for stability, so that it loses available chlorine if allowed to dry out on storage. Inclusion of borax, $Na_2B_4O_7 10H_2O$, to protect SPH by acting as a moisture reservoir has been patented. Various available oxygen bleaches, including sodium perborate and, particularly, a material which includes sodium monopersulphate and sodium bisulphate, have been tried, but chlorine bleaches have been found superior, at least in consumer preference in the UK.

The surfactant is sometimes added in the form of fines recovered as a by-product from a spray drying plant; it then contains other ingredients, such as sodium tripolyphosphate. It can also be added as a high active drum dried flake, which is usually ground in a ball mill with some of the mineral and the alkalis, so as to obtain good dispersion. Small manufacturers may find it more convenient to use a surfactant powder which does not have to be ground, particularly if a batch mixer is employed. If the scouring powder is coloured, it is commonly necessary to add the colour solution, or suspension, to another portion of the mineral and to grind in a ball mill to spread and develop the colour.

11.4 MACHINE DISHWASHING PRODUCTS

Machine dishwashers use the mechanical action of fast moving water, from jets or generated by agitators, but some detergent is also necessary.

Foam produced by the mechanical action can impair the machine performance, so that powders which have no tendency to foam, and, if possible, suppress the foam caused by some proteinaceous food residues, are needed. Normally the powders are based upon a combination of condensed phosphates, usually sodium tripolyphosphate, sodium silicates and, sometimes, sodium carbonate and sodium sulphate. 1–2% of a non-foaming nonionic surfactant, usually of the ethylene oxide/propylene oxide blocked polymer type, is often included, but products can be made without any organic surfactant. An available chlorine release compound, such as TCCA or SPH, is usually required; this helps the removal of some soils and the avoidance of spotting of articles. Available chlorine also makes the detergent solution germicidal, but this is probably not very important in view of the high temperatures normally employed in the machine dishwashing process.

The strong alkalis, particularly sodium metasilicate, the high temperatures, and the relatively long washing times, can cause damage to glassware and ceramics. Apart from the washing conditions, some ceramics are fired so as to produce articles with decoration and glaze which are more resistant to damage than other brands of ware; and the increasing use of machine dishwashers will probably encourage the wider production of resistant ceramics and glass. Ingredients such as borates and, particularly, sodium aluminate, are often added to minimize damage to this type of article. A tarnish inhibitor, usually benztriazole, is also used to reduce tarnishing of cutlery other than stainless steel which is not affected.

The formulation of a good machine dishwashing powder is not easy, as a careful balance is necessary to ensure that the washing performance is satisfactory under all likely conditions, whilst damage is minimal. A high proportion of sodium tripolyphosphate, say 40–50% is usually desirable, particularly when the water used is hard; but many machines now have built-in base exchange water softeners. With hard water, films of calcium carbonate, or other substantially insoluble salts, may tend to form which are particularly visible on glassware. Film deposition occurs especially during rinsing, when the ratio of phosphate to hardness salts is reduced. An article by Monsanto, *J. Am. Oil Chem. Soc.*, **56**, 910 (1979), gives a standard US formulation as:

	%
sodium tripolyphosphate (anhydrous)	40
liquid sodium silicate (42% solids, SiO_2/Na_2O ratio = 2.8)	24
Plurafac RA 43	2
sodium phosphate hypochlorite	24
sodium sulphate	10
	100

It also says that:

- attempts to reduce the washing temperature to 100–140°F (38–60°C) gave poor results, with spotting and filming of glassware. The adverse effect could be reduced by an increase in the polyphosphate content of the powder to 55%.
- reduction of polyphosphate content to 25–33%, to meet restrictions on the phosphorus content of detergent products, reduced product performance.
- products without phosphate gave such poor results that a domestic consumer testing panel soon stopped using them.

However, Kaneko (1979) reports that, at least in laboratory tests, good results have been obtained with the following non-phosphate formulation:

	%
surfactant Tetronic 90 Rl or 150 Rl	6
sodium citrate dihydrate	30
soda ash	20
sodium metasilicate pentahydrate	30
sodium sulphate	12.5
chlorinated cyanurate	1.5
	100

The Tetronics are compounds developed by the Wyandotte Corporation which are based on tetrahydroxylpropylethylenediamine to which ethylene oxide and propylene oxide are added sequentially.

Kaneko also says that:

- some nonionics used in dishwashing detergents foam if the temperature of the wash liquor falls below the cloud point of the surfactant.
- the new Tetronics are particularly good in the control of foam, including that due to proteinaceous soil.

Formulations commonly use sodium metasilicate powder, hydrated or anhydrous, instead of liquid silicate. Some care is necessary with regard to safety should the product be ingested if the anhydrous form is used.

Substances, such as sodium aluminate, which reduce damage to glass and china, operate by the deposition of a thin film on the articles being washed. It is the author's opinion that formulations which neither cause excessive damage to delicate china, nor produce visible films on glassware, probably achieve a subtle balance in which an invisible protective film is deposited.

Rinse aids are non-foaming non ionic surfactants intended to prevent the development of streaks, or spots, on articles being washed. With a good detergent powder, it is doubtful whether a rinse-aid is necessary to achieve good drainage of the rinse water and completely satisfactory results can often be obtained without rinse aid, particularly in machines which include a water softener.

Some granulation during the mixing process is desirable to avoid a dusty product; for this the use of some liquid sodium silicate may be helpful.

11.5 SOAP CHIPS, NOODLES AND FLAKES

Soap chips, noodles and flakes are produced and sold for industrial purposes:

- to small companies, often called remillers, which buy dried soap and produce from it specialized brands of toilet soap. Requirements are as already discussed in earlier chapters.
- to launderers and other users of soap, who usually dissolve the soap to make stock solutions. Requirements are agreed between the parties to suit the purchaser's needs. For laundries high titre soaps, based on tallow, are often selected.

Fine flakes were an important type of product in the domestic market, much used by housewives to wash delicate articles and baby things. Tonnages have declined greatly, but brands are still available in the UK. A good quality soap base, such as that used for toilet soaps, is dried to about 82–83% TFM and then milled. Common practice is to produce coarse flakes in a preliminary operation on a three-roll mill to provide some plastic working and homogenization; and then to make flakes which have a thickness of 3–5 thousandths of an inch on a five-roll mill. BS 1912:1973 specifies a maximum thickness of 0.2 m (8 thousandths approximately). The top rolls of the five roll mill are set very close together, normally with no static clearance, but so that they are forced apart slightly by the soap. Usually two cutting rolls with diagonal ridges are set on the top roll of the mill to produce diamond shaped flakes. If the mill is set correctly and the soap is in the right condition, the flakes are scored rather than cut, and leave the mill like sheets of postage stamps. The sheets then break up when the product is handled after it has cooled and become more brittle. Band driers were commonly used in the heyday of soap flakes; and some difficulty may be experienced in drying to the required level in a vaccum spray dry and chill type of plant, because of the hardness of the 83% TFM soap.

Soap flakes cannot be measured satisfactorily by volume and it is normal to use weighing heads on the filling machines.

A traditional toilet soap base made from about 80% tallow and 20% lauric oil is often used for fine flakes. Ready solubility in warm, rather than hot, water is desirable, because it makes easier the preparation of the wash liquor which housewives wish to use. This is helped by the thinness of the flakes, but for many years at least one major brand used a fat blend comprising roughly equal parts groundnut, or similar soft oil, and lauric oil, which gave a very soluble soap.

11.6 SHAVING SOAPS

A good shaving soap is required to:

- produce a copious, stiff, lather which holds as well as wets the beard; and which does not dry, or collapse, too quickly.
- be mild to the skin, which is inclined to be sensitive immediately after it has been shaved.

Shaving soap is produced as sticks or, less often, as bowls. Other shaving products, such as lather creams, brushless creams, and aerosol creams are rather outside the scope of this book. Shaving sticks are produced by drying, milling and plodding, as for toilet soaps, but a special type of base is used to give the required characteristics. The fat blend is usually a mixture of commercial stearin (predominantly stearic plus palmitic acids) and lauric oil, about 80%/20%. For the better grades, this fat is saponified with a mixture of about three parts potassium hydroxide to one part sodium hydroxide; for cheaper grades 1:1 mixture may be used. The process is 'all-in', usually carried out in a jacketted, stirred, pot. The glycerol liberated from the lauric oil is necessarily left in the soap; and additional glycerine is sometimes added. The soap is tested for free fat and/or alkali, and the pot adjusted with stearin, lauric oil, or alkali to finish with 1–2% free fat.

11.7 SOFT SOAPS

Soft soaps are jelly-like products which were very important, but have now declined greatly. In brief, they are, or were, commonly produced by a semi-boiled process from a range of fatty materials, but usually from highly unsaturated oils. The alkali can be caustic potash, caustic soda, or a blend. Potassium carbonate and potassium chloride may be added; and the blend of inorganic ingredients may be varied from season to season to maintain the required consistency. Details of various types of soft soaps and their manufacture are given in Davidsohn, Better and Davidsohn (1953).

Appendices

Appendix 1

NAMES AND ADDRESSES OF SOME COMPANIES AND ORGANIZATIONS

A1.1

This appendix includes names and addresses of various companies, and other organizations, which are likely to be of use to readers. It is concerned primarily with suppliers of plant and ingredients; but it is not comprehensive, particularly as regards suppliers of general chemicals. Some indications are given as to types of plant, or ingredient, supplied by the various companies which fall within the scope of this book, but this should not necessarily be taken to imply that they are the only suppliers of these items, or that they do no supply other things. Many of the companies are international and have addresses other than those listed here. Apologies are offered to organizations which have been omitted, but might have been included.

A1.2 MANUFACTURERS AND SUPPLIERS OF PROCESS PLANT

(1) AAF Limited (an Allis-Chalmers Company), Bassington Lane, Cramlington, Northumberland, NE23 8AF, England.
Dust Collectors.

(2) Alpha-Laval Company Limited, Great West Road, Brentford, Middlesex, TW8 9BT, England (sometimes called De Laval).
Centrifuges and soapmaking plant.

(3) APV Company Limited, PO Box No 4, Manor Royal, Crawley, West Sussex,RH10 20B.
Heat exchangers, evaporators and other plant.

(4) Ballestra SpA, Via Fantoli 21/17, 20138 Milano, Italy.
Plant for NSD products and the production of surfactants.

(5) Thomas Broadbent and Sons Limited, Huddersfield, HD1 3EA, England.
Centrifuges.

(6) Chemithon Corporation, 5430 W. Marginal Way SW, Seattle, WA 98 106, USA.
Sulphonation plant.

(7) CT (London) Limited, Walnut Tree House, Woodbridge Park, Guildford, Surrey, GU1 1EL, England (agents for Spraying Systems Company, North Avenue at Schmale Road, Wheaton, Illinois 60187, USA).
Spray nozzles.

(8) Feld and Hahn GmbH, Concordia Strasse 75, Bendorf 5413, West Germany.
Glycerine and other plant.

(9) Fratelli Gianazza SpA, 20025 Legnano (Milano-Italia), Viale Cadorna 80, Italy.
Plant for fatty acids and glycerine.

(10) Johnson-Progress Limited, Carpenters Road, Stratford, London E15 2DS, England.
Filtration plant.

(11) Lurgi (UK) Limited, No. 1 Kingsway, London WC2B 6XF, England; or, Lurgi Umwelt und Chemotechnik GmbH, D-6000 Frankfurt am Main, Gervinusstrasse 17/19, West Germany.
Plant for fatty acids, glycerine and soapless detergents (Lurgi also supplies Chemithon sulphonation plant).

(12) Luwa (UK) Limited, 36–38 Goldsworth Road, Woking, Surrey, GU21 1JT, England.
Scraped wall vaporizers.

(13) G. Mazzoni SpA, PO Box 421, 21052 Busto Arsizio, Italy.
Fatty acids, soap, soapless detergents and glycerine plants.

(14) Marchant Brothers Limited, 58–62 Verney Road, London, SE16 3DL, England.
Soap mills.

(15) Meccaniche Moderne, Corso Sempione 51, PO Box 388, 21052 Busto Arsizio, Italy.
Plant for soaps and soapless detergents (Meccaniche Moderne also supplies Allied Chemical Corporation sulphonation plants).

(16) Mikro Pul Ducon Limited, Towerfield Industrial Estate, Shoeburyness, Essex SS3 9QU, England.
Dust collectors.

(17) Peabody Holmes Limited, Brenchley House, 123–135 Week Street, Maidstone, Kent, ME14 1RF, England.
Air heaters and other furnaces.

(18) Pennwalt Limited, Doman Road, Camberley, Surrey, GU15 3DN, England.
Sharples centrifuges and soapmaking plant.

(19) John Thurley Limited, Ripon Road, Harrogate, HG1 2BU, North Yorkshire England.

Air heaters and other furnaces.

(20) Weber + Seelander, PO Box 1520, Braunschweiger Tor 17, D-3330 Helmstedt West Germany.
Soap plant.

(21) Wurster and Sanger, 222 West Adams Street, Chicago, Illinois 60606, USA
Fatty acid and glycerine plants.

A1.3 SUPPLIERS OF INGREDIENTS

(1) Albright and Wilson Limited, PO Box 15, Whitehaven, Cumbria CA28 9QQ, England.
Surfactants and phosphates.

(2) D. F. Anstead Limited, Victoria House, Radford Way, Billericay, Essex, England.
Colours.

(3) Allied Colloids, Bradford, England.
Polyelectrolytes, sequestrants.

(4) Akzo Chemie Nederland BV, Research Centre, Emmastraat 33, 7411 EK Deventer, Netherlands.
Oleochemicals.

(5) British Celanese (part of Courtaulds Group) PO Box 5, Spondon, Derby, DE2 7BP, England.
Sodium carboxymethyl cellulose.

(6) BASF United Kingdom Limited, PO Box 4, Earl Road, Cheadle Hulme, Cheadle, Cheshire SK8 6QG, England.
Surfactants including low foaming nonionics.

(7) British Titan Products, Tioxide Group plc, 10 Stratton Street, London W1A 4XP, England.
Titanium dioxide.

(8) Bush Boake Allen, Blackhorse Lane, Walthamstow, London E17 5QP, England.
Perfumes.

(9) Joseph Crosfield and Sons Limited (part of Unilever Group), PO Box 26, Warrington, Cheshire, WA5 1AB, England.
Silicates and siliceous desiccants.

(10) Ciba Geigy (ADP) Company, Clayton, Manchester, M11 4AR, England.
Optical brighteners, colours, germicides.

(11) Dow Chemical Company Limited, Meadowbank, Bath Road, Hounslow, TW5 9QY, England.
or
Dow Chemical Europe, Bachtobelstrasse 3, 8810 Horgen, Switzerland.
Surfactants, solvents.

(12) Fison Chemicals, Harvest House, Felixstowe, Suffolk 1P11 7LP, England.
Chlorcyanurates, etc.

(13) Givaudan and Company Limited, Whyteleafe, Surrey, CR3 0YE.
Perfumes.

(14) Gist Brocades NV, PO Box 1, NL-2600 MA, Delft, Netherlands.
Enzymes.

(15) Henkel KGaA, PO Box 1100, D-4000, Dusseldorf 1, West Germany.
Soap and detergent manufacturers. Suppliers of sequestrant 1000 SL (EHDP).

(16) Hickson and Welch Limited, Castleford, West Yorkshire, WF10 2JT.
Optical brighteners and colours.

(17) Hoechst UK Limited, Hoechst House, Salisbury Road, Hounslow, TW4 6JH, England.
Surfactants, including cationics, phosphates, and other sequestrants.

(18) IFF (Great Britain) Limited, Crown Road, Southbury Road, Enfield, Middlesex EN1 1TN, England.
Perfumes.

(19) Imperial Chemical Industries plc (ICI), Millbank, London, England.
Wide range of chemicals, including acids, alkalis, and anionic surfactants.

(20) Laporte Industries Limited (Interox), Hanover House, 14 Hanover Square, London, W1R 0BE, England.
Perborates.

(21) Lankro Chemicals Limited, PO Box 1, Eccles, Manchester M30 0BH, England.
Surfactants.

(22) Monsanto Limited, Monsanto House, Chineham Court, Chineham, Basingstoke, RG24 0UL, Hampshire, England.
Phosphates and other sequestrants. Phenols.

(23) Novo Industries A/F, R and D Division, Novo Alle, DK-2880, Bagsvaerd, Denmark.
Enzymes.

(24) Pfizer Europe Chemicals, Sandwich, Kent, England.
Citrates.

(25) PQ International Incorporated (formerly Philadelphia Quartz), PO Box 840, Valley Forge Executive Mall, Valley Forge, PA 19482, USA.
Silicates and related materials.

(26) Petresa International NV, Hilton Tower, Boulevard de Waterloo 39, B-1000, Brussels, Belgium.
Alkylates.

(27) PPF International Limited (part of Unilever Group), Ashford, Kent, TN24 0LT, England.
Perfumes.

(28) Rhone Poulenc UK Limited, Hulton House, 161–166 Fleet Street, London EC4A 2DP, England.
Silicates and siliceous desiccants.

(29) Rohm and Haas (UK) Limited, Lennig House, 2 Mason's Avenue, Croydon, CR9 3NB, England.
or
Rohm and Haas Company, Independence Mall, Philadelphia, PA 19105, USA.

Ion exchange resins and other chemicals.

(30) Sandoz Products Limited, Calverley Lane, Horsforth, Leeds, LS18 4RP, England.
Optical brighteners and colours.

(31) Shell International Chemical Company Limited, Shell Centre, Downstream Building, London SE1 7PG, England.
Alkylates, surfactants, solvents, preservatives.

(32) Unichema Chemie BV (Part of Unilever Group), PO Box 2, 2800 AA Gouda, Netherlands.
Fatty acids, glycerine, oleochemicals.

(33) Union Carbide UK Limited, Chemicals Division, Peter House, Oxford Street, Manchester, England.
Surfactants.

(34) Vista Chemical Company, PO Box 19029, Houston, Texas 77224, USA.
Alkylates, nonionic and other surfactants.

A1.4 LABORATORY EQUIPMENT AND CHEMICALS

(1) Griffin and George Limited, 285 Ealing Road, Alperton, Wembley, Middlesex, HA0 1HJ, England.
or
Griffin International, PO Box 14, Wembley, Middlesex, HA0 1HJ, England.

(2) Baird and Tatlock (London) Limited, PO Box 1, Romford, Essex, RM1 1HA, England.

(3) Townson and Mercer Limited, Croydon, Surrey, England.

A1.5 OTHER ORGANIZATIONS

(1) British Standards Institution, Linford Wood, Milton Keynes, MK 14 6LE, England,
or
2, Park Street, London W1A 2BS, England.

(2) FCRA (Fabric Care Research Association, which includes the former British Launderers Research Association), Forest House Laboratories, Knaresborough Road, Harrogate, North Yorkshire, HG2 7LZ, England.

(3) SDIA (Soap and Detergents Industry Association), P.O. Box 9, Hayes, Middlesex, UB4 0JD, England.

(4) Tropical Development and Research Institute, 127, Clerkenwell Road, London EC1R 5DB,
or
56/62 Gray's Inn Road, London WC1X 8LU, London.

Appendix 2

PRODUCTION AND MAINTENANCE OF VACUUM

A2.1

A plant which operates under vacuum is full of air when it is started up; and this has to be removed by some form of pump in order to establish a low absolute pressure, or vacuum. Some air leaks into the plant during normal operation and, in some cases, other non-condensible gases are evolved in the process. In most plants used in the detergents and related industries, water vapour, supplied as open steam or evaporated from the materials being processed, also needs to be removed continuously. Organic substances, such as fatty acids or glycerol, may be vaporized in the plant, but these are condensed substantially completely prior to the vacuum system, which will be assumed to have to deal only with water vapour and non-condensible gases described as air.

Vacuum, expressed in inches or millimetres of mercury, or other convenient units, is equal to the atmospheric pressure minus the absolute pressure at the relevant point in the plant. To avoid confusion, the discussion in the rest of this appendix will be in terms of absolute pressure rather than vacuum. It is, however, important to note that the absolute pressure at a given vacuum can vary considerably with the atmospheric pressure, if the latter differs widely from the approximate 760 mm Hg at sea-level, for example on the South African Rand which is some 5000 ft (1525 m) above sea-level.

A2.2

In principle, three systems can be used to raise the pressure of the gases and vapours leaving the plant to the level just above atmospheric pressure at which they can be discharged. These systems are shown diagrammatically in Fig. A2.1 (a), (b) and (c). In system (a), the whole of the gases and vapours are compressed in some form of pump from pressure P_2 at which they leave the plant to the discharge pressure P_A. This system is not normally used for the types of plant with which this book is concerned; it would be prohibitively expensive and probably impracticable.

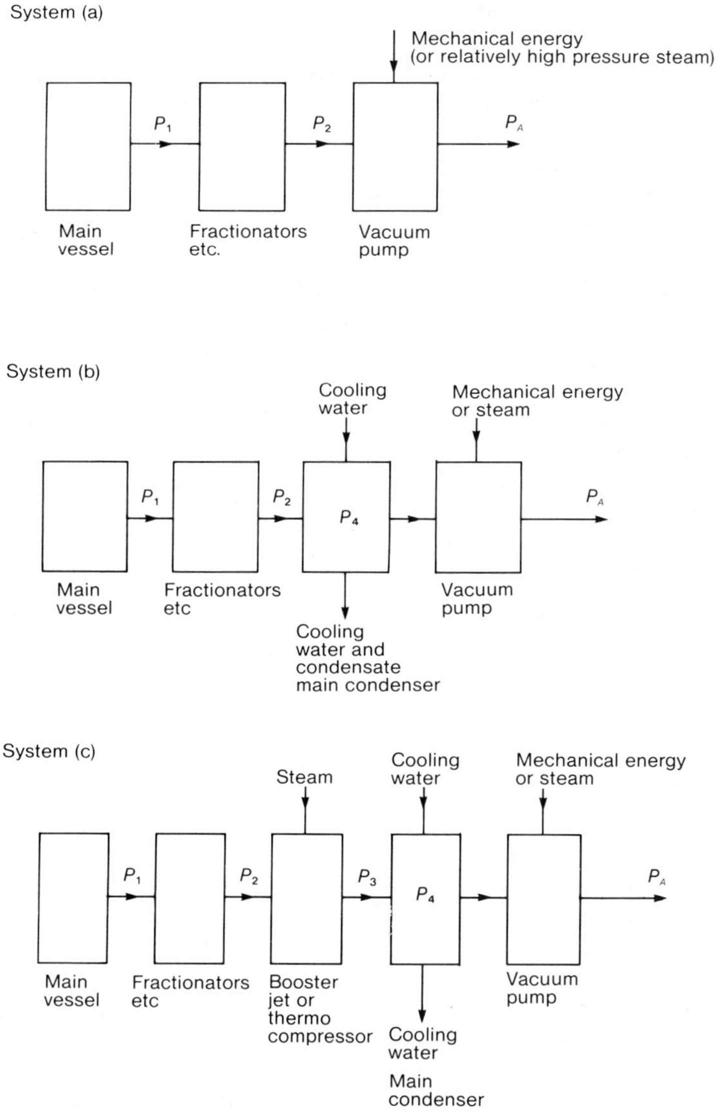

Fig. A2.1 – Systems for maintenance of vacuum.

In system (b) most of the water vapour is condensed by the use of cooling water. The condenser can be of the surface type in which the heat is transferred through walls and the condensate is recovered separately from the cooling water. Unless the condensate is required, a direct contact condenser, in which the cooling water and the vapours come into contact, and the condensate and cooling water leave together in a mixed stream, is normally more convenient and economical. A condenser should do two things, carried out in separate zones, although these are often in the same shell. Its primary duty is to condense the bulk of the water vapour, so that it can be removed from the system as liquid condensate. The secondary function of the condenser is to cool the residual air plus water vapour, which reduces the volume of gas to be removed by the air pump. Efficient cooling of the air requires counter-current contact, so that the process is completed by the coolest available water. This is achieved in the upper part of a counter-current direct contact condenser to which the gases enter near the base and the cooling water at the top. Surface condensers should have an air cooling section in which the tubes use the fresh cooling water, which then goes to the main condenser in a multi-pass water flow arrangement.

The temperature to which the main condensate is cooled in the condensing zone is of major importance, as it controls the minimum pressure which can be achieved with system (b). This minimum pressure is the vapour pressure of water at the highest temperature at which the condensate is in contact with the condensing vapours, that is the condensate exit temperature. In the case of a direct contact condenser, the relevant temperature is that of the mixed condensate and cooling water stream. The exit condensate temperature is controlled by the inlet temperature of the cooling water and the mass used per unit mass of water vapour to be condensed. In the case of a surface condenser, the area of cooling surface and the factors which affect heat transfer are important as these also influence the condensate temperature.

A vapour pressure curve for water is shown in Fig. 6.3 and a skeleton table is:

temperature, °C	*vapour pressure, mm Hg*
10	9
15	13
20	18
25	24
30	32

The total pressure in the condensing zone, P_4, equals $p_{w4} + p_{a4}$, where p_{w4} and p_{a4} are respectively the partial pressures of water vapour and air at point 4. p_{w4} is the vapour pressure of water at the highest temperature at which the condensate is in contact with the vapours in the condensing zone. With a given mass of air to the removed, p_{a4} is controlled by the performance of the air pump. A larger air pump reduces p_{a4}, which, in any case, should be relatively small, but it does not significantly reduce p_{w4}. Its only influence on p_{w4} is to carry away more water vapour with the air and so marginally to reduce the condensate temperature.

The total pressure in the air cooling zone is normally substantially the same as that in the condensing zone P_4; only a very small pressure difference being needed to induce flow from one zone to the other. The components p_w and p_a, however, vary, although $p_w + p_a$ continues to equal P_4. As the moist air is cooled, more water vapour is condensed. At any point

p_a/p_w = (moles air)/(moles water vapour),

or is proportional to

(mass air)/(mass water vapour).

As water is removed from the gas phase as liquid, the ratio increases, so that p_a rises and p_w falls to maintain P_4 constant. Cooling of the air reduces the volumetric load on the air pump in two ways:

- it reduces the mass of water vapour to be removed with the air as just described.
- by the direct effect of temperature on gaseous volumes.

In view of the discussion regarding p_{w4} in the condensing zone, it may seem surprising that p_w falls to a lower value in the air cooling zone. The reason is that the condensate which separates in the air cooling zone is at a lower temperature than the main condensate; and this is the temperature which controls p_w at a point in the air cooling zone. It is important for the zones to be separated sufficiently for water from the main condensate not to re-evaporate into the air in the cooling zone.

Various designs of counter-current direct contact condenser are possible, including types with perforated trays and others with devices to produce curtains of water. As for other parts of a vacuum system, condensers should be designed to minimize pressure drop.

Condensers can be low level or high level. The condensate, and in direct contact types also the cooling water, is extracted from a low level condenser by a pump. High level condensers, now usual, are installed at least some 10.5 m above ground level; and the water runs out through what is known as a barometric leg into an open seal tank, or hot well. This works as shown in Fig. A2.2. h_1 is the height of the column of water which, plus the absolute pressure in the condenser, just balances the atmospheric pressure on the water in the hot well.

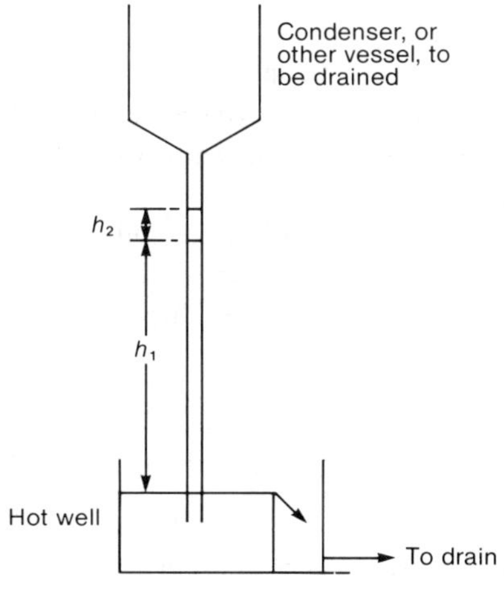

Fig. A2.2 – Principle of the barometric leg.

Some additional height, h_2, is needed to cause the water to flow down the leg at the required rate; and a little more is normally allowed to provide a margin of safety. If the barometric pressure is 760 mm Hg., the equivalent column of water equals 760/1000 × 13.6 = 10.4 m. If the pressure in the condenser is zero (the situation in a barometer) h_1 = 10.4 m, but if the pressure is, say 30 mm Hg, equivalent to 0.4 m water, h_1 = 10.4 – 0.4 = 10.0 m.

System (b) is normally used when it can achieve the required absolute pressure with the available cooling water. Under some circumstances a water chilling unit may be worthwhile, but this is more usual if very cold water is needed for duties such as plant cooling jackets.

When a lower pressure than that which could be achieved with system (b) is required, system (c) is used. In this, all the gases and vapours are compressed from P_2 to P_3, normally in a booster jet, or thermocompressor, in which mechanical energy is provided in the form of relatively high pressure steam. The booster jet discharges into a main condenser which operates as discussed for system (b); thus system (c) is a combination of system (a) and system (b). When the ratio P_3/P_2 to be provided by the booster jet is rather high, it may be necessary, or desirable, to use two, or even more, thermocompressor stages. The motive steam in the first booster stage then has to be compressed with the other gases and vapours in the next booster stage. The design of a booster jet is such that when the pressure required in the plant can be achieved using system (b), it is possible to shut off the steam to the booster jet, and to use the jet as a duct to the main condenser.

The vacuum pump used to compress to atmospheric pressure the moist air which leaves the condenser can be mechanical, or, now more usually, a two stage steam ejector unit, with an intermediate direct contact condenser to condense steam from the first stage. With either system (b) or system(c), a 'roughing jet' may be provided to reduce pressure more rapidly during start-up. Many years ago glycerine evaporators were sometimes equipped with large reciprocating pumps which extracted air, cooling water, and condensate together from a low-level condenser.

The vacuum maintenance system controls only the pressure at the point P_2. In many plants, such as glycerine stills, it is the pressure at P_1 which is particularly important. This depends upon the pressure change† between P_1 and P_2, as well as upon the performance of the vacuum maintenance system itself. Careful design of ducts, condensers and other plant items, is necessary; but, in some cases, certain types of fractionation, or other equipment, cannot be used if the desired pressure P_1 is to be achieved.

A2.3

If the pressure in a plant deteriorates, or is not up to specification, and the cause is not immediately evident, it may be helpful to diagnose the fault systematically along the following lines:

† Strictly, this discussion should consider total head rather than static head, or pressure, alone, because there can be some interchange between static and kinetic heads. The difference is usually small, except in jets, which do not affect pressures P_1 and P_2.

- check P_1 and P_2. If the difference is unduly large, there is some blockage (such as a dirty bed of packing), or the flow of gases and vapours is higher than expected.
- check the exit temperature of the condensate (the combined stream of cooling water and condensate in the case of a direct contact condenser) and note the corresponding vapour pressure of water. If the water temperature and the vapour pressure are high, the flow of water vapour to the condenser is high, or the supply of cooling water is sub-standard in quantity or temperature. This can be due to partial blockage of the water supply or the distribution system within the condenser, which is particularly liable to occur when the water is taken from a creek, or other dirty source. In the case of a surface condenser heat transfer can be impaired by deposits of scale or dirt.
- if the difference between the vapour pressure of water and the total pressure, that is the partial pressure of air, is high, either the vacuum pump system is not working properly, or, more likely, there is an additional intake of air through a leak in the plant. Such leaks can occur, for example, through holes, joints, faulty valves, or valve glands.
- if the system (c) is in use and the pressure rise P_2 to P_3 is low, check the steam supply and the nozzles in the booster jet or jets.

Appendix 3

ELEMENTS OF FAN ENGINEERING

A3.1

This appendix is concerned with fans used in process plants and similar situations, which work in ducts, rather than with open propeller fans used to move the air in a room.

As with centrifugal pumps, the performance of a fan under given circumstances depends upon the relationship between the characteristics of the fan and the resistance to flow through the system in which it is installed. The resistance of a system in terms of pressure requirement increases approximately with v^2, where v is the linear velocity ($v^{1.8}$ or $v^{1.9}$ may be more accurate). In a system with fixed geometry this can be expressed as pressure requirement is proportional to V^2, where V is the volumetric flow. As considered in more detail later, the performance of a fan at a given speed can be expressed in the form of a curve which relates the pressure produced to the volumetric flow. Fig. A3.1 shows curves for

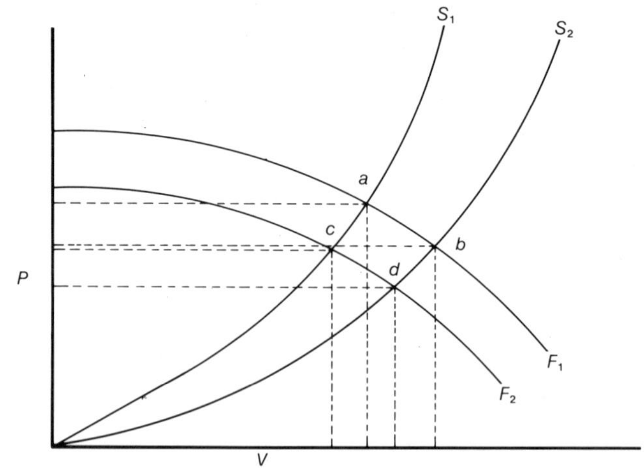

Fig. A3.1 – Operating points.

a fan working at two different speeds (*F1 and F2*) in two different external systems (*S1* and *S2*). An operating point is established where the two curves cut. Thus *a* is the operating point for fan speed 1 in system 1, *b* for fan speed 1 in system 2, *c* for fan speed 2 in system 1, and *d* for fan speed 2 in system 2. This shows how the pressure and volume produced by a fan are controlled by the external system in interaction with fan performance.

A3.2 DENSITY OF AIR

The performance of a fan is quoted, in curves or in tables, for standard air, usually at 20°C (68°F), 760 mm Hg. pressure and 50% relative humidity. Under these conditions air has a density, ρ, equal to about 0.074 lb/ft^3, or 1.18 g/1. Performance is significantly affected by the density of the air actually being handled; and this density can readily be calculated from that for standard conditions by application of the elementary gas laws. Air density is inversely proportional to the absolute temperature, so that ρ_t, density at t°C, = 0.074 × (273 + 20)/(273 + t) (or in Fahrenheit units $\rho_t = (460 + 68)/(460 + t)$.

Density is also directly proportional to the absolute pressure. Hence at p mm Hg and standard temperature $\rho_p = 0.074 \times p/760$ lb/ft^3. This is particularly important in relation to the altitude at which the fan is working. At 5000 ft. above sea level the normal barometric pressure is around 628 mm Hg; and the density of air at 20°C and 628 mm Hg is about 0.074 × 628/760 = 0.057 lb/ft^3 (see also section A3.6). Mass flow in lb/minute, M, equals volumetric flow V in ft^3/min × ρ in lb/ft^3.

A3.3 MEASUREMENT OF PRESSURES AND FLOW

Air pressures are measured using side gauges, facing gauges, and/or the two combined in pitotstatic tubes. As shown in Fig. A3.2, a side gauge is a tapping at right angles to the flow intended to measure static head, or pressure, alone, and not to be affected by the air movement. A facing gauge is placed directly facing the flow; and it measures total head, that is static head plus kinetic, or velocity, head. Kinetic head represents the energy due to the flow which is converted to pressure when the air is brought to rest, as in a facing gauge. The difference between the total head and the static head is thus the kinetic head, which is equal to $v^2/2g$ if the pressures are all expressed in terms of head of the flowing fluid, that is of the air of the density being handled. Consistent units must be used; in the English system that is:

v = ft/sec
g = acceleration due to gravity, 32.2 ft/sec^2 and kinetic head = ft of air.

Head in ft multiplied by density in lb/ft^3 equals pressure in lb/ft^2. It is often convenient to combine the side and facing tubes to give what is known as a pitotstatic tube. Such a tube must be designed carefully to minimize various errors; and it is usual to use a standard instrument such as one of the NPL (UK National Physical Laboratory) types. Technologists using English units usually express heads in terms of inches water gauge (WG), although inclined, and other sensitive, gauges are often used for the actual measurement, because of the small pressures involved. Velocities are also commonly expressed in terms of feet per minute, rather than per second. A practical version of the basic equation, kinetic head $= \frac{v^2}{2g}$, thus becomes v ft/min $= 1100\sqrt{\frac{\text{in WG}}{\rho_{\text{air}}}}$ approximately, or for standard air with $\rho_{\text{air}} = 0.074$, v ft/min $= 4000\sqrt{\text{in WG}}$, approximately.

For an accurate assessment of the total flow it is necessary to take measurements at appropriate points across the duct, but, as an approximation, it can be taken that the average linear velocity of gas in turbulent flow in a reasonably straight duct equals the velocity at the centre $\times$ 0.9. Hence the volumetric flow, V ft^3/min, equals linear velocity at the centre of the duct in ft/min $\times$ duct area in ft^2 $\times$ 0.9.

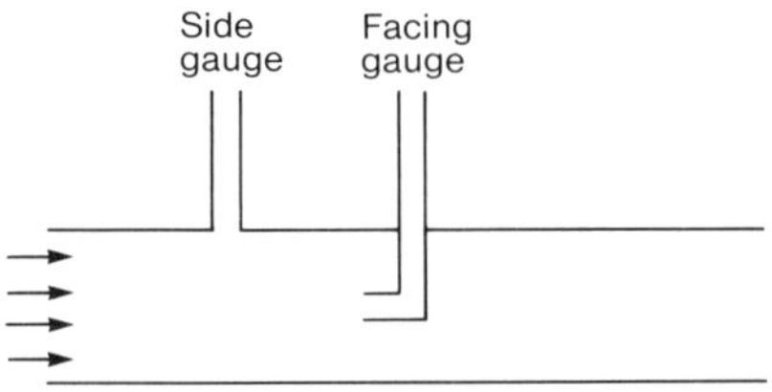

Basic arrangement in a duct

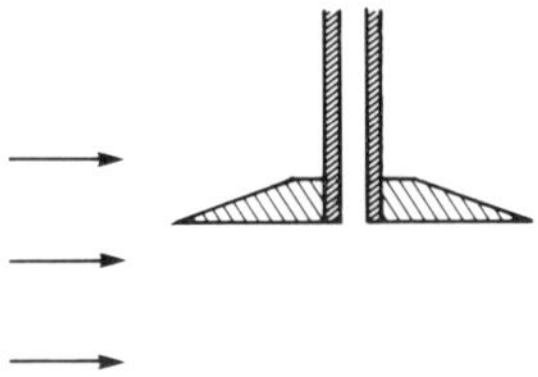

Side gauge tip for use when duct is not smooth

Fig. A3.2 – Side tubes and facing tubes.

Various precautions, as described in test codes and specialized books, are necessary when a fan is being tested; in particular, a straight length of duct is needed. An important point concerns the interpretation of the pressure gauge readings. With a pitotstatic tube, or separate side and facing tubes, in the discharge duct, both static and kinetic head are positive, and interpretation is strightforward. With similar instruments in the suction duct, the kinetic head, as always, is positive; but the static head is usually negative. Three cases are shown in Fig. A3.3, in which gauge S gives the side reading, gauge F the facing reading, and gauge C the combined reading. It will be noted that:

- when the entrance is blocked, so that there is no flow, gauge C reads zero, and gauges S and F both show the negative static pressure.

- when the suction is fully open, so that resistance to flow in the suction duct is negligible, the facing gauge *F* reads zero because the positive kinetic head just balances the negative static head recorded by side gauge *S*. The combined gauge *C* records the positive kinetic head, numerically equal to the negative static head shown by gauge *S*.
- when the entrance is partly blocked, that is there is some resistance to flow in the suction duct, the readings are as shown in (c). *S* shows the negative static head, and *C* the positive kinetic head. *F* shows the total head which is the algebraic sum of *S* and *C*; it is negative because some energy goes to overcome resistance to flow.

The total head produced by a fan operating with both suction and delivery ducts is given by the algebraic sum of the readings of a facing gauge in the delivery duct and a facing gauge in the suction duct. In principle, this can be measured directly by a manometer connected to facing gauges in the two positions, but this may not be convenient. To attempt to measure the total head produced by a fan by combining the readings of a facing gauge in the discharge duct with that of a side gauge in the suction duct is to include the kinetic head twice. The static head produced by a fan is the total head minus the kinetic head.

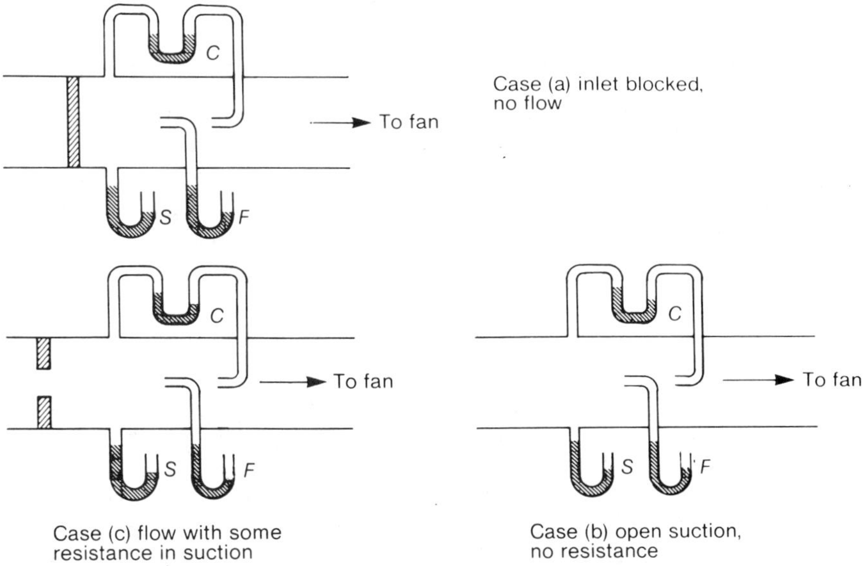

Fig. A3.3 – Gauge readings in the suction duct of a fan.

A3.4 POWER CONSUMPTION

The theoretical amount of energy, in ft-lb/min, contained in a given volume of air discharged at a certain pressure is found by multiplying the volumetric flow in ft^3/min by the head developed, measured in lb/ft^2. Division by 33 000 gives the theoretical, or air, horse-power; similarly division by 33 000/0.746 = 44 200 gives the theoretical power in kW. To convert to a formula using normal fan units.

$$\text{Theoretical power} = \underbrace{V}_{\text{ft}^3/\text{min}} \; \underbrace{\frac{\text{in WG} \times 62.4}{12}}_{\substack{\text{factor equal to pressure in}\\ \text{lb/ft}^2\text{, giving ft-lb/min.}}} \; \frac{1}{44\,200}\ \text{kW}$$

$$= V \times \text{in WG} \times 1.18 \times 10^{-4}\ \text{kW.}$$

To convert this figure to the electrical power needed to drive the fan, allowance must be made for aerodynamic losses due to friction and turbulence in the fan itself, and for mechanical and electrical losses in the drive and motor. It is commonly convenient to express the power requirement of a fan, and its efficiency, in terms of the brake power in HP, or kW, required to drive it. This leaves losses in the motor to be dealt with separately. Powers and fan efficiencies can be calculated on the basis of static head or total head. Both are often quoted; the main point is to be consistent throughout a calculation.

A3.5 FAN TYPES, CHARACTERISTIC CURVES, AND THE FAN LAWS

The results of manufacturers tests which provide information regarding the performance of a fan can be reported in tables, or in various types of graph. A common presentation is as a series of curves in which static, and/or total pressure, power required, and efficiency are all plotted against volume. The shapes of the curves can vary considerably with fan design, but the general form is roughly as shown in Fig. A3.4. The operating point is that where the fan pressure/volume curve cuts the system pressure curve as explained in section A3.1 and Fig. A3.1. It is desirable to work somewhere near the maximum of the efficiency curve.

Axial flow fans are now available, but centrifugal fans are usually used for the duties considered in this book. Centrifugal fans can have radial, forward, or backward facing blades as shown diagrammatically in Fig. A3.5. The velocity of the air leaving the fan impeller is the resultant of the flow component along the blade caused by centrifugal force and the tangential component caused directly by the rotation. Parallelograms of velocity are shown in the upper parts of the figures. Pressure is generated by centrifugal action and some of the kinetic energy can be converted to pressure in the fan scroll casing. The lower parts of Fig. A3.5 give some rough indication of the types of pressure/volume and power/volume curves typical of the three types of blade. The selection of a fan for a major duty is a matter for specialists; but it can be said that backward facing blades are commonly used in spray driers, whilst forward facing blades are often selected for ventilation. The motor which drives a forward bladed fan is liable to become overloaded if it is allowed to work against a low resistance.

When the air contains sticky powder, the self-cleaning properties of the fan are important. It is said that a compromise between radial and backward blades may be superior to fully backward blades as regards self-cleaning at the expense of some loss of efficiency. In spray drying, and similar, plants facilities should be provided to permit fans to be washed when necessary.

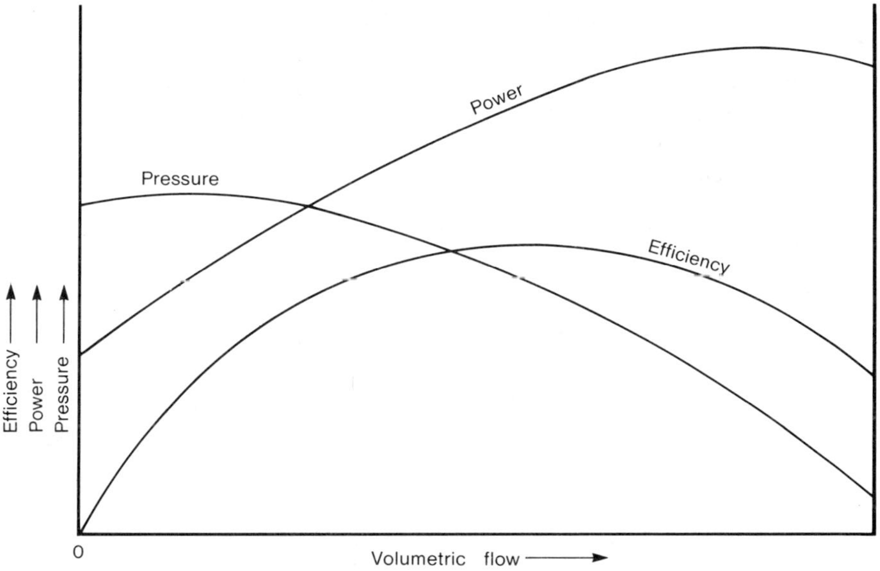

Fig. A3.4 – Fan characteristic curves.

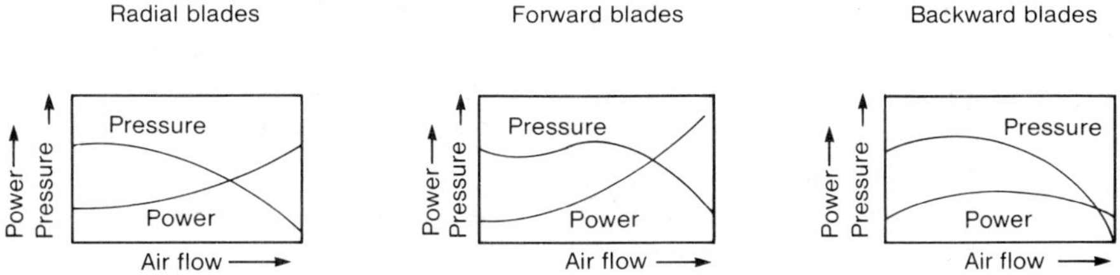

Fig. A3.5 – Centrifugal fan types and characteristics.

The following fan laws are generally applicable:

- *Law 1*. For a given fan with air of constant density,
  V (volume flow) and M (mass flow) vary directly with
  N (rotational speed).
  P (pressure) varies as N^2
  and Power varies as N^3
- *Law 2*. For a given fan working at a given speed, V is independent of air density, ρ, but M varies as ρ.
  P varies as ρ (the head of flowing fluid is constant)
  Power varies as ρ.
- *Law 3*. For geometrically similar fans of different sizes working with air of constant density with the *blade tip velocity constant*, V and M vary as d^2 (d = impeller diameter).
  P remains constant
  Power varies as d^2

Useful deductions from these laws include:

- *Relationship 4*. For geometrically similar fans of different sizes working with air of constant density with *rotational speed constant.*
  Blade tip speed varies as d and this is the fundamental variable.
  From law 3, V and M vary as $d^2.d = d^3$.
  From law 1, P varies as d^2, as at constant N blade tip speed varies as d, as just mentioned.
  Power varies as $d^3.d^2 = d^5$.
- *Relationship 5;* For geometrically similar fans working with air of constant density, when size and rotational speed are varied together:
  V and M vary as d^3N
  P varies as d^2N^2
  Power varies as d^5N^3

  For constant pressure development $P = d^2N^2$ = constant, so that N varies as d^{-1}.

  V for constant pressure development can be obtained by substitution of d^{-1} for N in the relationship V varies as d^3N, so that V varies as $d^3.d^{-1}$, or d^2.

With air at a constant density and with the geometry of the external system fixed, the external resistance/volume curve remains constant and the fan laws can often be applied directly. It is also possible, and usually preferable, to plot new estimated curves if it is proposed to consider, for example, a change in fan speed. This is essential if a change in air density, or in system geometry, for instance by the installation of different, or additional, equipment for dust collection, is involved, so that a new resistance curve can be plotted to give a revised operating point. Care must be taken to avoid mechanical problems due to excessive stresses, or to vibration, if an increase in speed is under consideration. If the capacity of a fan is unnecessarily high, it is usually more economical to reduce its speed rather than to restrict the flow by other means, such as dampers.

A3.6 EFFECT OF ALTITUDE

Table A3.1 indicates the way in which barometric pressure and air density are influenced by altitude. There are important implications summarized by Masters *Handbook of Spray Drying* (1979), and reprinted by permission of the publishers, Longman, as follows:

> With the decrease in air density with increase in altitude, higher air volumes must be handled by the fan for a given mass flow requirement. Care must be taken in sizing fans for operation at altitude. Data specifications are based upon sea level conditions and thus compensating factors must be applied to prevent fans and motors being undersized.
>
> A mass air flow at sea level is equivalent to a volume that would be 43% greater if the fan were operating at 3000 m (this is computed from taking the atmospheric pressure at sea level at 760 mm Hg and the atmospheric pressure at 3000 m at 530 mm Hg where increase in volume is in the ratio of 760/530 = 1.43 T_{ref} = 10°C). Pressure drop will increase substantially if a fan sized at sea level operates at altitude. Pressure drop varies as the square of velocity and for the above example a 43% increase in velocity will give a 205% increase in pressure drop. Larger fans and motor sizes are a characteristic of high-altitude working due to the additional fan work load. Actual system losses are lower with lower air density, and correcting sizing can hold velocities down to reasonable figures.

Table A3.1 – Effect of altitude on barometric pressure and air density

Altitude		Barometric pressure		Air density relative to that at sea level
ft	m	in Hg	mm Hg	
0	0	29.92	760	1.00
500	152	29.36	746	0.981
1000	305	28.80	732	0.962
1500	457	28.26	718	0.944
2000	610	27.72	704	0.926
2500	762	27.20	691	0.909
3000	914	26.68	678	0.891
4000	1219	25.68	652	0.858
5000	1524	24.72	628	0.826
6000	1829	23.79	604	0.795
7000	2134	22.90	582	0.766
8000	2438	22.04	560	0.737

Bibliography and other references

BOOKS OF GENERAL INTEREST

Bailey's (1979, 1982), *Industrial Oil and Fat Products,* 4th edn, Swern, D. (ed.), Vol. 1, 1979, Vol. 2, 1982, John Wiley.

Davidsohn, J., Better, E. J., and Davidsohn, A. (1953), *Soap Manufacture,* Vol. 1, Interscience.

Davidsohn, A., and Milwidsky, B. M. (1978), *Synthetic Detergents,* 6th edn, George Godwin/John Wiley.

Longman, G. F. (1975), *The Analysis of Detergents and Detergent Products,* John Wiley.

Milwidsky, B. M., and Gabriel, D. M. (1982), *Detergent Analysis,* George Godwin.

Masters, K. (1979), *Handbook of Spray Drying,* 3rd edn. George Goodwin. This is an updated version of Masters, K. (1972), *Spray Drying,* Leonard Hill.

Miner, C. S., and Dalton, N. N. (1953), *Glycerol,* Reinhold.

Newman, A. A. (1968), *Glycerol,* Morgan-Grampian (includes chapters by L. V. Cocks).

Rosen, M. J. (1978), *Surfactants and Interfacial Phenomena,* John Wiley.

Schwitzer, M. K. (1951), *Continuous Processing of Fats,* Leonard Hill.

Wigner, J. H. (1940), *Soap Manufacture: The Chemical Processes,* E. and F. N. Spon.

BOOKS TO WHICH REFERENCE IS MADE ON SPECIFIC POINTS

Jenson, V. G., and Jeffreys, G. V. (1963), *Mathematical methods in chemical engineering,* Academic Press.

Lawrie, J. W. (1928), *Glycerol and the glycols,* Chemical Catalog Company.
Levitt, B. (1967), *Oils, detergents and maintenance specialities,* Vol. 1, Chemical Publishing Company.
Markley, K. S. (1960, 1967), *Fatty acids,* 2nd edn. Part 1, 1960, Part 4, 1967. Interscience.
Patterson, H. B. W. (1983), *Hydrogenation of fats and oils,* Applied Science.
Sisley, J. P., and Wood, P. J. (1972), *Encyclopedia of surface active agents,* 2nd edn, Chemical Publishing Company.
Thomssen, E. G., and Kemp, C. R. (1937), *Modern soapmaking,* MacNair-Dorland.
Van Wazer, J. R. (1958, 1961), *Phosphorus and its compounds,* Vol. 1, 1958, Vol. 2, 1961, Interscience.
Webb, E. T. (1927), *Modern soap and glycerine manufacture,* Davis Brothers.

JOURNALS AND OTHER SOURCES

The following abbreviations are used in the list of references; as will be seen, some journals have changed names from time to time over the years.

Soap, became *Soap and Sanitary Chemicals – Soap Sanit. Chem.,* became *Soap and Chemical Specialities – Soap Chem. Spec.* became *Soap, Cosmetics and Chemical Specialities – Soap Cosmet. Chem. Spec.*
Soap, Perfumery and Cosmetics Trade Review became *Soap, Perfumery and Cosmetics – Soap Perfum. Cosmet.*
Oil and Soap, became *Journal of the American Oil Chemists Society – J. Am. Oil Chem. Soc.*
Chemical and Metallurgical Engineering – Chem. Metall. Eng., became *Chemical Engineering (New York) – Chem. Eng. N.Y.*
The Chemical Engineer (London) – Chem. Eng. (Lond.).
Transactions of the Institution of Chemical Engineers (London) – Trans. Inst. Chem. Eng. (Lond.).
Industrial and Engineering Chemistry – Ind. Eng. Chem. (this journal has several divisions).

Other journals cited only once are written in full in the list.

Useful articles may be found in various journals, particularly those just mentioned. Especially interesting review articles are sometimes presented to the short courses organized periodically by the American Oil Chemists Society and published in its Journal.

Anstead, D. F. (1973), Dyes in cosmetics, *Review of Progress in Coloration and Related Topics,* **4**, 13.
Berth, P. (1978), Recent developments in the field of inorganic builders, *J. Am. Oil Chem. Soc.,* **55**, 52.
Borchert, P. A., and Marino, R. A. (1980), Citrate as co-builder in zeolite detergents, *Soap Cosmet. Chem. Spec.,* **56**, 56.
Busby, G. W., and Grosvenor, D. E. (1952), Purification of glycerine by ion-exchange, *J. Am. Oil Chem. Soc.,* **29**, 318.

Cox, C. B. (1949), An approach to continuous fat splitting, *Trans. Inst. Chem. Eng.*, **27**, 123. (This paper is concerned with the Twitchell process.)

Crutchfield, M. M. (1978), Organic builders; a review of worldwide efforts to find organic replacements for detergent phosphates, *J. Am. Oil Chem. Soc.*, **55**, 58.

Davidsohn, A. (1965), Detergents for developing countries, *Soap,* **41**, 73.

D'Souza, G. B. (1979), The importance of glycerol in the fatty acid industry, *J. Am. Oil Chem. Soc.*, **56**, 812A.

Ginn, M. E., Steinhauer, R. C., Liebman, I., and Jungermann, E. (1968), Effects of tallow-coconut fatty acid ratios on properties of bar soaps, *J. Am. Oil Chem. Soc.*, **45**, 668.

Govan, W. J., Jr. (1946), Counter-current wash system in soapmaking, *Oil and Soap,* **23**, 229.

Govan, W. J. Jr. (1950), Glycerol distribution in soap boiling operations, *J. Am. Oil Chem. Soc.*, **27**, 44.

Griffin, W. C. (1949, 1955), Calculation of 'H.L.B' values of nonionic surfactants, *Journal of the Society of Cosmet Chemists,* **1**, 311 (1949) and *Am. Perfumer Essent. Oil Rev,* **65**, 26 (1955).

Hart, R. (1980), Chelating agents in detergents and speciality chemicals, *Soap Cosmet. Chem. Spec.*, **56**, 39.

Hollingsworth, M. W. (1976), *Household detergents,* published by Ciba-Geigy Limited, p. 12.

Hollingsworth, M. W. (1978), Role of detergent builders in fabric washing formulations, *J. Am. Oil Chem. Soc.*, **55**, 49.

Jeffreys, G. V., Jenson, V. G., and Miles, F. R. (1961), The analysis of a continuous fat hydrolysing plant, *Trans. Inst. Chem. Eng. (Lond.),* **39**, 389.

Jones, L. D. (1958), Saponification stressing the newer methods, *J. Am. Oil Chem. Soc.*, **35**, 630.

Jungermann, E. (1982), New trends in bar soap technology and the impact of liquid soaps, *J. Am. Oil Chem. Soc.*, **59**, 173A.

Kaneko, T. M. (1979), Low temperature non-phosphate home machine dish-washing formulations, *Soap Cosmet. Chem. Spec.*, **55**, 31.

Kapur, B. L., Solomon, J. M., and Bluestein, B. R. (1978), Summary of the technology for the manufacture of higher alpha-sulfo fatty acid esters, *J. Am. Oil. Chem. Soc.*, **55**, 549.

Lanteri, A. (1978), Sulfonation and sulfation technology. Part 1, *Soap Cosmet. Chem. Spec.*, **54**, 31 and 83.

Mallee, J. P. (1963), Tableting of detergents, *J. Am. Oil Chem. Soc.*, **40**, 621.

McBride, G. W. (1947), *Chem. Metall. Eng.,* **54**, Continuous process for soap, 94, 134.

Palmqvist, F. T. E. (1959), A new approach to continuous soapmaking – constant composition control, *J. Am. Oil Chem. Soc.*, **36**, 173.

Palmqvist, F. T. E. (1960a, 1960b), The De Laval Centripure process. A new method for the continuous production of soap, *Soap Perfum. Cosmet.*, **33**, 149; and *Soap Perfum. Cosmet.*, **33**, 279.

Palmqvist, F. T. E. (1961), Automatic control of the electrolyte content of soap

by means of variations in viscosity, *Soap Perfum. Cosmet.*, **34**, 353.
Palmqvist, F. T. E., and Harris, B. M. (1964), Neutral fat to finished soap in twelve minutes, *Soap Perfum. Cosmet.*, **37**, 865.
Platt, G. R. (1970), Manufacture of soap bases by a completely continuous process, *Soap Perfum. Cosmet.*, **43**, 233.
Rayner, A. (1938, 1939), Pan charge and physical properties, *Soap Perfum. Cosmet.*, **11**, 1086, and **12**, 49, 152, 324.
Rice, E. E. (1979), Materials of construction in the fatty acid industry, *J. Am. Oil Chem. Soc.*, **56**, 754A.
Schweiker, G. C. (1978), Sodium silicates and sodium aluminosilicates, *J. Am. Oil Chem. Soc.*, **55**, 36.
Schweiker, G. C. (1981), Detergent builders, *J. Am. Oil Chem. Soc.*, **58**, 170A.
Shen, C. Y., and Callis, C. F. (1965), Detergents by a new route, *Soap Chem. Spec.*, **41**, 64 and 109.
Sonntag, N. O. V. (1979a), Fat splitting, *J. Am. Oil Chem. Soc.*, **56**, 729A.
Sonntag, N. O. V. (1979b), New developments in the fatty acid industry, *J. Am. Oil Chem. Soc.*, **56**, 861A.
Sonntag, N. O. V. (1981), Current and future fat based raw materials for soap manufacture, *J. Am. Oil Chem. Soc.*, **58**, 155A.
Souririjan, S., and Kimura, S. (1967), Correlations of the reverse osmosis separation data for the system glycerol-water using porous cellulose acetate membranes, *Ind. Eng. Chem. Process design and development*, **6**, 504.
Souririjan, S., and Matsuura, T. (1982), Science of reverse osmosis – an essential tool for the chemical engineer, *Chem. Eng. (Lond.)*, 359.
Stavitsky, A. C. (1977), Utilization of type A zeolite as a laundry detergent builder, *Soap Cosmet. Chem. Spec.*, **53**, 29.
Sterace, C. A. (1981), Detergent enzymes: developments during the last decade, *J. Am. Oil Chem. Soc.*, **58**, 165A.
Stensby, Per, S. (1965), Optical brighteners in fabric softeners, *Soap Chem. Spec.*, **41**, 85.
Sturzenegger, A., and Sturm, H. (1951), Hydrolysis of fats at high temperatures, *Ind. Eng. Chem.*, **43**, 510.
Taylor, A. (1980), Soaps and detergents and the environment, *J. Am. Oil Chem. Soc.*, **57**, 859A.
Tuttle, J. B., and Woodward, E. R. (1946), Sodium chlorite is successfully used for bleaching tallow, *Chem. Metall. Eng.*, **53**, 114.
Woollatt, E. (1982), Evaporation and evaporators, *Processing*, **28**, 23.
Zilske, H. (1965), A new process for making fast dissolving detergent powders, *Soap*, **41**, 167.
Introduction to Laundry Chemistry (anonymous and undated), published by the Guild of Cleaners and Launderers.

Index

For many purposes readers may find it convenient to locate items by use of the detailed table of contents, pages 5 – 11. Only selected references are made in this index to some materials etc. which are mentioned frequently in the text.